Molecular Sieve Zeolites-I

The Second International
Conference co-sponsored by
the Division of Colloid and
Surface Chemistry, the
Division of Petroleum Chemistry,
and the Division of Physical
Chemistry of the American
Chemical Society and
Worcester Polytechnic Institute
at Worcester Polytechnic
Institute, Worcester, Mass.,
Sept. 8–11, 1970.

Edith M. Flanigen

Leonard B. Sand

Conference Co chairmen

ADVANCES IN CHEMISTRY SERIES **101**

AMERICAN CHEMICAL SOCIETY

WASHINGTON, D. C.   1971

Coden: ADCSHA

Copyright © 1971

American Chemical Society

All Rights Reserved

Library of Congress Catalog Card 77–156974

ISBN 8412–0114–5

PRINTED IN THE UNITED STATES OF AMERICA

Advances in Chemistry Series
Robert F. Gould, *Editor*

Advisory Board

Paul N. Craig

Thomas H. Donnelly

Gunther Eichhorn

Frederick M. Fowkes

Fred W. McLafferty

William E. Parham

Aaron A. Rosen

Charles N. Satterfield

Jack Weiner

FOREWORD

ADVANCES IN CHEMISTRY SERIES was founded in 1949 by the American Chemical Society as an outlet for symposia and collections of data in special areas of topical interest that could not be accommodated in the Society's journals. It provides a medium for symposia that would otherwise be fragmented, their papers distributed among several journals or not published at all. Papers are refereed critically according to ACS editorial standards and receive the careful attention and processing characteristic of ACS publications. Papers published in ADVANCES IN CHEMISTRY SERIES are original contributions not published elsewhere in whole or major part and include reports of research as well as reviews since symposia may embrace both types of presentation.

CONTENTS

Preface ix

INTRODUCTION

1. Recent Advances in Zeolite Science 1
   D. W. Breck

SYNTHESIS

2. Some Problems of Zeolite Crystallization 20
   S. P. Zhdanov

3. Stages of Zeolite Growth from Alkaline Media 44
   R. Aiello, R. M. Barrer, and I. S. Kerr

4. Zeolite Formation from Synthetic and Natural Glasses 51
   R. Aiello, C. Colella, and R. Sersale

5. Crystallization of Zeolites in the Presence of a Complexing Agent. Part II 63
   G. H. Kühl

6. Phosphorus Substitution in Zeolite Frameworks 76
   E. M. Flanigen and R. W. Grose

7. Reaction Process of Zeolite Formation in the System $Na_2O-Al_2O_3-SiO_2-H_2O$ 102
   F. Schwochow and G. Heinze

8. Synthesis and Structural Features of Zeolite ZSM-3 109
   G. T. Kokotailo and J. Ciric

9. Crystallization of Zeolitic Aluminosilicates in the System $Li_2O-Na_2O-Al_2O_3-SiO_2-H_2O$ at 100°C 122
   H. Borer and W. M. Meier

10. Synthesis of Lithium and Lithium, Sodium Mordenites 127
    M. L. Sand, W. S. Coblenz, and L. B. Sand

11. Synthesis of Beryllosilicate with the Structure of Analcite 135
    S. Ueda and M. Koizumi

12. Crystal Chemical Relationships in the Analcite Family. I. Synthesis and Cation Exchange Behavior 140
    W. D. Balgord and R. Roy

13. Synthesis of Thermodynamically Stable Zeolites in the $Na_2O-Al_2O_3-SiO_2-H_2O$ System 149
    E. E. Senderov and N. I. Khitarov

STRUCTURE

14. Zeolite Frameworks 155
    W. M. Meier and D. H. Olson

15. Faujasite-Type Structures: The Aluminosilicate Framework:
    Positions of Cations and Molecules: Nomenclature 171
    J. V. Smith

16. Infrared Structural Studies of Zeolite Frameworks 201
    E. M. Flanigen, H. A. Szymanski, and H. Khatami

17. Structural Studies on Erionite and Offretite 230
    J. A. Gard and J. M. Tait

18. Linde Type B Zeolites and Related Mineral and Synthetic Phases .. 237
    W. C. Beard

19. Crystal Structure of Gismondite, a Detailed Refinement 250
    K. F. Fischer and V. Schramm

20. Refinement of the Crystal Structure of Laumontite 259
    V. Schramm and K. F. Fischer

21. Crystal Structures of Ultrastable Faujasites 266
    P. K. Maher, F. D. Hunter, and J. Scherzer

MINERALOGY

22. Zeolites in Sedimentary Deposits of the United States—A Review .. 279
    R. A. Sheppard

23. Clinoptilolite from Japan 311
    H. Minato and M. Utada

24. Present Status of the Zeolite Facies 317
    D. S. Coombs

25. Graphical Analysis of Zeolite Mineral Assemblages from the Bay of
    Fundy Area, Nova Scotia 328
    A. B. Carpenter

26. Composition and Origin of Clinoptilolite in the Nakanosawa Tuff of
    Rumoi, Hokkaido 334
    A. Iijima

27. Present-Day Zeolitic Diagenesis of the Neogene Geosynclinal
    Deposits of the Niigata Oil Field, Japan 342
    A. Iijima and M. Utada

MODIFICATION AND GENERAL PROPERTIES

28. Cation Exchange on Zeolites 350
    H. S. Sherry

29. Infrared Spectroscopic Studies of Zeolites 380
    J. W. Ward

30. Kinetics of Ion Exchange from Partially Exchanged Starting
    Materials 405
    N. M. Brooke and L. V. C. Rees

31. Properties of Linde A in Aqueous, Nonaqueous, and Mixed Media .. 414
    R. B. Barrett and J. A. Marinsky

32. Thermodynamics of the Exchange of Alkylammonium Ions in
    Synthetic Faujasites 426
    E. F. Vansant and J. B. Uytterhoeven

33. Stability of Heteroionic Forms of the Synthetic Zeolite A 436
    A. Dyer, W. Z. Celler, and M. Shute

34. Evolution of the Structure and of the Texture of a Type 4A
    Molecular Sieve in the Course of Thermal Treatments Between
    400° and 800°C 443
    J. L. Thomas, M. Mange, and C. Eyraud

35. Recovery and Purification of Cesium-137 from Purex Waste Using
    Synthetic Zeolites 450
    L. A. Bray and H. T. Fullam

36. Dielectric Study of Synthetic Zeolites X and Y 456
    R. A. Schoonheydt and J. B. Uytterhoeven

37. NMR Relaxation of Water in Zeolite 13-X 473
    H. A. Resing and J. K. Thompson

38. Stereospecific Adsorption of Nitrous Oxide, Cyclopropane, Water,
    and Ammonia on the Co(II)A Synthetic Zeolite 480
    K. Klier

39. Thermal Decomposition Patterns in Methylammonium Cation-
    Exchanged Y-Type Faujasites 490
    E. L. Wu, G. H. Kühl, T. E. Whyte, and P. B. Venuto

40. Properties of Aluminum-Deficient Large-Port Mordenites 502
    W. L. Kranich, Y. H. Ma, L. B. Sand, A. H. Weiss, and I. Zwiebel

41. Investigation of Oxygen Mobility in Synthetic Zeolites by Isotopic
    Exchange Method 514
    G. V. Antoshin, Kh. M. Minachev, E. N. Sevastjanov,
    D. A. Kondratjev, and C. Z. Newy

Index 521

PREFACE

During his attendance at a Gordon Conference in the USA in 1962, Professor R. M. Barrer of Imperial College, London, expressed the opinion that periodic molecular sieve conferences would be worthwhile. On the premise that it was appropriate to hold the first conference in London to honor Professor Barrer's contributions, a group of British scientists was encouraged to initiate its organization. As a result, the first International Conference on Molecular Sieves was held successfully in London, April 4–6, 1967, under the chairmanship of Professor Barrer and under the sponsorship of the Society of Chemical Industry. The conference was attended by some 200 scientists, and 34 papers were presented. The proceedings, "Molecular Sieves," is published by the Society of Chemical Industry, 14 Belgrave Square, London, S.W. 1, 1968.

In 1968, Professor Barrer and one of us (LBS) initiated the continuation of the conferences on a triennial basis by inviting the second conference to be held at Worcester Polytechnic Institute, with the American Chemical Society invited to be the sponsoring society. It was believed appropriate to hold the second conference in the United States to recognize the pioneering commercialization of molecular sieve zeolites by Union Carbide Corporation and the petroleum refining industry. The organizing committee for the second conference included the co-chairman representing the American Chemical Society (EMF), Dr. W. L. Kranich, WPI, secretary-treasurer, Dr. P. K. Maher, Davison Chemical Division, W. R. Grace & Co., industrial liaison, and Prof. Barrer, honorary member. Dr. D. W. Breck participated helpfully in several discussions. The Petroleum Research Fund contributed to the expenses of invited distinguished speakers, and contributions for support of the conference were received from BP (North America) Ltd.–British Petroleum Co., Ltd., Esso Research & Engineering Co., W. R. Grace & Co.–Davison Division, Gulf Research & Development Co., Mobil Research & Development Corp., Nalco Chemical Co., Norton Co., Shell Development Co., Sun Oil Co., Texaco, Inc., Union Carbide Corp., Union Oil Co. of California, and Universal Oil Products Co. A number of special services were donated by the Union Carbide Corp.

Approximately 300 scientists from 18 countries participated in the conference. The meeting provided an opportunity for presentation of new scientific information and discussion and exchange of ideas both formally and informally, in an extremely pleasant and congenial environment. The success and enjoyment of the conference resulted partly from the careful planning and programming of the organizing committee, but especially from the untiring work of the local committees headed by: Dr. Y. H. Ma, Prof. W. B. Bridgman, Mr. W. Trask, Mr. G. Fuller, Mrs. A. H. Weiss, Mr. A. Begin, Mr. C. A. Keisling, Dr. I. Zwiebel, Dr. A. H. Weiss, Mrs. D. C. French, Mr. G. Gage, and Dr. R. E. Wagner.

On the final day of the conference, a permanent International Zeolite Conference Committee was elected, consisting of Miss Edith M. Flanigen, Chairman, Prof. R. M. Barrer, Dr. H. B. Habgood, Prof. A. V. Kiselev, Dr. P. K. Maher, Prof. W. M. Meier, Dr. C. Naccache, Prof. J. V. Smith, Prof. J. B. Uytterhoeven, and Dr. P. B. Venuto. It was voted to accept the invitation of Prof. Walter M. Meier to host the third conference in Zurich, Switzerland, in September 1973, under the chairmanship of Prof. Meier. Unfortunately, he could not attend the second conference because en route his plane was hijacked to the Jordanian desert where he was held hostage.

The Conference Committee would like to commend Robert F. Gould, Editor, Mrs. Colleen Stamm, Assistant Editor, and the staff of the ADVANCES IN CHEMISTRY SERIES for their efficient handling of an unexpectedly large number of manuscripts and their outstanding job in the publication of the Preprints and these two final volumes. The cooperation and contributions of the authors, discussants, session chairmen, panel chairmen and panelists, and other participants are also gratefully acknowledged.

The Proceedings are divided into two volumes following the session divisions at the Conference. Volume I contains the Introductory Lecture and 40 papers with discussion, presented under Synthesis, Structure and Mineralogy, and Modification and General Properties. Volume II includes 35 papers with discussion in Sorption and Catalysis, and the Concluding Remarks.

It appears from the contributions and participation in the first and second conferences that they serve a need in the rapidly growing field of molecular sieve science and technology, and they have been established accordingly on a triennial basis. May we take this opportunity to thank the many others who contributed to the success of the conferences and extend best wishes to those organizing the next. We can assure them that the pleasure of meeting and working with the many people involved in the conference far outweighs any burden of work in its organization.

EDITH M. FLANIGEN
LEONARD B. SAND

Tarrytown, New York
Worcester, Massachusetts
March 1971

Recent Advances in Zeolite Science

D. W. BRECK

Union Carbide Corp., Tarrytown Technical Center, Tarrytown, N. Y.

> *Zeolite properties are being studied by nearly every type of modern scientific discipline, and they are being utilized in many new chemical engineering processes. Important advances include detailed basic information on cations in zeolites, more understanding of the mechanism of zeolite formation, the formation and character of structural defects and hydroxyl groups, the role of zeolite structure in adsorption and catalysis, and the increasing technology of the use of molecular sieve zeolites in catalysis and adsorption.*

The discovery of synthetic crystalline zeolites has resulted in wide scientific interest and a great variety of applications in industry. The growth of scientific interest in zeolites may be measured by the increase in the number of related publications. From the early research on zeolite minerals in 1909 until 1969, over 4500 papers have been published in the open literature. The extensive interest is evidenced equally by over 1000 patents which have been issued in the United States alone. The fundamental properties and applications of molecular sieve zeolites involve many scientific disciplines and cross many of the traditional boundaries. Fields involved are inorganic and physical chemistry with emphasis on surface and colloid chemistry and catalysis, biochemistry, the geological sciences of geochemistry, geology, mineralogy, and physics, including crystallography, spectroscopy, and solid state physics.

Applications in the engineering sciences include adsorption separations, hydrocarbon catalysis, and purifications. The molecular sieve zeolites possess many unique properties which have resulted in their use in a great variety of applications such as hydrocarbon conversion catalysts, recovering radioactive ions from waste solutions, separating hydrogen isotopes, solubilizing enzymes, carrying active catalysts in the curing of plastics and rubber, transporting soil nutrients in fertilizers, and even filtering tars from cigarette smoke. Minato has presented an interesting review on industrial uses for mineral zeolites in Japan (65). Relatively few of the known zeolite types are utilized in major applications. A considerable number are sitting on the shelf, as it were, with perhaps hidden potential. In addition, developments in recent years lead to the unquestionable conclusion that the group of zeolite species can be much larger than is now known. The analysis of the crystal structures of most of the known zeolite minerals and some of the synthetic types has led to classifications based on structural principles. An extension of these principles suggests the formulation of hypothetical zeolite structures.

Some Recent Important Achievements in Zeolite Research

The growth in zeolite research is reflected in the growth in these conferences, from 31 papers at the first held in London 3 years ago, to 78 papers in this one. Recent developments in zeolite research fall into the following groups: zeolite structural chemistry, zeolite mineralogy, synthesis of zeolites, adsorption by zeolites, zeolite catalysts, and zeolite chemistry.

Zeolite Structural Chemistry. The elucidation of the intricate details of zeolite structures is intriguing in a scientific sense as well as important in providing an understanding for industrial applications. The basic framework types of zeolites can be grouped into classes by common subunits of their structure. Classifications of this type were proposed by Smith (71) and refined by Meier (59).

The basic structures of nearly all of the zeolite minerals appear to have been resolved, including the complex structures of laumontite (17), stilbite (39), heulandite (63), and yugawaralite (55). In addition, the relationship of the zeolites erionite and offretite has been delineated further by the work of Kawahara (51) and, in this symposium, by Gard. Zeolite minerals for which structural analyses have not been completed include the widely occurring mineral clinoptilolite and the rarer mineral stellerite, which apparently is related to stilbite.

Synthetic zeolites which have been resolved include the zeolite type L (14) and the zeolite Ω (13), both of which have open frameworks and should have important catalysis application. Ion exchange studies (71) have contributed to an understanding of the synthetic zeolite T (19), which appears to be structurally related to the minerals offretite and erionite.

The locations of cations in particular zeolites have been determined; in other zeolites, extensive studies have been completed on cations and the interaction of cations with other molecules in the structure. The extensive work carried out on faujasite and the topologically related synthetic zeolites type X and type Y is summarized by Smith, page 171.

Infrared spectroscopy has been a valuable technique for exploring zeolite structures. It is useful for studying the nature of hydroxyl groups in zeolites, the interaction of cations with adsorbed molecules, and the fundamental framework structures of zeolites.

Zeolite Minerals. Several zeolite minerals—i.e., clinoptilolite, ferrierite, erionite, analcime, mordenite, chabazite, phillipsite, laumontite—occur in extensive sedimentary deposits in many areas of the earth's surface (44). Recently, for example, erionite has been reported in the USSR (18). An unusual ferrierite with a high Si/Al ratio (about 7) was reported in California (82).

The mineralogy of yugawaralite (42), gonnardite (41), stilbite (43), ferrierite (2), offretite and erionite (70), and faujasite (50), have been reported.

No new zeolite minerals have been reported, and the total still stands at 34 recognized species (Table I). However, there is probably little chance of finding zeolite minerals on the surface of the moon.

Zeolite Synthesis. Research on zeolite synthesis has continued unabated. The objectives seem to be three-fold: the preparation of novel synthetic zeolites, the understanding of the mechanism of zeolite nucleation and growth, and the growth of larger single crystals of synthetic zeolites. The latter problem is important for understanding the zeolite crystallization process as well as for providing specimens which are needed urgently for many aspects of zeolite research—e.g., x-ray crystal structure studies and other physical measurements. Table II summarizes those synthetic zeolites reported recently which have been reasonably well characterized. The characterization of a new material as a zeolite must be done with considerable care and involves a variety of contributing techniques. Much of the early work on zeolite synthesis in the published literature is of little value owing to the lack of adequate and necessary details. In many cases, a synthesized material was loosely designated as a synthetic zeolite of some type based on preliminary x-ray powder patterns alone. Much of the earlier work should be reviewed, discredited if necessary, and updated if appropriate. In recent years, zeolites have been synthesized in somewhat different systems—e.g., lithium–sodium. Systems involving tetraalkylammonium bases have resulted in the formation of zeolite Ω and ZSM4.

Substitution of phosphorus in zeolite frameworks is reported in this symposium. The occlusion of phosphate in a zeolite with the type A framework was reported earlier by Kuehl in the synthetic zeolites ZK21 and ZK22 (56, 58). A contribution to an understanding of the

Table I.

No.	Name	Structure Group	Year Discovered
1	Analcime	1	1784
2	Bikitaite	6	1957
3	Brewsterite	7	1822
4	Chabazite	4	1772
5	Clinoptilolite	7	1890
6	Dachiardite	6	1905
7	Edingtonite	5	1825
8	Epistilbite	6	1823
9	Erionite	2	1890
10	Faujasite	4	1842
11	Ferrierite	6	1918
12	Garronite	3	1962
13	Gismondine	1	1816
14	Gmelinite	4	1807
15	Gonnardite	5	1896
16	Harmotome	1	1775
17	Herschelite	4	1825
18	Heulandite	7	1801
19	Kehoeite	1	1893
20	Laumontite	1	1785
21	Levynite	2	1825
22	Mesolite	5	1813
23	Mordenite	6	1864
24	Natrolite	5	1758
25	Offretite	2	1890
26	Paulingite	1	1960
27	Phillipsite	1	1824
28	Scolecite	5	1801
29	Stellerite	7	1909
30	Stilbite	7	1756
31	Thomsonite	5	1801
32	Viseite	1	1942
33	Wairakite	1	1955
34	Yugawaralite	1	1952

mechanism of zeolite crystallization and growth is presented by Aiello, Barrer, and Kerr, page 44.

The growth of zeolite X single crystals was reported by Ciric (30) and more recently in an interesting approach using nonaqueous solvents by Charnell (26), the latter resulting in 100-micron crystals of zeolite X.

Adsorption. During the past years, molecular sieve action has been reported for crystalline and noncrystalline solids other than zeolites. These include coal (64), special active carbons (59), porous glass (84), microporous beryllium oxide powder (47), and layer-type silicates which have been modified by exchange with large organic cations (11). How-

Zeolite Minerals

Typical Occurrence in Igneous Rocks	Occurrence in Sedimentary Rocks
Ireland, New Jersey, etc.	Western U.S., deep sea floor
Rhodesia	
Scotland	
Nova Scotia, Ireland, etc.	Arizona, Nevada, Italy
Wyoming	Western U.S., deep sea floor
Elba	
Scotland	
Iceland	
Rare; Oregon	Nevada, Oregon
Rare; Germany	
Rare; British Columbia	Utah
Ireland, Iceland	
Rare; Italy	
Nova Scotia	
France, Italy	
Scotland	
Sicily	Arizona
Iceland	New Zealand
Rare; South Dakota	
Nova Scotia, Faroe Islands	New Zealand, U.S.S.R., Calif.
Iceland	
Nova Scotia	
Nova Scotia	U.S.S.R., Japan, Western U.S.
Ireland, New Jersey	
Rare; France	
Rare; Washington	
Ireland, Sicily	Western U.S., Africa, deep sea floor
Iceland, Colorado	
Iceland, Ireland, Scotland	
Scotland, Colorado	
Rare; Belgium	
New Zealand	
Japan	

ever, only the crystalline zeolites have attained commercial significance as molecular sieves. Since zeolites are used in many diverse applications including catalysis and ion exchange, the term molecular sieve is not necessarily inclusive or consistent with this wide sphere of application. However, stereoselective catalytic and ion-sieve behavior are well known.

Studies of the fundamental adsorption phenomena on zeolites have continued in several different areas. These are reviewed by Kiselev and Barrer (Vol. II, p. 37). Adsorption studies have been confined largely to a few materials, chiefly those of commercial significance such as zeolites A, X, and Y. In recent years, interest in adsorption kinetics and diffusion

Table II. Some New Synthetic Zeolites

Name	Composition	Structure	Ref.
ZSM-2	$Li_2O \cdot Al_2O_3 \cdot 3.3$–$4.0$ $SiO_2 \cdot H_2O$	Tetragonal $a = 27.4, c = 28.1$A Adsorbs cyclohexane	(31)
ZSM-3	0.3–0.8 $Li_2O \cdot 0.7$–0.2 $Na_2O \cdot Al_2O_3 \cdot 2.8$–$4.5$ $SiO_2 \cdot 9H_2O$	Faujasite-type Adsorbs cyclohexane	(32)
ZSM-4	0.5–0.01 TMA $\cdot 0.5$–0.99 $Na_2O \cdot Al_2O_3 \cdot 6$–$15$ $SiO_2 \cdot 5H_2O$[a]	Cubic $a_o = 22.2$A Adsorbs cyclohexane	(22)
Zeolite N	0.83 ± 0.05 $Na_2O \cdot 0.03 \pm 0.01$ TMA $\cdot Al_2O_3 \cdot 1.8$–2.2 $SiO_2 \cdot YH_2O$	Cubic $a_o = 37.2$A Small pore	(1)
Zeolite Ω	0–0.7 TMA $\cdot 0.5$–1.5 $Na_2O \cdot Al_2O_3 \cdot 6$–$12$ $SiO_2 \cdot 10H_2O$	Hexagonal $a = 18.1, c = 7.59$ Large pore size ~ 11A	(38)
Zeolite α	0.2–0.5 TMA $\cdot 0.5$–0.8 $Na_2O \cdot Al_2O_3 \cdot 4.0$–$7.0$ $SiO_2 \cdot YH_2O$	Cubic, A-Type $a_o = 12.04$Å Adsorbs n-hexane	(78)
Zeolite β	$XNa(1.0 \pm 0.1$–$X)$ TEA $\cdot AlO_2 \cdot 5$–100 $SiO_2 \cdot 4H_2O$[b]	Cubic $a_o = 12.04$ Adsorbs cyclohexane	(77)
Zeolite N-A	TMA $\cdot Al_2O_3 \cdot 2.5$–6.0 $SiO_2 \cdot 7H_2O$	A-Type Cubic $a_o = 12.12$ Adsorbs n-hexane	(9)
ZK-19	$(Na_2O, K_2O) \cdot Al_2O_3 \cdot 3.0$–$6.25$ $SiO_2 \cdot \sim 5H_2O$	Phillipsite-type Absorbs H_2O	(57)
ZK-20	$(0.1$–$0.2)R_2O \cdot (0.8$–$0.9)$ $Na_2O \cdot Al_2O_3 \cdot 4$–$5$ $SiO_2 \cdot YH_2O$[c]	Levynite-type Adsorbs CH_4	(54)
ZK-21	$1.0 \pm Na_2O \cdot Al_2O_3 \cdot YSiO_2 \cdot ZP_2O_5$[d] $Y = 1.9$ to 4.5, $Z = 0.01$ to $\dfrac{Y+2}{48}$	A-type Cubic $a_o = 12.14 - 12.21$ Adsorbs n-hexane	(58)
ZK-22	(TMA, Na_2O)[d]	A-type Adsorbs n-hexane	(56)

[a] TMA = tetramethylammonium.
[b] TEA = tetraethylammonium.
[c]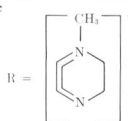

[d] Intercalated phosphate—up to 1 P/β cage. Thus, if $Y = 4$, $Z = 1/8$ unit cell; contents are $Na_8[(AlO_2)_8(SiO_2)_{16}] \cdot 1.0$ PO_2.

in zeolites has been pronounced. Another area of interest is the description of adsorption equilibria in terms of general relations such as the modified Polanyi relationship of Dubinin (35). In many cases, it appears that the Dubinin equation can describe adsorption equilibria quite satisfactorily. The interaction of the adsorbate with aspects of the zeolite structure such as cations with adsorbed molecules, the thermochemistry of zeolite adsorption, and the various components which contribute to the interaction of adsorbed molecules with the zeolite structure have been topics of considerable interest. Studies have been reported on the adsorption of hydrocarbons in zeolite L (10). The adsorption character of zeolite L originally delineated is amplified by these more recent measurements including heats of adsorption. The adsorption isotherms could not be described over the complete range of filling of the zeolite intracrystalline pores by any single model.

Adsorption studies have not been confined to organic vapors and inert gases. Studies of the adsorption of inorganic materials—in particular, mercury, sulfur, and phosphorus—were reported on several ion-exchanged forms of zeolite X, zeolite A, chabazite, and gmelinite (16). The adsorption of rare gases in the less stable minerals heulandite and stilbite were studied (12). These zeolites, with proper partial activation, adsorb the gases argon and krypton, although complete dehydration does collapse the structures of these minerals. Erionite received further attention by Eberly, who investigated the adsorption of normal hydrocarbons in the C_5–C_8 range and compared adsorption rates with similar measurements in zeolite 5A (36).

Simple gas molecules have been used to probe the electrostatic fields in zeolite Y exchanged with univalent and divalent cations (49). The large electric field which should be present near the polyvalent cations was not observed, and the heats of adsorption measured indicate that the field was obscured by strongly bound OH or O ions on the cation.

Adsorption studies on zeolite X and Y of varying sodium content (obtained by a variation in the Si/Al ratio and by NH_4^+ exchange) revealed that there are 2 types of surface cations in the zeolites which exhibit nearly equal adsorption energetics (66). These 2 cations are characterized by site III (which is preferentially removed from the structure) and site II. Adsorption heats for gases such as CH_4, Xe, and Kr suggested an effective charge of 0.66 esu for the surface sodium ions.

Zeolite Catalysts. During the past 3 years, interest in zeolite catalysts has intensified greatly. Zeolites X, Y, and mordenite have been the center of most of the attention, with some recent interest in synthetic zeolites related to erionite and offretite. Controversy still exists concerning the source of activity and the nature of the active sites in zeolites, such as the rare earth exchanged forms of zeolite Y for hydrocarbon conversion reactions. This is reviewed by Rabo (Vol. II, p. 284). At present, opinion seems to converge on the compromise that both the so-called Bronsted and Lewis acid sites are required, with optimum activity coinciding with equal numbers of each (68, 69). Infrared and thermal analytical studies have contributed greatly to an understanding of these materials (3, 25, 80). The diverse activities shown by zeolite catalysts in many different organic reactions is reviewed by Venuto. The relationship between activity and different species of rare earth cations in catalytic cracking are discussed by Scherzer (page 266). Little information on the catalytic properties of potentially important materials such as zeolite L and zeolite Ω has been published, although these materials are of considerable interest.

Zeolite Chemistry. The ion-exchange properties of zeolites X, Y (6, 73), and chabazite (7) have received intensive study in recent years and the ion-exchange properties of zeolite X and zeolite Y have been compared (8). The self-diffusion of sodium in zeolite X has been studied by Brown and Sherry (23). The exchange of ammonium and alkyl derivatives of ammonium ion in zeolites X and Y has been compared by Theng et al. (75). The ion-sieve character of the synthetic zeolite T toward the removal of potassium ions from the original zeolite has been reported by Sherry (71). In short, intensive studies of these several zeolites have revealed a diverse behavior. The relationship between structure, cation sites, and cation population as determined by Si/Al ratio and the ion-exchange behavior of zeolites has been clarified. Cation selectivity has also been related to structure (72). Zeolites offer an untapped source of unique materials for ion-exchange applications, although some applications have been proposed in a limited way. The large majority of available synthetic zeolites and zeolite minerals have received little, if any, attention in ion exchange.

Two additional developments in zeolite chemistry include the removal of tetrahedral aluminum atoms from the framework by complexing agents (53) and the increase in stability of a zeolite by essentially complete removal of the alkali metal cation (3). The latter process—ultrastabilization—is the center of some controversy. Two proposals for explaining the stability of these materials have been advanced; one is based upon the removal of tetrahedral aluminum, which results in an increase in the Si/Al ratio of the framework, the formation of additional O—Si—O linkages, and a decrease in the unit cell constant. The other is based upon the complete removal of alkali metal ion, which may act as a catalyst in perturbing the structure at elevated temperatures. Although there may be merit to one or both proposals, probably neither is the sole explanation.

Thermal stability results are significant in a commercial application of zeolites as catalysts. Since the zeolite must be able to withstand treatment with water vapor at quite high temperatures, it is also probable that a function of the rare earth ion in the hydrocarbon conversion cracking catalyst is to improve the thermal and hydrolytic stability of the zeolite as well as provide a catalytically active site.

From Mossbauer spectroscopy, the environment of Fe^{2+} ions in zeolites has been deduced (34). In summary, research in recent years on zeolites has been concerned with those materials of specific commercial importance. There seems to be a direct relation between the level of scientific interest and the area of major application. This, of course, may be prompted in part by the availability of materials since some zeolite minerals are rare and difficult to obtain. Many zeolites and many potentially interesting aspects of zeolite chemistry and zeolite properties have been neglected so far.

Nomenclature

A statement by Hey (45) that, "the development of mineralogical nomenclature has been rather haphazard and many species and varieties have received totally different names," is applicable to the zeolite nomenclature problem that exists today.

The differentiation of individual mineral species and mineral nomenclature is a continuing problem. In the zeolite group of minerals, the situation has been improved greatly as the result of recent structural analyses. At present, I believe the zeolite mineral group includes 34 different species. Five of these minerals have been discovered since 1950; garronite was reported in 1962 (79) (Table I).

In 1937, Hey (46) defined the zeolites as:
"a group of crystalline solids, hydrated aluminosilicates of mono- and divalent bases in which the ratio $(R'', R_2)O:Al_2O_3$ is unity, which are capable of losing a part or the whole of their water without change of crystal structure, and of adsorbing other compounds in place of the water removed, and which are capable of undergoing base exchange."

This definition is descriptive and still applicable. Today, x-ray powder diffraction methods are indispensable for describing new minerals. Without x-ray diffraction, synthetic zeolites would be hopelessly confused. All of the early attempts to synthesize zeolites are meaningless because x-ray diffraction data are not available.

The description of a new mineral should include accurate and complete determinations of the chemical, optical, and structural properties (60).

The designation of a synthetic zeolite as a separate species must be based on characteristic properties (20). The characterization and classification of synthetic zeolites have been hindered by the lack of a systematic chemical method for naming synthetic, complex aluminosilicates. At present, we have followed the historical "Law of Priority" and have assigned to each new synthetic zeolite an arbitrary letter or group of letters and numbers. The Greek alphabet also has been used. Different letters or code designations assigned to the same synthetic zeolite have led to confusion in the literature.

What are the alternatives? One might develop a system similar to the IUPAC rules for naming complex inorganic compounds, such as the heteropoly salts (*40*). Thus analcime, with typical unit cell contents of $Na_{16}[(AlO_2)_{16}(SiO_2)_{32}] \cdot 16\ H_2O$ could be named sodium 16-alumino-32-silicate-16-water. A nepheline of composition $Na_8(AlO_2)_8(SiO_2)_8$ would be named sodium 8-alumino-8-silicate. The synthetic zeolite type A, with unit cell contents of $Na_{12}[(AlO_2)_{12}(SiO_2)_{12}] \cdot 27H_2O$ (pseudo-cubic cell) could be named sodium 12-alumino-12-silicate-27-water. If half of the sodium ions were replaced by calcium ions, it could be named 6-sodium-3-calcium, etc. This is too unwieldy, and we would probably resort to a simpler designation.

The designation of a synthetic zeolite by an arbitrary code is no less subjective than the historical methods for naming minerals. The characteristic properties which are the basis for designating the synthetic zeolite as a separate species first must be determined carefully. It is unlikely that any proposed system for zeolite nomenclature will meet with complete approval by all concerned. However, the following practices are generally applicable.

(1) The synthetic zeolite is designated by the letter(s) assigned to the zeolite by the original investigator (Law of Priority), for example, zeolite A, zeolite K-G, zeolite α, zeolite ZK-5, etc.

(2) These letters designate the zeolite as synthesized. Thus, zeolite A designates the synthetic zeolite, $Na_{12}[(AlO_2)_{12}(SiO_2)_{12}] \cdot 27H_2O$ (pseudo-cubic unit cell) as prepared in the Na_2O, Al_2O_3, SiO_2, H_2O system, and zeolite L designates the zeolite $K_9[(AlO_2)_9(SiO_2)_{27}] \cdot 22H_2O$ as prepared in the K_2O, Al_2O_3, SiO_2, H_2O system. These refer to typical compositions. We also have used the terms "type A," "type X," etc., which have the same meaning.

(3) In some cases, various investigators have referred to a synthetic zeolite by the name of a related mineral—*e.g.*, "synthetic analcime," "synthetic mordenite," etc. This approach is inaccurate and inadequate. The terms "mordenite-type," "analcime-type," etc., are preferable for indicating that the synthetic zeolite is related structurally to a mineral. In the light of current knowledge of the effect of cation type and location, Si and Al distribution, and Si/Al ratio on the properties of the zeolite, a statement of identity is unwarranted when only similarity is established.

(4) Unfortunately, some confusion is unavoidable since we all use the same alphabet. Thus, Na-B refers to synthetic analcime-type zeolite (*15*), and Na-D indicates a mordenite-type. We have used the letter B to refer to the synthetic zeolite phases which Barrer and others have referred to as "P" (*5*). Since the use of P for these phases has precedence in the published literature, it is preferred. We have used D to refer to

a chabazite-type of synthetic zeolite. In summary, where a letter has been employed to refer to more than 1 synthetic zeolite, additional letter(s) are necessary; in these examples, the symbol for the predominant alkali metal is used.

(5) A problem arises if the synthetic zeolite contains tetrahedral atoms other than Al and Si—*e.g.*, P, Ga, Ge. We use the symbol as a prefix, as P-L, to indicate zeolite L which contains substantial P substitution in the framework; in this case, typical unit cell contents are: $K_{21}[(AlO_2)_{34}(SiO_2)_{25}(PO_2)_{13}]42H_2O$.

(6) It may be necessary to indicate a particular composition in cases where the letter(s) refer to a synthetic zeolite which can vary in framework composition. This is accomplished by giving the Si/Al ratio, unit cell contents, etc.

(7) We have adopted the use of "N" to refer to a synthetic zeolite which is prepared from systems that contain alkyl ammonium bases. Thus, "N-A" refers to a synthetic tetramethylammonium zeolite that has the type A framework (9).

(8) When a different cationic form of a synthetic zeolite is prepared by ion exchange, it may be so referred to—*i.e.*, calcium exchanged (zeolite) A abbreviated as $Ca^{ex}A$ or CaA (48). A hyphen between Ca and A, Ca-A, could refer to a completely different zeolite. Thus:

$$\text{Ca exchanged A} = Ca^{ex}A = CaA \neq Ca\text{-}A$$

Of course, these terms refer to an unspecified level of exchange. It is assumed that the major cation component is, in this example, Ca^{2+} (greater than 50% exchange). In most cases, it is necessary to specify additionally the degree of exchange as per cent of exchange equivalents, or in terms of the unit cell contents. Thus $Ca_2Na_8[(AlO_2)_{12}(SiO_2)_{12}] \cdot XH_2O$ is equal to 33% exchange.

(9) Obviously, any alteration of the parent zeolite by ion exchange, dealumination, decationization, etc., must be specified carefully.

Since the nomenclature problem is a matter of communication, a glossary is necessary. This leads naturally to the scheme for systematizing zeolites by structural classification.

Classification Scheme for Zeolites

Structural classifications of zeolites have been proposed by Smith (74) and Fischer and Meier (37, 61). The classification I propose is a modification of these. It is based on the framework topology of the zeolites for which the structures are known, 44 at present. The classification consists of 7 groups (Table III). Within each group, the zeolites have a common subunit of structure which is a specific array of $(Al,Si)O_4$ tetrahedra. The simplest units are the ring of 4 tetrahedra (4-ring) and 6 tetrahedra (6-ring) as found in many other framework aluminosilicates. These subunits have been called "secondary building units," sbu, by Meier. The primary units are the SiO_4 and AlO_4 tetrahedra. In many

Table III. Classification

Name	Typical Unit Cell Contents
Group 1 (S4R)	
Analcime	$Na_{16}[(AlO_2)_{16}(SiO_2)_{32}] \cdot 16\ H_2O$
Phillipsite	$(K,Na)_{10}[AlO_2)_{10}(SiO_2)_{22}] \cdot 20\ H_2O$
Paulingite	$K_{68}Na_{13}Ca_{3.6}Ba_{1.5}[(AlO_2)_{152}(SiO_2)_{520}] \cdot 700\ H_2O$
Yugawaralite	$Ca_2[(AlO_2)_4(SiO_2)_{12}] \cdot 8\ H_2O$
Group 2 (S6R)	
Erionite[f]	$(Ca,Mg,K_2,Na_2)_{4.5}[(AlO_2)_9(SiO_2)_{27}] \cdot 27\ H_2O$
Offretite[f]	$(K_2,Ca)_{2.7}[(AlO_2)_{5.4}(SiO_2)_{12.6}] \cdot 15\ H_2O$
Omega[g]	$Na_{6.8},TMA_{1.6}[(AlO_2)_8(SiO_2)_{28}] \cdot 21\ H_2O$
T	$Na_{1.2}K_{2.8}[(AlO_2)_4(SiO_2)_{14}] \cdot 14\ H_2O$
Group 3 (D4R)	
A	$Na_{12}[(AlO_2)_{12}(SiO_2)_{12}] \cdot 27\ H_2O$
P	$Na_6[(AlO_2)_6(SiO_2)_{10}] \cdot 15\ H_2O$
Group 4 (D6R)	
Faujasite	$(Na_2,K_2,Ca,Mg)_{29.5}[(AlO_2)_{59}(SiO_2)_{133}] \cdot 235\ H_2O$
X	$Na_{86}[(AlO_2)_{86}(SiO_2)_{106}] \cdot 264\ H_2O$
Chabazite	$Ca_2[(AlO_2)_4(SiO_2)_8] \cdot 18\ H_2O$
Gmelinite	$Na_8[(AlO_2)_8(SiO_2)_{16}] \cdot 24\ H_2O$
ZK-5	$(R_2,Na_2)_{30}[(AlO_2)_{30}(SiO_2)_{66}] \cdot 98\ H_2O$
L[h]	$K_9[(AlO_2)_9(SiO_2)_{27}] \cdot 22\ H_2O$
Group 5 (T_5O_{10})[i]	
Natrolite	$Na_{16}[(AlO_2)_{16}(SiO_2)_{24}] \cdot 16\ H_2O$
Thomsonite	$Na_4Ca_8[(AlO_2)_{20}(SiO_2)_{20}] \cdot 24\ H_2O$
Edingtonite	$Ba_2[(AlO_2)_4(SiO_2)_6] \cdot 8\ H_2O$

[a] Of the 5 space-filling solids of Federov, 3 (cube, hexagonal prism, and truncated octahedron) are found as polyhedral units in zeolite frameworks. The cube, or double four-ring (D4R) as shown here is shown in the list of zeolite polyhedra by Barrer (4) as the 6-hedron. The double six-ring (D6R) is the hexagonal prism or 8-hedron. The α cage is the Archimedean semiregular solid truncated cuboctahedron referred to also as a 26-hedron, Type I. The β cage is the truncated octahedron or 14-hedron, Type I. The γ cage is the 18-hedron and the ε cage the 11-hedron. Other polyhedral units are as given by Barrer.

[b] The framework density is based on the dimensions of the unit cell of the hydrated zeolite and framework contents only. Multiplication by 10 gives the density in units of tetrahedra/1000Å^3.

[c] The void fraction is determined from the water content of the hydrated zeolite.

[d] Refers to the network of channels which permeate the structure of the hydrated zeolite. Considerable distortion may occur in the Group 5 and 7 zeolites upon dehydration.

of Zeolites

Type of Polyhedral Cage[a]	Framework Density, G/Cc[b]	Void Fraction[c]	Type of Channels[d]	Free Aperture of Main Channels, Å[e]
	1.85	0.18	One	2.6
	1.58	0.31	Two	4.2–4.4
α, γ, δ (10-hedron)	1.54	0.49	Three	3.9
	1.81	0.27	Two	3.5
ε, 23-hedron	1.51	0.35	Three	3.6–4.8
ε, 14-hedron (11)	1.55	0.40	Three	3.6–4.8,∥a 6.3,∥c
γ, 14-hedron (11)	1.65	0.38	One	7.5
ε with D6R	1.50	0.40	Three	3.6–4.8
α, β	1.27	0.47	Three	4.2
	1.57	0.41	Three	3.5
β, 26-hedron (11)	1.27	0.47	Three	7.4
β, 26-hedron (11)	1.31	0.50	Three	7.4
20-hedron	1.45	0.47	Three	3.7–4.2
14-hedron (11)	1.46	0.44	Three	3.4–4.1,∥a 6.9,∥c
α, γ	1.46	0.44	Three	3.9
ε	1.61	0.32	One	7.5
	1.76	0.23	Two	2.6–3.9
	1.76	0.32	Two	2.6–3.9
	1.68	0.36	Two	3.5–3.9

[e] Based upon the structure of the hydrated zeolite.

[f] Erionite and offretite also may be considered to consist of double 6-rings linked by single 6-rings.

[g] Zeolite Ω may be considered to consist of single 6-rings linked by double 12-rings.

[h] Zeolite L consists of double 6-rings linked by single 12-rings.

[i] The T_5O_{10} refers to the unit of 5 tetrahedra as given by Meier for the 4–1 type of sbu (61).

[j] The T_8O_{16} unit refers to the characteristic configuration of tetrahedra shown in Figure 8b of Ref. 61.

[k] The $T_{10}O_{20}$ unit is the characteristic configuration of tetrahedra shown in Figure 8a of Ref. 61.

Table III.

Name	Typical Unit Cell Contents
Group 6 $(T_8O_{16})^j$	
Mordenite	$Na_8[(AlO_2)_8(SiO_2)_{40}] \cdot 24\ H_2O$
Dachiardite	$Na_5[(AlO_2)_5(SiO_2)_{19}] \cdot 12\ H_2O$
Epistilbite	$Ca_3[(AlO_2)_6(SiO_2)_{18}] \cdot 18\ H_2O$
Group 7 $(T_{10}O_{20})^k$	
Heulandite	$Ca_4[(AlO_2)_8(SiO_2)_{28}] \cdot 24\ H_2O$
Stilbite	$Ca_4[(AlO_2)_8(SiO_2)_{28}] \cdot 28\ H_2O$

cases, the framework may be considered to consist of larger polyhedral units (4) such as the truncated octahedron and truncated cuboctahedron (8, 33). Some of the sbu's proposed by Meier are complex—such as the 4–1, 5–1, and 4–4–1 units. I have designated the cage-like units with Greek letters as:

α cage = truncated cuboctahedron

β cage = truncated octahedron

γ cage = double 8 with 4-ring bridges

δ cage = double 8-ring

ε cage = "cancrinite" unit or 11-hedron

In earlier classifications, each group is named after a representative member, such as the "mordenite group," natrolite group, etc. I have adopted instead an arbitrary group designation by number. The 7 groups are based on the 7 sbu's of Meier.

The classification is shown in an abbreviated form in Table III with an example to illustrate each group. Included is information on the void fraction, expressed as cc per cc of dehydrated zeolite as derived from the water content for the fully hydrated zeolite, the framework density expressed in grams/cc, the approximate dimensions of the main channels (as derived from the structure of the hydrated zeolite), and the type of channel system, 1 for one-dimensional, etc. In many instances, of course, the dimensions of the apertures change substantially with dehydration. Consequently, the aperture dimensions are not necessarily consistent with the adsorption properties of the dehydrated zeolite. We need considerable information on the structures of dehydrated zeolites. In such a

Continued

Type of Polyhedral Cage[a]	Framework Density, G/Cc[b]	Void Fraction[c]	Type of Channels[d]	Free Aperture of Main Channels, A[e]
	1.70	0.28	Two	6.7–7.0,‖c 2.9–5.7,‖b
	1.72	0.32	Two	3.7–6.7,‖b 3.6–4.8,‖c
	1.76	0.25	Two	3.6–6.3,‖a 4.9–6.3,‖c
	1.69	0.39	Two	3.9–5.4,‖a 4.2–7.1,‖c
	1.64	0.39	Two	4.1–6.2,‖a 2.7–5.7,‖c

classification, the distribution of Si–Al and other tetrahedral atoms must be considered for individual species where possible. The zeolites natrolite, scolecite, and mesolite have the same framework topology but are different mineral species because of framework ordering of Si, Al, and/or cations. Other examples are phillipsite–harmotome, thomsonite–gonnardite, analcime–wairakite, and chabazite–herschelite.

Topics Which Need Attention

Several subjects which suffer from inadequate research need real effort to increase our knowledge and expand utilization of molecular sieve zeolites. Of course, many of these subjects are being studied in more than one laboratory because many have considerable industrial significance.

First, more complete information is needed on the structures of zeolites which have been resolved only partially in terms of framework. We need to know more about cation locations, how they change with cation species in terms of population, how they are altered by the removal of water, and about changes in the framework during removal of water. For the synthetic zeolites, the x-ray crystallographer will be aided by the ability to grow single crystals of sufficient size to study by single-crystal methods.

The changes which result from removal of tetrahedral aluminum ions by chemical methods are of considerable interest (52) but structural analyses are needed to determine these changes.

Second, several types of unknown zeolite structures have been postulated. The chemist presently is guided by qualitative concepts concerning the mechanism of zeolite formation. We would hope that in the future we will be able to synthesize a zeolite with a previously postulated structure—much like the practice of today's synthetic organic chemist. We now believe it is possible to synthesize zeolites with other framework atoms such as phosphorus. This leads to interesting possibilities, such as a framework which bears no charge or possibly a positive charge, thus containing exchangeable anions. Indeed, the zeolite synthesis area offers many intriguing possibilities.

Thirdly, chemical properties must be elucidated. For example, ion exchange was recognized early as a characteristic property, yet we know little about the ion exchange behavior of most of the zeolites. The ion exchange character of some zeolites, reviewed above, has been studied in detail and in depth. These in-depth studies have revealed unexpected cation selectivities and have shown that zeolites may make interesting ion separations. Optimization of these properties may lead to important applications in solving environmental problems. The novel application of the zeolite properties in nuclear chemistry has been demonstrated. The Szilard-Chalmers reaction to form actinides within the structure of zeolite X by the usual (n,γ) reaction is known. Recoil of the nuclide after neutron capture permits easy elution (24).

The behavior of zeolites in biological systems is a bit mysterious. For example, treatment of yeast mitochondria with zeolite A or X in the sodium or potassium cation forms disaggregates the mitochondria and solubilizes cytochrome oxidase in an aqueous system. However, the same zeolites in calcium or magnesium form showed little activity of this type (67). Similarly, zeolite A can disrupt bacterial cells (83).

Applications of the Future

Some urgent environmental problems may be solved or alleviated in the next decade by using the known properites of known zeolites and closely related materials or perhaps by preparing a new zeolite with desired chemical or physical properties. The ion exchange properties of certain zeolites may be used selectively to remove certain radioactive isotopes from nuclear wastes (21, 62). Also, clinoptilolite may be used to remove ammonia from secondary sewage effluents by ion exchange (28). Development of a catalytic process to remove sulfur dioxide from waste gases using a molecular sieve has been indicated (27, 29).

Relative to the behavior of zeolites in biological systems, a "biozeolitic" theory of sewage purification was proposed more than 30 years ago. It was reported then that sewage clarification by the so-called activated sludge process depends on the presence of an aluminosilicate complex which is said to be chemically the same as "zeolites" (76).

Literature Cited

(1) Acara, N. A., U. S. Patent **3,414,602** (1968).
(2) Alietti, A., Passaglia, E., Scarni, G., *Am. Mineralogist* **1967**, 52, 1562.
(3) Ambs, W. C., Flank, W. H., *J. Catalysis* **1969**, 14, 118.
(4) Barrer, R. M., *Chem. Ind. London* **1968**, 1203.
(5) Barrer, R. M., Baynham, J. W., Bultitude, F. W., Meier, W. M., *J. Chem. Soc.* **1959**, 195.
(6) Barrer, R. M., Davies, J. A., Rees, L. V. C., *J. Inorg. Nucl. Chem.* **1968**, 30, 3333.
(7) *Ibid.*, **1969**, 31, 219.
(8) *Ibid.*, **1969**, 31, 2599.
(9) Barrer, R. M., Denny, P. J., Flanigen, E. M., U. S. Patent **3,306,922** (1967).
(10) Barrer, R. M., Lee, J. A., *Surface Sci.* **1968**, 12, 341.
(11) Barrer, R. M., Millington, A. D., *J. Colloid Interface Sci.* **1967**, 25, 359.
(12) Barrer, R. M., Vaughan, D. E. W., *Surface Sci.* **1969**, 14, 77.
(13) Barrer, R. M., Villiger, H., *J. Chem. Soc.* (D), **1969**, 659.
(14) Barrer, R. M., Villiger, H., *Z. Krist.* **1969**, 128, 352.
(15) Barrer, R. M., White, E. A. D., *J. Chem. Soc.* **1952**, 1561.
(16) Barrer, R. M., Whiteman, J. L., *J. Chem. Soc.,* A **1967**, 13, 19.
(17) Bartl, H., Fischer, K. F., *Neues Jahrb. Mineral. Monatsh.* **1967**, 33.
(18) Belitskii, I. A., Bukin, G. V., *Dokl. Akad. Nauk SSSR* **1968**, 178, 169.
(19) Breck, D. W., Acara, W. A., U. S. Patent **2,950.952** (1960).
(20) Breck, D. W., Flanigen, E. M., *Proc. Intern. Symp. Mol. Sieves, 1st, London, 1967,* **1968**, 10.
(21) Bray, L. A., U. S. Energy Comm. **BNWL-288** (1967).
(22) British Patent **1,117,568** (1968).
(23) Brown, L. M., Sherry, H. S., Krambeck, F. J., *Abstr. Natl. Meeting ACS, 158th,* New York, Sept., 1969, Phys. 125, 126.
(24) Campbell, D., *Inorg. Nucl. Chem. Letters* **1970**, 6, 103.
(25) Cattanach, J., Wu, E. L., Venuto, P. B., *J. Catalysis* **1968**, 11, 342.
(26) Charnell, J. F., to be published.
(27) *Chem. Eng. News* **1968**, 46 (39), 22.
(28) *Ibid.*, **1968**, 46 (35), 39.
(29) *Chem. Week* **1967**, 101 (27), 37.
(30) Ciric, J., *Science* **1967**, 155, 689.
(31) Ciric, J., U. S. Patent **3,411,874** (1968).
(32) Ciric, J., U. S. Patent **3,415,736** (1968).
(33) Cundy, H. M., Rollett, A. P., "Mathematical Models," 2nd ed., Oxford Univ. Press, London, 1960.
(34) Delgass, W. N., Garten, R. L., Boudart, M., *J. Chem. Phys.* **1968**, 50, 4603.
(35) Dubinin, M. M., *Zh. Fiz. Khim.* **1965**, 39, 1305.
(36) Eberly, P. E., Jr., *Ind. Eng. Chem. Prod. Res. Develop.* **1969**, 8, 140.
(37) Fischer, K. F., Meier, W. M., *Fortschr. Mineral.* **1965**, 42, 50.
(38) Flanigen, E., Netherlands Patent **6,710,729** (1968).
(39) Galli, E., Gottardi, G., *Mineral. Petrogr. Acta* **1966**, 12, 1.
(40) "Handbook for Chemical Society Authors," p. 42, The Chemical Society, London, 1960.
(41) Harada, K., Iwamoto, S., Kihara, K., *Am. Mineralogist* **1967**, 52, 1785.
(42) Harada, K., Nagashima, K., Kinichi, K., *Am. Mineralogist* **1969**, 54, 306.
(43) Harada, K., Tomita, K., *Am. Mineralogist* **1967**, 52, 1438.
(44) Hay, R. L., GSA Special Paper No. **85**, Geological Society of America, New York, 1966.
(45) Hey, M. H., "An Index of Mineral Species and Varieties Arranged Chemically," p. *xvi*, British Museum, London, 1955.

(46) Hey, M. H., *Trans. Brit. Ceram. Soc.* **1937**, 36, 84.
(47) Horlock, R. F., Anderson, P. J., *Trans. Faraday Soc.* **1967**, 63, 531.
(48) Hoss, H., Roy, R., *Beitr. Mineral. Petrog.* **1960**, 7, 389.
(49) Huang, Y., Benson, J. E., Boudart, M., *Ind. Eng. Chem. Fundamentals* **1969**, 8, 346.
(50) Iiyama, K., Harada, K., *Am. Mineralogist* **1969**, 54, 182.
(51) Kawahara, A., Curien, H., *Bull. Soc. Franc. Mineral. Crist.* **1969**, 92, 250.
(52) Kerr, G. T., *J. Phys. Chem.* **1968**, 72, 2594.
(53) *Ibid.*, **1969**, 73, 2780.
(54) Kerr, G. T., U. S. Patent **3,459,676** (1969).
(55) Kerr, I. S., Williams, D. J., *Acta Cryst.* **1969**, Sect. B., 25, 1183.
(56) Kuehl, G. H., *Abstr. Natl. Meeting A.C.S., 158th*, **1969**, Inor. 37.
(57) Kuehl, G. H., *Am. Mineralogist* **1969**, 54, 1607.
(58) Kuehl, G. H., U. S. Patent **3,355,246** (1967).
(59) Lamond, T. G., Metcalfe, J. E., III, Walker, P. L., Jr., *Carbon* **1965**, 3, 59.
(60) McConnell, D., *Am. Mineralogist* **1948**, 43, 260.
(61) Meier, W. M., *Proc. Intern. Symp. Mol. Sieves, 1st, London, 1967*, **1968**, 10–27.
(62) Mercer, B. W., *Natl. Meeting Am. Inst. Chem. Engrs., 66th, Portland, Oregon, 1969*, **1968**, paper 22d.
(63) Merkle, A. B., Slaughter, M., *Am. Mineralogist* **1968**, 53, 1120.
(64) Metcalfe, J. E., III, Kawahata, M., Walker, P. L., Jr., *Fuel* **1963**, 42, 233.
(65) Minato, H., *Koatsu Gasu* **1968**, 5, 536.
(66) Neddenriep, R. J., *J. Colloid Interface Sci.* **1968**, 28, 293.
(67) Person, P., Zipper, H., Felton, J. H., *Arch. Biochem. Biophys.* **1969**, 131, 457.
(68) Richardson, J. T., *J. Catalysis* **1967**, 9, 182.
(69) *Ibid.*, **1968**, 11, 275.
(70) Sheppard, R. A., Gude, A. J., III, *Am. Mineralogist* **1969**, 54, 875.
(71) Sherry, H. S., Intern. Conf. Ion Exchange in the Process Industries, Soc. Chem. Ind., London, 1969.
(72) Sherry, H. S., *Ion Exchange* **1969**, 2, 89.
(73) Sherry, H. S., *J. Colloid Interface Sci.* **1968**, 28, 288.
(74) Smith, J. V., *Mineral. Soc. Am. Spec. Paper* No. 1, 1963.
(75) Theng, B. K. G., Vansant, E., Uytterhoeven, J. B., *Trans. Faraday Soc.* **1968**, 64, 3370.
(76) Theriault, E. J., McNamee, P. D., *Ind. Eng. Chem.* **1936**, 28, 79, 83.
(77) Wadlinger, R. L., Kerr, G. T., U. S. Patent **3,308,069** (1967).
(78) Wadlinger, R. L., Rosinski, E. J., Plank, C. J., U. S. Patent **3,375,205** (1968).
(79) Walker, G. P. L., *Mineral. Mag.* **1962**, 33, 173.
(80) Ward, J. W., *J. Phys. Chem.* **1968**, 72, 4211.
(81) Wells, A. F., "Third Dimension in Chemistry," Oxford Univ. Press, London, 1956.
(82) Wise, W. S., Nokleberg, W. J., Kokinos, M., *Am. Mineral.* **1969**, 54, 887.
(83) Wistreich, G., Lechtman, M. D., Bartholomew, J. W., Bils, R. F., *Appl. Microbiol.* **1968**, 16, 1269.
(84) Yastrebova, L. S., Bessonov, A. A., Khvoschchev, S. S., *Materialy Vses. Soveshch. po Tseolitam, 2nd, Leningrad* **1964**, 229.

RECEIVED February 18, 1970.

Discussion

W. M. Meier (Eidgenossische Technische Hochschule, Zurich): Most classification schemes which have been proposed so far are based on certain subunits of structure or secondary building units. The choice of these is frequently somewhat arbitrary. However, these secondary building units should always satisfy the following requirements: The unit cell must contain an integral number of secondary building units, and the secondary building units should be compatible with the observed Si/Al ratio(s). The units stated for Groups 6 and 7 are not in agreement with the first requirement. The yugawaralite framework (Group 1) is based on single 4-membered rings whereas analcime is not if requirement 2 is taken into account.

2

Some Problems of Zeolite Crystallization

S. P. ZHDANOV

I. V. Grebenshchikov Institute of Silicate Chemistry,
Academy of Sciences USSR, Nab.Makarova, 2, Leningrad, V-164, USSR

> *In some zeolite crystallization in heterogenous aluminosilicate systems, temperature is mainly a kinetic factor whereas the nature and composition of crystals are determined by concentrations and correlations of components. Studies of their chemical structure and the peculiarities of crystallization kinetics show that a solubility equilibrium exists between the solid and liquid phases of gels. Gel crystallization on heating is preceded by increased component concentrations in the liquid phase, which determines the composition of the zeolite crystals formed. Decreases in liquid component concentrations are compensated by dissolution of the solid phase of gels. This explains the linear rate of crystal growth during the first part of the crystallization process. Crystal size distribution for zeolite A and the linear rate of crystal growth indicate the autocatalysis of zeolite crystallization.*

Owing to systematic studies by Barrer in zeolite synthesis with the use of the autoclave technique (*2–12*), more than 50 species of zeolites differing in composition and structure were obtained. Temperature boundaries of zeolite crystallization were determined on $t°–n$ diagrams ($n = SiO_2/Al_2O_3$). Many Na-, K-, and Na,K-zeolites, including zeolites A, X, and Y, were synthesized by Breck and others (*13–17, 22*) at low temperatures without the autoclave.

Considerable progress achieved in the low-temperature synthesis of zeolites is connected with the use of highly reactive materials in the colloid-dispersed state (aluminosilica gels, aqueous silica sols, silica, alumina gels, etc.). The use of these materials in the majority of cases facilitates synthesis, but at the same time it involves certain difficulties. In many such cases, not only the kinetics of the process but also the actual final results appear to be dependent on the complicated chemical behavior of these materials during their preparation, in the course of crystallization, or on stirring. Initial materials can significantly influence the results of crystallization at elevated temperatures under conditions of the autoclave technique (1, 8, 19, 20, 23, 33). These dependences as yet are not understood, but must be explained in terms of the peculiarities of the mechanism and kinetics of zeolite crystallization.

Studies on the mechanism of zeolite crystallization have received extensive attention from many investigators (4, 16, 22, 24, 29, 30, 33, 37, 39). However, the main questions of the mechanism of crystallization referring to kinetics, such as the autocatalysis of the crystallization process, alkalinity effect on the rate of crystallization, the nature of the induction period, and seeding effects, have been insufficiently studied. The question of the role of solid and liquid phases in the formation of zeolite crystals in the heterogenous aluminosilicate systems also is still being discussed (16, 22, 25, 30, 31, 37, 39).

During recent years, new data have been obtained and published on the influence of temperature and composition of the initial mixtures as well as seeding and addition of complexing reagents upon the kinetics and final results of the process of zeolite crystallization (16, 20, 23, 31, 32, 33, 34, 35, 36, 37, 39, 41).

According to Refs. 6, 19, 20, 23, 32, and 35, the temperature of zeolite crystallization can be lowered significantly by increasing the alkalinity of initial mixtures. However, increases in temperature and alkalinity equally influence the component concentrations in the liquid phases of heterogenous mixtures. Therefore, the probability of formation of the zeolites in such cases is determined by the concentration of components rather than by the temperature. At the constant temperature, the nature and composition of crystals are determined mainly by chemical factors—i.e., by concentrations and correlations of components in the systems (27, 31, 35, 37, 39, 41) and more directly in the liquid phase (39). Therefore, investigations into the chemistry and kinetics of zeolite crystallization are important in studying the formation and growth of zeolite crystals in heterogenous aluminosilicate systems. Unfortunately, the composition of the liquid phase of such aluminosilicate systems remains insufficiently studied in connection with zeolite crystallization.

Chemical Structure of Alkali Aluminosilica Gels

Alkali aluminosilica gels from which zeolites usually are crystallized represent heterogenous colloidal systems consisting of liquid and solid phases which strongly differ in their chemical compositions. In those cases when aluminosilica gels are obtained under equal conditions and from the same initial materials—for example, from solutions of the silicates and aluminates—the results of their crystallization reproduce well,

Table I. Changes in Concentrations and Correlations of Components in Their General Compositions; Characteristics

Gel Samples	Concentrations and Correlations in Initial Mixtures			$\dfrac{SiO_2}{Al_2O_3}$ in Gel Skeleton
	$\dfrac{SiO_2}{mole/l}$	$\dfrac{Na\text{-}Al}{Si}$	$\dfrac{SiO_2}{Al_2O_3}$	
V	0.133	25.0	0.33	2.2
VIII	0.184	12.5	0.33	2.6
965	0.239	5.8	0.33	2.8
967	0.270	13.7	1.0	2.6
963	0.620	2.0	1.0	3.0
VII	0.710	3.5	2.0	2.3
467	0.770	2.0	2.0	2.8
III	0.470	9.2	5.0	2.5
966	0.930	3.1	5.0	3.4
196	1.340	1.2	5.0	3.5
660	1.210	0.9	36.8	6.6
104	2.140	0.9[a]	28.0	—
105	2.530	0.7[a]	25.6	8.9

[a] [(Na + K) − Al]/Si.
[b] Ch = K,Na-chabazite.
[c] Er = K,Na-erionite.

being determined only by gel composition at a given temperature (*31, 37–41*). Changes in the conditions of gel preparation—that is, the use of silica sols instead of the silicates, aging of the obtained gels at room temperature, or stirring—can lead to different results. The nature of crystallizing zeolites and kinetics of crystallization in such cases are not determined only by the common composition of the system and by temperature. They probably depend on the differences in concentrations and correlations of components in different phases of gels, on their distribution among the phases—*i.e.*, on the chemical structure of gels. Therefore, investigations into the chemical structure of aluminosilica gels in terms of their crystallization may appear useful for understanding the peculiarities of formation and growth of zeolite crystals in the heterogenous aluminosilicate systems.

Analyses of the solid phases of many samples of aluminosilica gels showed that the Si/Al ratio in the solid phase of gels (in the gel skeleton) always exceeds 1, whereas the Na/Al ratio is close to 1; *i.e.*, these relations in the gel skeleton are the same as those in the crystal lattice of zeolites. It was concluded from these results (*41*) that Al in the skeleton of alkali aluminosilica gels is coordinated four-fold within the common (Si,Al,O)-framework, whereas alkaline cations compensate excess negative charges of aluminum–oxygen tetrahedra. Analogous conclusions were drawn by the authors (*21*) who obtained the same correlations of comin the Liquid and Solid Phases of Aluminosilica Gels with Changes of Zeolite Crystals Obtained

Concentrations and Correlations in Liquid Phase of Gels				Zeolite Crystals	
$Al(OH)_4^-$, mole/l	SiO_2, mole/l	$[Al][Si] \times 10^2$	$Si/Al \times 10^3$	$\dfrac{SiO_2}{Al_2O_3}$	Type
0.466	0.030	1.40	64	1.8	A
0.800	0.026	2.08	33	—	A
1.096	0.027	2.96	24	1.8	A
0.218	0.054	1.17	25	—	A
0.570	0.011	0.63	19	1.9	A
0.066	0.062	0.41	940	—	A
0.064	0.023	0.15	356	2.0	A
0.061	0.263	1.57	4300	—	X
0.034	0.530	1.80	15600	2.6	X
0.020	0.548	1.09	27400	3.0	X
0.007	1.000	0.70	143000	5.1	Y
0.024	2.030	4.88	85000	5.1	Ch[b]
0.028	1.980	5.55	71000	7.2	Er[c]

ponents in the solid phase for many samples of aluminosilica gels of different compositions.

Consequently, the (Si,Al,O)-framework of aluminosilica gels can be similar to that of zeolites, at least in regard to the simplest structural elements forming the network. Apparently, the gel skeleton consists of more complicated structural elements such as the single and double 4- and 6-membered rings and their combinations. This can be suggested, for instance, from the data given below on the compositions of erionite-like crystals (perhaps erionite + offretite) and the aluminosilicate skeleton of gels yielding these crystals.

	Na_2O	K_2O	Al_2O_3	SiO_2
Initial mixture	0.71	0.28	1.0	26
Gel skeleton	0.27	0.73	1.0	8.9
Crystals	0.29	0.71	1.0	7.2

This may indicate that in both the (Si,Al,O)-framework of K,Na-gels and crystal lattice of erionite, similar structural elements are present, formation of which requires the participation of potassium ions. Potassium ions will play the dominant role in the erionite structure and in the gel skeleton on potassium deficit in the initial mixture. Such structural elements can be built from 6-membered rings of tetrahedra like columns of erionite structure. Potassium ions are blocked inside the elements of erionite structure and owing to this are nonexchangeable (37, 40).

Analytical data on the compositions of solid and liquid phases for different aluminosilica gels are given in Table I.

The data given in Table I show that the distribution of components between the liquid and solid phases of aluminosilica gels appears to be a complicated function of the initial solution compositions. A general tendency exists for the increase in Si/Al ratio in both the solid phases of gels and zeolite crystals with increasing SiO_2/Al_2O_3 ratio ($= n$) in the initial mixtures. At equal n, the Si/Al ratio in the solid phase of gels increases with decreasing concentration of the excess alkali (in respect to Al_2O_3 content); *i.e.*, with decrease of $Na-Al/SiO_2$ ratio. Such dependence was shown for zeolite crystals in Ref. 37.

This dependence can be explained in terms of the polycondensation mechanism of formation of aluminosilica gel skeleton from $Al(OH)_4^-$ hydroxoaluminate ions and $(OH)_{4-m}Si(O^-)_m$ silicate ions with different degrees of hydroxylation. At equal correlations and concentrations of the silicate and aluminate ions in solutions, the amount of silicate ions participating in the reaction of condensation should increase with the growth of their hydroxylation degree. The degree of hydroxylation of the silicate ions in alkaline solutions is determined by Equilibrium 1 and therefore it must increase with decrease in $NaOH/SiO_2$ ratio ($= m$) and with dilution.

$$Si(OH)_4 + m(Na^+ + OH^-) \rightleftarrows (OH)_{4-m}Si(O^-Na^+)_m + mH_2O \quad (1)$$

Increase in the Si/Al ratio in the crystals of synthetic faujasites (37, 41) and analcimes (35) with decreasing alkalinity of initial gels and mixtures can be explained on the basis of these considerations.

Growth of Si/Al ratio with increasing water content was observed in Ref. 37 for zeolite crystals obtained from usual Na- and K-aluminosilica gels and in Ref. 27 for zeolites obtained from gels with phosphate additions.

General dependences observed for the change in Si/Al ratio in both the amorphous (Si,Al,O)-framework of gel skeleton and zeolite frameworks (Table I) suggest the same mechanism of their formation in solution.

Correlations and concentrations of components in the liquid phase of gels appear to be quite different from those in the solid phase and initial mixtures. It is essential that in all cases both the silicate and aluminate ions are present in the liquid phase of gels, and the product of their concentrations is comparatively constant at significant changes in the absolute values for ion concentrations of each type. This can be an indication of the existence of an equilibrium similar to that of dissociation or solubility.

amorphous aluminosilicate ⇌ aluminate ions + silicate ions

(solid phase)   (solution)

aluminosilicate ions
(solution)

(2)

Though systems of this type cannot be regarded as thermodynamically true equilibrium states, the existence of quasi-equilibria between the solid and liquid phases of aluminosilica gels is confirmed by good reproducibility of the compositions of their solid and liquid phases in repeated syntheses.

In the case of gels prepared from solutions of the aluminates and silica sols, the concentrations and correlations of components in the liquid phase appear to be essentially different, as compared with gels of the same compositions obtained from solutions of the silicates and aluminates (Table II).

Table II. Concentrations in the Liquid Phase, Mole/L

Gel samples	Na_2O	K_2O	Al_2O_3	SiO_2	SiO_2/Al_2O_3
105 from silicate	0.72	0.21	0.018	2.12	120
105 from sol	0.86	0.25	0.020	0.05	2.6
237 from silicate	1.42	—	0.061	2.42	40
237 from sol	1.73	—	0.034	0.04	1.2

The compositions of liquid phases were analyzed at once after the gel formation. In both cases, the concentrations of silica in the liquid phases of gels obtained from silica sols appeared to be very low, and consequently the SiO_2 content in the skeleton must be higher. All colloidal silica in such gels is present in their skeleton and is not found analytically in the liquid phase. The skeleton of gels obtained from sols, unlike the gels prepared from the silicate solutions, does not consist of continuous (Si,Al,O)-framework only; it must contain inclusions of SiO_2 colloidal particles unstable in alkali media. Therefore, no equilibrium exists between the solid and liquid phases of such gels after their preparation. Differences in chemical structure of gels obtained from the silicate solutions and from silica sols naturally should affect the behavior of gels during their crystallization and the final results as well.

Thus, studies on the chemical structure of alkali aluminosilica gels utilized in zeolite synthesis reveal a complicated dependence in the distribution of components between the solid and liquid phases. At the same time, in the case of gels prepared from the homogenous solutions, the structural elements of disordered (Si,Al,O)-network in the gel skeleton and those of the regular (Si,Al,O)-frameworks in zeolites probably are similar. The distribution of components between the liquid and solid phases of gels depends on the method of gel preparation and initial materials used.

This dependence should be more evident for heterogenous aluminosilicate mixtures utilized in zeolite synthesis at elevated temperatures. Studies of the liquid phase compositions of such mixtures directly under the synthesis conditions appear to be extremely important for the explanation of the final results of crystallization.

As seen from the data of Table I, the nature and composition of zeolite crystals formed from aluminosilica gels are closely connected with the compositions of their liquid phases.

Chemical Changes in Aluminosilica Gels in the Course of Their Crystallization

Data on the compositions of liquid and solid phases of aluminosilica gels given in Table I refer to the equilibrium state at room temperature. Crystallization of gels proceeds, as a rule, on heating; therefore, it is essential to know what chemical changes take place in gels under these conditions. Table III shows the changes in composition of the liquid phase of gels after their heating at 90°C for 4 hours.

These data show that increases in concentrations of all components in the liquid phase of gels take place on heating. This may be connected with the growth of solubility of the aluminosilicate skeleton with increasing temperature. Consequently, increases in concentration of the silicate and aluminate ions in the liquid phase precede the beginning of gel crystallization.

Changes in concentrations in the liquid phase of Na,K gels during erionite crystallization are illustrated by Figure 1, where the continuous curves refer to the gel obtained from the mixtures of silicate and aluminate solutions, and the dashed curve represents the gel prepared from silica sol. In both cases, the total composition of initial mixtures is the same (7.1 Na_2O · 2.8 K_2O · Al_2O_3 · 25.6 SiO_2 · 433 H_2O). Points of the axis of ordinates correspond to the component concentrations in the initial mixtures.

Table III. Changes in Component Concentrations (Mole/L) in

Gel Samples	Na_2O	Al_2O_3	SiO_2	$[Al][Si] \times 10^2$
		20°C		
965	1.20	0.548	0.027	2.96
963	0.77	0.285	0.011	0.63
266	0.21	0.016	0.008	0.026
196	0.91	0.009	0.330	0.60
660	0.45	0.004	1.000	0.80

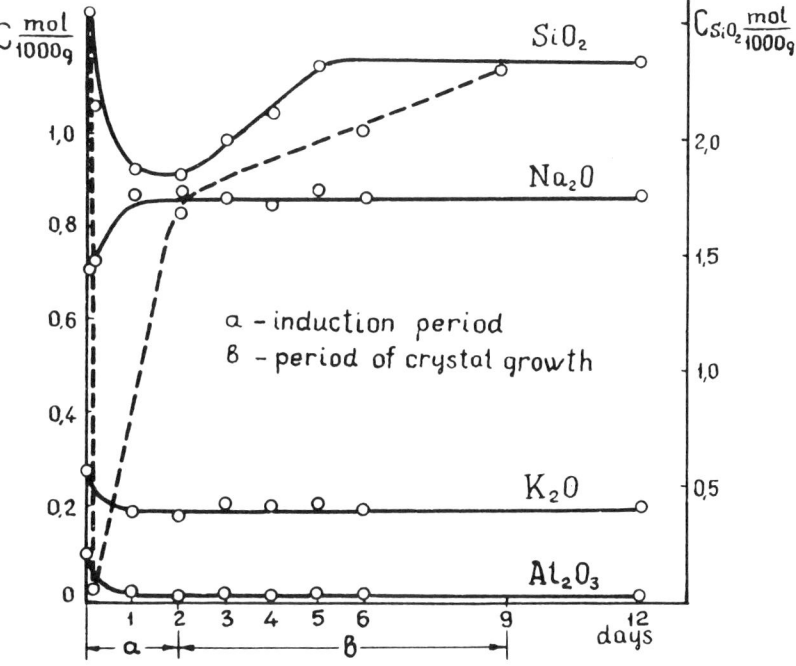

Figure 1. Changes in component concentrations in liquid phase of gel obtained from solution of silicates and aluminates and from silica sol (dotted curve) during crystallization

Formation of gel from the silicate and aluminate mixtures immediately leads to the decrease of concentrations in solution. Changes observed during the first 24 hours of heating can be owing to the establishment of solubility equilibrium at the temperature of the experiment (100°C). After this, the concentrations of all components except SiO_2 remain practically constant. Increase in the SiO_2 concentration observed after 48 hours of heating can be explained by the beginning of erionite crystallization, since the Si/Al ratio in erionite crystals is less than that in the solid phase of gels yielding these crystals (3.6 instead of 4.45).

the Liquid Phase of Aluminosilica Gels after Heating at 90°C

	90°C		
Na_2O	Al_2O_3	SiO_2	$[Al][Si] \times 10^2$
1.23	0.669	0.063	8.46
0.87	0.371	0.015	1.11
0.24	0.012	0.017	0.041
1.01	0.010	0.388	0.78
0.66	0.008	1.243	1.99

Changes in SiO_2 concentration in the liquid phase of gels obtained from silica sols are quite different. In this case, the liquid phase of freshly precipitated gel appears to be very poor in silica. A sharp increase in SiO_2 concentration during the first period of gel heating is caused by the dissolution in the alkaline intermicellar liquid of the SiO_2 colloid particles which form the inhomogenous skeleton of such gels together with (Si,Al,O)-framework. After 48 hours of heating, the SiO_2 concentration in the liquid phase reaches the values at which erionite crystallization proceeds in gel obtained from the silicate. Consequently, gels prepared from silica sols are really nonequilibrium systems in relation to SiO_2 concentration in the liquid phase, in contrast to gels obtained from solutions of the silicates and aluminates.

As seen from Table I, the Si/Al ratio in zeolite crystals in all cases is less than that in the skeleton of aluminosilica gels; therefore, the excess silica must pass from the gel skeleton into the liquid phase during crystal formation in gels.

When crystallization of aluminosilica gels is fully completed, the solution appears to be in contact not with amorphous but only with crystalline aluminosilicate. Therefore, other equilibrium concentrations should be established in the mother liquor. The data (37) show that the concentration of aluminate ions in the mother liquor is always lower than that in the liquid phase of gels, whereas the concentration of silicate ions is either higher or lower than that in the gel liquid phase. However, the decrease in the product of concentrations of the silicate and aluminate ions in mother liquors is a general regularity, as visibly evidenced by the data of Table IV obtained at 20°C.

These data show that the solubility of zeolite crystals, as could be expected, is lower than that of the amorphous aluminosilicate phase of gels from which the crystals were formed.

Thus, increases in concentrations of all components in the liquid phase of aluminosilica gels during the period preceding crystallization (Table III) are attributable to the growth of solubility of their solid amorphous phase with increasing temperature. As for the changes in concentrations occurring in the liquid phase during crystallization, they are connected with the formation of the crystalline phase in which composition and solubility differ from those of the amorphous aluminosilicate solid phase of gels.

Table IV. Aluminosilica Gel and Zeolite Crystal Solubility

Gel Samples	V	965	967	462	963	467	196	105
[Al] [Si] × 10^2								
In liquid phase	1.4	2.96	1.17	0.76	0.63	0.15	1.09	5.6
In mother liquor	0.48	0.70	0.24	0.63	0.14	0.07	0.28	5.3

On the Peculiarities of Kinetics of Zeolite Crystallization

The pecularities of zeolite crystallization reported in several works (*6, 13, 16, 20, 24*) are the induction period, autocatalytic nature of the process, growth of crystallization rate with increasing alkali concentration, and effects of seeding on the kinetics.

Induction Period. The causes of the induction period have been insufficiently cleared up. In this period which precedes rapid crystallization, the formation of nuclei and their growth to a size exceeding a critical one (*13, 16, 20*) take place. However, the induction period decreases with increasing temperature and alkali concentration in the mixtures (*16, 20, 24*), and it depends on the nature of the initial aluminosilicate materials utilized in the synthesis (*20*). These facts indicate that the induction period can be associated with the dissolution of components of the solid aluminosilicate phase during the period preceding crystallization.

As follows from our published data (*37, 39*) and the data of Table I on the crystallization of any zeolite in the heterogenous aluminosilicate systems, the nature and composition of crystals are determined directly by the composition of the liquid phase. Concentrations and correlations of components in the liquid phases of the heterogenous aluminosilicate mixtures needed for the beginning of crystallization can be reached by dissolving the solid phase. The rate of dissolution increases with increasing temperature and alkali concentration, reducing the induction period (*16, 20*).

In those cases when the initial materials are mixtures of alkaline solutions and crystalline aluminosilicates, the dissolution of crystalline material accompanied by the formation of amorphous aluminosilicate phase apparently precedes the beginning of crystallization (*6*). The induction period in such cases can be associated with the duration of dissolution of the initial crystalline material in the alkaline liquid phase. At a considerable degree of supersaturation of the solution, the formation of aluminosilica gels as an intermediate product appears to be necessary.

As for the gels prepared from colloidal aqueous silica and alkaline aluminate solutions, the dissolving of SiO_2 colloid particles of gel skeleton in the alkaline solution precedes the beginning of crystallization.

If, in the course of heating the gels up to a definite temperature, the concentrations of components in the liquid phase reach values allowing the formation and growth of nuclei, crystallization can start immediately after reaching this temperature without any induction period. The curves in Figure 2 show the time dependence of the size changes in zeolite A cubic crystals formed from gels of the same composition ($2.8\ Na_2O \cdot Al_2O_3 \cdot 1.9\ SiO_2 \cdot 427\ H_2O$) at different temperatures. Each point is an average result of 20–30 measurements of the edge length of the largest

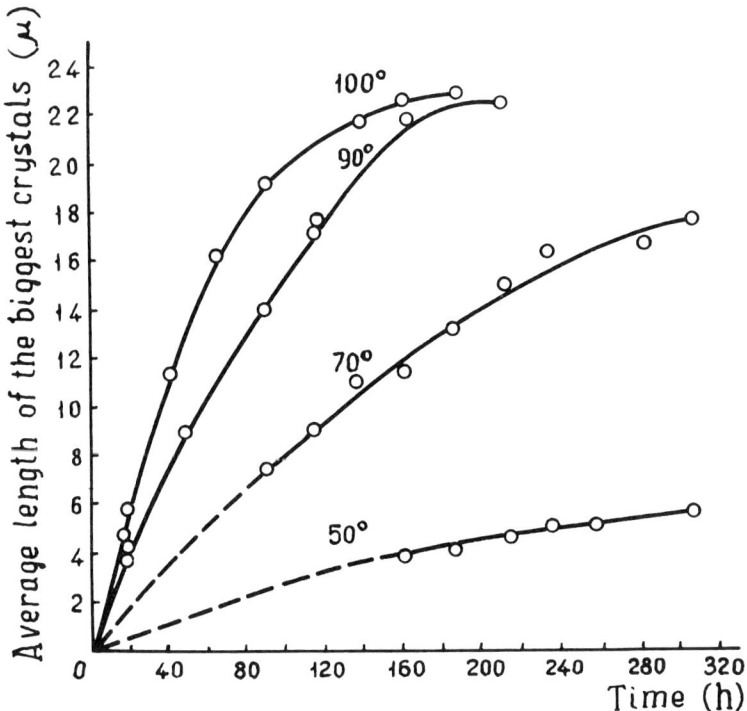

Figure 2. Growth of zeolite A crystals in aluminosilica gels of the same compositions at different temperatures

crystals under a microscope. For at least 2 temperatures (90° and 100°C), one can surely affirm the absence of a prolonged period preceding the beginning of crystal formation. The induction period here may be considered as the time required to achieve the temperature of crystallization. The absence of an induction period during the crystallization of zeolite A from gels was indicated by the authors (*18*).

Consequently, the induction period is the time required for providing the conditions of nuclei formation rather than the time which is necessary for the growth of nuclei to a critical size.

Autocatalytic Nature of Zeolite Crystallization. The autocatalytic nature of crystallization reported in Refs. *13*, *16*, *20*, and *24* suggests the existence of a period during which the acceleration of crystallization is observed at the constant temperature of the experiment.

Kerr (*24*, *25*) showed that the rate of zeolite formation increases proportionally to the quantity of crystalline product present.

$$\frac{dZ}{dt} = kZ \qquad (3)$$

The rate of any crystallization process is determined by the rate of nuclei formation and crystal growth.

Equation 3 indicates the increase in either the linear rate of crystal growth or the rate of nuclei formation during the period of crystallization described by the equation. Yet, there is a lack of data in the literature concerning the rate of growth of zeolite crystals and that of nucleation. Data given in Figure 2 show that the autocatalytic nature of the crystallization process cannot be attributed to a change in the rate of crystal growth in the course of crystallization. The rate of crystal growth is the greatest at the beginning of crystallization, gradually decreasing as the process develops.

The observed acceleration of crystallization evidently must be connected with the increase in the rate of nuclei formation after the beginning of crystallization. This finds confirmation in the results of the analysis of the curve for crystal size distribution of zeolite A obtained at 90°C (Figure 3) and the curve of crystal growth for the same zeolite under the same conditions (Figure 2).

The time passed from the beginning of crystallization to the formation of nuclei of crystals of each size can be calculated from the data represented by the curves in Figures 2 and 3.

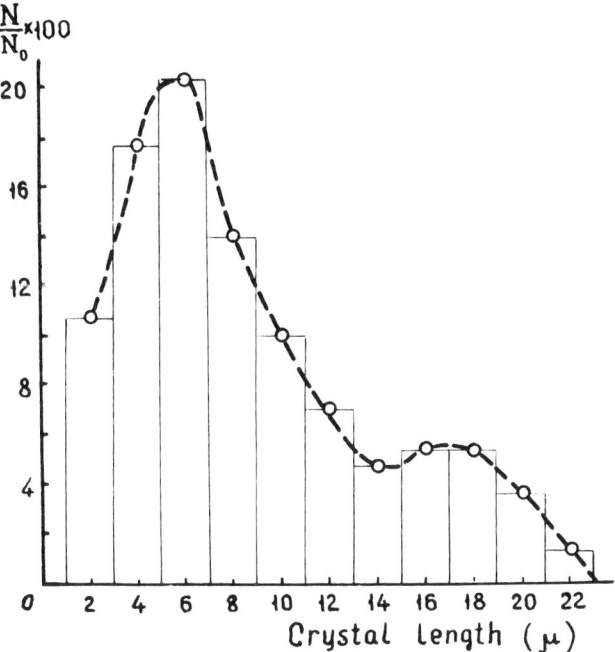

Figure 3. *Crystal size distribution of zeolite A in the product of crystallization (90°C).*

After the transformation of the scale of crystal size in Figure 3 into the scale of the time of crystal formation, it is easy to obtain the dependences shown in Figure 4. The only assumption made was that the linear rate of crystal growth is independent of its size. This assumption seems reasonable. As seen from the slope of the curves of Figure 2 (curves of 100° and 90°C), the rate of crystal growth remains practically constant up to 8–12μ. Decrease in the rate of crystal growth observed later can be caused by the decreasing component concentrations in solutions.

As follows from Figure 4, the formation of nuclei takes place during the entire process of crystallization, but the rate of nucleation is increasing only during its first period. This conclusion is confirmed by the results of mathematical analysis of the experimental curves for crystallization as a function of time obtained by Breck and Flanigen for zeolites A and X, and by Domine and Quobex for mordenite. In all cases, these curves are described by a common equation in the form $Z = kt^n$. Similar dependence was found by Ciric (*J. Colloid Interface Sci.* **1968**, 28, 315) for the crystallization of zeolite A. The continuous curves in Figure 5 are experimental curves, and the points denote the magnitudes of conversion calculated from this equation at the corresponding values of the constants k and n.

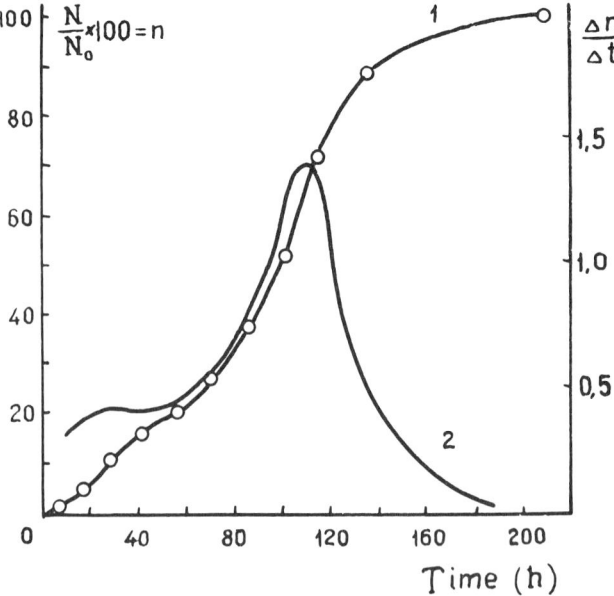

Figure 4. Growth of the number of nuclei (1) and change in the rate of their formation (2) during crystallization of zeolite A

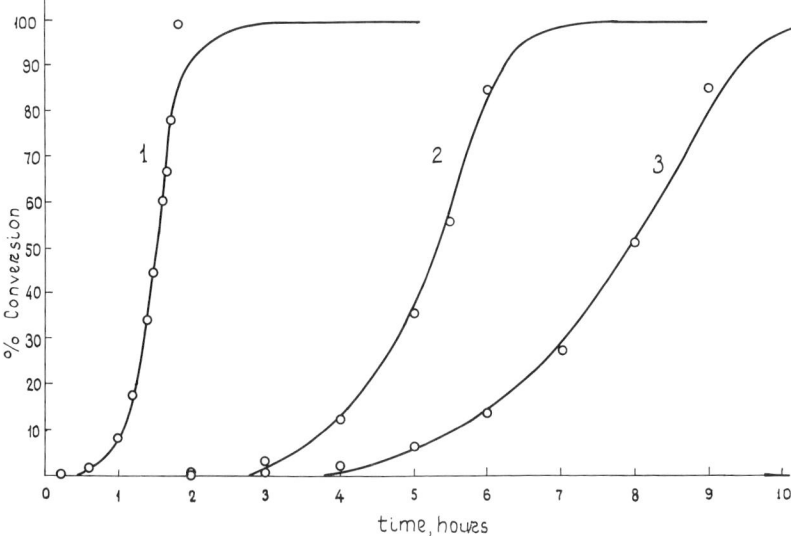

Figure 5. Crystallization of zeolites as a function of time

1. Zeolite A, 100°C
2. Zeolite X, 100°C (Breck and Flanigen)
3. Mordenite, 300°C (Domine and Quobex)
—— Experimental curves   ○ Points calculated from $Z = kt^n$

As Figure 5 shows, the calculated points agree quite well with the experimental curves. The calculations confirm that the autocatalytic growth of the mass of crystals can take place during the induction period. However, as the absolute quantities of crystals formed during this period are low, they cannot be found by existing methods.

The values of the constants k and n are easily calculated from the plot of $\log z$ vs. $\log t$ (Figure 6). It is essential that the constant values of n for all three cases are $n > 4$.

Theoretically, at the constant linear growth rate of crystals and at the constant nucleation rate, n must be equal to 4. The values of $n > 4$ obtained from the calculations are connected with the growth of the nucleation rate during the initial autocatalytic period of crystallization.

Such an increase in the nucleation rate can be explained by postulating that not only the aluminosilicate blocks formed in the liquid phase but also the similar blocks with ordered structure occurring in the gel skeleton can be the nuclei of crystals. The number of such blocks passing into solution and coming out at the surface of gel particles for a unit of time must increase with increasing dissolution rate of the gel skeleton during the autocatalytic stage of crystallization.

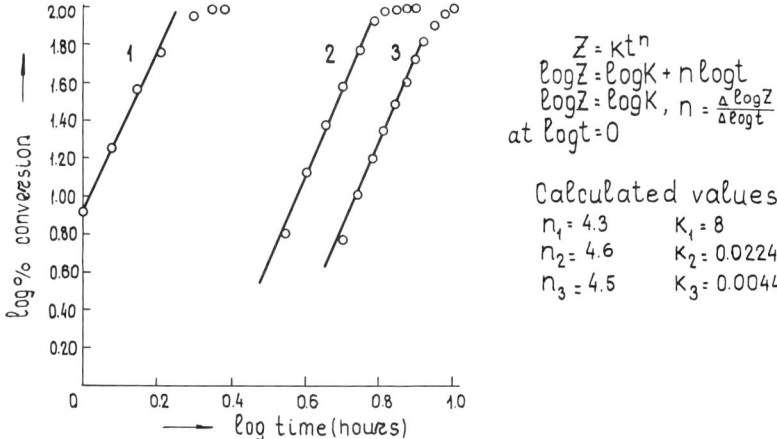

Figure 6. Kinetics of zeolite crystallization, log Z vs. log t
1. Zeolite A, 100°C
2. Zeolite X, 100°C (Breck and Flanigen)
3. Mordenite, 300°C (Domine and Quobex)

Temperature Dependence of the Rate of Crystal Growth of Zeolite A. The rate of zeolite crystallization is significantly dependent on temperature. Using the temperature–time dependence of crystallization, Breck and Flanigen (16) estimated the activation energy of the crystallization process for zeolites A, X, and Y. The values obtained are 11, 14, and 15 kcal/mole, respectively. Similarly, from kinetic data reported in Ref. 20, the activation energy of mordenite crystallization can be estimated at about 11 kcal/mole. The values obtained for activation energy are important as energetic characteristics for the zeolite crystallization process as a whole, yet they give no answer to the question as to which of the probable stages of the crystallization process (dissolution of the amorphous phase, formation of nuclei, or crystal growth) limits the rate of crystallization. In this light, studies on the temperature dependence of the rate of zeolite crystal growth are of interest.

The influence of temperature on the rate of crystal growth of zeolite A is illustrated by the curves in Figure 2. A linear dependence of the log of the rate of crystal growth agginst $1/T$ (Figure 7) was obtained from the results given in Figure 2. The value of 10.5 kcal/mole for activation energy of the zeolite A crystal growth was calculated from the slope of the straight line in Figure 7. It is in good agreement with the value of 11 kcal/mole found for activation energy of zeolite A crystallization in Ref. 16.

The closeness of the values for activation energy of the crystallization process as a whole and for one of its stages, crystal growth, suggests that in this case the rate of crystal growth but not the rate of nucleation and that of dissolution or diffusion limits the rate of crystallization at the temperature of crystallization. Really, the rate of nuclei formation in aluminosilica gels must be very high and, as seen from Figure 4, it increases during crystallization, accelerating the process. The activation energy of dissolution and diffusion in solutions can be expected (28) to be considerably lower than the 10.5 kcal/mole calculated above for crystal growth. According to our preliminary data, the activation energy of dissolution of aluminosilica gel studied above (Figure 2) in 0.5N NaOH is about 5 kcal/mole. However, at low alkali concentrations in gels, the rate of crystallization may be limited also by that of dissolution.

The conclusions made here concern the crystallization of aluminosilica gels, but Kerr (24) came to analogous conclusions in studying the formation of zeolite A from the amorphous aluminosilicate, treating it with sodium hydroxide solution—*i.e.*, under conditions different from those of the usual crystallization of aluminosilica gels.

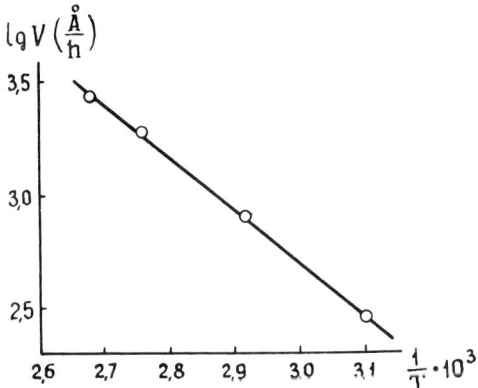

Figure 7. Log of the rate of crystal growth (Å/h) vs. reciprocal of the absolute temperature

Effect of Alkalinity on the Rate of Crystallization. The causes of the effect of alkalinity on the rate of crystallization have not been discussed in the literature. The authors (22) only reported that crystallization is catalyzed by excess alkali. Usually, an increase in alkali concentration in aluminosilica gels and aluminosilicate mixtures leads to a decrease in both the duration of crystallization and crystal sizes (32, 38). This can be explained by the growth of the rate of nucleation with increasing alkali concentration. Therefore, a greater number of nuclei should form during crystallization in a more alkaline medium.

Table V. Component Concentrations in the Liquid Phase of Gels (Mole/L) and Rate of Growth of Zeolite A Crystals at 90°C

Gel Samples	NaOH	Al_2O_3	SiO_2	$[Al][Si] \times 10^4$	V_{av} µ/Day
266–1	0.560	0.0266	0.0103	5.46	3.05
266–2	0.460	0.0258	0.0085	4.00	2.25
266–3	0.428	0.0208	0.0092	3.80	1.80
266–4	0.302	0.0104	0.0108	2.25	1.25

As seen from Table I, the change in alkali concentration in gels causes complicated changes in SiO_2 and Al_2O_3 concentrations in the liquid phase.

Low-alkali aluminosilica gels do not crystallize at all, even at 90°C. This indicates the existence of some minimum alkali concentrations at which crystallization at a given temperature becomes possible.

Figure 8 illustrates the unusual effect of changes in alkali concentrations in aluminosilica gels on the sizes of zeolite A crystals formed in the 4 sample gels with the same ratios of $Na_2O : Al_2O_3 : SiO_2 = 2.8 : 1 : 1.9$ (alkali concentration in gel was changed by dilution). In this case, the increase in alkali concentration led to the growth of crystal size, not to its decrease as in other cases (*38*).

To understand this apparent discrepancy, the compositions of liquid phases of the other samples of gels with the same oxide ratios were investigated together with the rates of crystal growth. The results obtained are given in Table V.

The data of Table V show that the average rate of crystal growth at a given temperature is approximately a linear function of the product of concentrations of the silicate and aluminate ions in the liquid phase of gel. Consequently, in the case shown in Figure 8, the change in crystal size can be caused by changes in concentrations of other components rather than a change in alkali concentration.

Seeding Effects. Crystallization of some zeolites (scolecite, natrolite) proceeds only in the presence of seeds (*26, 36*). The mechanism of seeding is insufficiently clear in such cases.

Breck and Flanigen (*16, 22*) observed the acceleration of zeolite A crystallization by the addition of seeds in aluminosilica gels, but found no significant increase in crystal size. This was indicated also by the authors (*18*).

Treating the amorphous powdered aluminosilicate with NaOH solution under dynamic conditions, Kerr showed (*24*) that the use of zeolite A crystals as seeds in the liquid phase considerably accelerates the conversion of the amorphous phase into the crystalline one. To understand the mechanism of aluminosilica gel crystallization, it is essential to know whether the seed growth occurs in the course of crystallization. As follows from Kerr's data obtained under special conditions of his experiments, the seed crystals must grow at the expense of soluble components of the liquid phase, though no measurements of crystal size were performed in his work (24).

Seeding effects of different synthetic zeolites on the results of aluminosilica gel crystallization have been investigated in our laboratory by Samulevich and Shubaeva. The greater the addition of seed crystals, the less is the duration of crystallization. Direct measurements of crystal size showed that the seed crystals grow during crystallization at a rate which differs only slightly from that of the growth of forming crystals. The maximum size of zeolite A crystals obtained from gels without seeds was 19μ, while the size of the seed crystals in the next test of the same gel amounted to 36μ. These results are of great importance for understanding the mechanism of zeolite crystallization.

On the Mechanism of Crystallization of Aluminosilica Gels

In the majority of investigations on zeolite synthesis, particular attention was paid to the effect of temperatures on crystallization. Recent works on mordenite crystallization (20, 32) indicate that temperature

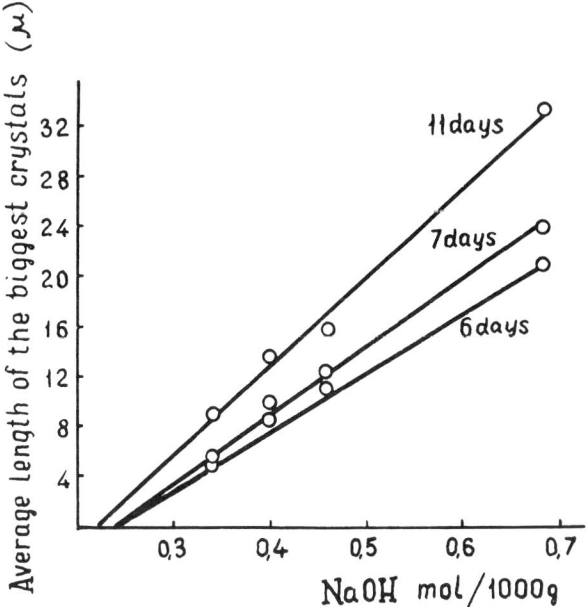

Figure 8. Dependence of crystal growth of zeolite A on alkali concentration in gel at 90°C

influences the kinetics of the process rather than its final results. The temperature of zeolite crystallization can be significantly reduced either by increase in alkalinity of the heterogenous aluminosilicate mixtures or by the use of more readily soluble initial materials (6, 20, 33, 34). Since the increase in both the temperature and alkalinity raises the solubility of the solid aluminosilicate phase, one can suggest that the composition of the liquid phase in such heterogenous systems has an essential effect on crystallization, yet there is a lack of information on this question in the literature.

Investigations carried out in our laboratory (37, 39) and results reported here refer to aluminosilica gels. No common opinion exists on the mechanism of crystallization of aluminosilica gels and the role of their solid and liquid phases. The authors (22, 29, 30) suggest that zeolite crystals are formed in the solid phase of gels. In addition, it is considered that no dissolution of the solid phase occurs during crystallization (22) and the components of the liquid phase do not directly participate in the formation of crystals (29, 30).

The experiments on the substitution of liquid phase in aluminosilica gels (37, 39, 41) clearly showed that the crystallization of aluminosilica gels cannot be considered as simple ordering of the structure of the gel skeleton without any participation of the liquid phase and without a transport of the components of the solid phase into solution during crystallization.

In our understanding (41), based on the results of investigation of the chemical structure of aluminosilica gels, the nuclei of zeolite crystals begin to form in the liquid phase of gels or at the interface of gel phases. The growth of crystal nuclei proceeds at the expense of aluminosilicate hydrated anions occurring in the solution. These anions represent different combinations of (Si,O)- and (Al,O)-tetrahedra, as it was postulated by Barrer et al. (4). These units can be the structural blocks of the growing crystals. Their compositions and structures are given by the compositions of liquid phases. The growth of crystals leads to the dissolving of the solid phase during all the period of crystallization.

To explain the autocatalytic nature of aluminosilica gel crystallization, it is assumed that the gel skeleton, being x-ray amorphous as a whole, contains together with the disordered (Si,Al,O)-network the simplest structural blocks—for example, in the form of the single and double 4- and 6-membered rings of (Si,O)- and (Al,O)-tetrahedra. Structurally, these aluminosilicate blocks are similar to those of the zeolite frameworks but they differ from them chemically by the presence of terminal Si–OH, Al–OH, and Si–O$^-$Na$^+$ (or generally Si–O$^-$R$^+$) groups which have not been used in polycondensation reactions of the formation of gel skeleton owing to the high reaction rate. The structure and Si/Al ratio of these

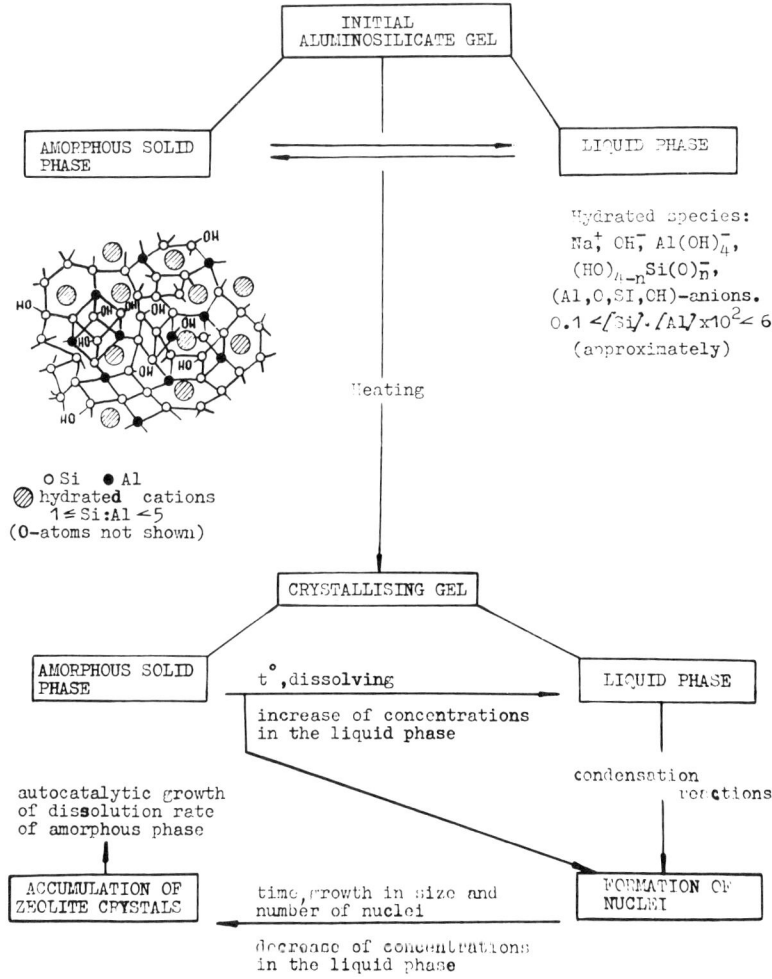

Figure 9. Schematic representation of aluminosilica gel crystallization

blocks as well as the composition of the gel skeleton as a whole and that of the liquid phase of gel are set by the composition of the initial mixture.

The solid and liquid phases of aluminosilica gels are connected by the solubility equilibrium. Owing to this, both the aluminosilicate and silicate ions are always present in the liquid phase of aluminosilica gels (Table I) and the equilibrium product concentrations of the ions (the solubility product of amorphous aluminosilicate) depend on the composition of amorphous aluminosilicate and temperature. In heating the gels, their solubility increases and Equilibrium 2 moves to the right. This leads to increased concentration of the silicate, aluminate, and aluminosilicate ions in the liquid phase. As a result, the probability of condensation reactions between the ions increases, giving rise to the formation of primary aluminosilicate blocks (4- and 6-membered rings) and crystal nuclei. The formation and growth of crystal nuclei lead to the exhaustion of liquid phase in the simplest silicate, aluminate, and aluminosilicate ions, and the equilibrium state is reached by permanent dissolving of the solid phase. Because of the lower solubility of zeolite crystals in comparison with that of the amorphous aluminosilicate skeleton of gels from which they are formed (Table IV), the crystallization process must continue to complete dissolution of the amorphous phase.

Closeness of the compositions between the solid phase of gels and zeolite crystals obtained from them (Table I) provides relative constancy of composition in the liquid phase during crystallization; i.e., stability of the conditions of nucleation. This may explain the ease of obtaining the pure zeolite phases in the crystallization of aluminosilica gels.

During the dissolving of the aluminosilicate skeleton of gels, both the simplest silicate and aluminate ions and aluminosilicate blocks transport into the liquid phase. These blocks may be regarded as prepared structural elements for growing crystals and nuclei. Consequently, a situation develops when structural elements of nuclei arise at the expense not only of reaction in solution but also of the breakdown of the gel skeleton during its dissolution. It has to lead to increases in the rate of nucleation and quantity of growing nuclei and crystals (Figure 4). At the constant rate of crystal growth at the expense of the components in the liquid phase (Figure 2), this must lead to an increase in the rate of dissolution of the solid phase—i.e., to the autocatalytic acceleration of crystallization.

Schematic representation of the crystallization of aluminosilica gels according to our understanding of the process is given in Figure 9.

As the nuclei and crystals grow at the expense of components of the liquid phase of gels, the composition of crystals should depend on that of the liquid phase. This is confirmed by the data given in Table VI.

Table VI. Dependence of Si/Al Ratio of Zeolite X on the Composition of Gel Liquid Phase

Gel Samples	Concentrations in Liquid Phase, Mole/1000 Grams of Solution			$\frac{SiO_2}{Al_2O_3}$	$\frac{Na_2O_{exc.}}{SiO_2}$	Si/Al in Zeolite X Crystals
	Na_2O	Al_2O_3	SiO_2			
196–1	0.702	0.0124	0.397	32.0	1.75	1.63
196–2	0.825	0.0114	0.304	26.6	2.68	1.38
196–3	0.926	0.0105	0.220	21.0	4.17	1.33
196–4	0.955	0.0116	0.174	15.0	5.42	1.15
196–5	1.200	0.0202	0.117	5.8	10.05	1.07

It is essential to note that both Si/Al ratio and NaOH content in all gels used was constant (Si/Al = 2.5; NaOH concentration = 2.61 mole/l). The data of Table VI indicates that the growth of the Si/Al ratio in the crystals of synthetic faujasite is directly associated with increases in the SiO_2 concentration and Si/Al ratio in the liquid phase of the gels.

As we concluded above, the activation energy of crystallization corresponds to that of crystal growth. The activation energies of different zeolites are close. For example, for mordenite and zeolite A, the calculated values mentioned above are 10 to 11 kcal/mole; *i.e.*, activation energy is equivalent to the energy of 2 hydrogen bonds. It can be connected with the necessity of dehydration of the silicate

$$\begin{array}{ccc} \overline{O} & OH & H \\ \diagdown & \diagup & \diagup \\ & Si & O \\ \diagup & \diagdown & \diagdown \\ O & OH & H \end{array}$$

and aluminate

$$\begin{array}{ccc} OH & OH & H \\ \diagdown & \diagup & \diagup \\ & Al & O \\ \diagup & \diagdown & \diagdown \\ OH & OH & H \end{array}$$

ions in solution before the condensation reactions between the ions could take place.

Much evidence can be presented in favor of the proposed mechanism of aluminosilica gel crystallization, some of which was reported earlier (*37, 39*). The most convincing arguments supporting this mechanism are the growth of seed crystals in gels, dependence of the rate of crystal growth upon the concentration of components in the liquid phase, and dependence of the Si/Al ratio in crystals on the composition of the liquid phase.

The probability of crystallization of aluminosilica gels with removal of a considerable part of the liquid phase (which is suggested as the main supporting evidence for the formation of zeolite crystals without any participation of liquid phase) is quite real from our point of view. The equilibrium between the solid and liquid phases of gels does not depend on the volume of solution but only on the common composition of gel and on the temperature. So, removal of some quantity of liquid phase will not lead to the change in equilibrium and should not influence the crystallization results of such gels.

In zeolite crystallization from the heterogenous aluminosilicate mixtures, the dissolving of some quantity of solids must precede the beginning of crystallization (induction period). For one specific case, it was well illustrated by Kerr (24).

Literature Cited

(1) Ames, L. L., *Am. Mineralogist* **1963**, 48, 1374.
(2) Barrer, R. M., *J. Chem. Soc.* **1948**, 2158.
(3) Barrer, R. M., Baynham, J. W., *J. Chem. Soc.* **1956**, 2822.
(4) Barrer, R. M., Baynham, J. W., Bultitude, F. W., Meier, W. M., *J. Chem. Soc.* **1959**, 195.
(5) Barrer, R. M., Baynham, J. W., McCallum, N., *J. Chem. Soc.* **1953**, 4035.
(6) Barrer, R. M., Cole, J. F., Sticher, H., *J. Chem. Soc.* **1968**, A, 2475.
(7) Barrer, R. M., Denny, P. J., *J. Chem. Soc.* **1961**, 971.
(8) *Ibid.*, **1961**, 983.
(9) Barrer, R. M., Marshall, D. J., *J. Chem. Soc.* **1964**, 485.
(10) Barrer, R. M., McCallum, N., *J. Chem. Soc.* **1953**, 4029.
(11) Barrer, R. M., White, E., *J. Chem. Soc.* **1951**, 1267.
(12) *Ibid.*, **1952**, 1561.
(13) Breck, D. W., *J. Chem. Educ.* **1964**, 41, 678.
(14) Breck, D. W., Eversole, W. G., Milton, R. M., *J. Am. Chem. Soc.* **1956**, 78, 2338.
(15) Breck, D. W., Eversole, W. G., Milton, R. M., Reed, T. B., Thomas, T. L., *J. Am. Chem. Soc.* **1956**, 78, 5963.
(16) Breck, D. W., Flanigen, E. M., "Molecular Sieves," p. 47, Society of the Chemical Industry, London, 1968.
(17) Breck, D. W., Smith, J. V., *Sci. Am.* **1959**, 200, 88.
(18) Budnikov, P. P., Petrovykh, N. M., *Zh. Prikl. Khim.* **1965**, 38, 10.
(19) Coombs, D. S., Ellis, A. J., Fyfe, W. S., Taylor, A. M., *Geochim. Cosmochim. Acta* **1959**, 17, 53.
(20) Domine, D., Quobex, J., "Molecular Sieves," p. 78, Society of the Chemical Industry, London, 1968.
(21) Fahlke, B., Wieker, W., Thilo, E., *Z. Anorg. Allgem. Chem.* **1966**, 347, 82.
(22) Flanigen, E. M., Breck, D. W., ACS, 137th Meeting, Cleveland, Ohio, 1960.
(23) Hawkins, D. B., *Mater. Res. Bull.* **1967**, 2, 951.
(24) Kerr, G. T., *J. Phys. Chem.* **1966**, 70, 1047.
(25) *Ibid.*, **1968**, 72, 1385.
(26) Koizumi, M., Roy, R., *J. Geol.* **1960**, 68, 41.
(27) Kuhl, G. H., "Molecular Sieves," p. 85, Society of the Chemical Industry, London, 1968.
(28) Kuznetsov, V. A., "Hydrothermal Synthesis of Crystals," p. 77, Nauka, Moscow, 1968.
(29) Myrsky, Ya. V., Mitrofanov, M. G., Dorogochinski, A. V., "New Adsorbents—Molecular Sieves," Grosnyi, 1964.
(30) Myrsky, Ya. V., Mitrofanov, M. G., Popkov, B. M., Bolotov, L. T., Ruchko, L. F., "Zeolites, Their Synthesis, Properties and Utilization," p. 192, Nauka, Moscow–Leningrad, 1965.
(31) Ovsepyan, M. E., Zhdanov, S. P., *Izv. Akad. Nauk SSSR, Ser. Khim.* **1965**, 11.
(32) Sand, L. B., "Molecular Sieves," p. 71, Society of the Chemical Industry, London, 1968.

(33) Senderov, E. E., "Geochemical Investigations on High Temperatures and Pressures," p. 173, Nauka, Moscow, 1965.
(34) Senderov, E. E., *Geochim.* **1963**, 820.
(35) Senderov, E. E., "Zeolites, Their Synthesis, Properties and Utilization," p. 165, Nauka, Moscow–Leningrad, 1965.
(36) Senderov, E. E., Khitarov, N. I., *Geochim.* **1966**, 1398.
(37) Zhdanov, S. P., "Molecular Sieves," p. 62, Society of the Chemical Industry, London, 1968.
(38) Zhdanov, S. P., Buntar, N. N., "Synthetic Zeolites," p. 105, Nauka, Moscow, 1962.
(39) Zhdanov, S. P., Egorova, E. N., "Chemistry of Zeolites," Nauka, Leningrad, 1968.
(40) Zhdanov, S. P., Novikov, B. G., *Izv. Akad. Nauk SSSR, Ser. Khim.* **1966**, 44.
(41) Zhdanov, S. P., Samulevich, N. N., Egorova, E. N., "Zeolites, Their Synthesis, Properties and Utilization," p. 129, Nauka, Moscow–Leningrad, 1965.

RECEIVED February 20, 1970.

Discussion

W. H. Flank (Houdry Laboratories, Marcus Hook, Pa. 19061): In crystallizing a given zeolite from a 2-phase nutrient system, the composition in the liquid phase, especially the effective hydroxyl ion concentration, and the temperature must be controlled so that a balance is maintained between the ratio of the various silica and alumina species being obtained from dissolution of the solid phase and the ratio of these species already present in the liquid phase. Degree of supersaturation should also be controlled. Since the rate dependence of the various reaction steps is not constant for all species, lack of such control may have a severe effect on the system. What was done to the gel samples in Table VI, which had constant Si/Al and Na/Si ratios, to produce the various ratios noted for the liquid phase?

S. P. Zhdanov: The changes in the liquid phase composition were reached by diluting the initial silicate and aluminate solutions. The concentrations of NaOH in the initial mixtures were constant but not the Na/Si ratios. The constancy of NaOH concentration was reached by diluting the solution.

3

Stages of Zeolite Growth from Alkaline Media

R. AIELLO, R. M. BARRER, and I. S. KERR

Physical Chemistry Laboratories, Imperial College, London S.W. 7, England

> *A study has been made of stages of aggregation from alkaline aluminosilicate solutions. Appropriate mixtures containing M_2O, Al_2O_3, SiO_2, and H_2O (where $M = Li$, Na, or K) were made with sufficient water and strength of alkali that the solids initially dissolved to give clear liquids. These liquids were heated at $80°C$ without stirring. Solids appeared, the evolution of which was followed by electron microscopy and electron and x-ray diffraction. The solid phase appeared initially as laminae, mostly amorphous. After longer heating, the solids contained thin laminae and thicker, broken laminae which seemed to be evolving into larger particles. At this stage, the x-ray pattern showed clear arcs caused by zeolites. Finally, the laminae were completely replaced by zeolites, nucleation of which appeared to be heterogeneous.*

Under suitably alkaline conditions and with a sufficient volume of water, it is possible to dissolve aluminosilicate gels to form clear liquids free of any residual solids. Such liquids are of interest because from them one may hope to examine the first stages of zeolite crystallization. The question of homogeneous vs. heterogeneous nucleation (7, 8) of zeolite crystals may be studied usefully when clear liquids provide the reaction mixture. For these reasons, we have examined the first solids to appear from such liquids and have followed their subsequent evolution.

Experimental

Limpid liquids were obtained by dissolving a small quantity of amorphous silica powder and freshly prepared aluminum hydroxide (total weight of oxides about 2 to 3 grams) in 1 liter of hot $1N$ Na, K, or Li hydroxide solution. The $SiO_2:Al_2O_3$ molar ratios were about 4. The liquids so prepared were filtered and each was divided into 4 parts, which were put in plastic bottles in an oven at 80°C without stirring. After a somewhat variable interval, usually of several hours, there was precipitation of solids. The bottles were removed, one at a time at intervals, from the oven, and the solids were separated from the liquid by centrifugation and washed. A little of the material suspended in water was placed on grids for subsequent electron microscopic examination, and the remainder was dried for x-ray study. Because so little was needed for the electron microscope, it was possible to obtain samples for this examination corresponding to shorter reaction times than were involved in collecting the first samples for x-ray examination.

Electron Microscopy

The first solids to form were thin bouyant particles which tended to circulate through the liquid by convection. As they aged, they became denser and settled at the bottom of the vessel. The freshly formed solids could, because of their bouyancy, be gel-like, with a considerable water content.

In all cases considered, these freshly formed particles appeared under the electron microscope to consist mainly of lamellae such as are shown in Figures 1A, 2A, and 3. As the reaction time increased, in addition to the thin flakes many thicker and broken ones were observed (Figures 4 and 5), as well as some thick particles or conglomerates which were

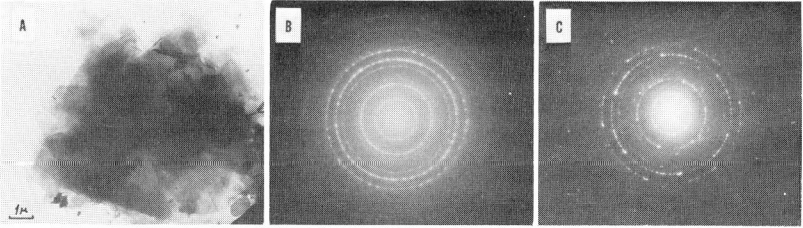

Figure 1. Laminae with the appearance of a smectite from solutions containing Na^+

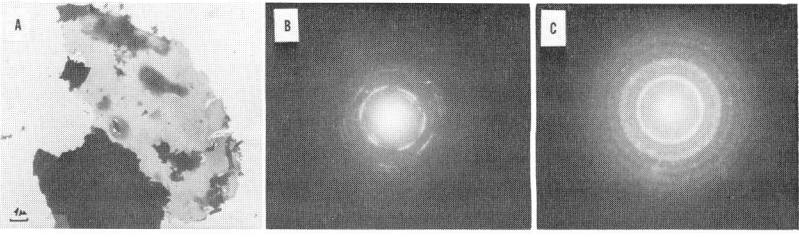

Figure 2. Crystallite laminae and electron diffraction patterns

Figure 3. Lamellae grown from liquids containing Li⁺

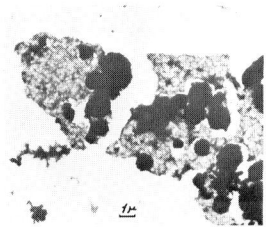

Figure 4. Gaps developing in lamellar material

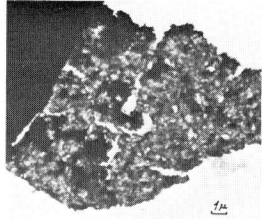

Figure 5. A further stage in the nucleation of zeolite

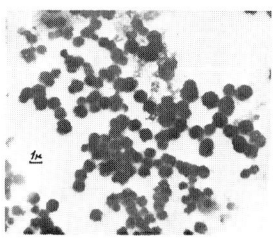

Figure 6. Final stage showing crystals of Na-P1 and basic sodalite

probably composed mainly of zeolites. At still longer reaction times, the thin laminae became less numerous, and the thick broken ones were predominant. Finally, at the longest times, the lamellae practically disappeared and were replaced by crystals of zeolite, such as those in Figure 6 which were identified by x-ray diffraction as Na-P1 (3) and basic sodalite (6). Although the formation and evolution of the lamellae seemed to be similar in all the cases examined, their nature appeared to be to some extent a function of the cation present, as shown below.

Electron Diffraction by Lamellae

Figure 3 shows typical freshly formed lamellae grown from the liquids containing Li⁺. All the lamellae examined showed only the diffuse halos characteristic of a nearly amorphous structure. A large majority of the laminae freshly grown from liquids containing K⁺ also were amorphous. However, a few of these, which already showed signs of the aging referred to in the previous section and contained small particles (Figure 7A), gave patterns such as that in Figure 7B, with a broad arc in the range $d = 2.88–2.93A$. This corresponds with the strongest diffraction found in the chabazite-like phase K-G of Barrer and Baynham

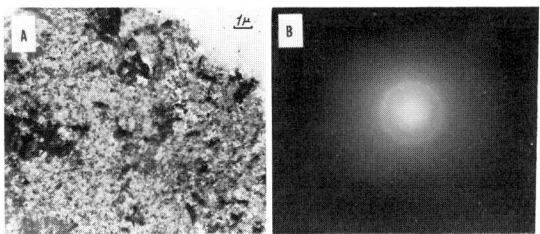

Figure 7. Laminae grown from liquids containing K^+

(2). X-ray diffraction showed that this phase was present, and it is concluded that K-G forms the dark particles seen in Figure 7A.

The lamellae grown from the liquids containing Na^+ also comprised mainly amorphous structures with diffuse halos, but there were 2 minor exceptions. A few laminae such as those in Figure 1A have the appearance of a smectite and in fact give electron diffraction patterns like those in Figures 1B and 1C, taken respectively from thin and thicker parts of the same lamella. The latter pattern also appears to show rotational slip. The d-spacings are given in Table I. They can be indexed as the $hk0$'s from a hexagonal unit cell with $a = 5.24$A or a monoclinic unit cell with $a = 5.24$A and $b = 9.08$A, similar to that for micas and clay minerals. Tilting the lamellae revealed heavy faulting in layers parallel to the basal plane, and it was therefore impossible to determine c.

A few other laminae produced patterns such as those in Figures 2B and 2C. Figure 2C shows the pattern normal for many such crystallites. Figure 2B shows the pattern from the laminae in Figure 2A, and indicates a type of electron diffraction pattern which occurs when there is a strongly preferred orientation parallel to the supporting film and this

Table I. Electron Diffraction Data[a] from the Lamina in Figure 1B

$Int.$	d_{obs}/A	hkl
vs	4.50	100
vs	2.61	110
vw	2.27	200
m	1.71	210
vs	1.52	300
s	1.31	220
w	1.26	310
vw	1.04	320
vw	0.99	410
vw	0.87	330

[a] Indexed on a hexagonal unit cell with $a = 5.24$A.

film is tilted with respect to the incident beam. d-Spacings from Figures 2B and 2C are shown in Table II. They correspond to the same as yet unidentified material having a hexagonal unit cell with $a = 7.05\text{A}$ and $c = 6.47\text{A}$.

X-Ray Examination

The x-ray diffraction patterns of the solids after they had begun to evolve towards the stage of thicker, broken lamellae having the appearance of those in Figures 4 or 7A showed lines which occur in the patterns of various zeolites. The reflections were initially very weak but became progressively more intense as the reaction time increased. In solids formed from liquids containing Na^+, zeolite Na-P and basic sodalite were the final crystalline constituents, but for short reaction times there were very weak reflections found in the patterns of Linde A and faujasite. Solids derived from liquids containing K^+ soon developed reflections of K-G, while those from liquids containing Li^+ developed reflections which corresponded with lines in the zeolite Li-A of Barrer and White (5). In general, these diffraction studies were made on products further evolved from the freshly formed lamellae than those used for the electron diffraction, though there was some overlap.

Discussion

The formation of the zeolites from clear liquids did not occur as a direct shower of crystals, as would be the case for example with rock salt crystallites from a supersaturated solution. Thus, homogeneous nucleation was not apparent. Instead, a rather complex evolution took place, always heralded by the appearance of amorphous lamellae. Crystallites which were not zeolites occasionally appeared as additional transient species—*e.g.*, layer silicates such as smectites, or the unknown phase of Table II. Zeolites then began to appear as crystallites in association with the lamellae, although fresh, zeolite-free lamellae also seemed to be forming at the same time. Where zeolite crystals were associated with lamellae, the latter soon developed holes and gaps and (Figure 4) gave the impression that the lamellar material was being consumed. From the foregoing evidence, the nucleation of zeolites is almost certainly heterogeneous under our conditions. The laminae may feed the alkaline solution and the solution the growing zeolites, until eventually the laminae disappear.

The sequence of the above processes recalls the principle first pointed out by Ostwald, that in all reactions the most stable state may

Table II. Electron Diffraction Data[a] from Unknown Material

Int.	d/Å	hkl
	Figure 2B	
m	6.14	100
m	3.52	110
vw	3.08	111
s	2.85	102
vw	2.40	112
vw	2.22	202
s	2.15	003
vs	2.02	300
m	1.89	212
vw	1.76	220
m	1.75	203
vw	1.34	313
vw	1.25	403
	Figure 2C	
vw	3.56	110
vw	3.23	002
vs	2.83	102
w	2.32	210
s, br	1.90	301, 212
m, br	1.50	400, 312, 303
w, br	1.29	411, 005, 322
w, br	1.15	311, 420

[a] Indexed on the hexagonal unit cell with $a = 7.05$Å, $c = 6.47$Å.

not be reached at once, but that first a succession of intermediate and less stable states tends to be traversed. Examples of this behavior in hydrothermal systems include the low-temperature formation of cristobalite instead of quartz from excess silica (4), the formation of high-temperature disordered potash felspar under conditions when the stable phase is the ordered felspar structure (2), and the initial formation of Na-mordenite from silica-rich highly alkaline aqueous aluminosilicate gels, which crystals then disproportionate into analcite and quartz (1). These latter species are stable under such alkaline conditions. This behavior signifies a general tendency for less stable species to nucleate more rapidly than stable ones so that kinetic considerations can initially outweigh thermodynamic ones. The Ostwald law of successive transformations correlates, at least in part, with the simplexity principle of J. R. Goldsmith (9), according to which phases tend to appear in the order of decreasing simplexity or entropy. For instance, the disordered felspar referred to above is in a state of higher simplexity than its ordered counterpart. Ready nucleation may be favored by high simplexity or entropy, but energy as well as entropy changes ultimately determine the relative stabilities of the phases and so the final product.

Literature Cited

(1) Barrer, R. M., *J. Chem. Soc.* **1948**, 2158.
(2) Barrer, R. M., Baynham, J. W., *J. Chem. Soc.* **1956**, 2882.
(3) Barrer, R. M., Baynham, J. W., Bultitude, F. W., Meier, W. M., *J. Chem. Soc.* **1959**, 195.
(4) Barrer, R. M., Marshall, D. J., *J. Chem. Soc.* **1964**, 485.
(5) Barrer, R. M., White, E. A. D., *J. Chem. Soc.* **1951**, 1267.
(6) Barrer, R. M., White, E. A. D., *J. Chem. Soc.* **1952**, 1561.
(7) Breck, D. W., Flanigen, E. M., "Molecular Sieves," p. 47, Society of the Chemical Industry, London, 1968.
(8) Ciric, J., *J. Colloid Interface Sci.* **1968**, 28, 315.
(9) Goldsmith, J. R., *J. Geol.* **1953**, 61, 439.

RECEIVED November 13, 1969.

Discussion

John Turkevich (Princeton University, Princeton, N. J. 08540): What were the conditions of centrifugation, and what types of particles could not be driven down?

R. Aiello: The conditions of centrifugation were about 4000 rpm for many hours, until the solution was clear and practically all the suspended particles were separated.

W. Sieber (Inst. fur Kristallographie und Petrographie, ETH, Zurich): Did you ever examine the behavior on heating of the filtrate from the amorphous lamellae? Is it excluded that formation of lamellae and formation of zeolites are independent phenomena?

R. Aiello: We did not examine the behavior of the filtrate from the lamellae on heating. We cannot exclude that formation of lamellae and zeolite are independent.

4

Zeolite Formation from Synthetic and Natural Glasses

R. AIELLO, C. COLELLA, and R. SERSALE

Istituto di Chimica Applicata, Facoltà di Ingegneria, Università di Napoli, Italy

> *The formation of zeolites from glasses has been investigated systematically. Synthetic sodic glasses with variable SiO_2/Al_2O_3 or Na_2O/Al_2O_3 ratios have been considered, together with synthetic, alkali-rich glasses obtained by melting natural products—i.e., leucite—with alkalis, and natural glasses (pumices). Zeolites such as zeolite A, Na-P, analcite, basic sodalite, and faujasite have been obtained. The influence of different factors affecting the transformation of glass to zeolite have been examined. The kinetics of zeolite formation have been followed, and the chemical composition of the mother liquors at different reaction times were determined. The existence of a gel as an intermediate stage of the glass–zeolite transformation has been admitted.*

For a long time, the Institute of Applied Chemistry of Naples University has been systematically investigating the zeolitization process of glasses, either volcanic (7) or synthetic, with composition near (8, 9) or different (4) from that of the natural ones.

This paper refers to the most recent results obtained from studying the zeolite formation from synthetic sodic glasses with variable SiO_2/Al_2O_3 or Na_2O/Al_2O_3 ratios, semisynthetic glasses prepared by melting leucite with sodium carbonate, and natural glasses (pumices).

Experimental

Synthetic glasses have been prepared by melting and subsequent quenching in distilled water of suitable oxide mixtures. The granulated glasses were oven-dried at 110°C and subsequently ground. In this way, 2 series of glasses were prepared, the first with Na_2O/Al_2O_3 ratio = 1 and SiO_2/Al_2O_3 ratio variable between 1 and 6, the second with SiO_2/Al_2O_3 ratio = 4 and Na_2O/Al_2O_3 ratio variable between 1 and 4. The composition of these glasses, determined by chemical analysis, is specified in the Na_2O–Al_2O_3–SiO_2 diagram in Figure 1.

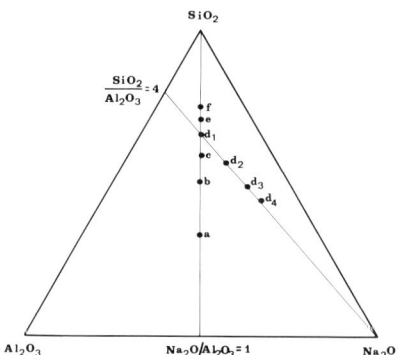

Figure 1. Molecular composition of synthetic glasses examined

None of the samples investigated, tested by x-ray, showed any presence of arcs.

The semisynthetic glasses were obtained by melting mixtures of leucite from Roccamonfina, Italy, and sodium carbonate. The samples tested have the molecular composition indicated in Table I.

Riolitic pumices from Lipari, Italy (Table II), served as natural glass samples.

All the glasses, when not otherwise specified, were sieved at 256 mesh. Glass samples were mixed with NaOH solutions of different concentration in perfectly closed Teflon containers. These containers were placed in an air-thermostated oven at variable temperature for a given time and rotated at 33 rpm. The ratio between the weight of the glass and the weight of the water contained in the alkaline contact solution was considered as the solid/liquid ratio (S/L). At the end of each experiment, the solids were centrifuged from the mother liquors, washed until pH $\cong$ 9, and oven-dried at 110°C.

The following abbreviations indicate the products obtained: X = faujasite-type zeolite (3); A = zeolite A (2); P = zeolite Na-P (1); B = analcite (1); I = basic sodalite (1); U denotes no crystallization.

Results and Discussion

Zeolitization of Synthetic Glasses. The formation areas of some products, obtained in different conditions from both series of glasses (Figure 1), are shown in Figures 2 and 3. Figure 2 refers to the series

Table I. Molecular Composition of Semisynthetic Glasses

Na_2O	K_2O	Al_2O_3	SiO_2	Symbol
1.98	0.97	1	4.05	s_1
2.95	0.94	1	4.03	s_2
3.94	0.95	1	4.02	s_3

Table II. Composition of Riolitic Pumices

SiO_2	70.85%
Al_2O_3	12.83
MnO	0.11
TiO_2	0.15
Fe_2O_3	1.02
FeO	1.35
CaO	0.83
MgO	0.55
Na_2O	4.46
K_2O	4.70
H_2O	3.71
	100.56

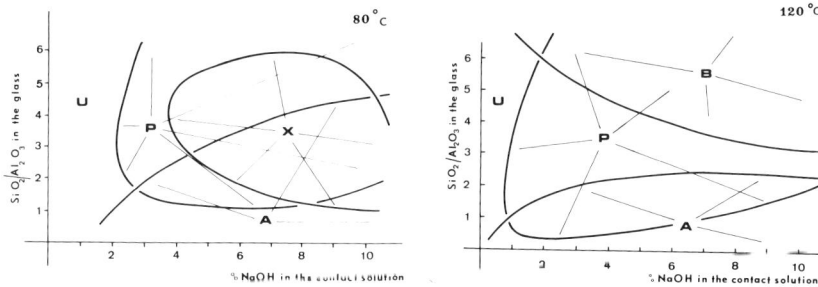

Figure 2. Formation areas of products obtained from a–f glasses treated for 36 hours with variably concentrated NaOH solutions and $S/L = 1/20$, at 80° and 120°C, respectively

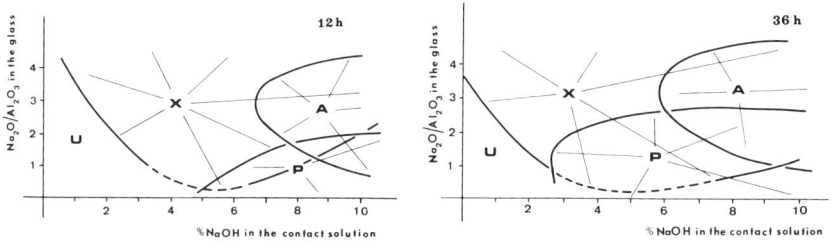

Figure 3. Formation areas of products obtained from d_1–d_4 glasses treated at 80°C with variably concentrated NaOH solutions and $S/L = 1/20$, for 12 and 36 hours, respectively

of glasses a–f, after treatment with NaOH solutions of variable concentration for 36 hours, with solid/liquid ratio = 1/20, at 80° and 120°C, respectively.

Zeolite P is predominant at both temperatures. This zeolite is normally present in cubic form. The tetragonal form of zeolite P has been obtained particularly from glasses with low SiO_2/Al_2O_3 ratio, with higher concentrations of the alkaline contact solution and at higher temperatures. Analcite appears only from starting glasses with high SiO_2/Al_2O_3 ratio and at higher temperatures. Low temperatures favor the formation of zeolite A, as well as zeolite X, which is completely absent at 120°C. At 80°C the greatest, although limited, yield of zeolite X has been obtained with ratio $SiO_2/Al_2O_3 = 4$. Considering the practical interest of the synthesis of faujasite group zeolites, we studied the zeolitization of more alkaline glasses (Na_2O/Al_2O_3 ratio > 1, SiO_2/Al_2O_3 ratio fixed and equal to 4; glasses d_2–d_4 in Figure 1) to investigate the influence of the glass alkalinity on the yield of zeolite X.

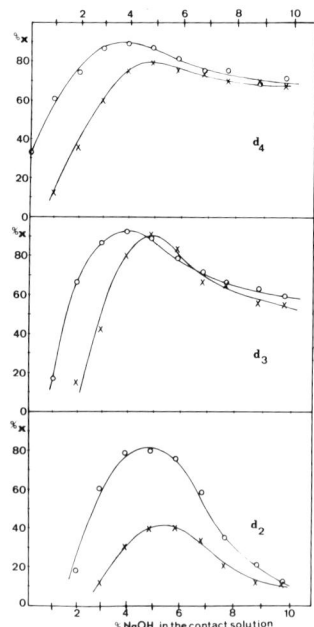

Figure 4. Zeolite X percentages from d_2–d_4 glasses treated at 80°C with variably concentrated NaOH solutions and $S/L = 1/20$ for 12 (x) and 36 hours (o)

The formation areas of some zeolites obtained by treating these glasses with NaOH solution of variable concentration, solid/liquid ratio = 1/20, at 80°C for 2 different lengths of time, are shown in Figure 3. The increase of alkalinity in the starting glass clearly promotes the faujasite formation; from the more alkaline glasses, zeolite X forms the single evolution phase in wide zones of the fields. Zeolite P formation, unlike zeolite A, seems to be promoted at the expense of zeolite X by the increase of the reaction time. The glass alkalinity also affects the yield of zeolite X from the glass; this is particularly evident for short reaction time. In Figure 4, the percentage of zeolite X in the solids after 12 and 36 reaction hours has been plotted against the NaOH concentration of the contact solution.

The percentage values were determined by x-ray quantitative analysis, using molecular sieve 13X of Union Carbide as reference. The maxima of the curves, corresponding to the highest values of the yield, move toward lower NaOH concentrations of the contact solution as the alkalinity of the starting glass increases. The more alkaline glasses (d_3 and d_4) also show a better reactivity than the less alkaline one (d_2); in fact, the yield of zeolite X obtained after 12 hours of reaction is comparable with the yield at 36 hours for the d_3 and d_4 glasses but is much inferior for d_2.

The a_o values of the zeolite X samples obtained from d_1–d_4 glasses correspond to the values of faujasites poor in silica (zeolites X, Si/Al ratio < 1.5) (3). For each of these glasses, the a_o value of zeolite X obtained increases with consequent decrease of the Si/Al ratio (3) as the NaOH concentration of the contact solution increases. Figure 5 shows the variation of a_o of the zeolite X samples obtained from the d_4 glass against the initial concentration of the contact solution. The interplanar spacings used to calculate the a_o values were corrected using $Pb(NO_3)_2$ as standard and employing a computer program. The a_o variation in the field examined is practically linear. By considering the diagram of Breck and Flanigen (3) which gives the a_o variation as a function of the Si/Al ratio in the faujasite–type zeolites, it is possible to extrapolate for the limiting value of the Si/Al ratio = 1 (6) an a_o value equal to 25.02 A. This shows that the Si/Al ratio of the sample corresponding to our highest a_o value (Figure 5) is around 1. The same glass d_4, after reaction with distilled water, again gives a good yield of zeolite X (Figure 4), whose a_o value is 24.78 A. This a_o value, which does not appear in Figure 5, lies widely apart from the other values obtained and corresponds to a zeolite Y (Si/Al ratio > 1.5) (3). This displacement may be related to the big pH variation in the contact solution passing from 0 to 1% of NaOH.

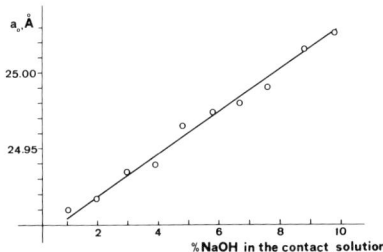

Figure 5. a_o variation as function of initial NaOH concentration in the contact solutions for zeolite X samples from d_1 glass treated at 80°C for 36 hours and $S/L = 1/20$

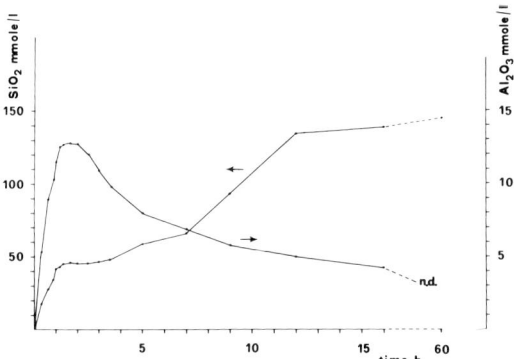

Figure 6. SiO_2 and Al_2O_3 concentrations in mother liquors obtained by treatment of d_3 glass at 80°C, with 4% NaOH solution and $S/L = 1/20$ at different reaction times; n.d. = not determinable

For a better understanding of the glass–zeolite transformation mechanism, the formation kinetic of zeolite X from d_3 glass has been followed with different investigation techniques. The treatment conditions of the glass have been 4% NaOH solution and solid/liquid ratio 1/20, at 80°C. For this purpose, a series of identical glass–solution mixtures was set to react; the moment the glass was added to each solution (already placed in the Teflon container and thermostated at 80°C) was assumed as time zero. Each container was extracted from the thermostat at the programmed time, and the solid was separated from the mother liquor. SiO_2 and Al_2O_3 were determined gravimetrically in the liquids; alumina was precipitated using 8-hydroxyquinoline. The solids were washed and a small part was used for electron microscopy investigations; the remaining part was dried at 110°C and subsequently equilibrated at 20°C for a week on $Ca(NO_3)_2$ saturated solution. The percentage of zeolite X (by x-ray quantitative analysis) and the percentage of water (by thermogravimetric analysis) present in these solids were then determined in the same thermostated room. The experimental results are summarized in Figures 6, 7, and 8.

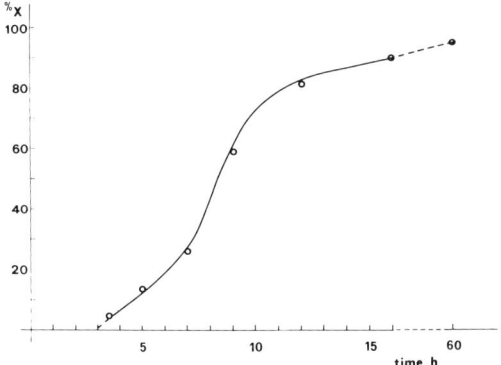

Figure 7. Crystallization kinetic of zeolite X from d_3 glass treated at 80°C with 4% NaOH solution and $S/L = 1/20$

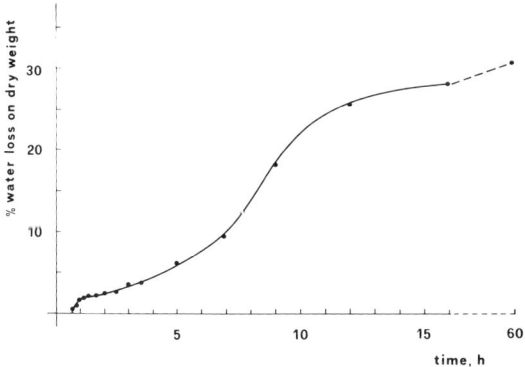

Figure 8. Percentages of water present in samples obtained from d_3 glass treated at 80°C with 4% NaOH solution and $S/L = 1/20$ at different reaction times

Figure 6 shows the variations with time of the silicate and aluminate concentrations, expressed as SiO_2 and Al_2O_3, in the mother liquor. The formation kinetic of zeolite X, obtained by x-ray quantitative analysis, using molecular sieve 13X of Union Carbide as reference, is reported in

Figure 7. The percentages of water present in the solids at different reaction times are reported in Figure 8.

The 2 curves in Figure 6 can be divided into 3 periods. In the first (up to about 1 hour) both curves are ascendent; in the second (up to about 2 hours) there is a leveling off in both cases, after which, in third period, the curves diverge until they reach practically constant values at the end of the reaction. Considering the curve in Figure 7, zeolite X becomes detectable by x-ray shortly after the beginning of the third period. A comparison of the curves in Figures 7 and 8 shows, however, a determinable amount of water in the solids before the third period. To evidence this finding, the water amounts in the solids at different reaction times not attributed to adsorption by zeolite X are indicated in Figure 9 as shaded area. This water is, in our opinion, related to the

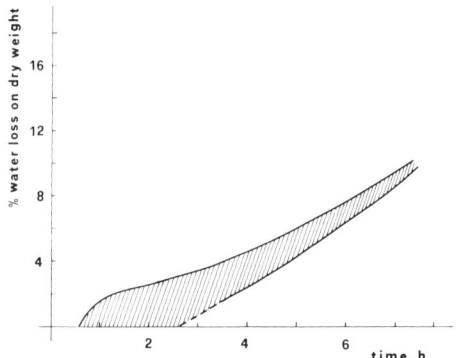

Figure 9. Upper curve as Figure 8. Lower curve: water percentages adsorbed by zeolite X in the same solids. Shaded area shows water percentages connected with gel

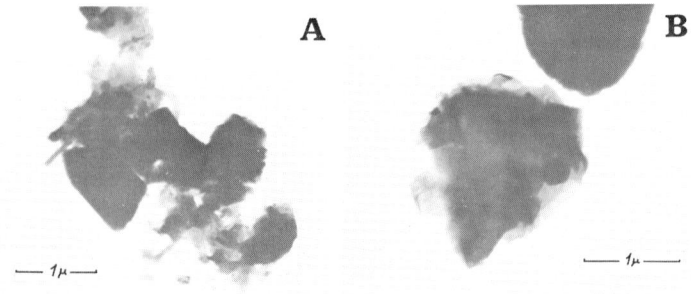

Figure 10. Electron micrographs. Samples obtained from d_3 at 80°C with 4% NaOH solution and $S/L = 1/20$ for 80 (A) and 120 (B) minutes

presence of a gel as intermediate stage in the crystallization process of zeolite X. The electron micrographs in Figure 10, together with the scales of nonreacted glass, show the presence of a semitransparent gel-like phase, amorphous to the electron diffraction. Referring again to the curves in Figure 6, the first period (up to about 1 hour) has been interpreted as the initial stage of the glass dissolution; the second (up to about 2 hours) as the period of gel formation. In the first period, the SiO_2/Al_2O_3 ratio in the mother liquor remains practically constant around values of 3.4–3.6 because of a probable differential dissolution of the glass. The gel formation begins about the end of the first period, following the saturation in SiO_2 and Al_2O_3 of the solution. Considering that in the second period the SiO_2 and Al_2O_3 concentrations remain almost constant and their ratio is near that found in the first period, we deduce that the SiO_2/Al_2O_3 ratio of the gel is around the values of 3.4–3.6 given above. Again referring to Figure 6, we notice that within 2 to 3 hours, the Al_2O_3 concentration begins to decrease considerably, while the SiO_2 concentration remains practically constant. This indicates the beginning of formation of zeolite X nuclei in the gel. The zeolite X in formation shows a ratio $SiO_2/Al_2O_3 = 2.3$, smaller than the value we attributed to the gel; the SiO_2/Al_2O_3 ratio in zeolite X does not seem to vary during the whole process of zeolite crystallization. In addition, the results shown in Figure 9 suggested to us the presence of gel in successive stages of the process. Sticher and Bach reached analogous conclusions studying the reaction between kaolinite and potassium hydroxide (10).

The use of a natural product with SiO_2/Al_2O_3 ratio = 4 as a raw material for the preparation of synthetic glasses has also been studied. Leucite was melted with sodium carbonate to obtain the s_1, s_2, and s_3 glasses (see Experimental). These glasses were submitted to the zeolitization process in sodic environment, with conditions similar to those used for the synthetic glasses considered above. The zeolites so far obtained from these glasses (A, X, P, and I) are not different from those obtained from glasses containing only sodium (Figures 2 and 3).

Good yields of zeolite X have been obtained especially from the glass richest in sodium (s_3); this is probably because of the higher alkalinity of the glass and the greater dilution in the glass of potassium present in the starting leucite.

Zeolitization of Natural Glasses. The research on the zeolitization of natural glasses has been devoted essentially to the investigation of the most favorable conditions for the crystallization of the various zeolites obtainable.

Volcanic glasses appear remarkably versatile for the crystallization of different zeolites, provided that the conditions of treatment in alkaline environment are chosen properly. We have paid particular attention to

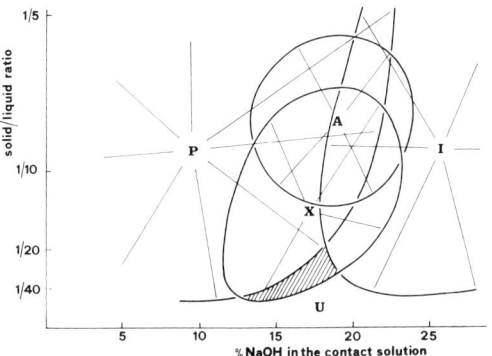

Figure 11. Formation areas of products obtained from pumices (Lipari) at 65°C for 60 hours with various S/L ratios and variably concentrated NaOH solutions

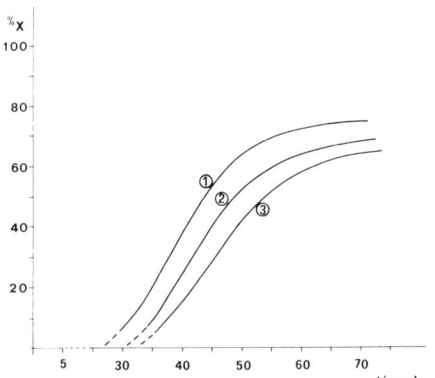

Figure 12. Crystallization kinetics of zeolite X from pumices at 3 different finenesses, treated at 65°C with $S/L = 1/25$ and 18% NaOH solution
Fraction passing at 320 mesh:
(1) 100%
(2) About 50%
(3) About 25%

the formation of faujasite-type zeolites. Starting from riolitic pumices (Table II), zeolite X, even if obtainable at about 80°C, is always accompanied at this temperature by other phases (5). At lower temperatures, the formation of zeolite X is favored in relation to the other phases; this is in agreement with the results of Barrer *et al.* (1) on the formation fields of faujasite-type zeolite from gels. As an example, the formation areas of some zeolites from riolitic pumices at 65°C are shown in Figure 11. In this case, because of the fixed chemical composition of the starting glass, the NaOH concentration in the contact solution and the solid/liquid ratio are considered variables. The shaded area refers to the conditions which permit zeolite X to be obtained unaccompanied by other phases. Figure 11 shows the influence of the solid/liquid ratio on the type of zeolite obtained. This ratio, evidently varying the SiO_2 and Al_2O_3 concentrations in the mother liquor, also influences the zeolite formation rate; with the lower solid/liquid ratios, the crystallization has not begun yet at 60 hours. Another factor which evidently affects the crystallization kinetic of the zeolites from glasses, considering that the glass dissolution is an essentially superficial reaction, is the fineness of the starting glass. For this purpose, the crystallization kinetics of zeolite X from pumices at 3 different finenesses, all treated at 65°C with 18% NaOH solution and with solid/liquid ratio = 1/25, are reported in Figure 12. The curves were obtained by x-ray quantitative analysis, always referring to molecular sieve 13X of Union Carbide.

Acknowledgment

The authors thank ENI (Ente Nazionale Idrocarburi) for financial help and A. Annetta and B. Terracciano for their help in carrying out some of the experiments.

Literature Cited

(1) Barrer, R. M., Baynham, J. W., Bultitude, F. W., Meier, W. M., *J. Chem. Soc.* **1959**, 195.
(2) Breck, D. W., Eversole, W. G., Milton, R. M., Reed, T. B., Thomas, T. L., *J. Am. Chem. Soc.* **1956**, 78, 5963.
(3) Breck, D. W., Flanigen, E, M , "Molecular Sieves," p. 53, Society of the Chemical Industry, London, 1968.
(4) Colella, C., Aiello, R., *Chim. Ind.* **1970**, 52, 151.
(5) Franco, E., Aiello, R., *Rend. Accad. Sci. Fis. Mat.* **1969**, 36, 1.
(6) Loewenstein, W., *Am. Mineralogist* **1942**, 39, 92.
(7) Sersale, R., *Ann. Chim. Rome* **1959**, 49, 1111.
(8) Sersale, R., *Rend. Accad. Sci. Fis. Mat.* **1961**, 28, 317.
(9) Sersale, R., *Silicates Ind.* **1960**, 11, 1.
(10) Sticher, H., Bach, R., *Helv. Chim. Acta* **1969**, 52, 543.

RECEIVED January 30, 1970.

Discussion

Harry Robson (Esso Research Laboratory, Baton Rouge, La.): You state that the composition of zeolite X produced is constant throughout the crystallization. Thermodynamically, we should expect that Si_2O/Al_2O_3 of the crystal should change as the composition of the mother liquor changes. Is it possible that the crystals initially formed are annealed during the remainder of the crystallization treatment to produce a uniform product?

R. Aiello: The cell constant of the initial and final crystals was practically identical. We therefore believe this to mean that our system was not in thermodynamic equilibrium.

G. Tsitsishvili (Academy of Sciences of the Georgian SSR, Tbilisi, USSR): We have synthesized zeolites P, A, and faujasite using natural obsidian aluminosilicates, reolite and perlite. By regulation of the crystallization process, it is possible to obtain products with different Si/Al ratios within one type of zeolite. Details of these results are published: *Izv. Akad. Nauk USSR, Inorg. Mater.* **1965,** 1, 285; **1966,** 2, 1306; **1969,** 5, 1848.

5

Crystallization of Zeolites in the Presence of a Complexing Agent

Part II[1]

GÜNTER H. KÜHL[2]

Mobil Research and Development Corp., Central Research Division Laboratory, Princeton, N. J.

> *A zeolite of increased SiO_2/Al_2O_3 ratio crystallizes when the ratio of silicate to tetrahydroxoaluminate in the reaction mixture is raised by complexing aluminum. The equilibrium between the resulting complex and tetrahydroxoaluminate is pH-dependent. All complexing agents investigated but sulfate are effective to varying degrees, but only phytate provides a buffering action comparable to phosphate. Attempts are made to explain the observed effect of these variables: SiO_2/Al_2O_3 ratio, pH, cation or cation ratio, silica source, time, concentration, and temperature. The explanation is based on the polymerization–depolymerization equilibrium, the stabilizing effect of certain cations on certain precursors, and the hypothesis that cations determine the way in which precursors are linked to form a zeolite structure. Simple precursors are probably 4-rings, double 4-rings, and double 6-rings.*

The effect of phosphate, arsenate, tartrate, and salicylate on the crystallization of zeolites has been reported previously (17). The influence of other complexing agents is described herein.

Zeolite synthesis has been quite empirical. Any information available will help, therefore, to understand the sequence of reactions occur-

---

[1] Part I presented as a monograph, "Molecular Sieves," Society of the Chemical Industry, London, 1968.
[2] Present address: Mobil Research and Development Corp., Applied Research and Development Division, Paulsboro, N. J. 08066

ring during crystallization. Information gained from zeolite synthesis in phosphate-containing reaction mixtures is particularly valuable because the pH is controlled well during the entire crystallization period, and the obscuring effect of a large excess of silicate can be avoided.

Experimental

Reagents. Sodium aluminate and sodium metasilicate were supplied by Allied Chemical Corp., General Chemical Division. The compositions were 43.3% Al_2O_3, 37.7% Na_2O; 21.9% SiO_2, 24.5% Na_2O. Sodium waterglass, obtained from Philadelphia Quartz Co., had the composition 28.7% SiO_2, 8.9% Na_2O. Colloidal silica sol (Ludox LS) was supplied by du Pont de Nemours; composition: 30.7% SiO_2, 0.23% Na_2O. Other sodium silicate solutions with different properties were obtained from Philadelphia Quartz Co. They can be described as follows:

Designation	SiO_2 %	Na_2O %	Molar Ratio SiO_2/Na_2O	Viscosity Centipoise
Star	26.5	10.6	2.48	60
D	29.4	14.7	2.06	350
C	36.0	18.0	2.06	70,000
B-W	31.4	19.7	1.65	7,000

Gluconic acid, as a technical grade 50% aqueous solution, was obtained from Matheson, Coleman, and Bell; phytic acid, as a 70% aqueous solution, from K&K Laboratory, Plainview, N. Y. Geigy Industrial Chemicals supplied ethylenediamine di(o-hydroxyphenylacetate) (Chel DP). Sodium diethanolglycinate (DEG Na) was obtained from Chas. Pfizer & Co., Inc. All the other chemicals were analytical grade reagents.

Preparation of Zeolites. Preparation was done as previously described (17). The temperature was in the range 90°–100°C. Polypropylene flasks were used throughout. The final pH was measured at ambient temperature after the crystallization. Although different from that at reaction temperature, it provides a convenient comparison between experiments.

Crystallization in the Presence of Complexing Agents

General Considerations. It has been reported previously (17) that phosphate aids in the preparation of high-silica zeolites. Aluminum is complexed by phosphate (15) and gradually released from this complex, thereby becoming available for reaction with the silicate. Under reaction conditions (pH, temperature) the complex must provide a sufficiently high concentration of hydroxoaluminate to enable the zeolite to crystallize. On the other hand, the stability of the complex must be high enough to provide a lower concentration of hydroxoaluminate than without the complexing agent.

Table I. Effect of Complexing Agents in the Preparation of Zeolite Y

Reaction Mixture Molar Ratios

No.	SiO_2/Al_2O_3	$C.A.^a/Al_2O_3$	$Na_2O/C.A.^a$	H_2O/Na_2O	Crystallization Time, Days	Final pH	SiO_2/Al_2O_3
				Diphosphate			
1	4.0	4.0	$(10.0)^b$	~65	5	12.8	3.22
2	4.0	4.0	$(9.0)^b$	~65	7	12.55	3.31
3	4.0	4.0	$(8.5)^b$	~65	14	11.85	3.43
4	4.0	4.0	$(8.0)^b$	~65	–	~10	Amorphous
				Sulfate			
5	4.0	5.6	2.05	~77	2	12.0	2.73
6	4.0	11.2	1.52	~52	6	12.0	2.80
				Oxalate			
7	4.0	2.0	2.48	~135	3	12.2	2.81
8	4.0	2.0	1.97	~170	4	12.2	3.12
9	4.0	3.0	1.97	~150	5	13.0(?)	2.98
10	4.0	3.0	1.47	~200	16	11.95	3.33
				Citrate			
11	4.0	2.12	3.77	~110	2	12.05	2.93
12	4.0	4.24	2.38	~90	3	12.35	3.23
13	4.0	5.29	2.09	~80	6	12.75	3.33
14	3.74	5.91	1.93	~75	–	10.9	Amorphous
				EDTA			
15	4.0	2.04	4.35	~125	5	12.4	3.04
16	4.0	3.06	3.24	~135	6	12.2	3.25
17	3.74	3.80	2.68	~140	12	12.5	3.19
18	3.74	4.75	2.34	~130	–	12.6	Amorphous
19	6.0	2.04	5.42	~100	2	12.2	3.27
20	6.0	3.06	3.95	~110	2	12.0	3.56
				Gluconate			
21	4.0	1.02	5.67	~115	4	12.5	2.96
22	4.0	3.06	2.06	~105	2	12.2	3.02
23	4.0	5.10	1.43	~90	5	13.0(?)	3.16
				Chel DP			
24	4.0	1.02	5.68	~230	6	12.6	3.11
25	4.0	1.60	4.00	~210	6	12.6	3.27
26	4.0	2.00	3.50	~190	11	11.0	3.40
				DEG-Na			
27	4.0	2.0	1.66	~180	6	12.8	3.15
28	4.0	3.0	1.33	~220	9	12.5	3.35
29	4.0	3.0	1.17	~190	23	11.7	3.61
30	4.0	3.0	1.00	~220	–	9.7	Amorphous
				Phytate			
31	4.0	1.0	10.28	~90	4	12.6	3.14
32	4.0	1.0	8.76	~110	30	11.0	3.48
33	6.0	1.0	10.44	~90	14	11.7	4.34
34	6.0	1.0	8.94	~105	45	10.7	Mainly amorphous

[a] Complexing agent.
[b] pH to which $NaAlO_2$–$Na_4P_2O_7$ solution was adjusted before addition of Na_2SiO_3.

It appears that the reaction mixture must have a pH of 11 ± 0.2 or higher for the faujasite structure to crystallize, which seems to be owing to the state of the silicate in solution. Therefore, a complexing agent for this application should provide a favorable equilibrium between its aluminum complex and hydroxoaluminate at a pH of 11 or higher.

The higher the pH, the more hydroxoaluminate will be in equilibrium with the aluminum complex. It will be helpful, therefore, if the complexing agent can buffer in the proper pH range. The ability of a salt–acid mixture to buffer a solution can be described best by the dissociation constants of the acid. However, the pK values may not always be valid in the presence of aluminum because complex acids may be formed.

Most experiments described herein were conducted with a 4.0 SiO_2/Al_2O_3 molar ratio of the mixture. Without complexing agent, the faujasite product of such a mixture had a SiO_2/Al_2O_3 ratio of 2.81.

Individual Complexing Agents. Experiments with the following complexing agents are summarized in Table I.

Diphosphate. Apparently, nothing is known of the existence of aluminum–diphosphate complexes in the alkaline range.

In our experiments with diphosphate as the complexing agent, part of the alkalinity was neutralized with HCl before sodium metasilicate was added (*see* column 4). Diphosphate does have an effect on the silica-to-alumina ratio of the product, but the pH has to be controlled very carefully.

Sulfate. Sulfate forms an anionic complex with aluminum (3), $Na[Al(SO_4)_2]$. The formation of this complex is slow and catalyzed by OH^- ions.

Zeolites prepared in the presence of sulfate have low SiO_2/Al_2O_3 ratios; no effect of sulfate can be seen (Table I).

Oxalate. Oxalatoaluminate complexes of the type $M_3[Al(C_2O_4)_3]$ are among the best known aluminum complexes (1). Lacroix (18, 19) found that oxalate can form 2 complex aluminate ions. The complexes $[Al(C_2O_4)_2]^-$ and $[Al(C_2O_4)_3]^{3-}$ are extremely stable in essentially neutral solutions; nothing is said about the stability of the complexes at higher pH values. The p_{K_2} value of oxalic acid (4.19) indicates that this reagent is not an effective buffer in the proper pH range. Because of the low solubility of $Na_2C_2O_4$, larger volumes than normal had to be used. Low-silica zeolite Y instead of zeolite X was obtained from a number of preparations (Table I), indicating that oxalate is capable of reducing the concentration of hydroxoaluminate to some extent.

Citrate. Cadariu, Goina, and Oniciu (4) report an inflection in the titration curve of an aluminum–citrate solution with hydroxyl ions at pH 11–12. This inflection diminishes as the ratio of citrate to aluminum increases, and disappears at citr./Al = 5. The complex acid formed at low citr./Al should be able to buffer in the desired pH range. Indeed, the best buffering action is observed at low citr./Al (Table I). Citric acid itself is a poor buffer ($p_{K_3} = 5.4$).

Ethylenediaminetetraacetate. An ethylenediaminetetraacetate complex of aluminum has been described by Schwarzenbach and coworkers (*22*). Saito and Terrey (*21*) assign the formula H[Al(H_2O)EDTA] to the acid form. Kerr (*12*) found that this complex is actually H_2[Al(OH)-EDTA] with 2 acid functions. The acid H_2[Al(OH)EDTA] forms a monosodium salt, NaH[Al(OH)EDTA], at pH 4.5 and a disodium salt, Na_2[Al(OH)EDTA], at pH 8.0 upon reaction with sodium hydroxide. The last acid hydrogen of ethylenediaminetetraacetic acid has a p_K of 10.26, which is quite high compared with most acids. It is approaching the value for arsenic acid (11.60) and may be capable of providing a limited degree of buffering in the pH range of interest.

Comparison of the example pairs 15,16 and 19,20 (Table I) shows that an increase in the ratio of EDTA/Al from 1 to 1.5 yields products of significantly higher SiO_2/Al_2O_3 ratios. A greater excess of EDTA appears to hinder the crystallization.

Gluconate. Grossmith (*8*) reported the preparation of aluminum complexes with gluconate.

The silica-to-alumina ratios of the faujasite-type products from gluconate-containing mixtures (Table I) were in the high X- or low Y-range. The final pH values were 12.2 or higher. When we tried to obtain lower alkalinities, zeolite B of $SiO_2/Al_2O_3 = 3.3$ crystallized, but the final pH remained the same. The crystallite size in the preparations with gluconate was generally larger than in the other examples.

Ethylenediamine di(o-hydroxyphenylacetate). Since Chel DP is effective for chelating iron in mildly alkaline solution (*7, 13*), it was hoped that a complex analog to the ferric complex is formed with aluminum. The faujasite products obtained with Chel DP were of a very high crystallinity and had compositions in the low zeolite Y range. The absence of buffering ability is evident from the fast decrease of the final pH when Na_2O/C.A. is lowered to 3.5. Decreasing alkalinity does increase the complexing ability of Chel DP for aluminum, as is evident from the higher silica-to-alumina ratio of the product. The amount of Chel DP may have been adequate only in Example 26.

Diethanolglycinate. The reagent forms very strong ferric ion chelates over a broad pH range, but is commonly used at pH 9.5 to 12.5. Although no data on complexing aluminum with this reagent have been found in the literature, the chemical similarity of iron and aluminum made it a candidate for this study.

The experimental results (Table I) indicate that a complex between aluminum and DEG is formed, as the silica-to-alumina ratio of the product increases with decreasing pH. The compound does not buffer the solution, so that control of the pH is difficult.

Phytate. Phytic acid is the phosphoric acid ester of inositol, hexahydroxycyclohexane. Every phosphoric acid group contains 2 acidic hydrogens, so that the compound is expected to buffer. Complex formation with a number of cations is known (24) but an aluminum complex has not been reported.

The results of crystallization in the presence of phytate (Table I) indicate that phytate is both a good complexing agent and a reasonably good buffer. Faujasites with silica-to-alumina ratios as high as those obtained in the presence of phosphate and tartrate can be prepared with this complexing agent. The buffering action of phytate is better than that of tartrate, but may not quite reach that of phosphate.

Variables in the Crystallization of Zeolites

Zeolites usually are prepared by mixing silicate and aluminate solutions and heating the resulting gel–solution mixture until a complete crystallization is obtained. Since zeolites have nonstoichiometric structures with Al substituting for Si, the composition of the product depends on the composition of the reaction mixture. Little is known about the mechanism of zeolite crystallization. Kerr (10) has shown that zeolite A crystallizes from solution rather than by solid–solid transformation. The same author (11) found induction periods in the crystallization of zeolites X and B and different growth rates of the respective nuclei. By seeding, the induction period (nucleation) could be eliminated. Ciric (5) found that the growth rate of zeolite A increases with rising alkalinity, but the induction period seems to remain unchanged.

No information is available on the composition and structure of soluble precursors. We feel that observations made in buffered reaction mixtures are of particular value because the degree of polymerization of silicate and aluminosilicate is better controlled. Most of our experiments have been carried out in phosphate-buffered mixtures, and it has to be taken into consideration that phosphate is also a complexing agent for aluminum.

SiO_2/Al_2O_3. The SiO_2/Al_2O_3 ratio (R) of the reaction mixture is one of the variables that control the monomer–polymer equilibrium and the composition of the crystallizing framework structure (example: X → Y). This ratio also facilitates formation of certain precursor structures that are required for certain zeolites to crystallize. The number of cations associated with a precursor can be regulated by choosing the appropriate R; example: crystallization of zeolite A at $R = 2$ and of zeolite X at $R = 4$. A reaction mixture made with phosphate-buffered sodium aluminate and sodium waterglass yields ZK-14 (chabazite structure) at $R = 4$ (final pH 11.85, $SiO_2/Al_2O_3 = 3.4$) and ZK-15 (similar to species S^2) at $R = 6$ (final pH 11.65; $SiO_2/Al_2O_3 = 4.1$).

pH. The pH influences the polymerization–depolymerization equilibrium of the soluble aluminosilicates—i.e., the average size of these species increases with decreasing pH. This shift in the size distribution of building blocks causes different structures to crystallize (example: sodalite → A).

When the pH is varied in the presence of phosphate, zeolites of the same crystal structure but different SiO_2/Al_2O_3 ratios crystallize (example: X → Y[17]). Further decrease of the pH can cause crystallization of a zeolite with a different crystal structure, evidently resulting from a change in the structure of the precursor (example: Y → ZK-15[17]) (Table II).

Table II. Effect of pH[a]

$\dfrac{Na_2O}{P_2O_5}$	Zeolites Obtained	$\dfrac{SiO_2}{Al_2O_3}$ (Product)
3.55	Y	3.8
3.11	Y (+ B + ZK-15)	4.3
2.87	Y + B (+ ZK-15)	4.8
2.72	ZK-15 + trace of Y	5.8
2.60	ZK-15 + trace of Y	6.0

[a] Sodium metasilicate as the silica source, $SiO_2/Al_2O_3 = 5.18$.

In the crystallization of gallosilicates at 90°C, we observed that every reaction mixture with a final pH higher than about 12 yielded a sodalite structure, whereas a faujasite-type material crystallized below pH 12.

It is likely that, barring any complications, a type of building blocks always will generate the densest structure possible. In the sodium system, the smallest building block that determines a crystal structure is conceivably a ring of 4 tetrahedra with associated cations. Such 4-member rings can combine to form the sodalite structure. Upon decrease of the pH, the predominant building block may be a dimer of the 4-member ring, a cube, also known as a double 4-ring, consisting of 8 tetrahedra and associated cations. Connecting these species in the same way as the single 4-rings yields the zeolite A structure. When the pH is lowered further, the most abundant species may be a trimer of the 4-member ring, a hexagonal prism double 6-ring, with associated cations. Connection of these precursors would form the faujasite structure. However, a complication arises here. It seems difficult for the faujasite structure to accommodate 96 hydrated sodium ions per unit cell. Therefore, this structure normally is observed only if either the number of cations is lowered by increasing the SiO_2/Al_2O_3 ratio of the reaction mixture or some of the sodium ions in the mixture are replaced by presumably smaller hydrated potassium ions (25).

However, we found that both zeolites, X and A, crystallize from a reaction mixture of $R = 2.0$ containing gluconate. Zeolite X formed in this way always has a low SiO_2/Al_2O_3 ratio (Table III). The lattice

Table III. Preparation of Low-Silica Faujasite

$SiO_2/$ Al_2O_3	$C_6H_{12}O_7/$ Al_2O_3	$Na_2O/$ $C_6H_{12}O_7$	$H_2O/$ Na_2O	Final pH	% X	a_0
2.0	3.06	1.67	~140	13.0	35	25.090
2.0	3.06	1.51	~200	12.75	65	25.132
2.06	3.06	1.37	~170	12.9	55	25.064
2.0	3.06	1.18	~200	12.5	50	25.029
2.0	3.57	1.01	~200	10.9	70	25.071
2.0	4.08	0.89	~200	9.4	Amorphous	

parameter of one of the faujasite structures from these preparations was 25.132 A, equivalent to $SiO_2/Al_2O_3 = 2.0$ (6). It is doubtful that the gluconic acid plays any other role than that of an acid.

The zeolite A structure has not been observed in gallosilicates. Possibly, a stable precursor of the cube type cannot be formed because of the larger gallium. Instead, the trimer is formed immediately, leading to the faujasite structure.

If the assumption is correct that the 3 precursors mentioned exist, then every single 4-ring has at least 1 Al, every double 4-ring at least 2, and every trimer at least 3 Al. Thus, the highest SiO_2/Al_2O_3 ratio derived from these precursors is 6.0. Indeed, no evidence has been found yet that A and X structures occur with lattice parameters smaller than those corresponding to $SiO_2/Al_2O_3 = 6.0$. However, preparation of high silica sodalite (16, 20) suggests that single 6-rings containing 1 Al and each associated with 1 tetramethylammonium cation are precursors of such a structure.

At higher R, decrease of the pH causes a higher degree of polymerization of the silicate and lower concentration of reactive species resulting in a slower crystallization of zeolite Y, frequently contaminated with zeolite B. Further decrease of the pH causes ZK-15 to crystallize.

Type of Cation and Cation Ratio. The polymerization–depolymerization equilibrium conceivably can be influenced by the stabilizing effect of certain cations on certain precursors. The cations present also may determine the way in which these building blocks are joined to give a framework structure. Reaction mixtures, identical except for the cation introduced with the phosphate, yield different products at the same pH.

When sodium is predominant, the faujasite structure crystallizes; when potassium is the major cation, a zeolite of chabazite structure (ZK-14) is formed. Both of these structures can be built from hexagonal prisms as the only building blocks. Apparently the presence of the presumably smaller hydrated potassium ions allows a denser structure (chabazite) to crystallize.

Tetramethylammonium ions appear to stabilize the precursors that cause the zeolite A structure to crystallize. Thus, a zeolite A structure (ZK-4, ZK-21, ZK-22) is obtained from a reaction mixture that would yield a zeolite of faujasite structure if the only cations present were sodium (9, 16).

A zeolite of phillipsite structure (ZK-19) (14, 17) is obtained from a mixture of $SiO_2/Al_2O_3 = 4$ to 6 when the fraction of sodium in a reaction mixture containing sodium and potassium is in the range of 0.5–0.8. This denser structure is thermodynamically more stable than either the faujasite or chabazite structure. It seems to play the same role as zeolite B in the pure sodium system (11).

When a zeolite crystallizes, the number of cationic charges incorporated in the zeolite structure is equal to the number of aluminum atoms in the zeolite framework. Obviously, if only a small number of cations are available, the amount of aluminum being included in the zeolite framework is limited. By using cations whose size prevents them from being as easily incorporated as sodium, one can limit the number of cations effectively available for zeolite crystallization without affecting the pH of the reaction mixture. If some of these large cations are incorporated in the crystallizing zeolite, they may further limit the number of cationic charges per unit cell, causing additional increase of the silica-to-alumina ratio of the product. It is important, of course, that the size and shape of the large cations be compatible with the zeolite structure desired.

In order to prepare a high-silica zeolite of faujasite structure ($SiO_2/Al_2O_3 > 4.5$) from a phosphate-containing mixture, the number of sodium ions can be lowered by substituting part of the sodium phosphate with the corresponding tetramethylammonium phosphate, thus retaining the pH (an example is given in Table VI).

Prepolymerization of Silicate. From phosphate-buffered mixtures of identical chemical composition, different zeolites crystallize depending on the degree of prepolymerization of the silicate. Polymeric silicates tend to give ZK-15 or zeolites of chabazite structure [low-silica ZK-14 (17)] in the same pH range where metasilicate causes zeolite Y to crystallize (Table IV). Thus, either the depolymerization of waterglass is slower than generally thought or the precursors formed with partially depolymerized silicate are stabilized sufficiently to prevent further rapid depolymerization.

Table IV. Effect of Silica Source

Silica Source	Mixture Ratios				Product	
	SiO_2/Al_2O_3	P_2O_5/Al_2O_3	Na_2O/P_2O_5	H_2O/Na_2O	Type	SiO_2/Al_2O_3
Metasilicate	4.0	4.22	3.35	63	Y	4.0
	6.47	6.46	3.33	80	Y + Trace B	4.1
Waterglass	4.0	4.22	3.3	64	ZK-14	3.4
	6.0	4.22	3.45	61	ZK-15	4.2
Colloidal silica sol	4.0	2.82	3.55	89	ZK-15	4.6
	6.0	2.82	3.55	89	ZK-15	4.8

Table V contains information on the crystallization of a certain aluminate–phosphate mixture with silicates of different degrees of polymerization. Although no correction was made for the different SiO_2/Na_2O ratios of the silicates, the final pH was always in a range that would have allowed zeolite Y to crystallize. A trend from Y to ZK-14 to ZK-15 to B is observed with increasing degree of polymerization of the silicate used.

Table V. Effect of Degree of Polymerization of the Silicate

$SiO_2/Al_2O_3 = 6.0$

Waterglass	Final pH	Zeolite Obtained	SiO_2/Al_2O_3
B-W	12.05	ZK-14 + ZK-15 + trace Y	4.09
C	12.1	ZK-15	4.35
D	11.9	ZK-15	4.39
Star	11.6	B + trace ZK-15	4.76

Time. Most zeolite structures are metastable; they are generally less hydrothermally stable the lower their density is. Therefore, when the more open zeolites are kept in their mother liquor for an extended time, they tend to recrystallize to denser structures. Zeolites of faujasite structure recrystallize in a sodium system to zeolite B, and those of chabazite structure are transformed in their mixed sodium–potassium mother liquor to products of the denser phillipsite structure.

The rate of crystallization of a certain zeolite crystal structure usually decreases with increasing SiO_2/Al_2O_3 ratio. As Kerr (11) pointed out, a longer crystallization time gives the denser material—e.g., zeolite B— a chance to nucleate. The crystallization rate of zeolite B is greater than that of zeolite X so that the remaining amorphous material crystallizes mainly to zeolite B. Furthermore, given sufficient time, the previously formed zeolite X recrystallizes also to zeolite B. The nucleation time of zeolite X can be eliminated by seeding.

Concentration. A greater reaction volume frequently results in a slightly higher SiO_2/Al_2O_3 ratio of the product. High concentrations of salts inhibit crystallization so that denser materials are obtained.

Table VI. Effect of Concentration on Crystallizing Structure

Constants: $\dfrac{SiO_2}{Al_2O_3} = 8.0$; $\dfrac{Na_2O + [(CH_3)_4N]_2O^a}{P_2O_5} = 3.40$;

$\dfrac{Na_2O}{Na_2O + [(CH_3)_4N]_2O} = 0.256$

$\dfrac{H_2O}{Na_2O + [(CH_3)_4N]_2O}$	Temp., °C	Final pH	Product X-Ray	SiO_2/Al_2O_3
53	90	12.3	X + 5% A	6.0
43	90	12.95	A + 10% X	4.9
26	95	12.6	Sodalite + 15% A	7.4
26	90	12.3	Sodalite	8.1

^a Only the $(CH_3)_4N^+$ added as hydroxide.

Table VI shows the influence of dilution in a Na-$(CH_3)_4N$ system containing $(CH_3)_4NBr$. The smaller precursors are formed in the presence of higher concentrations of TMA ions. Therefore, it seems that TMA ions stabilize not only the precursors of the A-type structure, presumably double 4-rings, but also, at higher concentrations, single 4-rings and 6-rings. The same number of TMA ions may be required for stabilization of either precursor; this explains the concentration effect.

Temperature. Temperature influences the polymerization–depolymerization equilibrium. Higher temperatures cause denser materials to crystallize. Selbin and Mason (23) reported that they had to use a lower temperature to obtain the gallosilicate analog of zeolite X. Temperatures above 70°C caused a gallosilicate of sodalite structure to crystallize. Countless examples for the temperature effect are given in the literature (see, e.g., Ref. 2).

Conclusions

Most complexing agents for aluminum do not have the buffering ability desired to aid in the crystallization of high-silica zeolites. Of the reagents described in this paper, only phytate is similar in performance to phosphate. Use of a nonbuffering complexing agent in combination with phosphate as buffer appears worthwhile.

Observations made in zeolite synthesis work can be explained by the sequence of reactions: Condensation of aluminosilicate (and silicate) to form polymers, ring closure, stabilization of certain ring structures by cations to form soluble precursors, combination of precursors under the influence of cations to give nuclei, and finally crystal growth.

Literature Cited

(1) Bailar, J. C., Jr., Jones, E. M., *Inorg. Syn.* **1939**, 1, 35.
(2) Barrer, R. M., Baynham, J. W., Bultitude, F. W., Meier, W. M., *J. Chem. Soc.* **1959**, 195.
(3) Behr, B., Wendt, H., *Z. Elektrochem.* **1962**, 66, 223.
(4) Cadariu, I., Goina, T., Oniciu, L., *Studia Univ. Babes-Bolyai, Ser. Chem.* **1962**, 7, 81.
(5) Ciric, J., *J. Colloid Interface Sci.* **1968**, 28, 315.
(6) Dempsey, E., Kühl, G. H., Olson, D. H., *J. Phys. Chem.* **1969**, 73, 387.
(7) Freedman, H. H., Frost, A. E., Westerback, S. J., Martell, A. E., *Nature* **1957**, 179, 1020.
(8) Grossmith, F., Brit. Patent **949,405** (1964).
(9) Kerr, G. T., *Inorg. Chem.* **1966**, 5, 1537.
(10) Kerr, G. T., *J. Phys. Chem.* **1966**, 70, 1047.
(11) *Ibid.*, **1968**, 72, 1385.
(12) Kerr, G. T., unpublished data.
(13) Kroll, H., Knell, M., Powers, J., Simonian, J., *J. Am. Chem. Soc.* **1957**, 79, 2024.
(14) Kühl, G. H., *Am. Mineralogist* **1969**, 54, 1607.
(15) Kühl, G. H., *J. Inorg. Nucl. Chem.* **1969**, 31, 1043.
(16) Kühl, G. H., 158th Meeting, ACS, New York, September, 1969.
(17) Kühl, G. H., "Molecular Sieves," p. 85, Society of the Chemical Industry, London, 1968.
(18) Lacroix, S., *Bull. Soc. Chim. France* **1947**, 408.
(19) Lacroix, S., *Ann. Chim.* **1949**, 4, 5.
(20) Meier, W. M., Baerlocher, Ch., *Helv. Chim. Acta* **1969**, 52, 1853.
(21) Saito, K., Terrey, H., *J. Chem. Soc.* **1956**, 4701.
(22) Schwarzenbach, G., Gut, R., Anderegg, G., *Helv. Chim. Acta* **1954**, 37, 937.
(23) Selbin, J., Mason, R. B., *J. Inorg. Nucl. Chem.* **1961**, 20, 222.
(24) Vohra, P., Gray, G. A., Kratzer, F. H., *Proc. Soc. Exptl. Biol. Med.* **1965**, 120, 447.
(25) Wolf, F., Fürtig, H., East German Patent **58,957** (1967).

RECEIVED February 4, 1970.

Discussion

Brian D. McNicol (Koninklijke/Shell Laboratorium, Amsterdam, Netherlands): Was there any evidence of tetrahedral framework substitution by arsenic or phosphorus?

G. H. Kühl: No.

John Turkevich (Princeton University, Princeton, N. J. 08540): Is the phosphate incorporated into the zeolite structure in those cases where phosphate is used to buffer?

G. H. Kühl: I have never encountered phosphate substitution in the framework. However, the zeolite A structure has been found with up to one P per sodalite cage intercalated.

J. A. Rabo (Union Carbide Corp., Tarrytown, N. Y. 10591): You suggested that large cations limit the Al content to lower levels. Do you have any examples for this effect with large alkali cations?

G. H. Kühl: The examples are given in the paper (Table VI, TMA ions, and possibly Ref. 25, where smaller hydrated K^+ ions appear to permit crystallization of low-Si zeolite X).

6

Phosphorus Substitution in Zeolite Frameworks

EDITH M. FLANIGEN and ROBERT W. GROSE

Union Carbide Corp., Linde Division Laboratory, Tarrytown Technical Center, Tarrytown, N. Y. 10591

Zeolites containing phosphorus in the tetrahedral site in the framework have been synthesized. Phosphorus incorporation in a variety of structural types of zeolite frameworks has been achieved: analcime, phillipsite, chabazite, Type A zeolite, Type L zeolite, and Type B (P) zeolite. The syntheses and properties of some of the new aluminosilicophosphate zeolites are described. The synthesis technique involves gel crystallization where incorporation of phosphorus is accomplished by controlled copolymerization and coprecipitation of all the framework component oxides, aluminate, silicate, and phosphate, into a relatively homogeneous gel phase. Subsequent crystallization of the gel is carried out at temperatures in the region of 80° to 210°C. Proof and mechanism of framework substitution of phosphorus is based on electron microprobe analysis, infrared spectroscopy, and other characterization.

The polymorphs of silica—quartz, tridymite, etc.—exist as infinite three-dimensional frameworks formed by corner-sharing $[SiO_4]$ tetrahedra. Tetrahedra of $[AlO_4]$ may, by isomorphous substitution, replace $[SiO_4]$ to form aluminosilicates with the excess negative charge neutralized by an alkali or alkaline earth cation. Phosphorus (P^{5+}) also exhibits tetrahedral coordination with oxygen anions to form $[PO_4]$ tetrahedra and complex three-dimensional frameworks. The space dimensions of $[PO_4]$ and $[SiO_4]$ are similar, and the oxygen chemistry of phosphorus is similar to that of silicon. It therefore seems possible that $[PO_4]$ could isomorphously replace $[SiO_4]$ in the silicate crystal framework.

Few instances of isomorphous replacement of $[SiO_4]$ by $[PO_4]$ have been found in mineral silicates. The mineral viseite, containing $[PO_4]$ in addition to [AlO$_4$] and [SiO$_4$], has a structure analogous to analcime (*10*). The mineral kehoeite is an analogue of the analcime structure but is a complex zinc aluminophosphate containing [AlO$_4$] and [PO$_4$] (*11*). Phosphate tetrahedra replace some [SiO$_4$] in griphite, a phosphate garnet (*12*), and [SiO$_4$] replace some [PO$_4$] in apatite (*9*). Barrer and Marshall reported unsuccessful attempts to synthesize phosphorus-containing zeolites (*2*). Kühl has reported the effect of phosphate complexing of aluminum in zeolite crystallization (*7*) and the synthesis of a zeolite ZK-21 containing "intercalated" phosphate with a crystal structure similar to zeolite A (*8*).

A number of compounds are related in their crystal structures to the polymorphs of silica (*6*). These substances can be arranged into 3 categories on the basis of structural relationship: simple analogues, stuffed derivatives, and coupled derivatives. An example of a simple analogue of silica is germanium dioxide. Germanium has been substituted for silicon successfully in zeolite frameworks by several researchers, as well as gallium for aluminum (*1*). Examples of the stuffed derivatives are NaAlSiO$_4$ (high-carnegieite) and KAlSiO$_4$ (kalsilite), where ions of appropriate size are introduced into vacant interstitial positions in the SiO$_2$ structure type.

The coupled derivatives are represented by AlPO$_4$, GaPO$_4$, AlAsO$_4$, FePO$_4$, and others which yield neutral frameworks with similarities to various silica crystal structures. The formation of an AlPO$_4$-type compound in the zeolite synthesis gel appeared to be a reasonable approach to the synthesis of phosphorus-substituted zeolites. Experiments therefore were initiated to attempt the isomorphous substitution of [PO$_4$] in zeolite structures using the low-temperature and pressure hydrothermal gel systems generally employed in zeolite synthesis in the Union Carbide laboratories (*4*).

These experiments were successful in synthesizing aluminosilicophosphate zeolites with the following types of zeolite frameworks: analcime, chabazite, phillipsite–harmotome, Type A zeolite, Type L zeolite, and Type B (P) zeolite, all of which contained significant amounts of phosphorus (5–25 wt % P$_2$O$_5$) incorporated in the crystal framework. Species related to the felspathoids, sodalite and cancrinite, were also synthesized and contained phosphate detected by chemical analysis. Framework substitution was not verified for these materials, as they normally contain various intercalated salts in their structure. Zeolites structurally analogous to gmelinite (Type S), mordenite, Type X, and Type Y zeolites were synthesized in the sodium aluminosilicophosphate system but contained only small amounts of phosphate (<5 wt % P$_2$O$_5$). In this work, characterization of the phosphorus-substituted species was performed only on those which contained appreciable amounts of phosphorus (>5 wt % P_2O_5) in an attempt to reduce possible misinterpretation of the data owing to the presence of extraneous phosphate.

The nomenclature used throughout this paper designates the structural type by the letter of the most closely related phosphorus-free synthetic zeolite as used in the Union Carbide laboratories (4, 5). Table II in Ref. 5 contains a description of the synthetic zeolites described here. For the aluminosilicophosphate zeolites, these letter designations are prefixed by P to indicate that the zeolite framework contains phosphorus substituted in the tetrahedral (Si, Al) site. For example, the phosphate zeolite structurally related to Type A zeolite is designated P–A; P–R zeolite is the phosphate zeolite related to zeolite R, a synthetic sodium aluminosilicate zeolite with a chabazite type framework structure (5). The aluminosilicophosphate zeolites will be abbreviated to phosphate zeolites or phosphorus-substituted zeolites throughout this paper.

Synthesis

The synthesis technique involves gel crystallization where incorporation of phosphorus is accomplished by controlled copolymerization and coprecipitation of all the framework component oxides, aluminate, silicate, and phosphate, into a relatively homogeneous gel phase. Subsequent crystallization of the gel is carried out at temperatures in the region of 80° to 210°C. Typical compositions of the gels in moles are given in Table I, and the synthesis procedure is described below for the major phosphate zeolite species.

Zeolite P-C (Analcime Structure Type). Zeolite P-C was crystallized from sodium aluminosilicophosphate gels prepared by simultaneously

Table I. Typical Synthesis Conditions for Crystallizing Phosphorus-Substituted Zeolites

Zeolite	Reactant Composition in Moles						Crystallization Temp., °C	Crystallization Time, Hrs.
	Na_2O^a	K_2O^a	Al_2O_3	SiO_2	P_2O_5	H_2O^a		
P-C	≥0.5	–	1.0	0.6	0.5	≥ 55	210	160
P-W	–	≥0.5	1.0	1.6	0.5	≥110	150	68
P-G	–	≥0.5	1.0	1.0	0.5	≥110	150	116
P-R	≥1.2	–	1.0	1.8	0.9	≥110	125	94
P-A	≥1.8	–	1.0	1.6	1.1	≥110	125	45
P-L	–	≥1.0	1.0	1.5	1.0	≥110	175	166
P-B (P)	≥0.4	–	1.0	0.6	0.7	≥110	200	70

[a] Values for Na_2O, K_2O, and H_2O are slightly higher than the values shown and undetermined because of unknown quantities absorbed on the precipitated hydrous aluminophosphate gel.

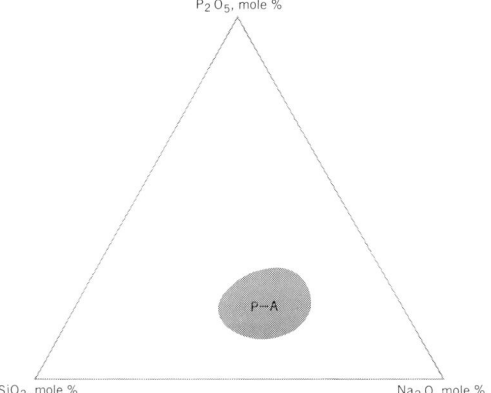

Figure 1. Partial reaction composition diagram. Projection of the Na_2O, Al_2O_3, SiO_2, P_2O_5, H_2O system at 100°–125°C. Na_2O + SiO_2 + P_2O_5 = 100 mole %. Al_2O_3 = 12–20 mole % of the total anhydrous gel composition; mole H_2O/Al_2O_3 is constant (= 110). A solid colloidal silica ("Cab-O-Sil") is the silica source.

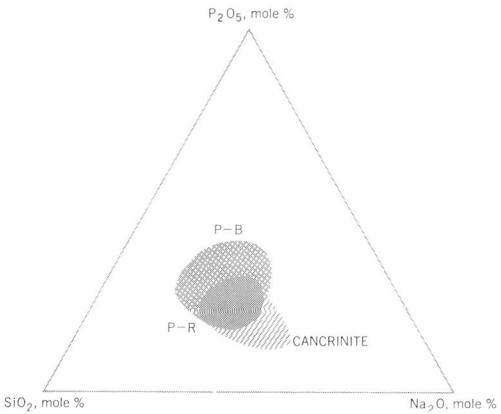

Figure 2. Same as Figure 1 with a crystallization temperature of 125°–150°C and an aqueous colloidal silica source ("Ludox"). Al_2O_3 = 17–23 mole %.

adding aqueous solutions of sodium metasilicate and phosphoric acid to an aqueous solution of $AlCl_3$ with agitation. The pH of the precipitating solution was adjusted to 7.5 by titration with concentrated NaOH solution. The resultant precipitate was separated by vacuum filtration, washed with distilled water, and placed in a Teflon-lined stainless steel autoclave. A predetermined amount of NaOH solution was added. The gel was crystallized at about 175°–210°C and autogenous pressure for 90–160 hours. Maximum phosphorus contents of 25 wt % P_2O_5 were achieved in zeolite P-C.

Zeolite P-W (Phillipsite–Harmotome Structure Type). The P-W synthesis gel was prepared by titrating an aqueous solution of $AlCl_3$ and phosphoric acid with concentrated KOH solution to a pH of 7.5. The precipitate was filtered, washed with distilled water, and then blended with "Ludox," an aqueous silica sol, and KOH solution. The reaction mixture was crystallized at 150°–175°C and autogenous pressure for about 72 hours. The use of a silica sol as the silica source appears to favor the formation of P-W zeolite over coexisting phases such as P-G or P-L zeolites. Phosphorus contents up to 20 wt % P_2O_5 were found in zeolite P-W.

Zeolite P-R and P-G (Chabazite Structure Type). Zeolite P-G was crystallized from potassium aluminosilicophosphate gels prepared by the same procedure used in the preparation of P-W gel. Zeolite P-R was crystallized in the sodium system by using NaOH in place of KOH. The reactant gel was crystallized at 125°–175°C and saturated water vapor pressure for 48–120 hours. Zeolite P-G containing up to 21 wt % P_2O_5, and zeolite P-R containing up to 16 wt % P_2O_5 were synthesized.

Zeolite P-A (A Structure Type). The reactant composition for P-A zeolite was prepared in the same manner as that for the synthesis of P-W or P-R zeolites using NaOH solution as the titrant and solvent. P-A zeolite also was crystallized from sodium aluminosilicophosphate gels prepared by simultaneous addition of $NaAlO_2$–NaOH solution and H_3PO_4 solution to an agitated aqueous slurry of "Cab-O-Sil," a colloidal silica powder. The synthesis gel was crystallized at 100°–150°C and autogenous pressure for 24–96 hours. The "Cab-O-Sil" silica source favors the crystallization of P-A zeolite over zeolites such as P-R or P-B. Zeolite P-A containing up to 10 wt % P_2O_5 was prepared.

Zeolite P-L (L Structure Type). Zeolite P-L was synthesized from aluminosilicophosphate gels prepared by the procedure described for zeolites P-W and P-G using KOH solution as the source of base. The use of a solid silica source such as "Cab-O-Sil" facilitates the formation of P-L

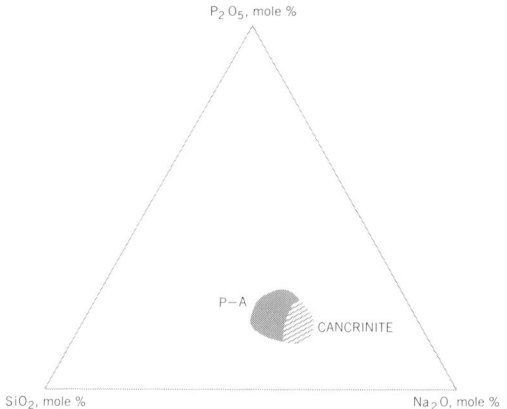

Figure 3. Same as Figure 1 with a crystallization temperature of 150°C and a "Cab-O-Sil" silica source. $Al_2O_3 = 14$–18 mole %.

zeolite rather than P-W or P-G zeolite. The gel was crystallized at 150°–175°C and autogenous pressure for 72–170 hours. Zeolite P-L containing up to 19 wt % P_2O_5 was synthesized.

Zeolite P-B (B or P Structure Type). Zeolite P-B was crystallized from coprecipitated gels using NaOH solution as the titrant and "Ludox" colloidal silica sol. The gel was crystallized at 150°–200°C for 45–72 hours. Phosphorus contents up to 23 wt % P_2O_5 were achieved in P-B zeolites.

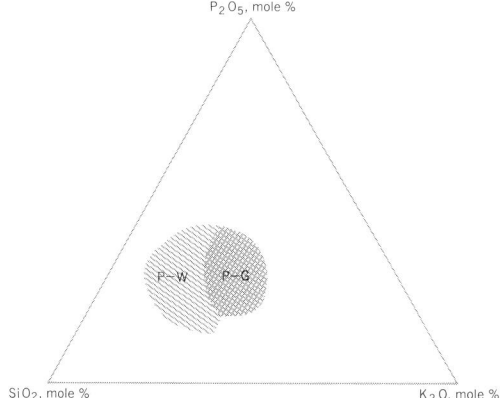

Figure 4. Partial reaction composition diagram. Projection of the K_2O, Al_2O_3, SiO_2, P_2O_5, H_2O system at 150°C. $K_2O + SiO_2 + P_2O_5 = 100$ mole %. $Al_2O_3 = 24$–42 mole % of the total anhydrous gel composition; mole H_2O/Al_2O_3 constant (= 110). "Ludox" is the silica source.

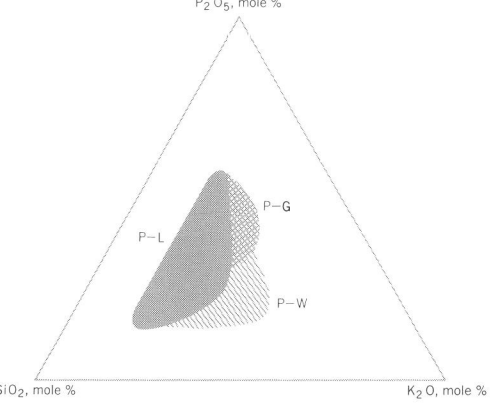

Figure 5. Same as Figure 4 with a crystallization temperature of 175°C and a "Ludox" or "Cab-O-Sil" silica source. $Al_2O_3 = 12$–25 mole %.

Table II. Typical X-Ray Powder Diffraction

P-C		P-W	
d, A	I/I$_0$	d, A	I/I$_0$
5.64	80	10.2	18
4.87	16	8.3	42
3.68	6	7.2	61
3.44	100	5.40	17
3.27	3	5.07	29
2.93	45	4.51	27
2.81	6	4.31	23
2.70	14	4.11	19
2.51	13	3.68	18
2.37	8	3.25	71
1.91	8	3.19	100
1.87	5	2.96	45
1.75	12	2.80	16
1.72	3	2.75	39
1.70	3	2.69	18
		2.57	23
		2.46	10
		2.20	8
		2.08	4
		1.79	5
		1.78	6
		1.73	8

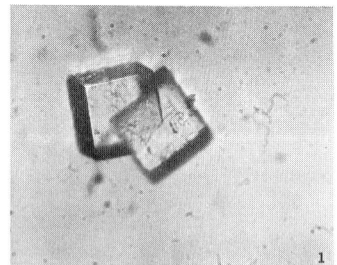
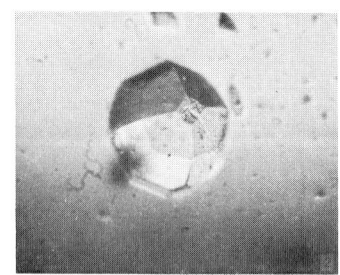
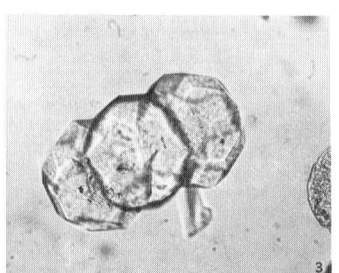

Figure 6. Optical photomicrographs of phosphate zeolites

Plate 1 Zeolite P-C (12.9 wt % P$_2$O$_5$) 200X   Plate 3 Zeolite P-B (21.1 wt % P$_2$O$_5$) 200X
Plate 2 Zeolite P-C (17.5 wt % P$_2$O$_5$) 150X   Plate 4 Zeolite P-W (14.3 wt % P$_2$O$_5$) 600X

Data for Phosphate Zeolites P-C, P-W, P-R, P-G

P-G		P-R	
d, A	I/I$_0$	d, A	I/I$_0$
9.46	100	9.46	100
6.97	21	6.94	29
5.61	17	5.59	10
5.10	21	5.09	32
4.72	10	4.72	9
4.53	5	–	–
4.36	74	4.35	76
4.15	7	4.11	5
4.02	9	4.00	8
3.90	43	3.90	24
3.62	28	3.63	32
3.48	16	3.47	16
3.25	12	3.25	10
3.14	10	–	–
2.95	95	2.94	82
2.92	53	–	–
2.71	9	2.71	5
2.64	19	2.62	11
2.54	14	2.53	8
2.33	9	2.32	5
2.11	7	2.10	8
1.89	8	1.89	8
1.82	14	1.82	9
1.73	12	1.73	6

The formation of the aluminosilicophosphate gel requires a reactive form of phosphorus such as phosphoric acid for its incorporation into the gel structure and zeolite framework. The mere presence of a phosphate salt such as sodium metaphosphate in the reactant gel will not result in phosphorus incorporation in the zeolite crystal lattice.

Ternary reaction diagrams showing the regions of gel composition for synthesizing the phosphate zeolites are shown in Figures 1–3 for the sodium aluminosilicophosphate zeolites, and in Figures 4 and 5 for the analogous potassium system.

Properties

Generally, the phosphate zeolites crystallize in the form of large, near-single crystals of the order of 100 μ in size. However, for zeolite P-A, the crystal size was of the order of 1 to several μ. Single-crystal measurements were therefore possible for most of the phosphate zeolites. Photomicrographs of several of the phosphate zeolite crystals are shown in Figure 6.

Typical x-ray powder diffraction data for the phosphorus-substituted zeolites are given in Tables II and III. Unit cell dimensions have been determined for several of the phosphate zeolites and compared with the structurally related phosphorus-free zeolites. The unit cell dimension of cubic P-C zeolite (13.2 wt % P_2O_5) was found to be 13.75 Å from single-crystal precession x-ray photographs. A single crystal of the analogous zeolite C has an $a = 13.73$ Å. The crystal structure of zeolite P-C (13.2 wt % P_2O_5) has been determined by Birle et al. and will be reported elsewhere (3).

A single crystal of P-W zeolite (16.5 wt % P_2O_5) was examined by precession x-ray photographs and found to be tetragonal, with unit cell

Table III. Typical X-Ray Powder Diffraction Data for Phosphate Zeolites P-A, P-L, P-B(P)

P-A		P-L		P-B	
d, Å	I/I_0	d, Å	I/I_0	d, Å	I/I_0
12.2	100	16.0	100	7.08	58
8.6	89	8.0	4	5.01	58
7.07	57	7.55	8	4.96	45
5.48	35	6.09	18	4.44	7
4.99	5	5.86	4	4.10	57
4.33	16	4.65	29	4.04	17
4.08	57	4.47	9	3.53	5
3.87	8	4.37	6	3.34	6
3.69	81	3.96	28	3.20	100
3.54	5	3.68	9	3.13	57
3.40	32	3.51	19	2.71	37
3.28	68	3.32	15	2.68	38
2.96	92	3.22	31	2.66	22
2.89	22	3.09	24	2.53	8
2.74	16	3.05	5	2.20	5
2.68	11	2.93	28	1.98	6
2.62	49	2.88	4	1.68	7
2.50	11	2.82	4		
2.45	8	2.69	24		
2.36	5	2.64	8		
2.24	3	2.53	9		
2.17	14	2.50	4		
2.14	8	2.46	4		
2.10	5	2.44	5		
2.07	5	2.32	3		
2.05	14	2.30	3		
2.02	2	2.22	9		
1.92	12	2.06	3		
1.89	7	1.96	2		
1.85	5	1.88	6		
1.83	5				
1.73	18				

Table IV. P-A Zeolite Unit Cell Dimensions, a vs. P_2O_5 Wt %

	Wt % P_2O_5	a, A
Na Form of P-A Zeolite	5.2	12.249 ± 0.005
	6.2	12.243 ± 0.005
	6.4	12.246 ± 0.005
	6.6	12.237 ± 0.005
	6.7	12.236 ± 0.005
	6.8	12.235 ± 0.005
	6.9	12.232 ± 0.005
	7.3	12.242 ± 0.005
	7.7	12.235 ± 0.005
	8.7	12.237 ± 0.005
	9.1	12.235 ± 0.005
	10.2	12.232 ± 0.005
Ca^{ex} Form of P-A Zeolite	6.3	12.237 ± 0.005
	8.0	12.251 ± 0.005
	8.6	12.220 ± 0.005

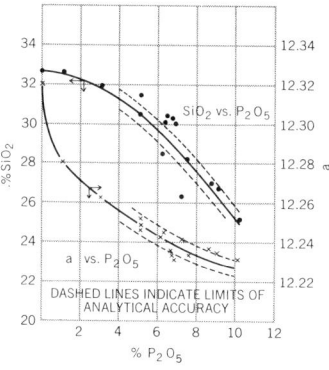

Figure 7. SiO_2 content vs. P_2O_5 content and unit cell dimension (a) vs. P_2O_5 content of P-A zeolite

dimensions of $a = 20.17$ A and $c = 10.03$ A. An alternate tetragonal cell with $a = 14$ A was also consistent with the x-ray data. A single crystal of phosphorus-free Type W zeolite was not available for comparison. Based on powder data, zeolite W has been indexed on a cubic unit cell with $a = 20$ A by D. W. Breck (unpublished results). Steinfink (14) reports an orthorhombic unit cell with $a = 9.96$ A, $b = 14.25$ A, and $c = 14.25$ A for the mineral zeolite phillipsite, and Sadanaga et al. (13) found a monoclinic unit cell for the related mineral zeolite harmotome which deviates only slightly from the orthorhombic phillipsite cell.

Table V. Typical Unit Cell

Phosphate Zeolite	Unit Cell	Tetrahedra/ U.C.
P-L	Hexagonal, $a = 18.75$, $c = 15.03^c$ $d_{meas} = 2.21$	72
P-W	Tetragonal, $a = 20.17$, $c = 10.03^d$	64^d
P-G P-R	Rhombohedral, $a = 9.44$, $\alpha = 94°28'$ (chabazite)	12
P-C	Cubic, $a = 13.73$ $d_{meas} = 2.30 \pm 0.02$	48
P-B	Tetragonal,^{e} $a = 10.1$ $c = 9.8$	16
P-A	Cubic, $a = 12.24$ (pseudo cell) $d_{meas} = 2.11$	24

^{a} Calculated from wet chemical analysis for zeolites shown in Table VII except for zeolite P-C, where the average of wet chemical and microprobe analyses (3) was used. All calculations were normalized to the appropriate number of tetrahedra/unit cell. Unit cell dimensions are in A, and densities in grams/cc at 25°C.

The cubic unit cell constants, a, of a number of P-A zeolites were determined over a range of phosphorus content (0 to 10 wt % P_2O_5). The unit cell dimension decreases with increase in phosphorus content (Table IV, Figure 7). This reduction in unit cell constant is consistent with the substitution of phosphorus in the zeolite framework by the mechanism discussed later, and is attributed to the smaller tetrahedral P—O bond distance of 1.54 A, compared to the tetrahedral Si—O bond distance of 1.61 A and Al—O bond distance of 1.75 A. Unit cell compositions for P-A zeolites calculated on the basis of the pseudo cubic unit cell of Type A zeolite, $a \sim 12$ A, are included in Table V. Measured densities for the sodium forms and calcium-exchanged forms of zeolite A and zeolite P-A are given in Table VI.

Typical chemical compositions of the phosphate zeolites are found in Table VII. The phosphorus content of the zeolites may be varied by use of suitable reactant gel compositions to give the desired phosphorus substitution. Single-crystal electron microprobe analysis was employed to verify the P_2O_5 content of several of the phosphate zeolites where crystals of sufficient size were available. The electron microprobe analyses were carried out by J. V. Smith and C. R. Knowles at the University of Chicago (unpublished results). A comparison of the electron microprobe results of Smith and Knowles and the wet chemical analyses of the bulk

Compositions for Phosphate Zeolites[a]

Composition	Charge Deficiency[b]
$K_{22.9}[(AlO_2)_{33.1}(SiO_2)_{26.3}(PO_2)_{12.6}] \cdot 42\ H_2O$	-2.6
$K_{15.6}[(AlO_2)_{28.8}(SiO_2)_{24.5}(PO_2)_{10.7}] \cdot 55\ H_2O$	$+2.5$
$\{K_{3.0}[(AlO_2)_{5.6}(SiO_2)_{4.3}(PO_2)_{2.0}] \cdot 10.7\ H_2O$	$+0.6$
$\ Na_{4.4}[(AlO_2)_{5.5}(SiO_2)_{4.5}(PO_2)_{2.0}] \cdot 10.9\ H_2O$	-0.9
$Na_{15.3}Ca_{0.6}[(AlO_2)_{2.30}(SiO_2)_{18.2}(PO_2)_{6.7}] \cdot 18.6\ H_2O$	-0.2
$Na_{4.7}[(AlO_2)_{7.9}(SiO_2)_{4.3}(PO_2)_{3.8}] \cdot 10.4\ H_2O$	-0.6
$Na_{11.5}[(AlO_2)_{11.5}(SiO_2)_{9.7}(PO_2)_{2.8}]\ OH_{2.8} \cdot 24.8\ H_2O$	0

[b] The charge deficiency values listed represent the additional charge required to balance the unit cell charge. Although the presence of OH^- and $H_3O_2^-$ groups are postulated in some cases in the substitution mechanism, their assignment to balance charge seems arbitrary except in the case of P-A zeolite (see discussion). Also, the values of charge deficiency in most cases are well within the errors in analysis.
[c] Cell determined from single-crystal electron diffraction studies on P-L.
[d] Unit cell from single-crystal x-ray precession studies; tetrahedra/U.C. chosen on the basis of the tetrahedra density in the related harmotome–phillipsite structures. The density was not determined.
[e] Tetragonal cell that of Taylor and Roy (15) for related P_t zeolite.

Table VI. Density of Hydrated Na and Ca^{ex} Forms of P-A Zeolites Compared with Hydrated NaA and Ca^{ex} A Zeolites

	Measured Density[a] 25°C, Grams/Cm³
Na form of P-A zeolite (8.7 wt % P_2O_5)	2.112 ± 0.005
Na A zeolite	2.047 ± 0.005
Ca^{ex} form of P-A zeolite (8.6 wt % P_2O_5)	2.129 ± 0.005
Ca^{ex} A zeolite	2.078 ± 0.005

[a] Measured by water displacement.

samples is given in Table VIII. The good agreement between single-crystal analysis (microprobe) and bulk chemical analysis supports the proposed framework substitution of phosphorus.

The phosphate zeolites can be ion-exchanged by use of the usual methods employed for aluminosilicate zeolites. The cation composition of some cation-exchanged phosphorus-substituted zeolites is given in Table IX. Essentially complete exchange (>90%) of the cation content

Table VII. Typical Chemical Compositions

Chemical Analysis, Wt %

Zeolite	Na_2O	K_2O	Al_2O_3	SiO_2	P_2O_5	H_2O
P-C	12.7	–	32.3	21.2	22.0	11.8
P-W	–	13.7	27.7	28.3	14.3	14.6
P-G	–	13.9	27.9	25.3	14.1	18.8
P-R	13.2	–	25.8	29.1	12.5	18.8
P-A	16.3	–	26.7	26.8	9.1	20.4
P-L	–	17.8	27.8	25.9	14.8	12.4
P-B	11.1	–	31.2	18.1	22.6	17.2

Table VIII. Single-Crystal Electron Microprobe Analysis of Phosphorus-Substituted Zeolites

Zeolite	Wet Chemical Analysis (Composite Sample), Wt % P_2O_5	Electron Probe Analysis (Single-Crystal), Wt % P_2O_5
P-C	22.0	19.5
P-W	19.8	21.7
P-G	14.1	15.7
P-B	15.6	16.1

was demonstrated for zeolites P-W, P-G, P-R, P-A, P-L, and P-B, with at least one of the exchanging cations studied.

The characteristic adsorption capacities of phosphorus-substituted zeolites and their ion-exchanged analogues are given in Table X. The adsorption properties of the phosphate zeolites in most cases do not differ significantly from those of phosphorus-free zeolites. Zeolite P-C exhibits small-pore (<3 A) and low-capacity (1–4 wt % H_2O) adsorption characteristics, as do phosphorus-free analcime structures. Zeolite P-W and its Na^+-exchanged form have an apparent adsorption pore size of about 3.8 A, which is slightly larger than that of phosphorus-free zeolite W (∼3 A), and are thermally stable to activation at 350°C in vacuum. Calcium-exchanged P-W zeolite lost about 75% of its x-ray crystallinity after activation at 350°C in vacuum.

Zeolite P-R and its Ca^{2+}- and NH_4^+-exchanged forms have an apparent adsorption pore size of about 3.8 A. The K^+-exchanged form of P-R zeolite does not adsorb oxygen and has an apparent pore size of <3.5 A. Considerable loss of crystallinity was observed for zeolite P-R and its ion-exchanged forms when activated above 200°C in vacuum, with the most severe degradation occuring with the NH_4^+-exchanged material. Zeolite P-G and its Na^+- and Ca^{2+}-exchanged forms are thermally stable to activation at 350°C in vacuum; the NH_4^+-exchanged form is not stable above 200°C. Zeolite P-G and Na^+-exchanged P-G have an apparent pore of Phosphorus-Substituted Zeolites

Oxide Composition, Mole

Na_2O	K_2O	Al_2O_3	SiO_2	P_2O_5	H_2O
0.65	–	1.00	1.11	0.49	2.07
–	0.54	1.00	1.73	0.37	2.98
–	0.54	1.00	1.54	0.36	3.80
0.84	–	1.00	1.92	0.35	4.13
1.00	–	1.00	1.71	0.24	4.32
–	0.69	1.00	1.59	0.38	2.53
0.58	–	1.00	0.99	0.52	3.12

Table IX. Cation Composition of some Cation Exchanged Phosphorus-Substituted Zeolites[a]

Zeolite	Initial Cation, $Wt\%$		Exchange Cation, $Wt\%$	% Exchange[b]
	K_2O	Na_2O		
P-W	13.7	–	–	–
Na+-exchanged	1.3	–	8.0 Na_2O	88
Ca2+-exchanged	2.2	–	7.6 CaO	93
P-G	13.9	–	–	–
Na+-exchanged	0.5	–	7.6 Na_2O	83
Ca2+-exchanged	0.7	–	8.1 CaO	99
P-R	–	13.2	–	–
K+-exchanged	–	0.8	18.2 K_2O	91
NH4+-exchanged	–	0.4	9.6 $(NH_4)_2O$	87
Ca2+-exchanged	–	0.8	11.5 CaO	96
P-A	–	15.1	–	–
K+-exchanged	–	0.8	20.2 K_2O	88
Ca2+-exchanged	–	1.7	12.8 CaO	93
P-L	17.8	–	–	–
Na+-exchanged	10.9	–	4.7 Na_2O	40
NH4+-exchanged	3.2	–	8.9 $(NH_4)_2O$	90
Ca2+-exchanged	10.5	–	4.5 CaO	42
La3+-exchanged	14.5	–	2.9 La_2O_3	5
P-B	–	12.1	–	–
K+-exchanged	–	0.1	18.4 K_2O	100

[a] Phosphate zeolites were treated with concentrated chloride solutions of the exchange ion at 25°–90°C for 4–24 hours.
[b] % Exchange = equivalents of exchange cation/equivalents of initial cation × 100.

size of about 3.8 A. Na+-exchanged P-G zeolite exhibits a lower capacity for CO_2, O_2, and N_2 than would be expected for a small cation form (0.95 A ionic radius for Na+ vs. 1.33 A for K+). This may be caused by a shift in cation position near the pores. Ca2+-exchanged P-G exhibits

Table X. Characteristic Adsorption Capacities

Zeolite	Adsorbate = H_2O Kinetic Diameter, A = 2.7 Pressure, Torr = 20 Temperature, °C = 25
C	1.3
P-C	4.4
W	14.0
P-W	19.1
Na^+-exchanged	23.1
Ca^{2+}-exchanged	10.6
R	22.4
P-R	19.0
Ca^{2+}-exchanged	11.1
NH_4^+-exchanged	0.8
K^+-exchanged	18.0
G	21.1
P-G	23.3
Na^+-exchanged	26.1
Ca^{2+}-exchanged	26.2
NH_4^+-exchanged	5.7
A	28.5
P-A	26.7
K^+-exchanged	20.7
Ca^{2+}-exhanged	27.8
L	20.5
P-L	14.4
Na^+-exchanged	15.9
NH_4^+-exchanged	14.8
Ca^{2+}-exchanged	16.4
La^{3+}-exchanged	15.6
B	22.0
P-B (P)	–
K^+-exchanged	4.1

[a] Activation conditions: 350°C, ~ 10^{-5} torr, ~ 16 hours.

the expected increase in adsorption capacity owing to decrease in cation density, and an increased pore size to ≥4.3 A reflecting the smaller cation radius (0.9 A vs. 1.33 A).

The phosphate analogues of Type A zeolite exhibit adsorption properties comparable with the phosphorus-free Type A zeolite. The P-A(Na) zeolite has an apparent pore size of about 4 A, K^+-exchanged P-A zeolite

of Phosphorus-Substituted Zeolites[a]

Grams/Gram Activated Zeolite × 100

CO_2	O_2	N_2	n-C_4H_{10}	iso-C_4H_{10}	neopentane	$(C_3H_7)_3N$
3.3	3.46	3.64	4.3	5.0	6.2	8.1
750	750	750	750	750	750	2
25	−183	−196	25	25	25	50
0	–	–	–	–	–	–
–	–	–	–	–	–	–
–	0.3	–	–	–	–	–
–	10.1	8.3	1.7	–	–	–
–	9.9	9.6	1.3	–	–	–
–	9.0	8.2	1.5	–	–	–
19.7	22.7	–	6.6	–	–	–
11.6	6.6	–	0.8	–	–	–
9.0	9.6	–	0.8	–	–	–
–	0.2	–	–	–	–	–
11.2	0	–	0	–	–	–
16.1	0	0	0	–	–	–
11.0	11.2	9.1	1.6	–	–	–
9.6	9.8	8.8	1.8	–	–	–
–	28.4	24.6	3.5	–	–	–
–	4.0	–	–	–	–	–
18.7	24.5	0.1	–	–	–	–
–	23.8	–	–	–	–	–
13.2	3.0	–	–	–	–	–
21.4	26.7	22.6	11.5	1.8	–	–
–	21.4	–	–	7.7	9.0	11.9
–	8.9	–	–	3.3	2.0	2.2
–	5.3	–	–	2.2	1.2	1.4
–	12.1	–	–	1.7	0.6	0.6
–	14.0	–	–	3.1	1.1	0.8
–	11.6	–	–	1.7	0.4	1.2
1.7	–	–	–	–	–	–
–	–	–	–	–	–	–
2.4	–	–	–	–	–	–

about a 3.4-A pore, and Ca^{2+}-exchanged P-A zeolite about a 5-A pore. These materials are stable to thermal activation at 350°C in vacuum.

The substitution of phosphorus in the framework structure of Type L zeolite results in reduced adsorption capacity (∼50%) and a reduction in apparent pore size from about 10 to 6–7 A. Na^+-exchange of P-L zeolite results in a further reduction in adsorption capacity except for water. Ca^{2+}-exchange of P-L results in an increased adsorption capacity for H_2O and O_2 to about 90% of that of phosphorus-free L zeolite, but reduces

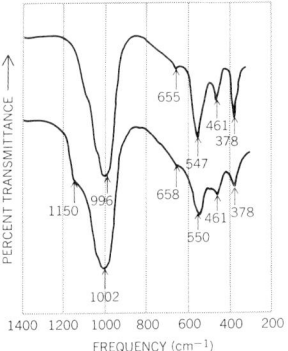

Figure 8. Infrared spectra for the sodium forms of P-A zeolite (lower curve) and type A zeolite (upper curve)

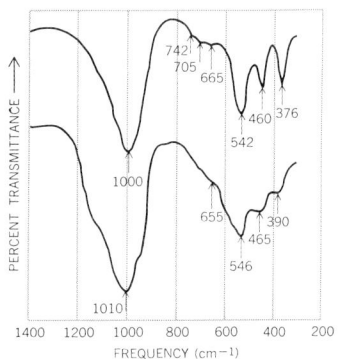

Figure 9. Infrared spectra for the calcium-exchanged forms of P-A zeolite (lower curve) and type A zeolite (upper curve)

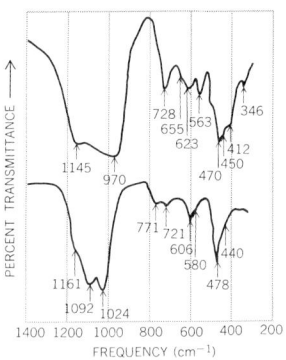

Figure 10. Infrared spectra of P-L zeolite (upper curve) and type L zeolite (lower curve)

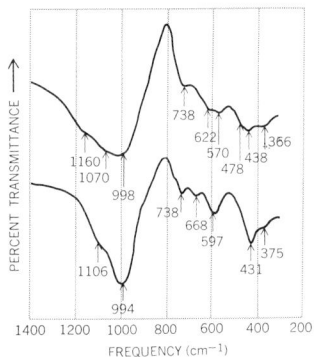

Figure 11. Infrared spectra for P-B zeolite (upper curve) and type B zeolite (lower curve)

the capacity for large molecules such as neopentane. Zeolite P-L and its Na- and Ca-exchanged forms are stable to thermal activation at 350°C. The NH_4^+-exchanged P-L is not stable to activation at 400°C.

Phosphorus-substituted Type B zeolite is not stable to thermal activation. Potassium-exchanged P-B was activated at 350°C and exhibited small-pore (<3 A) and low-capacity adsorption properties. The K^+-exchanged material is crystalline after thermal activation but powder x-ray diffractograms indicate a shift of the lattice spacings.

Infrared spectra of the phosphate zeolites in the mid-infrared region (1300–200 cm^{-1}) also were used to characterize the framework composi-

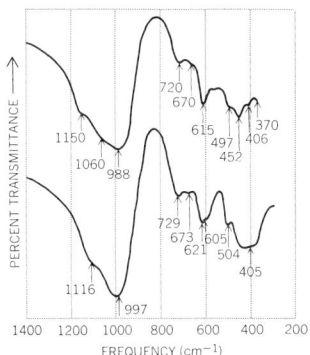

Figure 12. Infrared spectra for P-R zeolite (upper curve) and type R zeolite (lower curve)

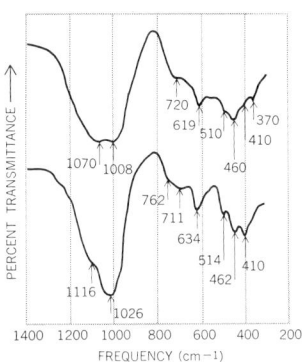

Figure 13. Infrared spectra for P-G zeolite (upper curve) and type G zeolite (lower curve)

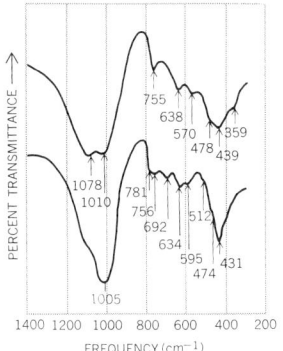

Figure 14. Infrared spectra for P-W zeolite (upper curve) and type W zeolite (lower curve)

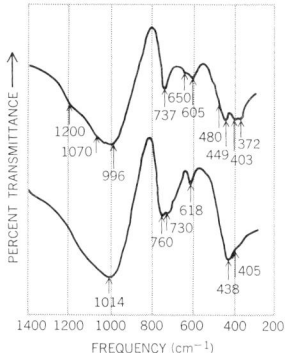

Figure 15. Infrared spectra for P-C zeolite (upper curve) and type C zeolite (lower curve)

tion and establish proof of phosphorus substitution. In another paper in this conference (5), the use of infrared spectroscopy to characterize zeolite frameworks is discussed.

As pointed out there and elsewhere, some infrared vibrations are sensitive to the Si, Al framework composition. In the case of stretch frequencies, substitution of Al for Si in the framework causes a shift to lower frequency owing to the longer Al—O bond distance and decreased bond order compared with Si—O. Substitution of phosphorus in the zeolite framework should show similar changes, but because of the shorter tetrahedral P—O bond distance (see above) the shift in the main asymmetric stretch band near 1000 cm^{-1} should move toward higher frequencies.

Table XI. Framework Compositions for Phosphorus Substitution Mechanisms

	Anhydrous Unit Cell Composition[a]				
	M	$[(AlO_2)$	(SiO_2)	$(PO_2)]$	OH^-
P-C	Na				
Zeolite C	16	16	32	–	–
Hypothetical, Al "saturated"	24	24	24	–	–
Zeolite P-C	14.3	23.3	15.9	8.8	–
Hypothetical, substitution ratio Al:P = 1:1, Al "saturation" via Mechanism 1	16	24	16	8	–
Zeolite P-C	16.0	22.0	12.8	13.2	5.2
Hypothetical, substitution ratio Al:P = 2:3, Al "saturation" via Mechanism 1, excess P via Mechanism 3	16	24	12	12	4
P-W	K				
Zeolite W	24	24	40	–	–
Hypothetical, Al "saturated"	32	32	32	–	–
Zeolite P-W	16.8	30.4	18.8	14.8	–
Hypothetical, Al "saturation" via Mechanism 1, excess P via Mechanism 2	20	32	20	12	–
Zeolite P-W	15.2	28.4	24.8	10.8	–
Hypothetical, Mechanism 1 but not to Al "saturation", excess P via Mechanism 2	16	28	24	12	–
P-R	Na				
Zeolite R	4.5	4.5	7.5	–	–
Hypothetical, Al "saturated"	6	6	6	–	–
Zeolite P-R	4.5	5.5	4.6	1.9	–
Hypothetical, Al "saturation" via Mechanism 1	4	6	4	2	–
Zeolite P-R	4.1	5.5	4.2	2.3	–
Hypothetical, Mechanism 1, and Mechanism 3	4.5	5.5	4.5	2.0	1.0
P-G	K				
Zeolite G	3.5	3.5	8.5	–	–
Hypothetical, Al "saturated"	6	6	6	–	–
Zeolite P-G	3.0	5.6	4.3	2.1	–
Hypothetical, Al "saturation" via Mechanism 1	4	6	4	2	–

Table XI. Continued

	Anhydrous Unit Cell Composition[a]				
	M	[(AlO$_2$)	(SiO$_2$)	(PO$_2$)]	OH$^-$
P-A	Na				
Zeolite A	12	12	12	–	–
Zeolite P-A	10.9	11.1	10.8	2.1	2.1
Zeolite P-A	11.5	11.5	9.7	2.8	2.8
Hypothetical, Mechanism 3	12	12	10	2	2
	12	12	9	3	3
P-L	K				
Zeolite L	18	18	54	–	–
Hypothetical, Al "saturated"	36	36	36	–	–
Zeolite P-L	22.9	33.1	26.3	12.6	–
Hypothetical, Mechanism 1	18	30	30	12	–
P-B	Na				
Zeolite B	6	6	10	–	–
Hypothetical, Al "saturated"	8	8	8	–	–
Zeolite P-B	4.7	7.9	4.3	3.8	–
Hypothetical, Al "saturation" via Mechanism 1, excess P via Mechanism 2	4.5	8.0	4.5	3.5	–

[a] Unit cell compositions for the phosphate zeolites were calculated from wet chemical analysis; the unit cells for the aluminosilicate zeolites are typical. Unit cells used are those shown in Table V; M represents exchangeable cation. The hypothetical cells were calculated on the basis of the substitution mechanisms proposed (*see* discussion section).

The infrared spectra of the phosphate zeolites are shown in Figures 8 to 15, along with their phosphorus-free synthetic zeolite analogues. Comparison of the spectra of the phosphate zeolites with those of the phosphorus-free analogues shows increased absorption in the higher frequency portion of the asymmetric stretch band because of the presence of phosphate groups in the framework. Similarly, where the substitution of phosphorus is accompanied by enrichment of the framework with aluminum (*see* Table VII and discussion section), an increase in absorption in the lower frequency portion of the asymmetric stretch is observed. Differences caused by phosphorus substitution also are observed in other bands in the spectra.

Discussion

The proposed mechanisms of phosphorus substitution in zeolites are:
(1) $(AlO_2)^- + (PO_2)^+ = 2(SiO_2)$
(2) $(PO_2)^+ = (K^+)Na^+ + (SiO_2)$
(3) $(PO_2)^+ + (OH)^- = (SiO_2)$

(4) $(PO_2)^+ + (H_3O_2)^- = 2(SiO_2)$
(5) $(PO_2)^+ + (OH)^- = (K^+)Na^+ + (AlO_2)^-$

In most experiments, the synthesis of phosphate zeolites has taken place in alumina-rich gels so that some substitution of alumina for silica $[(AlO_2)^- + Na^+ \rightarrow (SiO_2)]$ is probable. Mechanism 1 is proposed to occur in most of the phosphate zeolites through the formation of an aluminophosphate complex in the reactant gel which requires the presence of a reactive phosphate species such as phosphoric acid. This requirement would explain the failure to synthesize phosphate zeolites from gels containing a phosphate salt as the source of phosphorus. The combined effects of alumina substitution for silica and the increase in alumina content owing to the "$AlPO_4$" substitution results in alumina-rich zeolites which may have the inherent thermal instability observed in many low SiO_2/Al_2O_3 ratio zeolites. This may explain, at least in part, the poor thermal stability of phosphorus-substituted P-R and the NH_4^+-exchanged forms of P-G and P-L zeolite.

Typical unit cell compositions for the phosphate zeolites are given in Table V. Compositions for hypothetical phosphate zeolite frameworks calculated on the basis of the substitution reactions above are compared in Table XI with the unit cell compositions for the phosphate zeolites (calculated from chemical analysis). The substitution mechanism assigned to each phosphate zeolite species below can be understood best by referring to Table XI.

More than one mechanism may introduce phosphorus into the framework of most of the phosphate zeolites. Analysis of P-C zeolite indicates that $(PO_2)^+$ and $(AlO_2)^-$ simultaneously substitute for (SiO_2) (Mechanism 1) and neutralize each other until the analcime type framework is saturated with respect to $(AlO_2)^-$ in observance of the accepted structure rule that 2 alumina tetrahedra cannot be adjacent in a zeolite framework. In some cases, however, the $(PO_2)^+$ substitution exceeds this $(AlO_2):(PO_2)$ ratio, and the $(PO_2)^+$ must be balanced by $(OH)^-$ (Mechanism 3). The presence of $(OH)^-$ groups postulated in Mechanism 3 is based on charge balancing considerations and has not been verified experimentally. The role of the $(OH)^-$ ion would be analogous to the extra-framework cations balancing the charge of framework $(AlO_2)^-$ groups in aluminosilicate zeolites. Evidence of structural hydroxyl groups in zeolites is well established.

Tetrahedra of $(PO_2)^+$ and $(AlO_2)^-$ substitute for (SiO_2) (Mechanism 1) in zeolite P-W with the degree of $(PO_2)^+$ substitution greater than that of the $(AlO_2)^-$ substitution, resulting in neutralization of normal $(AlO_2)^-$ in the framework by $(PO_2)^+$ rather than K^+ (Mechanism 2). The decreased cation density is evident in the improved adsorption capacities.

zeolite W indicates a 36% increase in H_2O capacity for the phosphate analogue and an increase in pore size sufficient to show significant adsorption of oxygen.

Simultaneous substitution of $(PO_2)^+$ and $(AlO_2)^-$ (Mechanism 1) occurs in P-R zeolite until the framework is almost saturated with $(AlO_2)^-$. The additional $(PO_2)^+$ substituted for (SiO_2) is neutralized then by $(OH)^-$ (Mechanism 3). Phosphorus substitution in P-G zeolite also occurs by Mechanism 1. The analyses of some P-G zeolite samples indicate a slight excess of $(PO_2)^+$ over that balanced by $(AlO_2)^-$ and a deficiency of K^+; this suggests the possibility of $(PO_2)^+$ replacing $K^+ + (SiO_2)$, as in Mechanism 2.

The unit cell of Type A zeolite is saturated with respect to $(AlO_2)^-$ since it has a Si/Al near 1. The coupled $(AlO_2)-(PO_2)$ substitution (Mechanism 1) therefore should not occur in P-A zeolite. The substitution is explained by Mechanism 3, where the $(PO_2)^+$ is neutralized by $(OH)^-$. The analyses of P-A zeolites tend to show slight deficiency in $(AlO_2)^-$ and Na^+ ion, indicating some $(PO_2)^+$ or (SiO_2) substitution for $(AlO_2)^- + Na^+$ [$(PO_2)^+$ by Mechanism 5]. The lack of significant change in alumina content and cation density provides the P-A zeolite with similar adsorption and stability characteristics, as observed in the phosphorus-free Type A zeolite.

The coupled substitution of $(PO_2)^+$ and $(AlO_2)^-$ for $2(SiO_2)$ (Mechanism 1) appears to be the predominant substitution mechanism in P-L and P-B zeolites. In both cases, additional phosphorus substitution appears to occur by Mechanism 2.

Mechanism 4, $(PO_2)^+ + (H_3O_2)^- \rightarrow 2(SiO_2)$, has not been identified with any specific zeolite because of the difficulty in proving the existence of a tetrahedral configuration such as $(H_3O_2)^-$. This tetrahedral unit has been proposed as occupying tetrahedral sites in the structure of viseite and kehoeite, the mineral phosphate analogues of analcime (*10, 11*). Mechanism 5 is a possible means of substitution utilizing the replacement of alumina rather than silica, but is not probable because of the excess alumina (or silica deficiency) contained in the reactant compositions described here.

The shifts and changes in infrared spectra for the phosphate zeolites compared with their related phosphorus-free analogues are in agreement with the above proposed substitution mechanisms.

In conclusion, we propose that the properties and characteristics of the phosphate zeolites presented, including single-crystal microprobe analysis, infrared spectra, zeolite framework stoichiometry, and unit cell data, establishes beyond doubt the substitution of phosphorus in zeolite frameworks.

Acknowledgment

The authors thank J. V. Smith, Univ. of Chicago, for the electron microprobe analyses and for many helpful discussions on structural aspects. The expert assistance of Barbara A. Bierl in the infrared determinations and interpretation, L. G. Dowell in the x-ray studies, and R. G. Pankhurst in chemical analyses is gratefully acknowledged. We thank D. W. Breck for his encouragement, support, and helpful advice during the course of these studies and in the preparation of the manuscript.

Literature Cited

(1) Barrer, R. M., Baynham, J. W., Bultitude, F. W., Meier, W. M., *J. Chem. Soc.* **1959**, 195.
(2) Barrer, R. M., Marshall, D. J., *J. Chem. Soc.* **1965**, 6616.
(3) Birle, J. D., Knowles, C. R., Smith, J. V., Dowell, L. G., in preparation.
(4) Breck, D. W., Flanigen, E. M., "Conference on Molecular Sieves," p. 47, Society of the Chemical Industry, London, 1968.
(5) Flanigen, E. M., Szymanski, H. A., Khatami, H., ADVAN. CHEM. SER. **1971**, 101, 201.
(6) Frondel, C., "Dana's System of Mineralogy," 7th ed., Vol. III, p. 5, Wiley, New York, 1962.
(7) Kühl, G. H., "Conference on Molecular Sieves," p. 85, Society of the Chemical Industry, London, 1968.
(8) Kühl, G. H., U. S. Patent **3,355,246** (1967).
(9) McConnell, D., *Am. Mineralogist* **1937**, 22, 977.
(10) *Ibid.*, **1952**, 37, 609.
(11) McConnell, D., *Mineral. Mag.* **1964**, 33, 799.
(12) McConnell, D., Verhoek, F. H., *J. Chem. Educ.* **1963**, 40, 512.
(13) Sadanaga, R., Marumo, F., Takeuchi, Y., *Acta. Cryst.* **1961**, 14, 1153–63.
(14) Steinfink, H., *Acta. Cryst.* **1962**, 15, 644.
(15) Taylor, A. M., Roy, R., *Am. Mineralogist* **1964**, 49, 656.

RECEIVED March 4, 1970.

Discussion

W. M. Meier (Eidgenössische Technische Hochschule, Zurich): Work done by Liebau and coworkers on silicon phosphates (*Z. Anorg. Allgem. Chem.* **1968**, 359, 113) should perhaps be mentioned here. At least five different crystalline phases of composition SiP_2O_7 have been synthesized by these investigators using reaction mixtures containing silica gel and phosphoric acid at temperatures ranging from 250°C upwards and autogenous pressures. The crystalline products thus obtained are quite remarkable since the silicon was shown to be octahedrally coordinated (*Acta Cryst.* **1970**, 26, 233).

H. Villiger (Martinswerk GmbH, West Germany): 1) The cell constant of P–A is given in Table V as $a = 12.24$ Å (pseudo cell). Are there any extra lines visible which would indicate the presence of a larger cell? Table III does not contain such lines. I would like to know in particular whether the line 531 ($d \approx 4.05$ Å, $a \approx 24.5$ Å) is clearly absent, a state of affairs which might point to a disordered distribution of Si and Al caused by phosphorus substitution.

2) The unit cell dimension of P–L (a-direction) is compatible with the high alumina content observed. However, there is a drastic drop in adsorption capacity which I believe is owing to occlusion of phosphate. Do you think that the larger number of cations explains the lower sorption capacity?

E. M. Flanigen: 1) We did not observe any extra lines in P–A zeolite compared with NaA zeolite. The major difference in x-ray is the decrease in unit cell dimension with increasing phosphorus content in P–A. The [531] superstructure reflection is observed as a weak shoulder in P–A zeolites and in NaA zeolite ($d \approx 4.13$ Å, not listed in Table III).

2) One of the main problems here is the simultaneous substitution of Al with P. It is difficult to separate the changes in properties expected with increased Al content from those owing to phosphorus substitution. The establishment of proof of framework substitution is very difficult and cannot be based on one single property such as adsorption. We have tried to consider carefully all of the interrelated complex effects and properties and characterize the phosphate zeolites by as many techniques as possible before proposing phosphorus framework substitution. We are convinced that the only explanation for all of the observed properties and characteristics is framework incorporation of phosphorus. The reduced adsorption capacity and pore size of the P–L zeolite may be related to cation content since we report that cation exchange alters the adsorption properties.

With respect to the structure of P–L, J. M. Bennett has done some initial electron diffraction work from which the unit cell data given in the paper were taken.

J. M. Bennett (Union Carbide Corp., Tarrytown, N. Y. 10591): There are differences between the space group of L and P–L. In P–L, the c dimension is doubled by electron diffraction, and systematic absences show the presence of two c glide planes.

G. H. Kühl (Mobil Research and Development Corp., Paulsboro, N. J. 08066): You postulate OH$^-$ along with Na$^+$. In what way is this different from occluded NaOH?

E. M. Flanigen: If by occluded NaOH, you mean NaOH molecules which are extraneous to the zeolite structure, we propose that the Na ion is not directly associated with the hydroxyl ion, but that each serves as a framework charge-compensating moiety, the Na$^+$ charge compensating for AlO$_2^-$ tetrahedra, and the OH$^-$ for PO$_2^+$ tetrahedra.

R. M. Barrer (Imperial College, London SW7, England): 1) Barrer and Marshall (*J. Chem. Soc.* **1965**, 6616), working under apparently similar conditions of temperature, obtained aluminophosphates and aluminosilicates from the system Base–Al_2O_3–P_2O_5–SiO_2–H_2O, often as co-precipitates of crystals. Can you clarify the differences in the crystallizations and/or the experimental conditions which could yield these different results?

2) On the question how much of the phosphate was in the framework and how much was merely intercalated, it is to be observed that in the tight analcite framework, there is no room for intercalation. However, analcite is already known to have a modification with (presumably) framework phosphorus, which is viseite. Therefore, have you made any other tight network zeolite (possibly natrolite) which contains phosphorus? This could shed considerable light on the site of the phosphorus.

E. M. Flanigen: 1) We can only attribute differences in the results of Barrer and Marshall and those reported here to variation in the method of preparing the gel network, and in some cases perhaps to different crystallization temperatures. We believe the critical point in achieving framework substitution is incorporation of all of the phosphate in the insoluble gel network, rather than in a soluble phosphate form. This is controlled by the exact method of coprecipitation and gelation.

2) No, we have not.

D. E. W. Vaughan (W. R. Grace and Co., Clarksville, Md. 21029): Although the data convincingly demonstrate P substitution in the zeolite frameworks, substantial occlusion of phosphate is also evident on the basis of the stoichiometry presented in the tables (*i.e.*, Al_2O_3–$P_2O_5 \neq Na_2O$). The sorption data for A would seem to indicate such occlusion reflected in reduced sorption capacity, compared with zeolite W where the expected enhanced capacity is observed, reflecting lower cation content of the zeolite.

E. M. Flanigen: Since different mechanisms have been proposed for phosphorus substitution, the stoichiometry of substitution cannot be generalized but would have to be discussed for each phosphate zeolite species. However, it is not correct to assume that framework substitution imposes a stoichiometric requirement of Al_2O_3–$P_2O_5 = Na_2O$. In addition, the proposed phosphorus substitution is based not only on stoichiometry but on a combination of at least several other characteristics. In the case of phosphate zeolites containing 15 to 23 wt % P_2O_5, considering your suggestion that a substantial portion of that is occluded, the changes expected in the properties in many cases would be greater than any changes observed or reported.

In the case of P–A zeolite, although some reduction in adsorption capacity is observed in the case of some cation forms of P–A and with some adsorbates, in the case of Ca-exchanged P–A, no reduction was observed. It does not seem likely that Ca-exchange should remove occluded material, and indeed, chemical analysis showed no change in framework composition before and after exchange.

L. Moscou (Ketjen N.V., Amsterdam, Netherlands): I wonder whether the accuracy of the microprobe analysis you mentioned is good enough to be able to study homogeneity in composition over separate crystals.

E. M. Flanigen: Since the electron microprobe analyses were carried out by J. V. Smith and C. R. Knowles, I wonder if Professor Smith would comment.

J. V. Smith (University of Chicago, Chicago, Ill. 60637): The reproducibility of the instrument is 1 or 2%, but because it is a zeolite, you have to use something like a 10- to 20-μ beam, with crystals of the order of 50 to 100 μ; therefore, you cannot take many points across a crystal. Although the analytical reproducibility cited applies to any crystal and between crystals, you cannot prove to a 1-μ level that there is not a variation in phosphorus content.

7

Process of Zeolite Formation in the System $Na_2O-Al_2O_3-SiO_2-H_2O$

FRIEDRICH E. SCHWOCHOW and GERHARD W. HEINZE

Farbenfabriken Bayer AG, Leverkusen, West Germany

This paper is aimed at clarification of the change of concentration with time in the liquid phase before crystallization starts. To find optimum conditions for the commercial production of pure zeolites of the types A and faujasite, the reaction of fine-particle amorphous silica with sodium aluminate solution was studied at 20°, 40°, and 75°C. The liquid phase separated by filtration nucleates the zeolite types A, sodalite, phillipsite, and faujasite, depending on stirring time before liquid–solid separation. Quite similar conditions are observed in precipitated sodium aluminosilicate gels and mother liquor.

Interest in understanding the reaction mechanism of the formation of synthetic zeolites grows with their technical importance. The experience gathered for the hydrothermal synthesis of defined zeolites remains restricted to a specific case, until an understanding of the mechanism can be presented. The type of the crystallizing zeolite is predetermined not only by the concentrations of the reaction partners, but also by additional factors—SiO_2 source, precipitation step, digestion of the gels. According to Zhdanov (5), it first is necessary, in order to understand the synthesis, to replace the customary over-all ratios of the reactants by separate concentration data for the 2 phases in the heterogeneous system. Experiments in the system precipitated sodium aluminosilicate gels–mother liquor make it probable that the lattice type of the crystallizing zeolite is predetermined mainly by the composition of the liquid phase. The importance of the liquid phase is emphasized also by Kerr (2) using zeolite A crystallization as an example. Our investigations refer to the time-dependent concentration change in the 2 phases before crystallization starts and to zeolite formation in the mother liquors separated from the respective solid phases.

Experimental

For our experiments, we selected the 2 following SiO_2 sources: amorphous precipitated silica (specific surface area: 200 m^2/gram according to BET) and technical water glass solution, of density 1.33 grams/ml, containing 1.7 moles of Na_2O/l and 5.78 moles of SiO_2/l. Sodium aluminate, 1.36 grams/ml, was used as a solution containing 3.4 moles of Na_2O/l and 2.0 moles of Al_2O_3/l. The separation of solid–liquid phases was effected *via* filter paper (Schleicher & Schull No. 589).

Within the separated liquid phases, SiO_2 and Al_2O_3 were analyzed according to conventional chemical methods. Na_2O was determined by flame photometry. Crystalline reaction products were identified by Debye-Scherrer diagrams; the composition of the mixtures was determined by comparison with diagrams of test samples.

Table I. System Amorphous Silica–Sodium Aluminate

Conditions Prior to Separation of Liquid Phases		Composition[a] of Separated Liquid Phases, Mole Ratios, Na_2O: Al_2O: SiO_2			Zeolite Crystallizing from Separated Liquid Phases	
Temp., °C	Stirring Time, Hours				Main Product[b]	Impurities
20	24	3.67:	1	: 0.04	Zeolite A $a_o = 12.29_7$A	No
20	48	4.08:	1	: 0.06	Zeolite A $a_o = 12.29_1$A	No
20	72	6.01:	1	: 0.09	Zeolite A $a_o = 12.30_1$A	No
40	2	3.67:	1	: 0.06	Zeolite A $a_o = 12.30_1$A	2% Sodalite
40	4	8.92:	1	: 0.22	Zeolite A $a_o = 12.29_6$A	10% Sodalite
40	8	67.82:	1	:55.58	Faujasite $a_o = 25.02_1$A	10% Phillipsite 3% Zeolite A
40	16	50.00:	1	:74.17	—	—
75	2	57.59:	1	:79.13	Faujasite $a_o = 24.77_9$A	8% Phillipsite

[a] The composition of the liquid phase depends on temperature and stirring time, as well as the corresponding zeolite type crystallizing from the separated liquid phases.
[b] a_o = unit cell constant.

Concentration Change in the Liquid Phase

For our investigations on sodium–aluminium–silicate–water mixtures, we have selected concentration ratios in which crystallization of faujasite is basically possible. First, the reaction of amorphous precipitated silica with sodium aluminate solution was investigated, the mole ratios being $SiO_2/Al_2O_3 = 5$, $Na_2O/SiO_2 = 0.6$, and $H_2O/Na_2O = 40$ (based on the total batch). In this system, faujasite is obtained in high purity after a few hours at 100°C. If the reaction batch is stirred at 20°, 40°, and 75°C and subsequently the solid phase is separated from the liquid phase by filtration, one obtains—depending on temperature and stirring time before the separation of phases—either sodium aluminate solutions or sodium silicate solutions as the liquid phase (Table I, Figure 1). These concentration points are passed in the liquid phase during the continuous course of the synthesis—for example, during predigestion and heating. While the SiO_2/Al_2O_3 ratio in the solid phase decreases, the liquid phase becomes poor in Al_2O_3 content until there is practically nothing left but sodium hydroxide solution. The solution then changes over from the aluminate side to the silicate side, which means that increasing quantities of SiO_2 are being dissolved in the sodium hydroxide solution.

Zeolite Crystallization from the Separated Mother Liquors

From the liquid phases present in the amorphous SiO_2–sodium aluminate system, it is possible to obtain crystalline zeolites without the

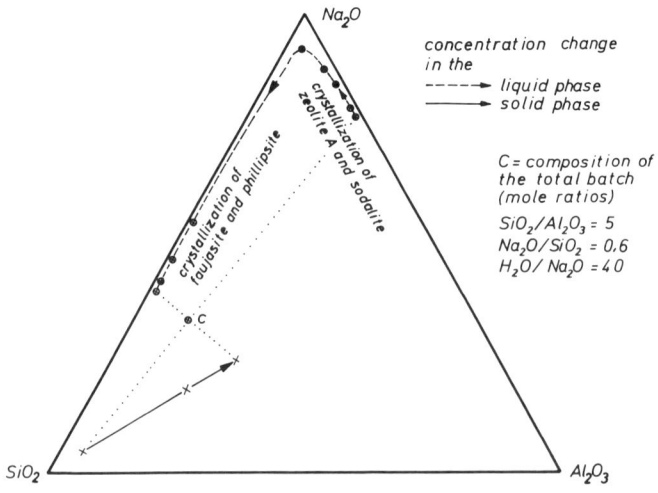

Figure 1. Concentration change in the liquid and solid phases during the course of zeolite synthesis

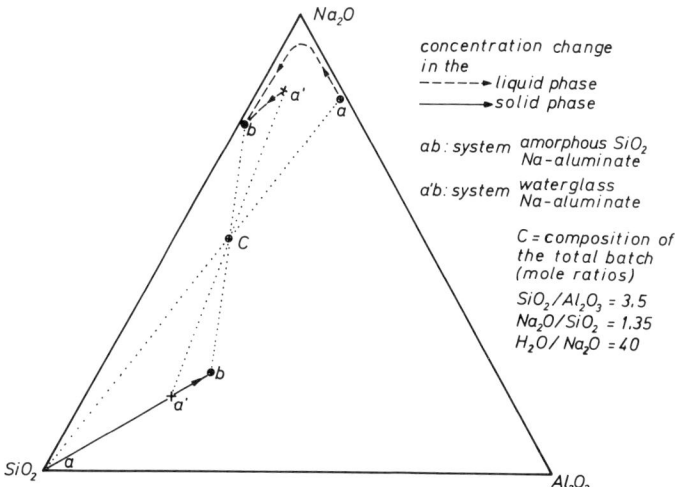

Figure 2. *Change of concentration with time and temperature in the 2 phases, using different SiO_2 sources*

simultaneous presence of the appropriate solid phases. If the respective liquid phases separated from the batch are heated to 85°C (24 hours), the different zeolite types A, sodalite, faujasite, and phillipsite will crystallize, depending on temperature and stirring time prior to the solid–liquid separation. These experiments show conclusively that zeolite A and sodalite, which are poorer in silica, are obtained from the liquid phase if the solutions remain on the aluminate side of the concentration diagram. It is only after the changeover of the liquid phase from the aluminate side to the silicate side that faujasite or phillipsite, both richer in silica, are obtained. The SiO_2 content of the faujasite rises with the SiO_2/Na_2O ratio in the liquid phase.

Concentrations in the Liquid Phase for Different SiO_2 Sources

Our findings pertaining to the change of composition in the liquid phase with time, as discussed in the first section, are confirmed in essence within reaction mixtures which cause the formation of faujasites poor in silica—smaller SiO_2/Al_2O_3 and larger Na_2O/SiO_2 mole ratios in the total batch.

Figure 2 shows the change of concentration with time and temperature in the liquid phase for a batch with the mole ratios $SiO_2/Al_2O_3 = 3.5$, $Na_2O/SiO_2 = 1.35$, and $H_2O/Na_2O = 40$. Prior to crystallization, the reaction mixtures containing amorphous silica or water glass approach the same final concentrations of SiO_2 in the liquid phase. Both, however, may differ with respect to the polymerization state of the dissolved silicate. The figures in Table II show that only a part of the available water glass is precipitated in gel precipitation. As such, the liquid phase contains at least a part of the polymeric silicates from the water glass unchanged. On the other hand, the formation of dissolved silicate from amorphous silica and sodium hydroxide results in essentially monomeric species.

Crystallization of Faujasite or Phillipsite from Mother Liquors of the Same Analytical Composition

Within the phase diagram (Figure 2), our findings show a region within which faujasite as well as phillipsite may crystallize. Which of the 2 actually predominates in the crystallization mixture is influenced by the choice of the SiO_2 source. For example, amorphous SiO_2 dissolved in sodium hydroxide solution (10% NaOH) and commercially available water glass solution are precipitated with sodium aluminate solution at 25°C under equal conditions (mole ratios $SiO_2/Al_2O_3 = 5$, $Na_2O/SiO_2 = 2$, and $H_2O/Na_2O = 40$ in the total batch). After heating to 100°C (within 30 minutes), the 2 liquid phases contain the same concentrations of Na_2O, Al_2O_3, SiO_2, and H_2O (Table III). After separation

Table II. Change of Silica Concentration with Time in the Liquid Phases Using Different SiO_2 Sources

Stirring Time Prior to Liquid–Solid Separation, Temp., 40°C	SiO_2 Concentration (g/l) in Separated Liquid Phases	
	System: Amorphous SiO_2–Sodium Aluminate	System: Water Glass–Sodium Aluminate
< 5 Minutes	0	8.14
15 Minutes	1.78	15.28
2 Hours	1.42	16.80
8 Hours	4.82	17.12
24 Hours	19.00	–

Table III. Differences in Crystallization Behavior of Liquid Phases of Equal Chemical Composition

SiO_2 Sources Used for Gel Precipitation	Composition of Liquid Phases at 100°C			Zeolite Type Crystallizing From Separated Liquid Phases	
	Na_2O, g/l	Al_2O_3, g/l	SiO_2, g/l	Main Product	Impurities
Silica dissolved in sodium hydroxide solution, 10% NaOH	79.20	1.86	20.04	Faujasite	15% Phillipsite
Technical water glass solution	76.50	1.70	20.50	Phillipsite	25% Faujasite

of the respective solid phases, the 2 mother liquors are further heated to 100°C (48 hours). Faujasite crystallizes predominantly from one solution and phillipsite from the other. By applying the method of Thilo and Wieker (4), we found differences in the paper chromatograms which we attribute to different states of polymerization. Based on the R_f ratios, it seems likely that solutions containing higher polymeric anions result in phillipsite.

Discussion of the Results

The above findings and conclusions point rather obviously toward 2 prerequisites for a technical faujasite synthesis:

(1) At the time of nucleation, the composition of the liquid phase —not that of the total batch—must correspond to a very high SiO_2/Al_2O_3 ratio (>20).

(2) In addition, the dissolved silica must be present in a monomeric state predominantly, because otherwise phillipsite crystallizes as the main product.

The favorable influence of a predigestion step on the faujasite synthesis is known. If dissolvable silica is used, the effect of predigestion is clearly recognizable from the experiments reported in Table II, building up a sufficiently high SiO_2 concentration in the liquid phase. In the case of rapid precipitation of gels from water glass solutions, the effect of the predigestion step cannot be the one just described. Predigestion causes the silicate micels to equilibrate within the gel in such a manner that afterward monomeric silica is split off during dissolving. Hence, it is possible to avoid the time-consuming step either by using reactive silicates for gel precipitation, such as $Na_2SiO_3 \cdot 5\ H_2O$ (1), or by converting *a priori* unsuitable silicate solutions, such as technical water glass, *via* depolymerization to solutions with reactive silica (3), carrying out the gel precipitation thereafter.

Literature Cited

(1) Andrews, E. B., Kerr, J., Whittam, T. V., Ger. Patent **1,269,111** (1964).
(2) Kerr, G. T., *J. Phys. Chem.* **1966**, 70, 1047.
(3) Schwochow, F. E., Heinze, G. W., unpublished work.
(4) Wieker, W., *Z. Anorg. Allgem. Chem.* **1969**, 366, 139.
(5) Zhdanov, S. P., "Molecular Sieves," p. 62, Society of the Chemical Industry, London, 1968.

RECEIVED January 29, 1970.

Discussion

Hans Villiger (Martinswerk GmbH, Bergheim, Germany): You mention phillipsite as a frequently-occurring impurity. Is this really so, or do you use a different nomenclature? To my knowledge, Barrer-P1 or Linde B is observed during the synthesis of Linde X. Phillipsite tends to grow in potassium-bearing systems rather than the pure sodium system.

Friedrich Schwochow: Our designation "phillipsite" means a crystalline zeolite with the structure of the mineral phillipsite, which in our reaction system necessarily occurs in the sodium form.

8

Synthesis and Structural Features of Zeolite ZSM-3

G. T. KOKOTAILO and J. CIRIC

Mobil Research and Development Corp., Research Department, Paulsboro, N. J. 08066

> *A new large-pore zeolite has been crystallized from lithium and sodium aluminosilicate gels at 60° to 100°C. Its composition is $(0.05-0.8)Li_2O \cdot (0.2-0.95)Na_2O \cdot Al_2O_3 \cdot (2-6)SiO_2 \cdot (0-9)H_2O$. This material adsorbs 17–18% cyclohexane (at 20 mm Hg) and 29–30% H_2O (at 12 mm Hg). The zeolite grows as hexagonal platelets which exhibit characteristic twinning. Electron diffraction patterns show hexagonal symmetry. The x-ray powder diffraction patterns were indexed in the hexagonal system. The c parameter is determined from the growth steps and is a function of the stacking sequence. This zeolite is a member of a family of structures based on the hexagonal form of faujasite with varying stacking sequences and c parameters.*

Early work in the synthesis of zeolites was summarized in a review by Morey and Ingerson (30). Research into the hydrothermal synthesis of zeolites was put on a systematic basis by Barrer and resulted in a number of synthetic zeolites (1, 2, 3, 4, 5, 6, 8, 9, 10, 11). Linde A and X (16), ZK-5 (25), and Linde L (15) were other prominent zeolites synthesized.

Hypothetical framework structures possible of synthesis were postulated (12, 24, 27, 29, 37). These are of interest as they provide insight into the many possible arrangements of linked (Si,Al) tetrahedra and tetrahedral units. All these are of interest not only because of their structural features but because of their industrial applications as sorbents and molecular sieves.

Using a controlled crystallization in the $Li_2O-Na_2O-Al_2O_3-SiO_2-H_2O$ system, a new zeolite, ZSM-3, was synthesized. The basic building block, the truncated octahedron, is common to sodalite, Linde A, faujasite,

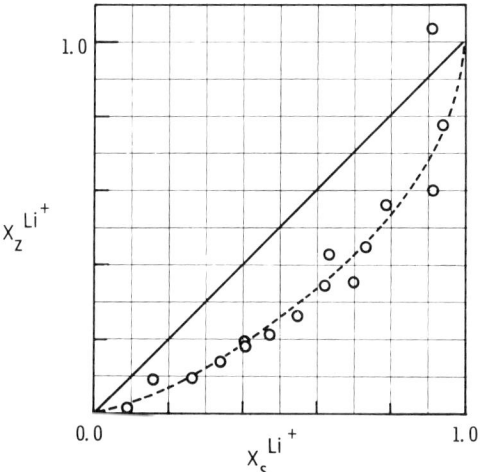

Figure 1. Ion-exchange isotherm for ZSM-3, Li^+-Na^+

Total normality = 0.1
Temperature = 25°C
Sample No. CSZ-330, SiO_2/Al_2O_3 = 3.25
Contact time = 48 hours
$X_z^{Li^+}$ = equivalents of lithium in zeolite/gram atom Al
$X_s^{Li^+}$ = equivalents of lithium in solution/total equivalents in solution

and ZSM-3. The mode of linking to form the framework structure of ZSM-3 consistent with its properties and x-ray data is proposed.

This paper describes the synthesis, physical properties, crystal habit, and structural features of ZSM-3.

Synthesis

The zeolite ZSM-3 was prepared from aluminosilicate hydrogels containing sodium and lithium cations. The crystallization technique consists of first preparing a "precursor" solution of concentrated sodium aluminosilicate and then mixing it with aqueous sodium silicate and aluminum chloride solutions to form the starting hydrogel slurry. This slurry is filtered to remove excess soluble sodium silicate. Lithium is added to this filter cake as lithium hydroxide solution. This mixture is held at temperatures of 60° to 100°C until ZSM-3 crystals form. At 60°C, crystallization requires 5 days while at 100°C, crystals are formed in 16 hours. In order to obtain the desired SiO_2/Al_2O_3 ratio in the crystalline product, the aluminum chloride content is varied.

Direct mixing without precursor direction results in the crystallization of Barrer's P zeolite and sodalite, resulting in a lower yield of ZSM-3.

The chemical composition of ZSM-3 in terms of the constituent oxides is $(0.05-0.8)Li_2O \cdot (0.2-0.95)Na_2O \cdot Al_2O_3 \cdot (2-6)SiO_2 \cdot (0-9)H_2O$ (18). The zeolite is synthesized from systems containing both sodium and lithium, the major cation component being sodium.

Physical Properties

Ion exchange isotherms for the ion pairs Li^+-Na^+ and $Ca^{2+}-Na^+$ are given in Figures 1 and 2. The experimental technique was essentially the same as that described by Sherry (35). A known weight of zeolite was tumbled in polyethylene bottles containing 0.1 normal solutions. The contact time was 48 hours; water bath temperature, 25°C. The slurries were filtered and solutions analyzed for sodium, lithium, and calcium by flame photometry. The points on the graph were calculated from the material balance of salts in the solution before and after exchange. These ion exchange isotherms are like those for zeolites X and Y.

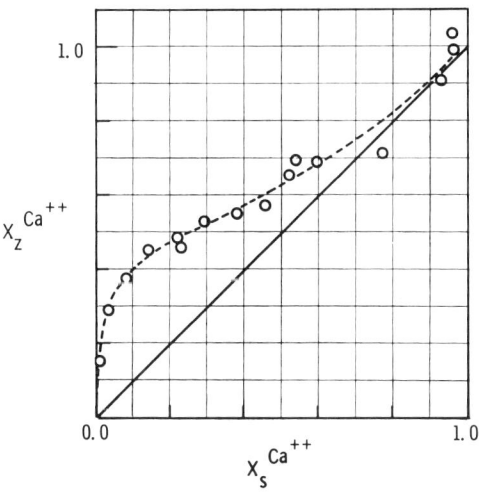

Figure 2. Ion-exchange isotherm for ZSM-3, $Ca^{2+}-Na^+$
Total normality = 0.1
Temperature = 25°C
Sample No. CSZ-330, $SiO_2/Al_2O_3 = 3.25$
Contact time = 48 hours
$X_z^{Ca^{2+}}$ = equivalents of calcium in zeolite/gram atoms Al
$X_s^{Ca^{2+}}$ = equivalents of calcium in solution/total equivalents in solution

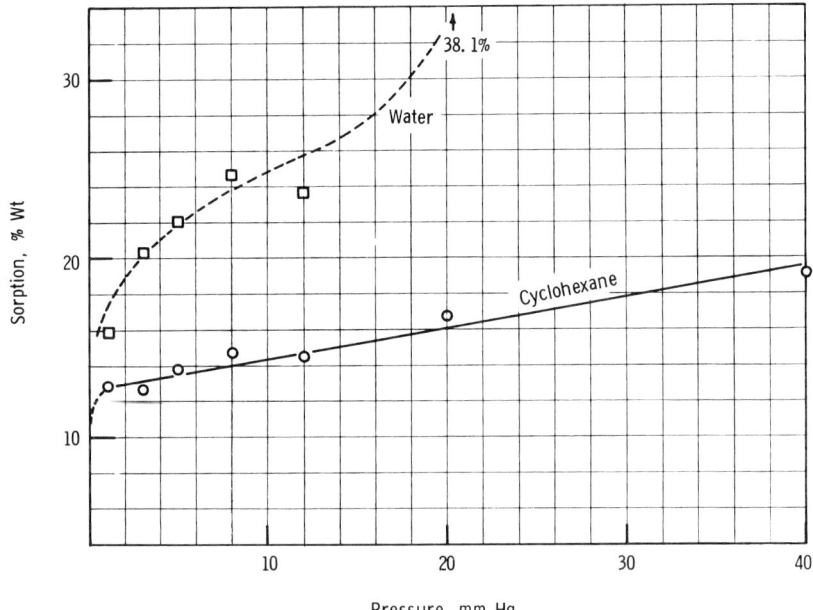

Figure 3. Sorption of water and cyclohexane by ZSM-3

Batch CSZ-330, $SiO_2/Al_2O_3 = 3.25$
$Na_2O/Al_2O_3 = 0.66$
$Li_2O/Al_2O_3 = 0.26$

Table I. Comparison of Sorptive Properties of ZSM-3

ZSM-3	Composition	% Cyclohexane	% Water
CSZ-353A	$0.13Li_2O$–$0.85Na_2O$–Al_2O_3–$3.05SiO_2$	17.6	30.2
CSZ-353B	$0.14Li_2O$–$0.86Na_2O$–Al_2O_3–$2.99SiO_2$	18.0	29.5
CSZ-315C	$0.11Li_2O$–$0.88Na_2O$–Al_2O_3–$3.6SiO_2$	18.9	29.1

Sorption isotherms for water and cyclohexane are shown in Figure 3 and for nitrogen in Figure 4. They have a characteristic shape typical of zeolites. Cyclohexane and water sorption measurements (at 20 and 12 mm Hg and 25°C) are shown in Table I.

An electron micrograph of a carbon replica of ZSM-3 (CSZ-168), Figure 5, shows a hexagonal platelet morphology and evidence of growth steps. There is also evidence of vertical growths parallel to the 100 direction in the ZSM-3 crystals. This crystal shape is a distinguishing property of this zeolite. The hexagonal platelets are of the order of 1 μ in diameter and about 0.1 μ thick. A comparison of the Linde X morphology is shown in Figure 6; the octahedral morphology is quite evident.

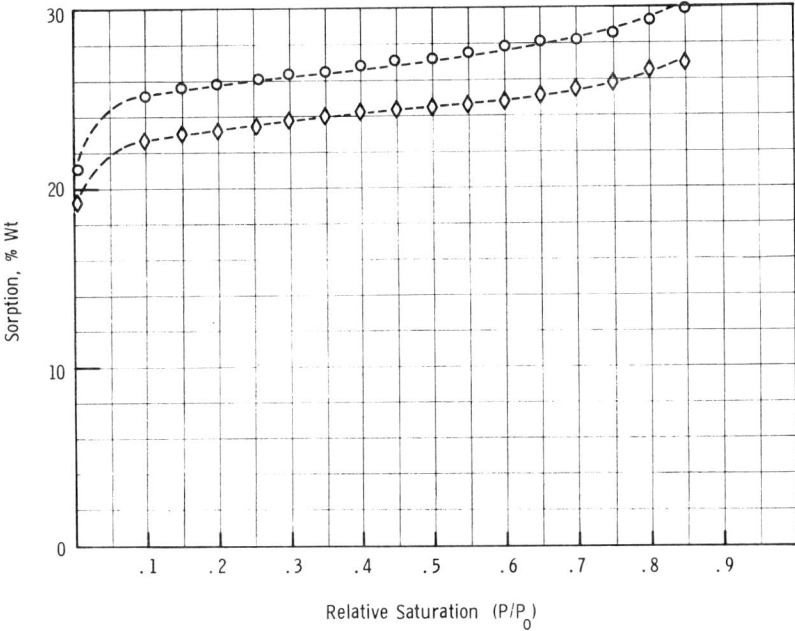

Figure 4. Nitrogen sorption isotherms for ZSM-3

$Temperature = 25°C$
○ Sample CSZ-256C, $SiO_2/Al_2O_3 = 3.4$
◇ Sample CSZ-263C, $SiO_2/Al_2O_3 = 3.5$

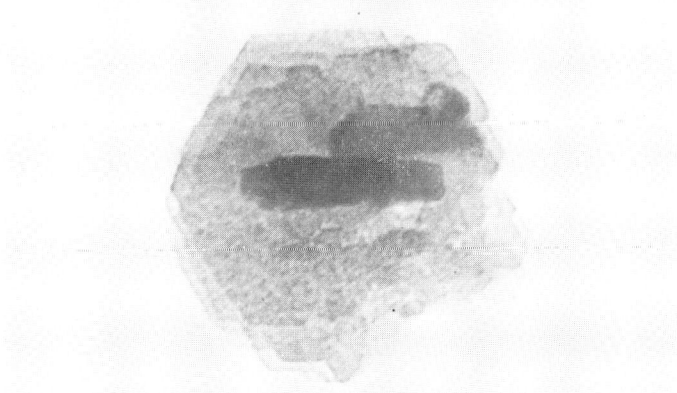

Figure 5. Electron micrograph of a single particle of ZSM-3 (CSZ 168), magnification 41,760 ×

Certain crystals grow in a normal manner for part of their growth, and then at a certain plane the orientation of the lattice changes discontinuously so that on opposite sides of this plane the crystal axes are

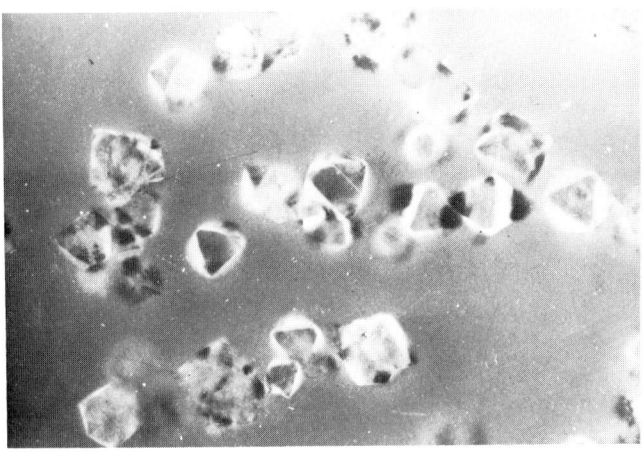

Figure 6. Optical micrograph of large crystals of zeolite X, magnification 230 ×

Figure 7. Electron micrograph of carbon replica of zeolite X crystals, magnification 3480 ×

not parallel. This extensive twinning of ZSM-3 crystals poses a problem as to the ability to grow large single crystals. The twinning occurs across a plane such that the twin growth direction is perpendicular to the original growth direction. This type of twinning across the 111 plane in the cubic form has been observed. An electron micrograph of a carbon replica of a twinned Linde X crystal, such that the twin growth is perpen-

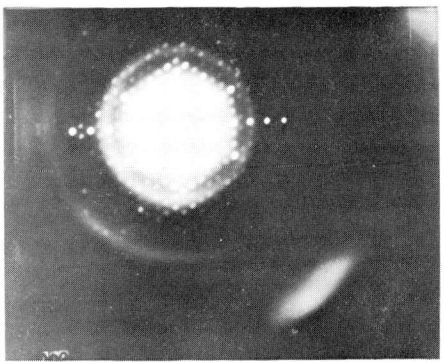

Figure 8. Electron diffraction pattern of a single crystal of ZSM-3 (CSZ 168)

Table II. X-Ray Data for ZSM-3 (CSZ-168)

d, Å	I	d	I
15.26	62	4.49	5
14.16	124	4.38	51
13.19	26	4.19	14
11.86	4	4.16	6
9.21	2	4.11	13
8.76	44	3.99	1
8.00	3	3.96	2
7.61	9	3.93	7
7.41	32	3.85	14
7.22	7	3.78	19
7.03	15	3.72	23
6.89	2	3.52	1
5.94	4	3.47	10
5.72	37	3.40	10
5.62	25	3.31	42
5.48	6	3.22	16
5.15	4	3.18	17
5.02	7	3.02	60
4.90	3	2.98	11
4.82	2	2.92	31
4.78	2	2.87	10
4.75	18	2.85	8
4.70	1	2.80	9
4.62	3	2.73	11
4.58	1	2.70	11

dicular to the direction of growth of the original crystal, is shown in Figure 7.

An electron diffraction pattern of one of the ZSM-3 platelets, obtained with the electron beam parallel to the c-axis, shows hexagonal symmetry (Figure 8). The a parameter, as determined from this pattern, is 17.5 A. X-ray data for ZSM-3 (CSZ-168) are given in Table II and may be indexed in the hexagonal system with $a = 17.5$ A. The pattern cannot be indexed with $c = 28.6$. A multiple of $c/2$ is required. These lattice parameters will vary somewhat with different Li/Na and Si/Al ratios.

An electron micrograph of a replica of ZSM-3 platelets was obtained, focusing the beam at the top of the twin growth, in order to study this edge in more detail (Figure 9). A uniform step growth of 125 A is evident. It has been shown (26, 39) that the growth steps of hexagonal crystals are equal to the height of the unit cell along the c parameter, as determined by electron microscopy and multiple beam interference. Then, from x-ray diffraction, electron microscopy, and electron diffraction, it was determined that the crystal class of ZSM-3 is hexagonal with lattice parameters $a = 17.5$ and a maximum value of $c = 129$ A. The distance between adjacent layers is 14.3 A.

Figure 9. Electron micrograph of a carbon replica of ZSM-3 (CSZ 168), magnification 83,520 ×

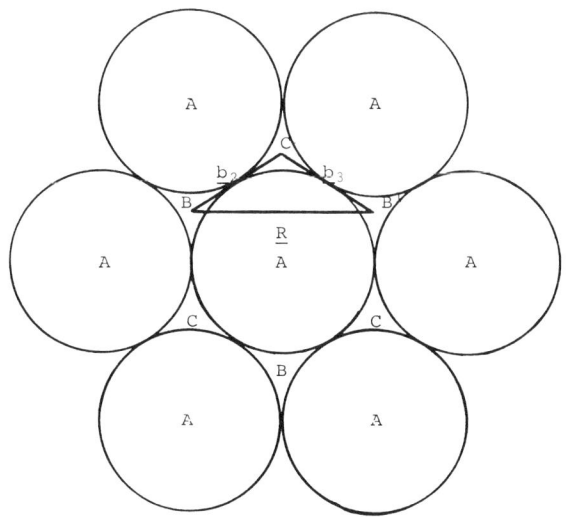

Figure 10. Displacement of atoms owing to a stacking fault

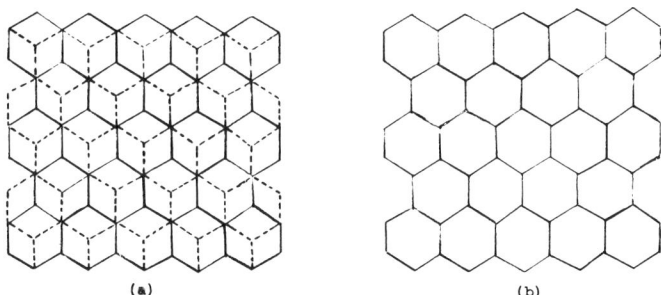

Figure 11. 001 Projections. Each line represents a sodalite cage

(a) AB layer, with dotted layer rotated through 60°
(b) Hexagonal array of linked sodalite cages

From a topological viewpoint, the truncated octahedra of faujasite may be represented by the positions occupied by spheres in the cubic close packing arrangement. An infinite number of structures can be produced from hexagonal layers of close packed spheres conventionally represented by A, B, and C (*34, 36, 40*). The arrangement of spheres in the layer stacked over the A layer (Figure 10) may occupy either position B or C. The c parameter and the number of stacking sequences is dependent on the number of layers in the identity period. All these arrangements of layers may be described in terms of the ideal cubic and primitive hexagonal structures. If a layer has layers which are the same above and below ($ABAB$), we designate it as hexagonal (h); if the layers above and below are different (ABC), it is cubic (c). Now any sequence of layers may be designated as series of h and c. The layers in the AB hexagonal structure are stacked along 001 while in the ABC cubic structure they are stacked along 111 of the unit cell.

The aluminosilicate framework of faujasite (*14, 17, 31*) consists of truncated octahedra linked through double six-membered rings to form a diamond structure. The structure may be considered a cc stacking of layers of truncated octahedra along 111 of the unit cell. The 111 projection of the cubic framework structure of faujasite and the hexagonal array of linked truncated octahedra (sodalite cages) in a layer are shown in Figure 11. The stacking sequence may be altered to give an hh stacking sequence with parameter

$$a_H = \frac{a_c}{\sqrt{2}} \text{ and } c_H = \frac{2a_c}{\sqrt{3}}$$

By altering the h and c stacking sequence, a large number of structures are possible with the c_H parameter determined by the number of

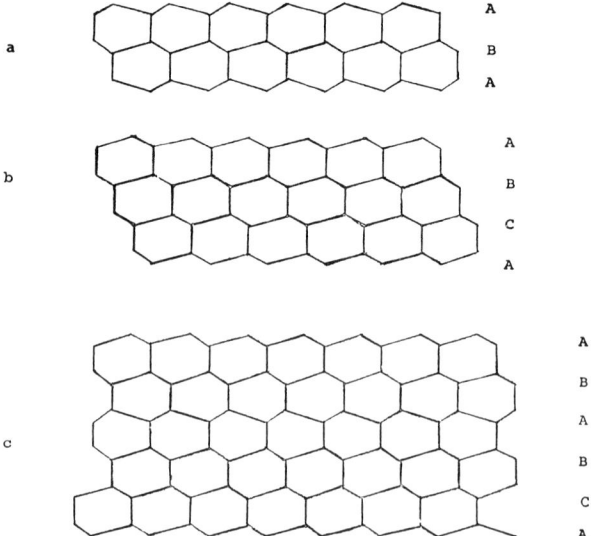

Figure 12. Stacking of layers in structures related to faujasite. Sodalite cages are represented by the line junctions. View perpendicular to c-axis, 110 projection

(a) Hexagonal AB
(b) Faujasite ABC
(c) ABABC

layers in the identity period. The 110 projection of a series of stacked layer sequences is shown in Figure 12. The faujasite framework forms a channel and pore system consisting of 4 twelve-membered ring openings to each large cage.

The framework of the hexagonal structure (hh) has a channel and pore system similar to that of faujasite, with 5 twelve-membered ring openings to the large cavities, one channel direction being parallel to c^*. The linking of truncated octahedra through double six-membered rings remains the same as for the cc stacking in the cubic faujasite structure.

The framework of chabazite (7, 19, 32) is related to that of gmelinite by rotating every second layer of double six-membered rings and altering the AB stacking sequence of gmelinite to the ABC of chabazite (7). The a parameter remains the same, with the c parameter increasing by a factor of 1.5. A series of hypothetical structures were predicted by further altering the stacking sequence (13). A similar relationship exists between the frameworks of erionite (7, 19, 23, 32, 38) and offretite (13).

A stacking fault divides the crystal, and the parts on each side of the stacking fault are displaced relative to each other by an amount equal to the partial dislocation. This displacement causes a phase difference between electrons scattered on either side of the fault and results in a corresponding change in the intensity of the diffracted beam. Thus, stacking faults in crystals are characterized in electron micrographs by contrast lines running parallel to the intersections of the fault with the surfaces. Such contrast lines, indicative of random stacking faults, were observed in crystals of erionite (28).

The x-ray diffraction data, adsorption isotherms, and ion exchange properties led us to believe that there was a relationship between the structures of ZSM-3 and faujasite.

Crystals of cubic SiC are octahedral, with the 111 and $\overline{111}$ faces having the least energy (21, 22). These faces will preserve their stability if displacements owing to edge dislocation occur. Thus, at high saturations, growth stops on the 111 and $\overline{111}$ faces and continues on the others. This makes the crystals grow as platelets. Frank (20) also speculated that the cubic SiC is the thermodynamically stable structure. Similar growth habits and stability data are noticed in Linde X and ZSM-3 crystals.

If the framework structure proposed for ZSM-3 is a nine-layer stacking sequence of truncated octahedra, there are 256 possible structures. Many of these are identical, and in some cases limitations are imposed by space group symmetry but the number of possible framework structures is still large.

Patterson (33) showed that when stacking faults are introduced in a cubic (cc layer sequence) crystal, the reflections with $H-K = 3N$ and $L = 3N'$, where N and N' are any integer, remain sharp, but the other reflections broaden and their maxima shift in opposite directions.

Many of the line profiles in the x-ray diffraction pattern of ZSM-3 are asymmetric. There is also some line broadening, some of it caused by crystallite size, and a shifting of some lines. This makes the structure determination of these polytype materials extremely difficult, especially from powder data. The large number of possible stacking arrangements and the existence of random stacking faults make it difficult to arrive at the correct structure by trial and error methods.

Acknowledgment

We thank S. L. Lawton for the valuable discussions and J. B. Milliken, S. Sawruk, and E. E. Jenkins for their help in the experimental work.

Literature Cited

(1) Barrer, R. M., *J. Chem. Soc.* **1948**, 2158.
(2) Barrer, R. M., Baynham, J. W., *J. Chem. Soc.* **1956**, 2882.
(3) Barrer, R. M., Baynham, J. W., Bultitude, F. W., Meier, W. M., *J. Chem. Soc.* **1959**, 195.
(4) Barrer, R. M., Baynham, J. W., McCallum, N., *J. Chem. Soc.* **1953**, 4035.
(5) Barrer, R. M., Denny, P. J., *J. Chem. Soc.* **1961**, 971.
(6) *Ibid.*, **1961**, 983.
(7) Barrer, R. M., Kerr, I. S., *Trans. Faraday Soc.* **1959**, 55, 1915.
(8) Barrer, R. M., Marshall, D. J., *J. Chem. Soc.* **1964**, 485.
(9) Barrer, R. M., McCallum, N., *J. Chem. Soc.* **1953**, 4029.
(10) Barrer, R. M., White, E. J., *J. Chem. Soc.* **1951**, 1267.
(11) *Ibid.*, **1952**, 1561.
(12) Barrer, R. M., Villiger, H., *Z. Krist.* **1969**, 128, 352.
(13) Bennett, J. M., Gard, J. A., *Nature* **1967**, 214, 1005.
(14) Bergerhoff, G., Baur, W. H., Nowacki, W., *Neues Jahrb. Mineral. Monatsch.* **1958**, 9, 193.
(15) Breck, D. W., Acara, N. A., U. S. Patent **3,216,789** (1969).
(16) Breck, D. W., Ebersole, W. G., Milton, R. M., *J. Am. Chem. Soc.* **1956**, 78, 2338.
(17) Broussard, L., Shoemaker, D. P., *J. Am. Chem. Soc.* **1960**, 82, 1041.
(18) Ciric, J., U. S. Patent **3,415,736** (1968).
(19) Dent, L. S., Smith, J. V., *Nature* **1958**, 181, 1795.
(20) Frank, F. C., *Phil. Mag.* **1951**, 42, 1014.
(21) Jagodzinski, H., *Acta Cryst.* **1954**, 7, 300.
(22) Jagodzinski, H., *Neues Jahrb. Mineral. Monatsch.* **1954**, 10, 209.
(23) Kawahara, A., Curien, H., *Bull. Soc. Franc. Mineral. Crist.* **1969**, 92, 250.
(24) Kerr, G. T., *Science* **1963**, 140, 1412.
(25) Kerr, I. S., *Nature* **1963**, 197, 1194.
(26) Kohn, J. A., Eckart, D. W., Cook, C. F., *Mater. Res. Bull.* **1967**, 2, 55.
(27) Kokotailo, G. T., Lawton, S. L., *Nature* **1964**, 203, 621.
(28) Kokotailo, G. T., Sawruk, S., unpublished data.
(29) Meier, W. M., Kokotailo, G. T., *Z. Krist.* **1965**, 121, 211.

(30) Morey, G. W., Ingerson, E. J., *J. Econ. Geol.* **1937**, 32, 607.
(31) Nowacki, W., Bergerhoff, G., *Schwiez. Mineral. Petrog. Mitt.* **1956**, 36, 621.
(32) Nowacki, W., Koyami, H., Mladeck, K., *Experientia* **1958**, 14, 396.
(33) Patterson, M. S., *J. Appl. Phys.* **1952**, 23, 805.
(34) Pauling, L., "The Nature of the Chemical Bond," 3rd ed., Cornell University Press, Ithaca, N. Y., 1960.
(35) Sherry, H. S., *J. Phys. Chem.* **1966**, 70, 1158.
(36) Smith, J. V., *Mineral. Soc. Am. Spec. Paper 1,* **1963**.
(37) Smith, J. V., Rinaldi, F., *Mineral. Mag.* **1962**, 33, 202.
(38) Staples, L. W., Gard, J. A., *Mineral. Mag.* **1959**, 32, 261.
(39) Verma, A. R., *Phil. Mag.* **1951**, 42, 1005.
(40) Wells, A. F., "Structural Inorganic Chemistry," 3rd ed., Clarendon Press, Oxford, 1952.

RECEIVED March 23, 1970.

9

Crystallization of Zeolitic Aluminosilicates in the System $Li_2O–Na_2O–Al_2O_3–SiO_2–H_2O$ at 100° C

H. BORER and W. M. MEIER

Institut für Kristallographie und Petrographie, Eidgenössische Technische Hochschule, Sonneggstrasse 5, Zürich, Switzerland

> *Crystallization sequences have been determined for over 400 compositional points in the Li,Na-aluminosilicate system at 100°C and reaction times of up to 14 days. A total of 9 zeolitic species has been observed thereby. Crystallization times and the nature of the first appearing solid phase depend largely on the lithium and sodium concentrations. The same crystallization sequences occur within certain fields, and these sequences are nowhere reversed. A simple scheme illustrating the relative stabilities of the solid phases in the system is presented.*

The most open zeolites crystallize at relatively low temperatures from highly reactive alkali aluminosilicate gels. These zeolitic phases are metastable, and usually several phases are formed in succession from a particular reaction mixture. This paper describes crystallization sequences observed in mixed Li,Na-aluminosilicate gels at 100°C. The details of our investigation of this system at 100°C are reported elsewhere (8, 9).

Crystallization fields of zeolites growing from homocationic Na gels at around 100°C have been studied extensively at a number of laboratories (*cf.* 3, 11, 21). Sand and coworkers recorded the coexisting phases for different reaction times (19). The kinetics of zeolite crystallization in the Na system also has received attention (12, 14, 15). Studies of low-temperature crystallizations from Li gels, on the other hand, have been much more limited (13).

Experimental

Well-defined starting materials were chosen which were combined according to a standard procedure in order to ascertain maximum reproducibility. Highly reactive gels were prepared thus from reagent grade lithium and sodium hydroxide, tetramethoxysilane, and analyzed sodium aluminate containing excess caustic. Measured amounts of 2M NaOH, 2M freshly prepared Na-aluminate solution, deionized water, $(CH_3O)_4Si$, and 2M LiOH were combined in this order at room temperature and vigorously stirred to produce homogeneous gels. A number of test experiments showed there was no need to remove the small amounts of methanol which formed in the reaction mixtures.

The dry weight of the constituents was about 65 mg/ml in all experiments. The molar ratios of the components were based on

$$n(Li_2O) + n(Na_2O) + n(Al_2O_3) + n(SiO_2) = 1$$

The concentrations of all components except water were varied systematically in steps of $\Delta n = 0.05$ within the following limits:

$$n(SiO_2)/n(Al_2O_3) \geq 0.5$$

$$n(Na_2O)/n(Al_2O_3) \geq 1.25 \text{ and } n(Al_2O_3) \geq 0.05.$$

A total of 448 compositional points of the system had to be examined as a consequence.

All crystallizations were carried out in sealed polypropylene tubes of 30–40-ml capacity. Glass was avoided purposely in all our experiments since it is attacked readily by the reactants used and the crystals nucleate mostly on the glass walls. In particular, we have noted that reproducibility of Li,Na-zeolite crystallizations is exceedingly poor when carried out in borosilicate glass even at relatively low temperatures. The polypropylene tubes containing the reactants were rotated slowly (about 8 revolutions per minute) at 100°C in thermostatically controlled ovens.

Samples of the reaction mixtures were taken after 3 hrs, 18 hrs, 112 hrs, and 14 days. The solid products were filtered off and washed thoroughly with distilled water. The dried solids were all examined under the microscope and identified by means of x-ray powder patterns using a Guinier camera. Selected samples were characterized further by means of chemical analyses, density and sorption measurements, electron microscopy, and thermal analyses (DTA and TGA).

Description of the Solid Phases

A total of 9 zeolites and 3 nonzeolitic solids could be observed. These species, which are summarized in Table I, can be divided into the following groups: species appearing in the Na system (A,P,S,T,X,Z), in the Li system (H,I,N), and those requiring both Li and Na (C,E,O). Species X and Z were obtained only in relatively few experiments in the high Na_2O/Li_2O region.

P and T pick up only minor amounts of lithium. P is always spherulitic, whereas T normally forms very small prisms showing well-developed

Table I. Crystalline Species and Notation

Symbol	Type	Other Designations and References
A[a]	Linde A	A(*10,16,19*), Q(*3*)
X[a]	Faujasite	X(*17*), F(*19*), R(*3*)
P[a]	Na-P1	cubic P(*3*), P1(*4*), P_c(*19*), B(*18*)
T[a]	Na-P2	tetrag. P(*3*), P2(*4*), P_t(*19*)
S[a]	Sodalite hydrate	S(*19*), I(*3*), Zh(*20*)
C[a]	Cancrinite hydrate	C(*7,19*)
Z[a]	Chabazite-like	S(*3*), E(*20*)
H[a]	Li-A (Barrer)	A(*6*)
I	Li-metasilicate	(*13*)
N	Li-aluminate	(*13*)
E[a]	K-F (Barrer)	F(*2*)
O	new	

[a] Zeolite.

faces. DTA-curves of P and T (of similar composition) are also remarkably different. Structurally, P is an isotype of gismondite and not truly cubic (*1*).

The fibrous crystals of C take up lithium preferably. Pure samples of C can be obtained readily in good yield from Li,Na gels. E, an important phase in the Li,Na system, appears as needle-like crystals or aggregates with a constant Si/Al ratio of 1. The potassium-exchanged form is identical with K-F, which was thought to be a typical product in the potassium field (*5*). Unexchanged crystals of E contain appreciable amounts of both Li and Na.

High Li concentrations are required for the formation of H and O. Species O, which crystallizes very slowly as tiny needles, has not been reported before. A representative sample sorbed 4.8% H_2O reversibly and contained a large excess of alkali (mostly Li) which could not be removed. For this reason, O does not appear to be a typical zeolite.

Crystallization Sequences

Typical examples of crystallization sequences are given in Table II. High alkalinity brings about faster crystallization rates and the fast-forming A and/or E are frequently the first appearing phases. The nature of the phase forming initially depends mostly on the lithium and sodium concentrations, as indicated in Table II. As a rule, E appears first above $n(Li_2O) = 0.35$.

The observed crystallization sequences are shown in Figure 1. They are all based on recorded changes in the powder patterns. The frequency of a particular sequence is indicated by the width of the respective arrow. Species which may appear first are enclosed in circles. In some areas,

Table II. Some Typical Crystallizations

Composition of Reaction Mixture				Products of Crystallization[a]			
$n(Li_2O)$	$n(Na_2O)$	$n(Al_2O_3)$	$n(SiO_2)$	3 hrs	18 hrs	112 hrs	14 days
0.05	0.80	0.05	0.10	A	A	st	T
0.10	0.70	0.10	0.10	A	A	A	as
0.30	0.50	0.10	0.10	ae	ce	C	C
0.35	0.45	0.10	0.10	ae	ce	C	ch
0.45	0.35	0.10	0.10	E	E	H	H
0.40	0.40	0.05	0.15	E	E	ho	ho
0.20	0.55	0.10	0.15	ae	ae	E	E
0.60	0.15	0.05	0.20	—	—	O	O
0.35	0.35	0.10	0.20	ce	ce	ch	H
0.25	0.45	0.10	0.20	A	aec	ce	C
0.05	0.60	0.10	0.25	—	at	T	T
0.05	0.50	0.25	0.20	A	A	A	A
0.05	0.45	0.10	0.40	—	P	P	pi
0.10	0.35	0.20	0.35	—	A	ac	ac
0.05	0.35	0.10	0.50	—	—	pt	T
0.15	0.25	0.15	0.45	—	P	P	hp

[a] Small letters denote components of mixture.

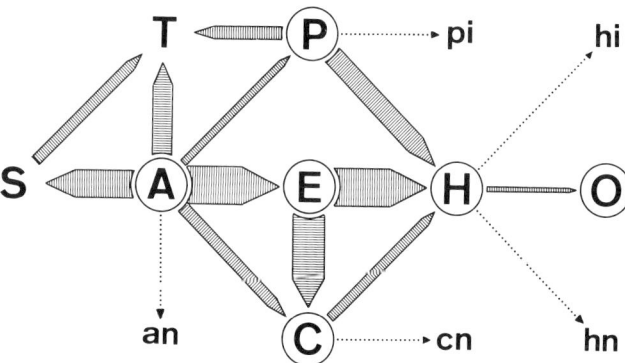

Figure 1. Crystallization sequences

phases I and N start forming (notably in long runs) without diminishing the main product. These less significant changes are represented by dotted lines in Figure 1.

Well-established sequences involving over 2 zeolitic species are A–E–C, A–S–T, and E–C–H. Each of the recorded sequences can be observed within a certain composition field. These fields are better defined than the "crystallization fields" which depend on factors controlling the relative rates of growth (15).

All reactions proceed in one direction; *i.e.*, the sequences are never reversed and reactions such as A–C and A–T do not occur at the same time. Figure 1, therefore, represents a scheme of the relative stabilities of the species in the system.

Acknowledgment

This study was supported by the Schweizerische Nationalfonds and a grant from the Mobil Oil Corp.

Literature Cited

(1) Baerlocher, Ch., Meier, W. M., to be published.
(2) Barrer, R. M., Baynham, J., *J. Chem. Soc.* **1956**, 2882.
(3) Barrer, R. M., Baynham, J., Bultitude, F. W., Meier, W. M., *J. Chem. Soc.* **1959**, 195.
(4) Barrer, R. M., Bultitude, F. W., Kerr, I. S., *J. Chem. Soc.* **1959**, 1521.
(5) Barrer, R. M., Cole, J., Sticher, H., *J. Chem. Soc. (A)* **1968**, 2475.
(6) Barrer, R. M., White, E. A. D., *J. Chem. Soc.* **1951**, 1267.
(7) *Ibid.*, **1952**, 1561
(8) Borer, H., Ph.D. Thesis, ETH Zurich, 1969.
(9) Borer, H., Meier, W. M., to be published.
(10) Breck, D. W., Eversole, W. G., Milton, R. M., Reed, T. B., Thomas, T. L., *J. Am. Chem. Soc.* **1956**, 78, 5963.
(11) Breck, D. W., Flanigen, E. M., *S.C.I. Monograph Mol. Sieves* **1968**, 47.
(12) Ciric, J., *J. Colloid Interface Sci.* **1968**, 28, 315.
(13) Gusseva, I. V., Liliev, I. S., *Zh. Neorgan. Khim.* **1965**, 10, 92.
(14) Kerr, G. T., *J. Phys. Chem.* **1966**, 70, 1047.
(15) *Ibid.*, **1968**, 72, 1385.
(16) Milton, R. M., U. S. Patent 2,882,243 (1959).
(17) *Ibid.*, 2,882,244 (1959).
(18) *Ibid.*, 3,008,803 (1961).
(19) Regis, A. J., Sand, L. B., Calmon, C., Gilwood, M. E., *J. Phys. Chem.* **1960**, 64, 1567.
(20) Zhdanov, S. P., *Izv. Akad. Nauk SSSR, Ser. Khim.* **1965**, 950.
(21) Zhdanov, S. P., *S.C.I. Monograph Mol. Sieves* **1968**, 62.

RECEIVED February 4, 1970.

10

Synthesis of Lithium and Lithium, Sodium Mordenites

MICHAEL L. SAND, WILLIAM S. COBLENZ, and L. B. SAND

Department of Chemical Engineering, Worcester Polytechnic Institute, Worcester, Mass. 01609

First syntheses are reported for lithium mordenite, lithium analcime, and lithium phillipsite. Lithium mordenite was synthesized in the temperature range 150°–200°C under autogenous pressure. Coexisting phases found with the mordenite were lithium analcime, lithium phillipsite, quartz, opaline silica, and lithium silicate. The best yields of Li-mordenite were obtained using aqueous colloidal silica sol, aluminum hydroxide, and lithium carbonate as the reactants. Li,Na-mordenite was synthesized in the system $(0.5Li_2O \cdot 0.5Na_2O) \cdot Al_2O_3 \cdot SiO_2 \cdot H_2O$ in the temperature range 150°–190°C with analcime, phillipsites, quartz, opaline silica, and alkali silicates found as coexisting phases. The Li,Na-mordenites contain the same Li/Na ratio as in the batch composition. Best yields were obtained using silicic acid, aluminum hydroxide, lithium hydroxide, and sodium hydroxide as reactants. Comparative data are given for the reactant conditions producing the Li-, Li,Na-, and Na-mordenites.

This paper reports the first syntheses of lithium mordenite, lithium analcime, and lithium phillipsite. A systematic study of the system lithia–alumina–silica–water has been in progress for several years and has proven so far to be a unusually challenging system in which to determine phase relationships of the zeolite phases. In this preliminary note, the authors will present only the reactant conditions to synthesize lithium mordenites and lithium, sodium mordenites, and compare these syntheses with that of sodium mordenites previously reported in the literature.

No lithium-containing mordenites have been found in natural occurrences; these are calcium, sodium mordenites with varying but small contents of potassium. The first probable synthesis of mordenite in 1927 by Leonard (10) was the autoclaved product of reacting sodium carbonate solution with spodumene. Although the product, identified tentatively as "mordenite(?)" was not analyzed, the results of our studies suggest that he synthesized a Na,Li-mordenite.

Since the original work of Barrer (3) in 1948 on the syntheses of sodium mordenites, a relatively large number of investigators (2, 5, 7, 8, 13), including later contributions by Barrer, have reported on the synthesis of mordenites from a variety of starting materials in other systems as well as those containing sodium. Conspicuously absent from the list of synthetic mordenites (considering ion exchanged varieties as derivatives of synthetic mordenites) has been synthetic lithium mordenite and synthetic potassium mordenite. Separate from these relatively high-temperature syntheses or mordenites was the discovery of conditions to synthesize large-port sodium mordenites (12). The authors use the designation "large-port" to distinguish between those mordenites which sorb large molecules into its 12-membered ring channels from those designated "small-port" or with no designations, which sorb only the smaller molecules such as methane (12).

The synthesis of 2 new zeolite-like phases reported by Barrer and White (6) in 1951 and the synthesis of a clinoptilolite-like phase by Ames (1) in 1963 represent the reported hydrothermal zeolite syntheses in the lithia system. In 1960, Hoss and Roy (9) reported the hydrothermal conversion of lithium-exchanged gmelinite to bikitaite at 250°C and 1000 atm. Barrer used lithium hydroxide, amorphous aluminum hydroxide, and silicic acid as reactants in which the batch compositions were 2 grams of gel ($Li_2O \cdot Al_2O_3 \cdot nSiO_2 \cdot nH_2O$, in which n varied between 1 and 10) dried at 120°C and 10cc of water or $\sim N$ lithia solution. In the temperature range 130°–450°C using these reactants, no mordenite, analcime, or phillipsite was reported as synthesized. In 1953, Barrer, Baynham, and McCallum (4) reported on Na-, K-, Rb-, Tl-, and Cs-analcimes synthesized from gels but that "Li-analcime" was made only indirectly by ion-exchange procedures with Ag-exchanged Na-analcime as an intermediate. Ames synthesized the clinoptilolite-like phase at 295°C in the lithia system using silica gel, aluminum hydroxide, and lithium hydroxide as reactants and batch compositions ranging from $0.6Li_2O \cdot Al_2O_3 \cdot 8SiO_2 \cdot 5H_2O$ to $Li_2O \cdot Al_2O_3 \cdot 10SiO_2 \cdot 8.5H_2O$. He synthesized mordenite at 295°C in the lithia–soda system using the same reactants with the addition of sodium hydroxide and a batch composition of $Li_2O \cdot Na_2O \cdot Al_2O_3 \cdot 10SiO_2 \cdot 8.5H_2O$.

Experimental

Reactants used were reagent grade lithium carbonate, lithium hydroxide, and sodium hydroxide, aluminum hydroxide (Grades C-31, C-730, Hydral 705, and 710 ALCOA), precipitated silicic acid (Fisher Scientific), and ammonium-stabilized aqueous colloidal silica sol (Ludox "AS," DuPont). Other reactants, which were used and found unsatisfactory to produce mordenite as a phase under the experimental conditions investigated, were fumed silica, silica gel, silica–alumina gels, diatomite, and sodium aluminate.

Modified Morey-type reactor vessels, 12-ml capacity, silver-lined and silver-sealed, were used in all runs. New, separate sets of vessels were used for reactions in the lithia and lithia–soda systems to avoid contamination. Reactants were mixed in a mortar and pestle and loaded into the autoclaves to 6 mm from the top. When the mixtures were of a "dry" consistency, they were tamped firmly into the vessel. The charged vessels were weighed before and after runs to detect small leaks. The vessels were placed in controlled-air ovens at temperature and upon completion of the run were quenched under cold tap water. The contents were extracted, dispersed gently in a mortar and pestle, and washed to near neutral on a Buchner funnel before oven-drying at 80°C. The phases were identified and data taken from x-ray diffractograms obtained on a G.E. XRD-5 unit and from films taken with a Norelco 114.6-mm diameter powder camera using aluminum powder as an internal standard and copper, nickel-filtered radiation. To obtain mean indices of refraction of the 3 synthetic mordenites, the samples first were water-saturated and equilibrated at room temperature in a controlled atmosphere ($P/P_s = 0.75$). Crystallization curves were obtained by placing a number of identically charged autoclaves in the pre-set oven, removing each at successive time intervals, and determining the per cent crystallization by the summation of peak intensities on the x-ray diffractograms using a reference mordenite standard. Crystallization was checked by microscopic examination as the zeolites crystallized as euhedral crystals in the 5–25μm range and could be differentiated from the unreacted material.

Table I. Typical Runs in the Lithia System

Batch Composition $Li_2O/Al_2O_3/SiO_2/H_2O$	T, °C	Time, Hrs.	Products
6/1/40/314	180	65	Phillipsite (major) Mordenite (moderate) Analcime (weak)
3/1/40/314	180	65	Mordenite (major) Phillipsite (moderate) Analcime (weak)
0.2/1/20/158	174	21	Analcime
1/1/4/34	152	67	Analcime
12.0/1/34.33/270	150	370	Phillipsite (major) Mordenite (weak)

Results

The first syntheses of lithium mordenites, phillipsites, and analcimes are the result of a systematic investigation started on this system several years previous to determine phase relationships. The first phase of the program was an evaluation of reactant materials. This parameter was much more critical than in other systems producing zeolites and probably is the reason these common zeolites had not been synthesized in earlier studies in the lithia system. The most successful combination of reactants to produce these zeolites was lithium carbonate, aluminum hydroxide (Hydral 710), and ammonium-stabilized aqueous colloidal silica sol (Ludox AS); 100% yields of Li-analcime were obtained, but to date Li-mordenite and Li-phillipsite have been obtained up to 75% yield, each with the other present as a co-existing phase. Typical runs are given in Table I. The Li-mordenite crystals occur with the same morphology as synthetic Na-mordenite crystals with 010, 100, and 001 pinacoids predominating and some crystals with prism termination, all in the size range 5–25μm. The mean index of refraction is 1.470. The Li-phillipsite crystals

Table II. Typical Runs in the Lithia–Soda System

Batch Composition $Li_2O/Na_2O/Al_2O_3/SiO_2/H_2O$	Reactants[a]	T, °C	Time, Hrs.	Products
1.25/1.25/1/10/25	A	185	84	Mordenite (major) Analcime (mod.)
0.5/0.5/1/5/50	A	187	48	Analcime (major) Mordenite (minor)
0.5/0.5/1/10/50	B	200	24	Mordenite
0.5/0.5/1/10/50	C	190	48	Mordenite (major) Quartz (mod.)
4.9/4.9/1/39.2/490	D	150	336	Mordenite (mod.) Phillipsite (mod.)
0.5/0.5/1/10/100	E	190	16	Analcime

[a] A = Grade C-730 aluminum hydroxide.
B = Grade Hydral 710 aluminum hydroxide.
C = Grade C-31 aluminum hydroxide.
D = Grade Hydral 705 aluminum hydroxide.
E = Grade Hydral 705 aluminum hydroxide and Ludox AS.

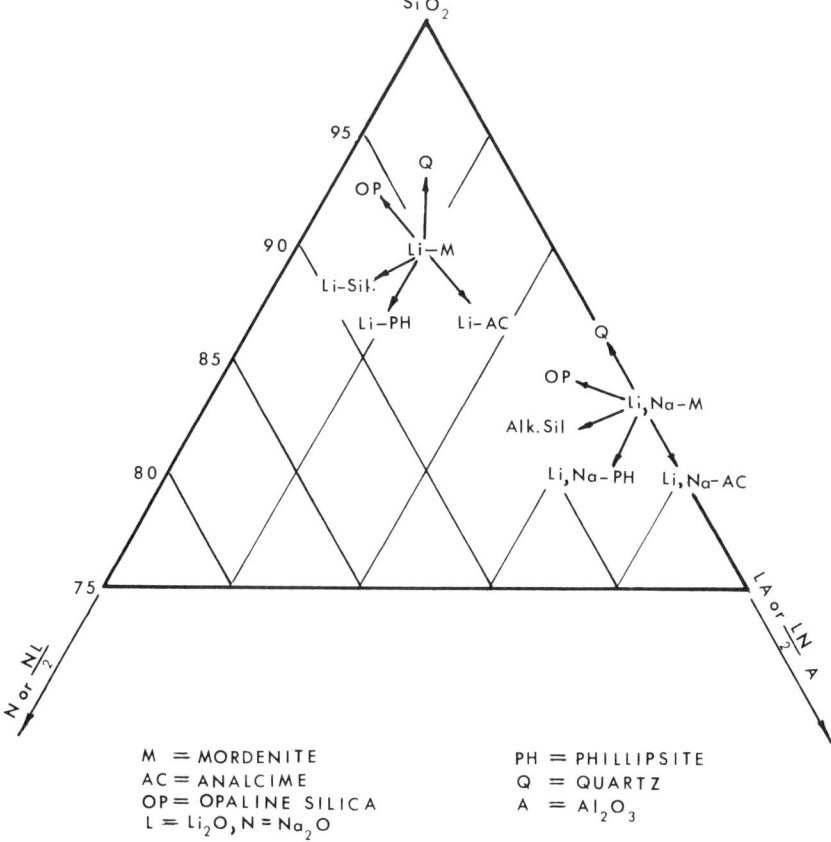

Figure 1. Schematic of compositional areas (mole basis) producing lithium and lithium, sodium zeolites

are acicular to 15μm length, mean index of refraction 1.504. The Li-analcime crystals occurred as equant crystals in the 20–80μm size range or as crystal composites. As has been reported by Saha (*11*) for Na-analcime synthesized from glasses, a substitutional series of Al for Si obtains also in the synthetic Li-analcimes. A range was found of $a_0 = 13.64$A, $n = 1.483$ for Li-analcime synthesized from siliceous batch compositions to $a_0 = 13.69$A, $n = 1.488$ for Li-analcime synthesized from low-silica batch compositions.

When the lithium phases were discovered, investigations were started in the lithia–soda system to determine if zeolites of intermediate composition could be synthesized. A 1:1 ratio of lithia to soda was used in all batch compositions. Silicic acid, lithium hydroxide, sodium hydroxide, and aluminum hydroxide (Hydral 710) were the best combinations of reactants to produce the Li,Na zeolites, mordenite, phillipsite, and analcime. Typical runs are given in Table II. A compositional diagram is given in Figure 1 to show the general areas of synthesis of mordenite and co-existing phases in the lithia and lithia–soda systems. The Li,Na-mordenite, as is the case with small-port Na-mordenite, is synthesized on composition, whereas the Li-mordenite could not be synthesized on composition. A 90% yield (10% opaline silica) of Li,Na-mordenite was obtained, and a crystallization curve on this synthesis is shown in Figure 2. Its chemical analysis is as follows: $70.12SiO_2$, $11.69Al_2O_3$, $2.55Na_2O$, $1.18Li_2O$, 13.60L.O.I. (W. H. Gerdes, analyst), which calculates to a formula of $0.345Li_2O–0.360Na_2O–Al_2O_3–10.2SiO_2–6.6H_2O$. The deficiency in exchangeable alkali probably is owing to hydrogen exchange. Its mean index of refraction is 1.471.

X-ray diffraction data for Na-, large-port Na-, Li-, and Li,Na-mordenites are given in Table III.

The small-port characteristics of the mordenites were determined by benzene sorption capacities, which were negligible, at low partial pressures of benzene (0.01–0.06) at room temperature using a quartz spring balance.

Discussion

Phase studies in the siliceous portion of the lithia system and the lithia–soda system resulted in the synthesis of mordenites with the same coexisting phases—analcimes, phillipsites, quartz, opaline silica, and crystalline alkali silicates—as had been found in the soda system. Whereas the starting materials used as reactants are not critical parameters in the synthesis of these zeolites in the soda system, the choice of reactants is a predominant factor in lithia-containing systems to produce these phases. The mechanism is not understood yet but the sensitivity of these

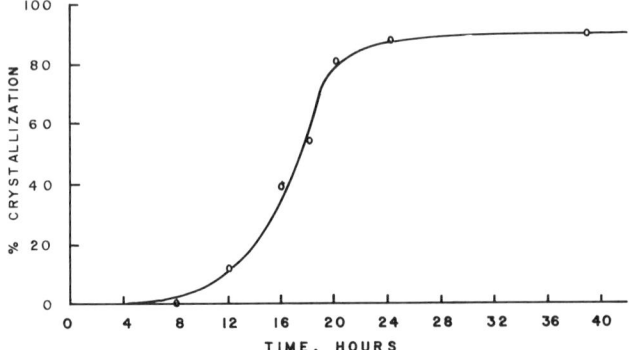

Figure 2. Crystallization curve for small-port lithium, sodium mordenite at 200°C

Table III. X-Ray Diffraction Data for Synthetic Mordenites

Synthetic Na-mordenite (8)		Synthetic Large-Port Na-Mordenite		Synthetic Na, Li-Mordenite		Synthetic Li-Mordenite	
d, A	I	d, A	I	d, A	I	d, A	I
13.53	40	13.4	40	13.5	40	13.5	30
10.24		10.2	10	10.2	10	10.3	10
9.06	50	9.02	70	9.02	80	9.02	75
6.57	55	6.50	50	6.51	50	6.50	40
6.39		6.32	30	6.32	25	6.32	15
6.08		6.02	10	6.01	10	5.98	10
5.80	15	5.75	20	5.75	25	5.75	25
5.05		5.03	2	5.03	2	5.05	2
4.83		4.84	2	4.82	5	4.80	2
4.52	25	4.50	35	4.50	40	4.49	45
4.15		4.12	5	4.11	5	a	
4.00	60	3.97	70	3.96	70	3.97	60
3.84		3.81	15	3.80	10	3.80	20
3.76		3.73	10	3.72	10	3.73	10
3.53		3.52	10	3.52	5	a	
3.47	100	3.45	100	3.44	100	3.45	60
3.39		3.37	60	3.37	65	3.37	50
3.29		3.28	10	3.28	10	a	
3.22		3.21	55	3.20	65	3.20	100
3.16		3.13	10	3.13	10	3.13	10

a Masked by phillipsite.

systems to starting materials provides an excellent opportunity to correlate the reactivity of interacting chemicals in studying the kinetics of zeolite crystallization. Varying the grade of crystalline aluminum hydroxide alone made a very large difference in the reaction products of both the lithia and lithia–soda systems. In the lithia system, in which aqueous colloidal silica sol was effective, the use of stirred autoclaves probably would eliminate part of the difficulty in determining the phase relationships. Although these studies were carried out primarily in the siliceous compositional region, the extension of the investigation into several lower silica compositions to evaluate the solid solution effects in the analcimes showed excellent crystallization of phases in this region. Research on this compositional area is in progress, particularly on the Al/Si and Li/Na substitutions in the crystalline phases. Also to be confirmed is whether a complete series exists between Li- and Na-mordenites.

Acknowledgment

Financial aid from the National Science Foundation Undergraduate Research Program GY 6076, and the U.S. Department of Health Education and Welfare for Mr. Coblenz is gratefully acknowledged.

Literature Cited

(1) Ames, L. L., Jr., *Am. Mineralogist* **1963**, 48, 1374–80.
(2) Ames, L. L., Jr., Sand, L. B., *Am. Mineralogist* **1958**, 43, 476–480.
(3) Barrer, R. M., *J. Chem. Soc.* **1948**, 2158–63.
(4) Barrer, R. M., Baynham, J. W., McCallum, N., *J. Chem. Soc.* **1953**, 4035–41.
(5) Barrer, R. M., Denny, P. J., *J. Chem. Soc.* **1961**, 983–1000.
(6) Barrer, R. M., White, E. A. D., *J. Chem. Soc.* **1951**, 1267–78.
(7) Coombs, D. S., Ellis, A. J., Fyfe, W. S., Taylor, A. M., *Geochim. Cosmochim. Acta* **1959**, 17, 53–107.
(8) Domine, D., Quobex, J., "Molecular Sieves," p. 78–84, Society of the Chemical Industry, Spec. Publ., 1968.
(9) Hoss, H., Roy, R., *Beitr. Mineral. Petrog.* **1960**, 7, 389–408.
(10) Leonard, R. J., *Econ. Geol.* **1927**, 22, 18–43.
(11) Saha, P., *Am. Mineralogist* **1959**, 44, 300–13.
(12) Sand, L. B., "Molecular Sieves," p. 71–78, Society of the Chemical Industry, Spec. Publ., 1968.
(13) Senderov, E. E., *Geokhimiya* **1963**, 9, 848–59.

RECEIVED February 10, 1970.

Discussion

Charanjit Rai (Cities Service Research, Cranbury, N. J. 08512): How do the properties of various lithium and lithium,sodium mordenites reported in your paper change as a function of the temperature at which the synthesis has been carried out?

M. L. Sand: We have not yet determined the properties of these mordenites. Mr. Coblenz has undertaken this work as his M.S. thesis.

D. E. Vaughan (W. R. Grace & Co., Clarksville, Md. 21029): In view of the report of widespread occurrence of mordenite lake bed deposits (Sheppard, page 279), the low-temperature synthesis of mordenite is of some interest and importance. Would you care to comment on and correlate the experimental and field evidence?

L. B. Sand: As mentioned in the text, these lithium-containing synthetic zeolites do not occur in nature, so experimental data do not apply to the problem of genesis. The synthesis of Na-mordenite at low temperature (75°C) could be of some significance, but a great deal of experimental data are required to apply to the interpretation the conditions for formation in a natural deposit.

J. Ciric (Mobil R & D Corp., Paulsboro, N. J. 08066): Are the Li-Ph crystals with the acicular habit birefringent?

L. B. Sand: Slightly, with a mean index of refraction similar to natural phillipsite.

11

Synthesis of a Beryllosilicate with the Structure of Analcime

S. UEDA and M. KOIZUMI

The Institute of Scientific and Industrial Research, Osaka University, Suita, Osaka 565, Japan

A crystalline phase of beryllosilicate with analcime structure was obtained under hydrothermal conditions from starting material of composition $Na_3Be_{1.5}Si_5O_{13} \cdot NaCl$. An analcime-like phase was obtained as a nearly pure phase at 200°C. A small amount of sodium chloride coexisted with the above phase. The lattice constant was $a = 13.35 \pm 0.01$ Å. The refractive index was 1.519 ± 0.002, higher than that of normal analcime. The broad endothermic peak, indicating the dehydration of zeolitic water, was observed in the temperature range from 150° to 500°C on a DTA curve. The specimen showed little change of structure on heating to 500°C. These results indicate that a beryllosilicate with the analcime structure was prepared by direct synthesis.

In naturally occurring zeolites, the extensive isomorphous replacement of aluminum, which is situated in the tetrahedral sites of the framework, by cations other than silicon is quite rare.

Two examples of substitution by phosphorus have been reported by McConnell (4) for viseite $[Ca_{10}Na_2(Al_{20}Si_6P_{10}(H_3)_{12})O_{96}(H_2O)_{16}]$ and kehoeite $[Zn_{5.5}Ca_{2.5}(Al_{16}P_{16}(H_3)_{16})O_{96}(H_2O)_{32}]$, which are isostructural with analcime ($NaAlSi_2O_6 \cdot H_2O$). On the basis of the occurrence of these phosphate zeolites in nature, Barrer et al. (2) attempted to synthesize zeolites consisting of alumino- and silicophosphates hydrothermally, but no crystalline phase with zeolitic structure was obtained in their experiments. On the other hand, they (1) earlier had succeeded in obtaining various species of gallogermanates with zeolitic structure in place of aluminosilicate. Their results demonstrated the isomorphous replacements of Ga ⇌ Al and Ge ⇌ Si between gallogermanate and aluminosilicate. In spite of this fact, the natural occurrence of gallogermanate zeolites has never been observed.

An example of Al $\rightleftarrows$ Be substitution in the aluminosilicate with zeolitic structure is given for helvine. The structure of helvine ($Mn_4Be_3Si_3O_{12}S$), similar to sodalite ($Na_4Al_3Si_3O_{12}Cl$), is composed of the three-dimensional framework with bonding of (BeO_4) and (SiO_4) tetrahedra (5). In order to keep the electrostatic charge balance, Al^{3+} is replaced by Be^{2+}, Na^+ by Mn^{2+}, and Cl^- by S^{2-}, although each substitution pair involves changes of both size and charge. In the case of so-called tugtupite [$Na_8Al_2Be_2Si_8O_{24}(Cl,S)_2$], formerly named beryllosodalite by Semenov and Bykov (8), the structure consists of linked MO_4 tetrahedra with M = Be, Al, and Si as the metallic cations (3). The result proves that the type of isomorphous replacement described below takes place between sodalite and so-called tugtupite.

$$2Al^{3+} \rightleftarrows Be^{2+} + Si^{4+}$$

From these facts, it is reasonable, then, that some beryllosilicates may be disposed to form a framework with large zeolite-like cavities.

The purpose of this study has been to synthesize new crystalline phases of beryllosilicate with zeolitic structure. The first attempt was made to obtain a phase with the analcime-like structure, because it is known that analcime is the most easily crystallized zeolitic phase under hydrothermal conditions in the system $Na_2O-Al_2O_3-SiO_2-H_2O$.

Experimental

Starting Material. The composition of beryllosilicate equivalent to ideal analcime may be written $NaBe_{1.5}Si_2O_6 \cdot aq$, if Al in analcime ($NaAlSi_2O_6 \cdot aq$) is replaced by Be to keep the electrostatic balance as shown by the formula of $2Al^{3+} \rightleftarrows 3Be^{2+}$.

Saha (7), who systematically studied the phase stability of analcime, concluded that they form as a stable phase over the wide range from nepheline composition ($NaAlSiO_4$) to albite ($NaAlSi_3O_8$). Since the sodalite composition corresponds to that of 3 nepheline ($Na_3Al_3Si_3O_{12}$) + sodium chloride (NaCl), it is possible for the analcime with nepheline composition to form in association with sodium chloride from the starting material with sodalite composition. If the replacement $2Al^{3+} \rightleftarrows Be^{2+} + Si^{4+}$ takes place in sodalite, as seen in the case of so-called tugtupite, the formula of the aluminum-free end member of sodalite will be given as $Na_4Be_{1.5}Si_{4.5}O_{12}Cl$.

Consequently, starting materials with the composition of $NaBe_{1.5}Si_2O_6$ and $Na_3Be_{1.5}Si_{4.5}O_{12} \cdot NaCl$ were prepared as described below. Calculated amounts of Snowtex-O silica sol (Nissan Chemical Industry Co., Ltd.), 1N solutions of NaOH and NaCl, and powdered BeO were mixed. The mixture was stirred vigorously and evaporated to dryness. The product obtained was ground to fine powder to make it homogeneous. No diffraction peaks were observed in the x-ray powder pattern of the mixture except those for beryllia.

Techniques of Hydrothermal Synthesis. In every run, the materials were contained in small sealed silver tubes to prevent selective leaching. A small amount of distilled water was added before the tube was sealed. The type of pressure vessel used was the test tube (6).

Most of the runs were made in the pressure–temperature range of 1000 bars (H_2O) and 150°–400°C for 1–3 weeks. The charge was quenched after the run.

The products were identified by x-ray diffraction techniques, petrographic methods, and differential thermal analysis.

Results and Discussion

From the starting material with the composition of $NaBe_{1.5}Si_2O_6$, fine-grained crystals with analcime-like structure were obtained only in the temperature range from 175° to 225°C, associated with beryllia and unknown phases. Excluding the x-ray reflections for the latter 2 phases, the x-ray powder data for the product are tabulated in Table I.

The lattice constant: $a = 13.35 \pm 0.01A$; refractive index: 1.519 ± 0.002. The specimen was obtained from the starting material of $NaBe_{1.5}Si_2O_6$, at 200°C and 1000 bars (H_2O) for a week.

The diffraction pattern was almost identical with that of normal aluminosilicate analcime. The lattice constant was calculated to be $a = 13.35 \pm 0.01A$ by indexing the x-ray reflections, assuming that the phase has cubic symmetry. The value is slightly smaller than the average one, 13.6A, for normal analcime (7). The crystals were optically isotropic and the refractive index was 1.519 ± 0.002, higher than that of normal analcime, 1.48. These differences in the value of lattice constant and refractive index may result from the existence of beryllium occupying a

Table I. X-Ray Data for the Beryllosilicate with Analcime Structure

hkl	d(A)	I	hkl	d(A)	I
211	5.43	31	633	1.817	23
200	4.71	5	642	1.787	8
321	3.56	6	732	1.696	46
400	3.33	100	800	1.670	17
332	2.845	70	741	1.646	18
422	2.724	17	820	1.621	11
431	2.618	27	822	1.573	9
521	2.436	24	831	1.551	24
440	2.361	14	842	1.458	7
611	2.167	32	761	1.440	20
620	2.114	5	664	1.424	12
631	1.968	7	754	1.408	16
543	1.888	6	932	1.378	29
640	1.853	28	941	1.347	11

part of the tetrahedral sites replacing silicon and aluminum in the framework of analcime.

From the starting materials with the composition of $Na_3Be_{1.5}Si_{4.5}O_{12}$ · NaCl, the phase with analcime-like structure was obtained also in the temperature range from 175° to 225°C. The data of x-ray diffraction and refractive index correspond with those for the beryllium-bearing analcime indicated in Table I. From this starting material, however, the 2 phases, sodium chloride and chkalovite ($Na_2BeSi_2O_6$) were observed in addition to the beryllium-bearing analcime.

In order to make the chkalovite phase disappear, runs were made on the starting materials with compositional ranges from $Na_3Be_{1.5}Si_{3.0}O_{10}$ · NaCl to $Na_3Be_{1.5}Si_{7.5}O_{18}$ · NaCl. At 200°C, beryllium-bearing analcime associated with a small amount of sodium chloride was obtained from the starting material, $Na_3Be_{1.5}Si_{5.0}O_{13}$ · NaCl, after 1–2 weeks. As the silica content of starting material was reduced, chkalovite appeared with the analcime. Quartz was observed instead of chkalovite for higher silica contents.

Since the zeolitic water in normal aluminosilicate analcime is removed gradually from the crystal lattice by heating, a broad endothermic peak is observed on the DTA pattern of the mineral, as indicated in Figure 1(A).

The DTA curve for the synthetic product obtained from the composition $Na_3Be_{1.5}Si_{5.0}O_{13}$ · NaCl, including beryllium-bearing analcime and a minor amount of sodium chlorides, is illustrated in Figure 1(B). The broad endothermic peak in the temperature range between 150° and 500°C indicates dehydration of zeolitic water contained in the beryllium-bearing analcime. The endothermic peak at 800°C resulted from the melting of sodium chloride. Similar thermal data were obtained for the products consisting of beryllium-bearing analcime and either chkalovite or quartz.

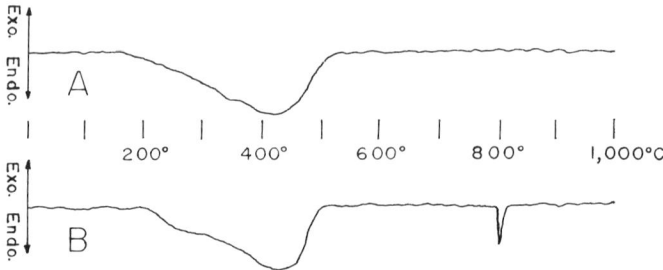

Figure 1. DTA curves for: (A) analcime from Maze, Niigata, Japan and (B) beryllosilicate obtained from the starting material with the composition $Na_3Be_{1.5}Si_{5.0}O_{13}$ · NaCl at 200°C, 1000 bars (H_2O) for 2 weeks

Sodium chloride and chkalovite were thermally stable up to their melting points, which were determined to be about 800°C for sodium chloride and 1050°C for chkalovite by the DTA method. Since a broad peak indicating the dehydration of zeolitic water was not altered by other endothermic peaks resulting from melting of sodium chloride and chkalovite, the existence of zeolitic water could be inferred from the DTA curve.

As for the beryllium-bearing analcime, there is little difference between the diffraction patterns of the original specimens and those heated to 500°C, except for the lattice constant change from 13.35 to 13.02A. Therefore, one may assume little change in the crystal structure of the analcime during gradual dehydration.

The accurate structural formula of the beryllium-bearing analcime cannot be given, because the data of chemical analysis are not available at the present stage of the investigation. Assuming that the substitution $2Al^{3+} \rightleftharpoons Be^{2+}Si^{4+}$ takes place in normal analcime as seen in the case of so-called tugtupite, however, the supposed formula of the beryllium-bearing analcime is set down as $Na_{16}Be_8Si_{40}O_{96}(H_2O)_{16}$.

It is uncertain whether chlorine is situated in the cavity of beryllium-bearing analcime lattice. However, since there is no difference in lattice constant and refractive index between the beryllium-bearing analcime synthesized from the Cl-free material of $NaBe_{1.5}Si_2O_6$ and that from $Na_3Be_{1.5}Si_{4.5}O_{12} \cdot NaCl$ and from $Na_3Be_{1.5}Si_{5.0}O_{13} \cdot NaCl$, it is unlikely for chlorine to occupy the cavity. The chlorine may, however, play a role as mineralizer in the crystallization process of beryllium-bearing analcime.

Conclusion

A beryllosilicate with the analcime structure can be prepared by direct synthesis. The lattice constant, refractive index, and DTA data are presented for the compound. These results indicate that beryllium may occupy the tetrahedral sites, replacing aluminum in the zeolite structure.

Literature Cited

(1) Barrer, R. M., Baynham, J. W., Bultitude, F. W., Meier, W. M., *J. Chem. Soc.* **1959**, 195.
(2) Barrer, R. M., Marshall, D. J., *J. Chem. Soc.* **1965**, 6616.
(3) Dano, M., *Acta Cryst.* **1966**, 20, 812.
(4) McConnell, D., *Mineral. Mag.* **1964**, 33, 799.
(5) Pauling, L., *Z. Krist.* **1930**, 74, 213.
(6) Roy, R., Tuttle, O. F., "Physics and Chemistry of the Earth," Ch. 6, Pergamon, London, 1956.
(7) Saha, P., *Am. Mineralogist* **1961**, 46, 859.
(8) Semenov, E. I., Bykov, A. V., *Dokl. Earth Sci. Sect. English Trans.* **1961**, 133, 812.

RECEIVED January 29, 1970.

12

Crystal Chemical Relationships in the Analcite Family

I. Synthesis and Cation Exchange Behavior

WILLIAM D. BALGORD[1] and RUSTUM ROY

Materials Research Laboratory, The Pennsylvania State University, University Park, Pa.

Systematic investigation of analyzed analcites ($NaAlSi_2O_6 \cdot H_2O$, ideal formula) of normal, high, and low Al/Si ratios prepared by hydrothermal synthesis and cation exchange showed that the anionic framework common to all members of the family—analcite, leucite, wairakite, pollucite, viseite, and certain other artificially prepared cation derivatives not yet found in nature—possesses a distinct robustness with respect to resisting major reconstructive transformations over broad ranges of composition, temperature, and p_{H_2O}. However, detectable second-order structural changes and deviations from cubic symmetry were brought about by variation of Al/Si ratio and cation population.

The unit cell of stoichiometric analcite contains 16 $NaAlSi_2O_6 \cdot H_2O$ formulas. The structure of analcite has been described in some detail, first by Taylor (*14*) and later by Coombs (*7*). Of direct concern here are the cavities within the structure which lie collinear with 3 sets of nonintersecting channels. The cavities are of 2 types: a set of 16 sites (1/8, 1/8, 1/8) occupied by H_2O, as in analcite, or by K or Cs, as in leucite or pollucite, coordinated by 12 framework oxygens, and a set of 24 smaller sites (1/8, 0, 1/4) occupied statistically by 16 Na in normal analcite.

Since 1950, Barrer and coworkers (*3, 4, 5, 6*) have reported on the properties of various cation exchanged forms of analcite having the ideal

---

[1] Present address: Division of Laboratories and Research, New York State Department of Health, Albany, N. Y.

Al/Si ratio—*viz*, 1/2. However, there have been no ion exchange investigations of the systems, Na-K, Na-Ca, or K-Ca in analcites of higher or lower than normal Al/Si ratios. Whereas earlier attempts to effect cation exchange of normal analcite by Li, Cs, Mg, Ca, and Ba were met with only limited success, indirect methods of exchange or direct synthesis from gels provided means whereby Li, K, Ca, Cs, and Pb^{2+} forms were said to have been obtained (*1, 3, 4, 5*).

In the present study, synthetic analcites having fixed Al/Si ratios of 2/3, 1/2, and 1/3 were subjected to ion exchange with a series of cations of various size, charge, and polarizability to fix limits of crystalline solubility and to determine the effects of compositional change on structure. Structure data as a function of temperature and p_{H_2O} (to be presented in a future publication) are available in a doctoral thesis by Balgord (*2*).

Experimental

Parent materials were synthesized hydrothermally in multigram quantities from gels according to methods described by Roy (*10*), Saha (*11*), and Luth and Ingamells (*9*). Detailed procedures used in preparing both parent analcites and cation-exchanged derivatives are obtainable also from the doctoral thesis by Balgord (*2*).

All samples were examined by optical microscopy, powder x-ray diffraction, and chemical analysis to determine phase composition, homogeneity, morphology, changes of symmetry, Al/Si ratio, cation population, and H_2O content. Precise lattice parameters were obtained at controlled p_{H_2O} using internal standards and computer least squares refinement. Chemical analyses were performed by flame photometry, x-ray fluorescence, emission spectrography, direct-reading emission spectrometry, and thermogravimetry.

Results

Modifications in the methods used by Saha (*11*) to prepare milligram quantities of analcites of several Al/Si ratios led to successful preparation of multigram quantities of optically homogeneous materials having the following characteristics:

$Al/Si = 2/3 \quad Na_{19.4}^{VI}(Al_{19.6}Si_{28.4})^{IV}O_{96} \cdot 14.2H_2O^{XII}$

$a_o = 13.74_8 \pm 0.017$ A, $n_D^{25} = 1.494 \pm 0.002$

$Al/Si = 1/2 \quad Na_{15.6}^{VI}(Al_{15.9}Si_{32.1})^{IV}O_{96} \cdot 16.4H_2O^{XII}$

$a_o = 13.72_4 \pm 0.017$ A, $n_D^{25} = 1.486 \pm 0.002$

$Al/Si = 1/3 \quad Na_{12.4}^{VI}(Al_{12.2}Si_{35.8})^{IV}O_{96} \cdot 18.4H_2O^{XII}$

$a_o = 13.65_9 \pm 0.017$ A, $n_D^{25} = 1.470 \pm 0.002$

Symmetry, as determined from x-ray powder patterns, is consistent with the space group *Ia3d*. Birefringence was absent or at most extremely weak.

Ion exchange runs were carried out under various conditions using salts of several monovalent and divalent cations with the dual objectives of preparing materials for dehydration and stability studies (to be described in a subsequent paper) and defining the limits of true crystalline solubility in the analcite structure.

As may be inferred from Table I, only certain ions are accommodated readily by the analcite structure. Found among this group are Li^+, Ag^+, K^+, NH_4^+, and Rb^+. But not so readily apparent from Table I is that extensive substitution involving K^+, NH_4^+, Rb^+, or Tl^+ invariably gives rise to the exsolution of a tetragonal phase seen both by x-ray diffraction and by microscopy as concentric reaction zones of contrasting relief. Ion exchange of high and low Al/Si analcite with K, on the other hand, yields products containing appreciable exchanged K within a single cubic phase: 62 and 35%, respectively.

Much more difficult is the exchange of divalent ions for Na^+ in analcite. Only with some effort were the writers able to achieve exchange

Table I. Results of Cation Exchange of Analcite

Al/Si	Cation	Conditions °C	PSI	Cell Edges, A	Refractive Indices
1/2	Li(81)[a]	100	[b]	13.53	1.501
	Ag(100)	100	[b]	13.68	1.560
	K(93)[c]	22	[b]	13.79	1.490
	NH_4^+(100)	100	[b]	13.12, 13.70	1.524
	Rb(100)	100	[b]	13.2, 13.6	1.520
	Tl(92)[d]	250	[b]	13.5, 14.6	1.640
	Mg(32)	225	4000	13.7	1.491
	Ca(82)	250	5000	13.64	1.492
	Sr(55)	225	4000	13.64	1.502–1.512
	Co(3)	100	[b]	13.7	1.486
	Ni(11)	100	[b]	13.7	1.486
2/3	K(62)	8	[b]	13.80	1.500
	Ca(100)	250	5000	13.62	1.514
1/3	K(35)	8	[b]	13.62	1.472–1.478
	Ca(81)	250	5000	13.64	1.474–1.481

[a] Numbers in parentheses indicate percentage exchange.
[b] Autogenous pressure.
[c] Metastable species, exsolution gives rise to leucite + K-saturated analcite.
[d] Taylor (*13*).

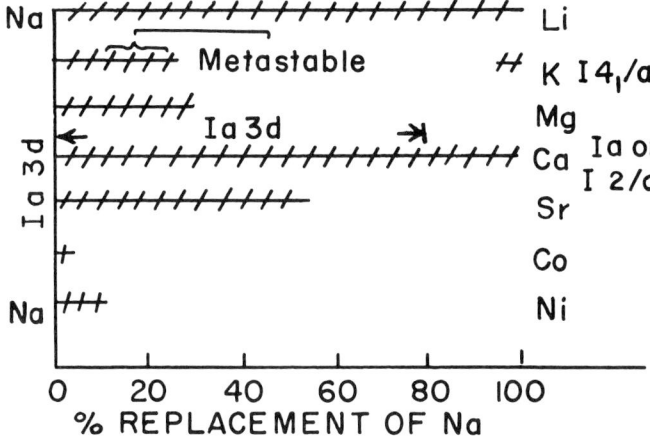

Figure 1. Extent of cation exchange in 1/2 analcite

with Ca^{2+}, Mg^{2+}, and Sr^{2+}. Two successive 4-day treatments of analcite in saturated $CaCl_2$ at 250°C and 5000 psi produced 82% exchange while maintaining cubic symmetry. But repeated treatments with salts of Sr, Mg, Ni, and Co produced cubic analcites containing lesser populations of altervalent ions decreasing in the stated order.

Both high and low Al/Si analcites readily undergo exchange with Ca at 250°C, 5000 psi during successive 4-day treatments.

Analcite partially exchanged with Ag^+ underwent a photosensitized redox reaction believed to involve Ag^+ with H_2O within the cavities, and giving rise to $Ag°$. The phenomenon was manifested first by the appearance of a yellow discoloration of the bulk material, suggesting the presence of color centers, and later by opaque metallic silver disseminated along grain boundaries within analcite crystallites.

Discussion

Formerly, all zeolites were presumed to possess the ability to exchange their "exchangeable" cations readily and without structural change. A study by Taylor and Roy (12) on the P-type zeolite demonstrated emphatically that this assumption is a gross oversimplification. In the present study, despite a greater degree of "openness" of the interstices of analcite relative to the P-zeolites, the extent of exchange with many cations is very much limited. In Figure 1, data are presented on the extent to which a given cation can substitute for sodium in the parent "analcite" structure. In the case of K^+ exchange, no more than 25% K^+ (and probably as little as 10%) is tolerated in the crystalline solution of the Na phase at equilibrium. Subsequent exchange resulting in further replacement of Na by K only succeeds in bringing about a double decomposition reaction giving rise to the exsolution of a K-rich second phase, leucite.

Included also in Figure 1 is information regarding symmetry changes induced by cation exchange. In view of these data, derived from powder diffraction data, the analcites, in marked contradistinction to the P-zeolite family, show rather strong and easily recognizable (by powder x-ray pattern) familial affinities despite any cation changes.

The difficulty of replacement of Na^+ by divalent ions is demonstrated amply in the observation that all previous workers failed to achieve substitution of Ca^{2+} for Na^+ to any significant extent by straightforward exchange methods. Although Ca substitution was achieved in this study under hydrothermal conditions, it is not certain whether the partial replacement of Na by Mg, Ni, or Co is limited kinetically or represents equilibrium. The relative facility with which Ca^{2+} replaces $2Na^+$ in the high-Al analcite may provide evidence that Al/Si ordering exercises some degree of control over the extent of exchange. A high proportion of the Na^+ sites actually occupied by Ca^{2+} presumably are coordinated by at least 2 of 4 framework oxygens which themselves are part of Al-containing tetrahedra.

The relation of the number of water molecules per unit cell to the size of the exchangeable cation and free volume (here defined as unit cell volume less the volume occupied by the framework and cations), is presented in Figure 2. Two groups of phases emerge, fully hydrated

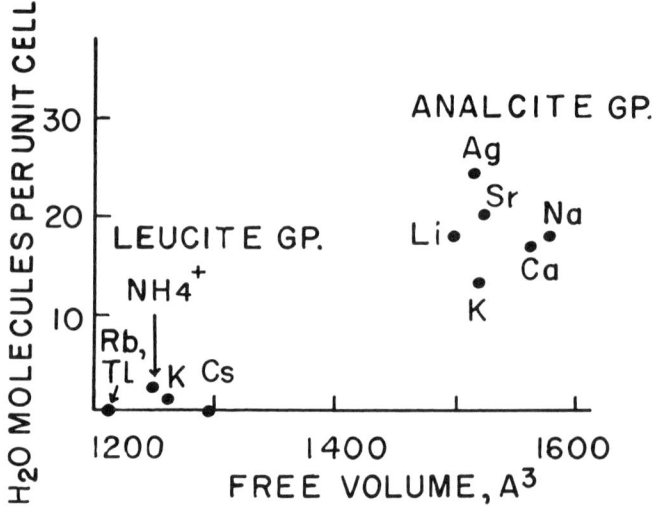

Figure 2. Relationship between free volume and H_2O content of 1/2 analcite unit cell exchangeable cation population

structures having an expanded cell, designated "analcite," and contracted structures containing little or no water, designated "leucite." The situation may be rationalized in terms of Figure 3, adapted from Deer *et al.* (8), which shows schematically a view of the sites in the analcite struc-

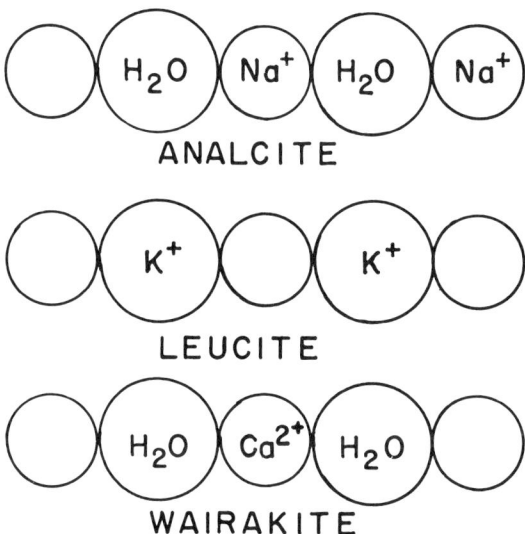

Figure 3. Schematic representation of H_2O and cation sites in the analcite structure

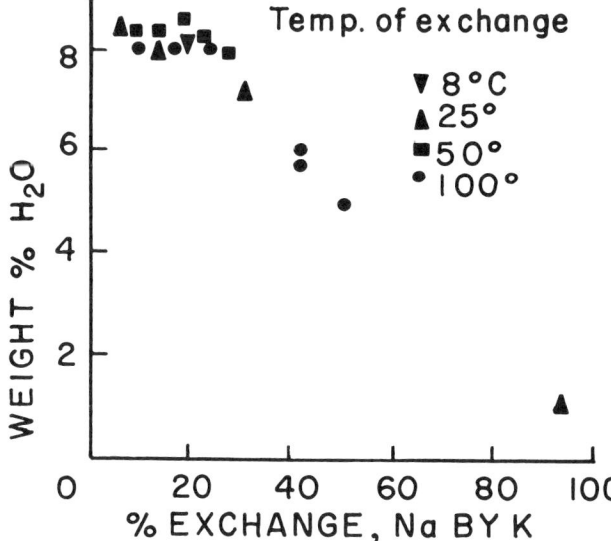

Figure 4. Dependence of H_2O content on K in 1/2 analcite

ture normally occupied by Na^+ and H_2O, respectively. Depicted in Figures 1 and 3 are the relationships on the limits of solubility and the location of the cations and the H_2O molecules. In principle, the divalent cationic species should admit more water. That the extra "space" in the channels is not occupied to any appreciable extent by water (at given p_{H_2O} and T) implies an exclusion of the H_2O molecules from the Na^+ site. In a K^+-saturated cubic analcite, the water content is constant (Figure 4); here a random (Na^+, K^+) distribution over the Na^+ sites pertains. In leucite, however, the K^+ (Figure 3) occupies the H_2O sites, thus excluding the H_2O molecule. H_2O in turn cannot occupy the vacated Na^+ sites because part of the occupied channel volume of leucite is taken up by a contraction of the unit cell.

Finally, a decrease of Al/Si ratio from 2/3 to 1/3 occasions a marked increase in water content. This fact, however, is consistent with the lower cation density in the Na^+ sites of the low-Al analcite.

Literature Cited

(1) Ames, L. L., Sand, L. B., *Am. Mineralogist* **1958**, 43, 477.
(2) Balgord, W. D., Ph.D. Thesis, Pennsylvania State University, 1966.
(3) Barrer, R. M., *J. Chem. Soc. (London)* **1950**, 2344.
(4) Barrer, R. M., Baynham, J. M., *J. Chem. Soc. (London)* **1956**, 2888.
(5) Barrer, R. M., Hinds, L., *J. Chem. Soc. (London)* **1953**, 1883.
(6) Barrer, R. M., McCallum, N., *J. Chem. Soc. (London)* **1953**, 4029–4031.
(7) Coombs, D. S., *Mineral. Mag.* **1955**, 30, 699–708.
(8) Deer, W. A., Howie, R. A., Zussman, J., "Rock-forming Minerals," Vol. 4, p. 350, Wiley, New York, 1964.
(9) Luth, W. C., Ingamells, C. O., *Am. Mineralogist* **1965**, 50, 255–258.
(10) Roy, Rustum, *J. Am. Ceram. Soc.* **1956**, 39, 145–146.
(11) Saha, Prasenjit, *Am. Mineralogist* **1959**, 44, 300–313.
(12) Taylor, A. M., Roy, Rustum, *Am. Mineralogist* **1964**, 49, 656–682.
(13) Taylor, H. F. W., *J. Chem. Soc. (London)* **1949**, 1256.
(14) Taylor, W. H., *Z. Krist.* **1930**, 74, 1–19.

RECEIVED January 21, 1970.

Discussion

Brian D. McNicol (Koninklijke/Shell Laboratorium, Amsterdam, Netherlands): With respect to your comment on the formation of $Ag°$ within the cavities by a photosensitized redox reaction, if the $Ag°$ atoms are monatomically dispersed, then they could be identified by electron spin resonance. $Ag°$, of course, is paramagnetic.

W. D. Balgord: Yes, if they stay that way (monatomic) long enough to measure them.

J. A. Rabo (Union Carbide Research Institute, Tarrytown, N. Y. 10591): The existence of Ag° atoms—upon reduction of Ag$^+$—in zeolites has been extensively investigated with X and Y zeolites using ESR without success. The smallest reduced species found so far are Ag$_2^+$ ions, which exist up to $\sim$ $-80°$C in Y zeolite.

W. D. Balgord: Several questioners seem to have interpreted our comments on p. 143 to mean that we actually observed monatomic Ag in the analcite. No effort was made to detect Ag atoms. In fact, the paragraph does not mention them as such. It does seem reasonable, however, that silver, if initially present as individual Ag$^+$ ions in the restricted analcite cavities, may have existed as discrete Ag° atoms, if only momentarily, at one step in the mechanism by which the metallic silver aggregated.

Douglas S. Coombs (University of Otago, Dunedin, New Zealand): In connection with the discussion on nonlinearity of cell dimensions plotted against Al atoms per formula unit, it may be commented that Saha's plot was for cell edge against Al/Si ratio. If this latter gives a straight-line relationship, a plot of cell edge against number of Al atoms must be curvilinear.

When H$_2$O exceeds 16 per unit cell, where is this extra water accommodated? If it is distributed through two lattice sites, is this reflected in dehydration phenomena?

W. D. Balgord: A replot of cell edge and IR data *vs*. Al/Si ratio does not reveal the linear relationship implied in the first of Dr. Coombs' questions. From the unit cell composition data presented, it may be observed that a decrease of 4 Na$^+$ ions, associated with a corresponding

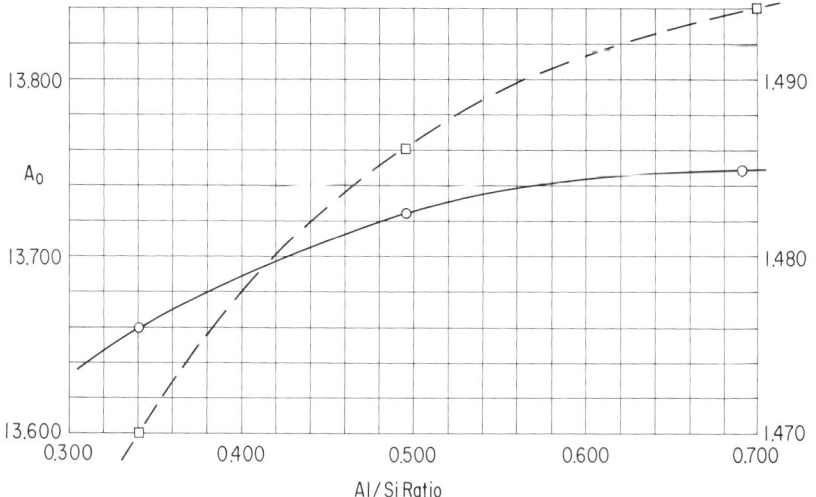

decrease of 4 Al^{3+} ions from Al/Si ratio of 1/2 to 1/3, is accompanied by an increase of 2 H_2O molecules. The inverse relationship between 2 Na^+ and H_2O suggests that the concurrence of vacant adjoining Na^+ sites provides sufficient space to accommodate the additional H_2O molecule.

13

Synthesis of Thermodynamically Stable Zeolites in the $Na_2O-Al_2O_3-SiO_2-H_2O$ System

E. E. SENDEROV and N. I. KHITAROV

V. I. Vernadsky Institute of Geochemistry and Analytical Chemistry, USSR Academy of Sciences, Moscow, V-334, Vorobiovskoe shosse, 47-A, USSR

> *Zeolites natrolite and analcime (composition of the latter is close to its ideal formula, $NaAlSi_2O_6 \cdot H_2O$) have been obtained in the $Na_2O-Al_2O_3-SiO_2-H_2O$ system under conditions hampering the formation and conservation of metastable crystals. Among sodium zeolites, only these 2 phases appear truly stable; formation of the others is a result of metastable growth from highly reactive starting materials.*

In the $Na_2O-Al_2O_3-SiO_2-H_2O$ system, a great number of artificial zeolites may be obtained. Their synthesis is mainly carried out below 200°–300°C using amorphous starting materials, gel-like mixtures and glasses. High supersaturation of a solution arising in a reactor with such a charge and slow rates of crystal formation under low-temperature zeolite synthesis conditions make development and conservation of metastable states a rule rather than an exception (2, 17).

When starting with highly reactive gel-like mixtures, the probability of metastable crystallization, as Fyfe (6) pointed out, is particularly high. In that case, a system should have an initial maximum free energy excess with respect to the final stable state. This increases the number of possible intermediate metastable states (and phases corresponding to them) through which the system passes to the end state according to Ostwald's law. Following Goldsmith's idea (7), during this transition the "simplicity" of crystal structure decreases in successively formed solid phases. Nucleation is the slowest for the most complex and ordered lattices, which form after the others.

Prolongation of a run makes preservation of the intermediate phases possessing only relative stability more difficult. Mixing a charge in a reactor should cause a similar effect, as well as the use of catalysts which accelerate stable phase nucleation. The appearance of crystals with a slow nucleation rate should be favored by introducing its seeds, and that of intermediate phases should be averted by using less reactive crystalline starting materials instead of amorphous ones. In the latter case, however, the time necessary for new phase formation becomes considerably longer.

Catalyzing influence on reactions in silicate systems may be excited by an increase of crystal-forming solution alkalinity. This is evidenced by zeolite synthesis experience (*17*) and by direct experiments of Campbell and Fyfe (*3*) and Kerr (*8*). However, an increase of pH may affect not only growth rate, but displacement of desilication reactions towards products poorer in silica (*15, 16*). These 2 effects must be distinguished when analyzing the influence of alkalinity growth.

In spite of the synthesis in the Na_2O–Al_2O_3–SiO_2–H_2O system of a great number of zeolites of different structural groups, reliable communications on artificial natrolite, $Na_2Al_2Si_3O_{10} \cdot 2H_2O$, were lacking until recently (*18*). This might seem strange because of extensive investigation of the system and wide abundance of the zeolite in nature. Difficulties with natrolite synthesis were supposed to be conditioned by the complexity of its structure (*12*), particularly by ordered Si and Al distribution in it. The ordering also explains the constancy of natrolite composition, unlike the majority of artificial sodium zeolites. Natrolite appears to differ by the slowness of its nucleation from disordered, possibly less stable phases. The factors preventing metastable growth should be taken into account in the formation of natrolite.

Gels of $(1-2)Na_2O \cdot Al_2O_3 \cdot 3SiO_2$ + aq composition seeded with natrolite were prepared to synthesize it (*18*). Natrolite had been crystallized in the range of about 100°–200°C under saturated vapor of crystal-forming solution pressure (Table I). In this range, the zeolite (commonly with analcime) appeared instead of chabazite and garronite, which crystallize from the same mixtures in absence of the seeds (*17*). No difference was noticed between the synthetic and natural natrolite. Chabazite and garronite also were formed in runs with the seeds but increased alkalinity resulting from concentration of relative Na_2O contents in the initial mixture caused natrolite substitution for them. The desilication reactions did not cause this substitution because chabazite and garronite may be as poor in silica as natrolite.

Another way to obtain natrolite is the use of natural sodium aluminosilicates—nepheline, $(Na,K)AlSiO_4$, and albite, $NaAlSi_3O_8$—separately and in mixtures, in a starting charge which was treated with neutral (H_2O and $0.2N$ NaCl) and alkaline ($0.2N$ NaOH) solutions. The charge was placed in a rocking autoclave in which a stainless steel ball mixed the contents. Pressure was approximately 300 atm; duration ranged from

Table I. Natrolite Synthesis from Seeded Gels

Na_2O/Al_2O_3 in a Gel	$Na_2O + Al_2O_3 + SiO_2$ Conc., Wt. %	Duration, Days	Products[a]
	120°C		
2	5	103	nt + an
1.5	20	103	am + nt
2	20	103	am + nt
1.5	39	103	an + nt
2	39	103	nt + an
	180°C		
1.5	5	19	an + nt
2	5	19	an + nt
1.2	20	19	an + nt
1.5	20	19	an + nt
1	40	19	nt + an

[a] am = amorphous matter, an = analcime, nt = natrolite.

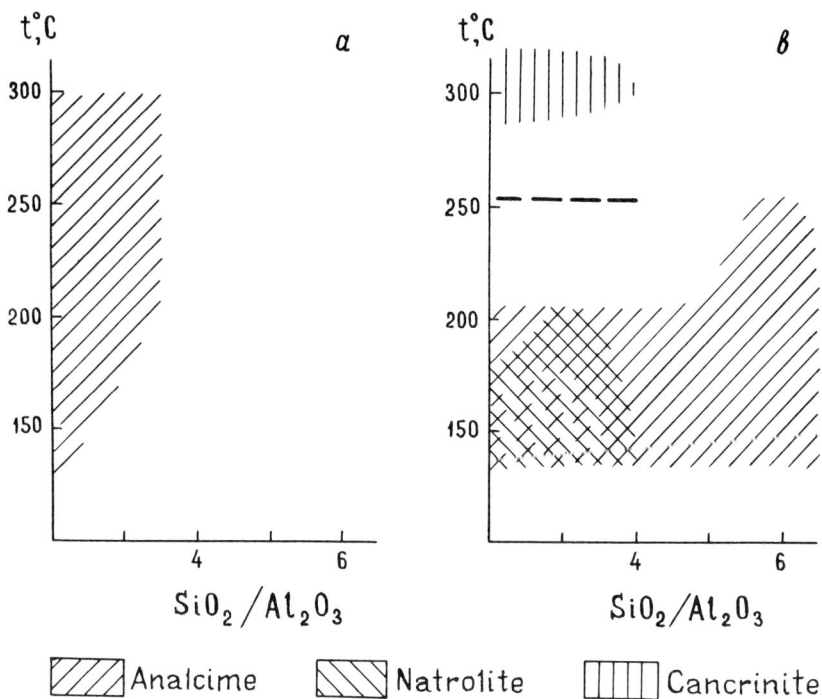

Figure 1. Recrystallization of mineral mixtures treated with neutral (a) and alkaline (b) solutions

several weeks to about 100 days. Reliably identified new phases were obtained only for the limited fields shown in Figure 1. Alkaline media favored natrolite crystallization, which arose in approximately the same temperature range as from the seeded gels. Natrolite as starting material decomposes at nearly 300°C, giving rise to hydroxycancrinite. This permits demarcation of the boundary of natrolite's stability field at roughly 250°C.

Thus, natrolite replaces garronite and chabazite with increasing alkalinity when crystallizing from seeded gels, and the latter 2 zeolites are not formed from minerals. All this suggests that natrolite is more thermodynamically stable.

The composition of analcime formed at recrystallization of the mineral mixtures can not vary as widely as in synthesis from amorphous gels and glasses. This suggests that analcimes of some compositions might grow as metastable phases. To determine this possibility, special experiments were conducted in which calcined gels, $Na_2O \cdot Al_2O_3 \cdot (2.5–13)SiO_2$, were treated with water and NaOH solutions of different concentrations (*10*). Mixing of a charge may be used during the experiments, as in the case of natrolite. However, this factor combined with changing the crystallization time over a range of several weeks did not prove as great an influence as variation of the solution alkalinity.

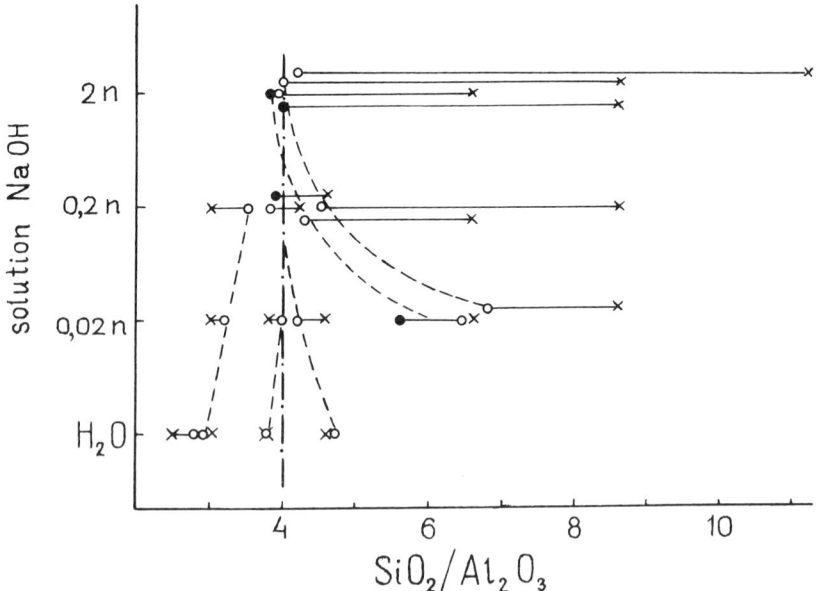

Figure 2. SiO_2/Al_2O_3 ratio in initial gels (×) and analcimes crystallized at 250°C (circles). Open circles = SiO_2/Al_2O_2 ratio in analcime in experiments with mixing, closed ones = without mixing. Each horizontal line connects the point of an initial gel with the point of analcimes grown from this gel. Points of crystal compositions formed from the same gels in different solutions are connected by dashes. A dot-and-dash line shows the SiO_2/Al_2O_3 ratio in the ideal formula of analcime.

Some results of the determination of analcime compositions are shown in Figure 2. The initial SiO_2/Al_2O_3 ratio in gels and in analcimes grown from them at 250°C is indicated on the abscissa. Analcime composition was measured with an accuracy in SiO_2/Al_2O_3 value of ±0.2 using the Saha (14) and Coombs and Whetten (5) method based on x-ray data. Composition of a solution acted on gels is shown by the ordinate.

Analcimes crystallized in more concentrated NaOH solutions are closer to the ideal formula, where the SiO_2/Al_2O_3 ratio is 4. If in the experiments NaOH influenced displacement of equilibrium, silica-deficient species, where $SiO_2/Al_2O_3 \approx 3$, would arise. Limits for SiO_2/Al_2O_3 in analcimes according to our new data are 2.8 and 8.2 (10). With an alkalinity increase, the displacement is limited, and not only high-SiO_2 species but low-SiO_2 ones disappear. Instead of the latter, feldspathoids were crystallized. The alkalinity change influenced the kinetics of the process and caused the disappearance of less stable analcimes of fringe compositions.

The ideal stoichiometric composition is a limit to which analcimes tend from both sides in experiments up to 400°C. At that and higher temperatures, prolongation of runs gives rise to analcimes with lower contents of silica, even when water reacts with gels enriched in SiO_2. This is illustrated by data on crystallization of $Na_2O \cdot Al_2O_3 \cdot 4.6\ SiO_2 + H_2O$ (Table II). Absence of sodium hydroxide in the solution with which the gels are treated allows neglect of the role of the desilication reactions.

The most probable reason for the constancy of composition is the ordering distribution of Si and Al in the lattice of such analcime (11). Here the analogy with natrolite and other groups of framework silicates, feldspars, for which low-temperature ordered varieties are known, is reasonable. Analcime of the $NaAlSi_2O_6 \cdot H_2O$ composition appears to be the low-temperature variety, stable up to 400°C. Deviation from a definite composition indicates the beginning of disorder. The order must

Table II. SiO_2/Al_2O_3 Ratio in Analcime Formed from $Na_2O \cdot Al_2O_3 \cdot 4.6\ SiO_2 + H_2O$

Temp., °C	Duration, Days	SiO_2/Al_2O_3
250	14	4.7
300	10[a]	4.4
350	15[a]	4.5
400	2	4.6
400	14[a]	3.7
450	3	4.4
450	14[a]	3.7

[a] Experiments with mixing.

remain unchanged at all temperatures lower than 400°C. It further means that, if the definite composition reflects the ordering, it must remain stable down to the lowest temperature of analcime formation which is likely to reach ambient temperature.

Thus, the factors making metastable growth difficult force the phases easily formed from gels—garronite, chabazite, analcime solid solutions—to disappear. Mordenite crystallization also is connected with metastable growth and equilibria, not with quartz but with less stable forms of silica (4). All this suggests that among the zeolites in the Na_2O–Al_2O_3–SiO_2–H_2O system the only thermodynamically stable phases are natrolite and the $NaAlSi_2O_6 \cdot H_2O$ analcime (up to 400°C). This is indirectly supported by the very wide abundance of the 2 zeolites in nature.

For the other sodium zeolites, grown metastably, it is possible to define sequences of increasing relative stability. Replacement of one phase by another may be influenced by the increase of crystallization time that was noticed by many investigators (1, 9, 13, 19) and by the increase of solution alkalinity and temperature when the latter 2 parameters accelerate rates of reactions. Here the sequences are: garronite (NaP) → analcime (solid solution), tetragonal P2 (more ordered) being more stable than cubic P1; in the poorest in silica field, NaX → NaA → NaP.

Literature Cited

(1) Barrer, R. M., Baynham, J. M., Bultitude, F. W., Meier, W. M., *J. Chem. Soc.* **1959**, 195.
(2) Breck, D. W., Flanigen, E. M., *Proc. Conf. Mol. Sieves, London,* **1967**.
(3) Campbell, A. S., Fyfe, W. S., *Am. Mineralogist* **1960**, 45, 463.
(4) Coombs, D. S., Ellis, A. F., Fyfe, W. S., Taylor, A. M., *Geochim. Cosmochim. Acta* **1959**, 17, 53.
(5) Coombs, D. S., Whetten, J. T., *Bull. Geol. Soc. Am.* **1967**, 78, 153.
(6) Fyfe, W. S., *J. Geol.* **1960**, 68, 553.
(7) Goldsmith, J. R., *J. Geol.* **1953**, 61, 439.
(8) Kerr, G. T., *J. Phys. Chem.* **1966**, 70, 1047.
(9) *Ibid.,* **1968**, 72, 1385.
(10) Khundadze, A. G., Senderov, E. E., Khitarov, N. I., *Geokhim.* **1970**, in press.
(11) Knowles, C. R., Rinaldi, F. F., Smith, J. V., *Ind. Mineralogist* **1965**, 6, 127.
(12) Meier, W. M., *Z. Krist.* **1960**, 113, 430.
(13) Regis, A. J., Sand, L. B., Calmon, C., Gilwood, M. E., *J. Phys. Chem.* **1960**, 64, 1567.
(14) Saha, P., *Am. Mineralogist* **1959**, 44, 300.
(15) Senderov, E. E., "Zeolites, Their Synthesis, Properties, and Applications," p. 165, Nauka, 1965.
(16) Senderov, E. E., *Geokhim.* **1966**, N 5, 600.
(17) *Ibid.,* **1968**, N 1, 3.
(18) Senderov, E. E., Khitarov, N. I., *Geokhim.* **1966**, N 12, 1398.
(19) Taylor, A. M., Roy, R., *Am. Mineralogist* **1964**, 49, 656.

RECEIVED January 21, 1970.

14

Zeolite Frameworks

W. M. MEIER

Institut für Kristallographie and Petrographie der ETH, Zürich, Switzerland

D. H. OLSON

Mobil Research and Development Corp., Princeton, N. J. 08540

> *A collection of stereopairs showing the presently known framework structures of zeolites is presented. Only well-established structures have been incorporated in this survey which includes crystal data, information on channel geometry, and possible fault planes.*

Most zeolite structures are fairly complex and cannot be visualized readily. For this reason, stereoscopic drawings of 27 well-established zeolite frameworks have been prepared as a general aid (Figures 1–27). These skeletal framework drawings are based on the T-atoms (Si,Al) only and T–O–T bridges are represented by straight lines. In general, the viewing direction has been chosen in such a way that the main channels are clearly visible. The idealized cell contents, crystal system, space group, and unit cell dimensions have been summarized in the figure captions. In many cases, the listed space group represents a pseudo-symmetry which does not account for Si,Al ordering. The unit cell has been indicated in all cases where this appeared feasible.

The present atlas of zeolite frameworks includes only reasonably well-established structures which have been at least partially refined. Mere proposals have been excluded since past experience has shown that all too frequently these have been incorrect. As a rule, the references in the captions have been limited to the first correct description of the framework structure and to its subsequent refinement.

Zeolites do not represent an easily definable group of crystalline aluminosilicates. There are obvious borderline cases like some sodalite-type species which have been included in this survey. On the other hand, nepheline hydrate, the scapolites, osumilite (*12*), and buddingtonite, an ammonium feldspar with zeolitic water (*17*), have not been considered here.

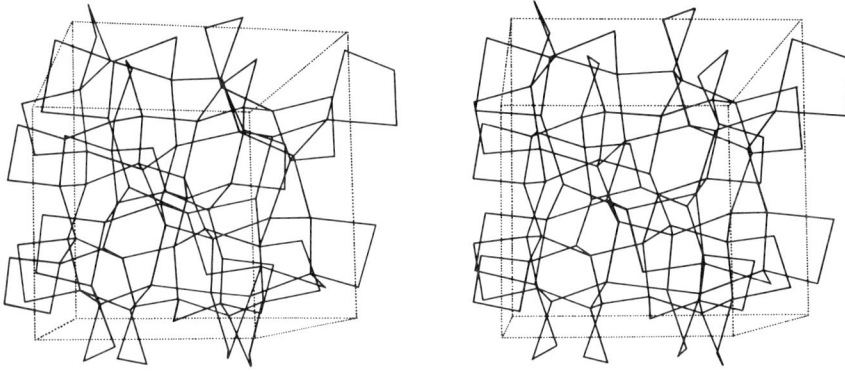

Figure 1. Analcime (29, 55), $Na_{16}Al_{16}Si_{32}O_{96} \cdot 16\ H_2O$, viewed along [100]
cubic, Ia3d, a = 13.73 Å
Isotypes: wairakite (14), leucite (62), pollucite (40), viseite (31), kehoite (32)

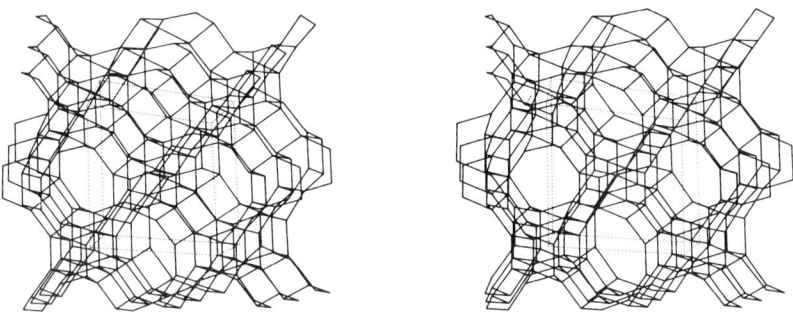

Figure 2. Laumontite (6), $Ca_4Al_8Si_{16}O_{48} \cdot 16\ H_2O$, viewed along [100]
monoclinic, Am, a = 7.57, b = 14.75, c = 13.10, γ = 112.0°
CH: [100] **10** 4.0 × 5.6*
Isotype: leonhardite (13)

General accounts of zeolite structures and classification schemes can be found in several recent articles (20, 35, 50) and tables (54).

Structure Types

Species which are based on topologically equivalent frameworks represent the same structure type irrespective of composition, distribution of the framework atoms, cell dimensions, and symmetry. Marked differences with respect to these properties frequently can be observed for isotypic species. A number of zeolitic isotypes have been listed in the figure captions together with appropriate references, usually to structural work.

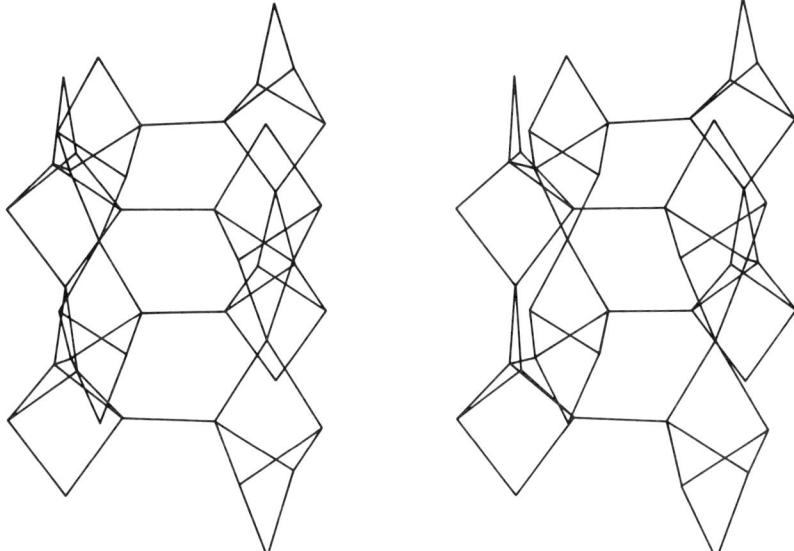

Figure 3. Natrolite (33, 42, 58), $Na_{16}Al_{16}Si_{24}O_{80} \cdot 16\ H_2O$, viewed along [110]

orthorhombic, Fdd2, a = 18.30, b = 18.63, c = 6.60 A
CH: ⊥ [001] **8** 2.6 × 3.9**
FP: (110). *Isotype: scolecite* (57)

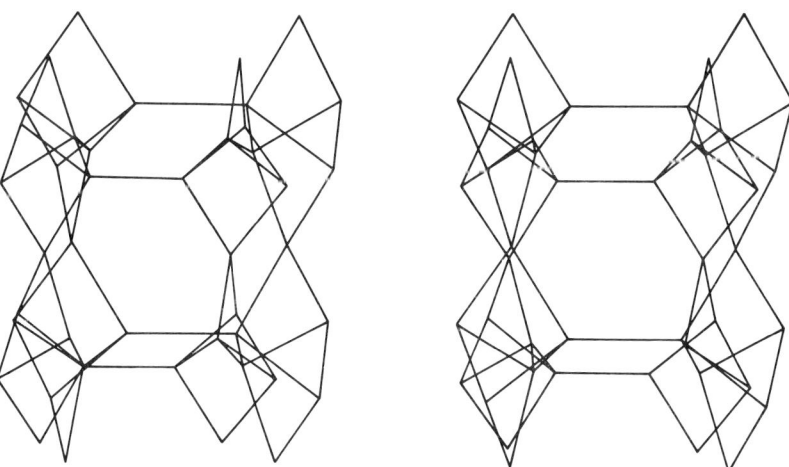

Figure 4. Thomsonite (57), $Na_4Ca_8Al_{20}Si_{20}O_{80} \cdot 24\ H_2O$, viewed along [100]

orthorhombic, Pnn2, a = 13.07, b = 13.08, c = 13.18 A
CH: ⊥ [001] **8** 2.6 × 3.9**
FP: (100), (010)

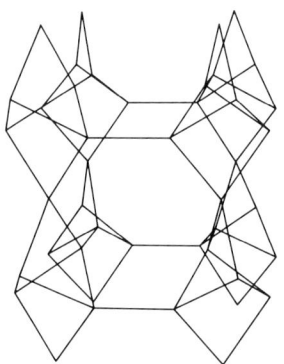 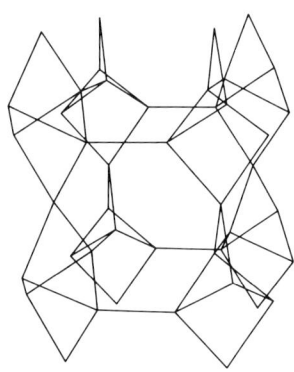

Figure 5. Edingtonite (56), $Ba_2Al_4Si_6O_{20} \cdot 8\ H_2O$, viewed along [1$\bar{1}$0]

orthorhombic, $P2_12_12$, a = 9.54, b = 9.65, c = 6.50 A
CH: ⊥ [001] **8** 3.5 × 3.9**
FP: (110)

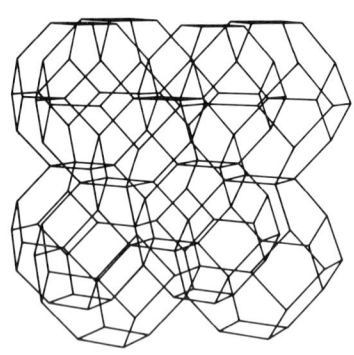 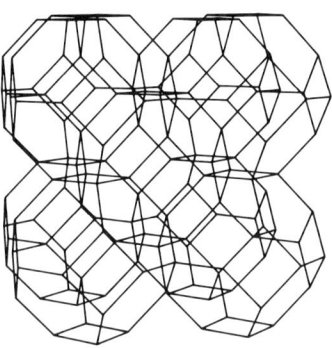

Figure 6. Sodalite (30, 43), $Na_6Al_6Si_6O_{24} \cdot 2\ NaCl$, viewed along [100]

cubic, $P\bar{4}3n$, a = 8.87 A
FP: (111)
Isotypes: sodalite hydrate or Zhdanov G (48, 49), TMA-sodalite (2),
tugtupite (15)

Synthetic zeolite NaP1 (or Linde B) is apparently an isotype of gismondine (Figure 14) according to a recent study by Baerlocher and Meier (4). The cubic structure which was proposed earlier has been ruled out in this work. The symmetry of the NaP1 framework is noncubic (despite the fact that the unit cell is compatible with cubic symmetry), and the phase appears to consist of mimetic twins.

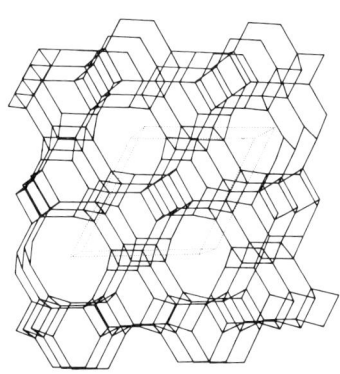 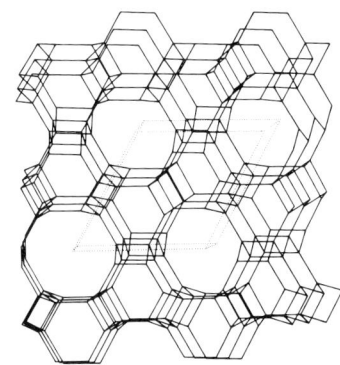

Figure 7. Cancrinite (24, 42), $Na_6Al_6Si_6O_{24} \cdot CaCO_3 \cdot 2\ H_2O$, viewed along [001]

hexagonal, P6$_3$, a = 12.75, c = 5.14 A
CH: [001] **12** 6.2*
FP: (001). Isotype: cancrinite hydrate (61)

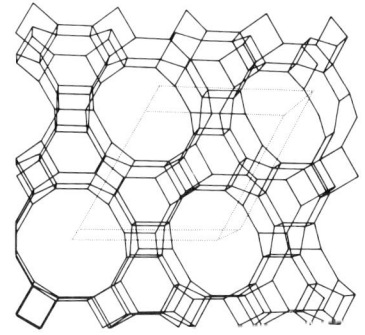 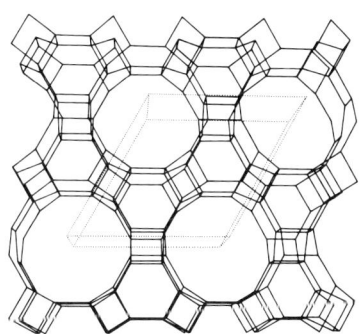

Figure 8. Gmelinite (19), $(Na_2,Ca)_4Al_8Si_{16}O_{48} \cdot 24\ H_2O$, viewed along [001]

hexagonal, P6$_3$/mmc, a = 1⸱.8, c = 10.0 A
CH: [001] **12** 7.0* ⟷ ⊥[001] **8** 3.6 × 3.9**
FP: (001)

Channel Geometry

The figure captions include information on the channels (CH). A shorthand notation has been used for the description of the channels in the various frameworks. Each system of equivalent channels has been characterized by the channel direction, the number of tetrahedra forming the smallest rings of the channels, and the crystallographic free diameters

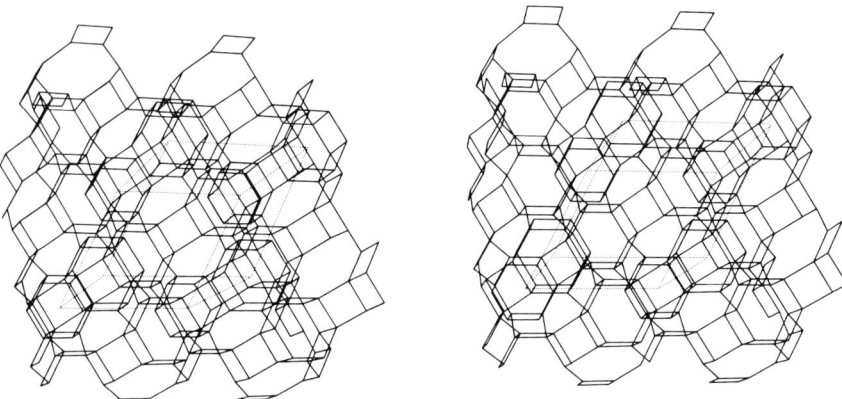

Figure 9. Chabazite (16, 51), $Ca_6Al_{12}Si_{24}O_{72} \cdot 40\ H_2O$, *viewed along* [001]

hexagonal, $R\bar{3}m$, a = 13.17, c = 15.06 A
CH: ⊥ [001] **8** 3.6 × 3.7***
FP: (001)

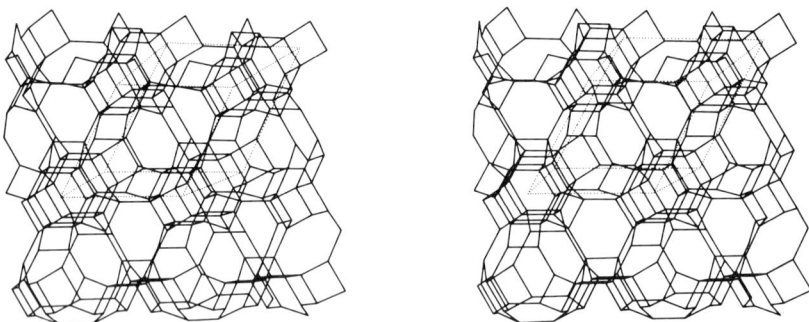

Figure 10. Erionite (52), $(Na_2,Ca,\ etc.)_{4.5}Al_9Si_{27}O_{72} \cdot 27\ H_2O$, *viewed along* [001]

hexagonal, $P6_3/mmc$, a = 13.26, c = 15.12 A
CH: ⊥ [001] **8** 3.6 × 5.2***
FP: (001)

of the channels. The free diameter values are based on 1.35 A for the oxygen radius, and both minimum and maximum values are given for noncircular apertures. The number of asterisks following these figures indicates whether the channel system is one-, two-, or three-dimensional. Only those apertures have been taken into account which are more open than regular six-membered rings. In most cases, these smaller openings

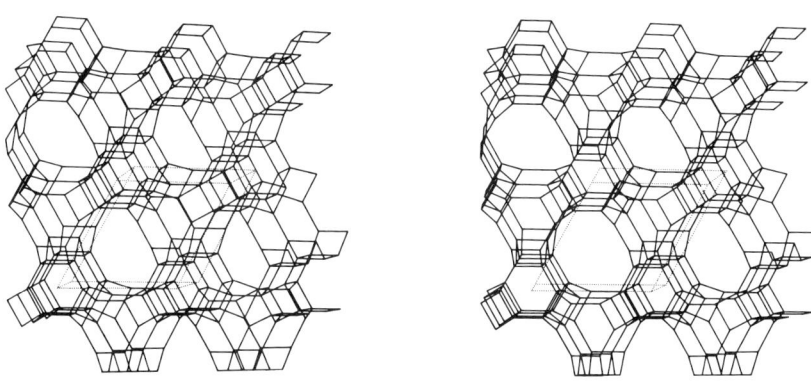

Figure 11. Offretite (8), (Na$_2$,Ca, etc.)$_2$Al$_4$Si$_{14}$O$_{36}$ · 14 H$_2$O, viewed along [001]

hexagonal, P̄6m2, a = 13.3, c = 7.6 Å
CH: [001] **12** 6.4* ⟷ ⊥ [001] **8** 3.6 × 5.2**
FP: (001)

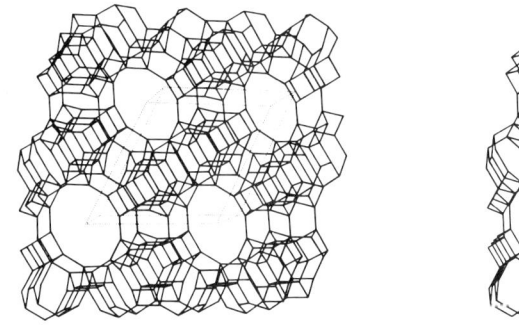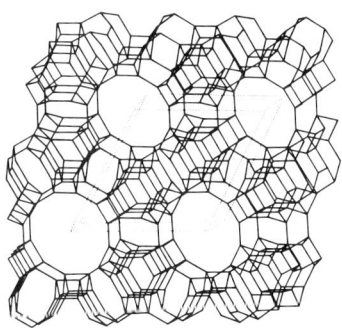

Figure 12. Linde L (5), K$_6$Na$_3$Al$_9$Si$_{27}$O$_{72}$ · 21 H$_2$O, viewed along [001]

hexagonal, P6/mmm, a = 18.4, c = 7.5 Å
CH: [001] **12** 7.1*
FP: (001)

form simple windows (rather than channels) connecting larger cavities. Interconnecting channel systems are separated by a double arrow (⟷). A vertical bar (|) means that there is no direct access from one channel system to the other.

Crystallographic free diameters depend on the state and the composition of the zeolite. The channel dimensions of isotoypic species can differ appreciably, particularly in the case of nonrigid frameworks.

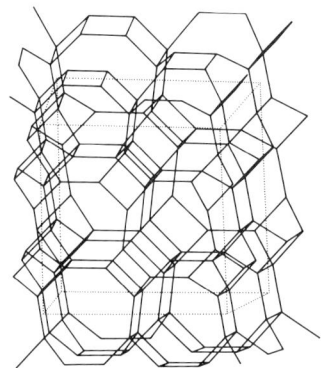 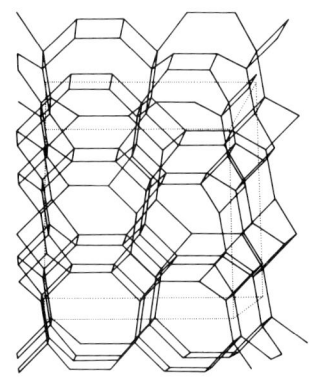

Figure 13. Phillipsite (53), $(K,Na)_{10}Al_{10}Si_{22}O_{64} \cdot 20\ H_2O$, viewed along [100]

orthorhombic, B2mb, a = 9.96, b = 14.25, c = 14.25 A
CH: [100] **8** 4.2 × 4.4* ⟷ [010] **8** 2.8 × 4.8* ⟷ [001] **8** 3.3*
FP: (010). Isotype: harmotome (47)

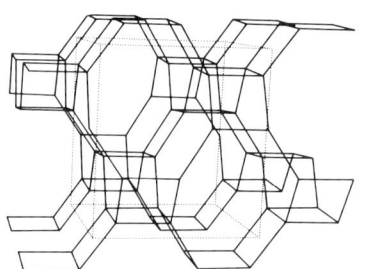 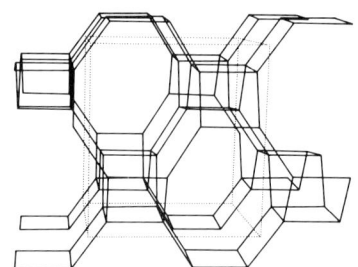

Figure 14. Gismondine (18), $Ca_4Al_8Si_8O_{32} \cdot 16\ H_2O$, viewed along [010]

monoclinic, P2$_1$/a, a = 9.84, b = 10.02, c = 10.62 A, γ = 92.4°
CH: {[100] **8** 3.1 × 4.4 ⟷ [010] **8** 2.8 × 4.9}***
FP: (101), (011)
Isotypes: TMA-gismondite (3), Barrer P1 or Linde B (4)

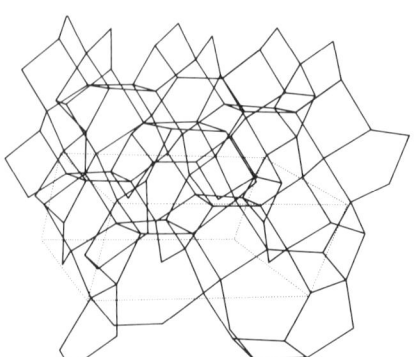 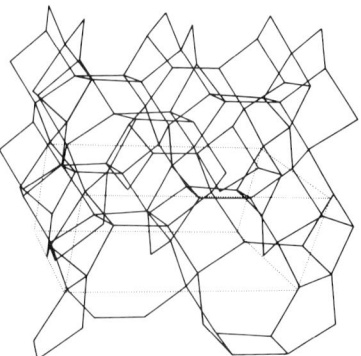

Figure 15. Yugawaralite (27, 28), $Ca_2Al_4Si_{12}O_{32} \cdot 8\ H_2O$, viewed along [001]

monoclinic, Pc, a = 6.73, b = 13.95, c = 10.03 A, β = 111.5°
CH: [100] **8** 3.1 × 3.5* ⟷ [001] **8** 3.2 × 3.3*

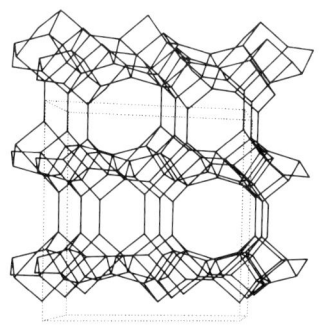 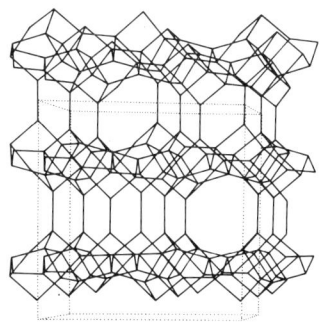

Figure 16. Heulandite (37, 38), $Ca_4Al_8Si_{28}O_{72} \cdot 24\ H_2O$, viewed along [001]

monoclinic, Cm, a = 17.73, b = 17.82, c = 7.43 A, $\beta = 116.3°$
CH: [100] **8** 4.0 × 5.5* ⟷ {[001] **10** 4.4 × 7.2* and **8** 4.1 × 4.7}*
FP: (010)

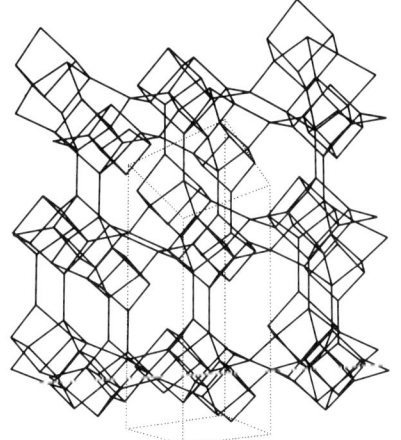 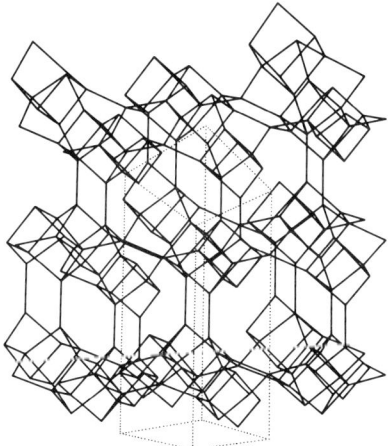

Figure 17. Stilbite (21), $Na_2Ca_4Al_{10}Si_{26}O_{72} \cdot 28\ H_2O$, viewed along [001]

monoclinic, C2/m, a = 13.64, b = 18.24, c = 11.27 A, $\beta = 128.0°$
CH: [100] **10** 4.1 × 6.2* ⟷ [001] **8** 2.7 × 5.7*
FP: (010)

Fault Planes

A number of zeolite frameworks have been described in terms of stacking sequences of certain building blocks such as six-membered single or double rings (cf. 35, 50). Structural relationships among members of

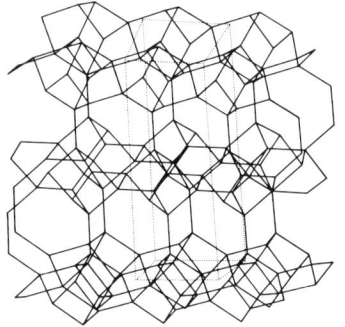
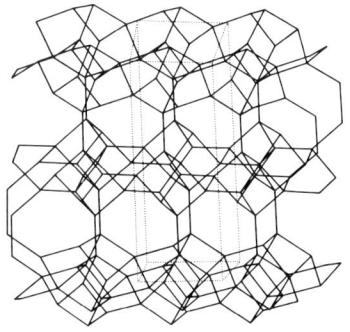

Figure 18. Brewsterite (45), $(Sr,Ba,Ca)_2Al_4Si_{12}O_{32} \cdot 10\ H_2O$, viewed along [001]

monoclinic, P2$_1$/m, a = 6.77, b = 17.51, c = 7.74 A, β = 94.3°
CH: [100] **8** 2.3 × 5.0* ⟷ [001] **8** 2.7 × 4.1*
FP: (010)

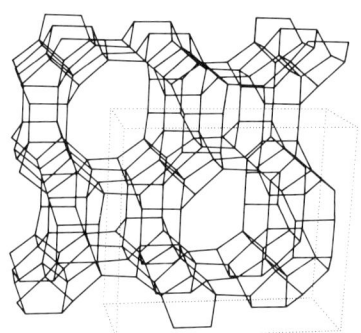
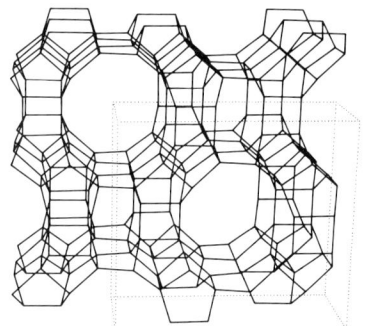

Figure 19. Mordenite (34), $Na_8Al_8Si_{40}O_{96} \cdot 24\ H_2O$, viewed along [001]

orthorhombic, Cmcm, a = 18.13, b = 20.49, c = 7.52 A
CH: [001] **12** 6.7 × 7.0* ⟷ [010] **8** 2.9 × 5.7*
FP: (010)

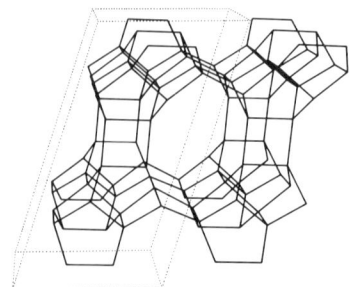
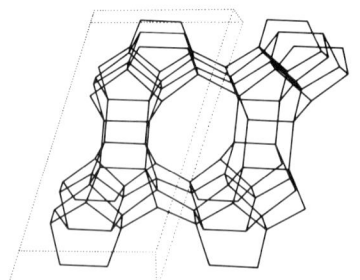

Figure 20. Dachiardite (23), $Na_5Al_5Si_{19}O_{48} \cdot 12\ H_2O$, viewed along [010]

monoclinic, C2/m, a = 18.73, b = 7.54, c = 10.30 A, β = 107.9°
CH: [010] **10** 3.7 × 6.7* ⟷ [001] **8** 3.6 × 4.8*
FP: (010)

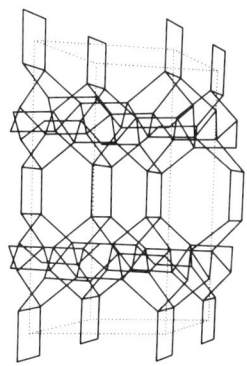 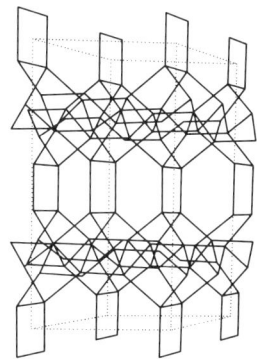

Figure 21. Epistilbite (44), $Ca_3Al_6Si_{18}O_{48} \cdot 16\ H_2O$, viewed along [001]

monoclinic, C2/m, a = 8.92, b = 17.73, c = 10.21 A, β = 124.3°
CH: [100] **10** 3.2 × 5.3* ⟷ [001] **8** 3.7 × 4.4*

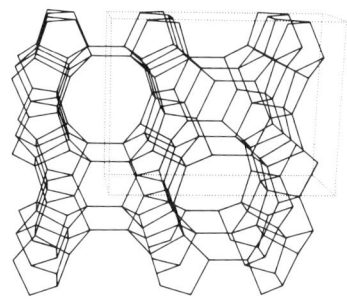 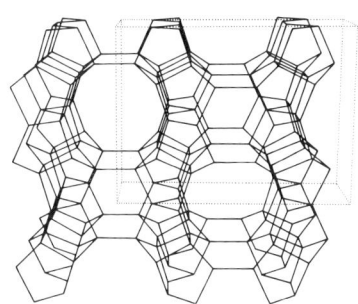

Figure 22. Ferrierite (59), $Na_2Mg_2Al_6Si_{30}O_{72} \cdot 18\ H_2O$, viewed along [001]

orthorhombic, Immm, a = 19.16, b = 14.13, c = 7.49 A
CH: [001] **10** 4.3 × 5.5* ⟷ [010] **8** 3.4 × 4.8*

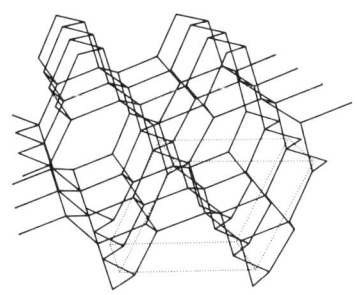 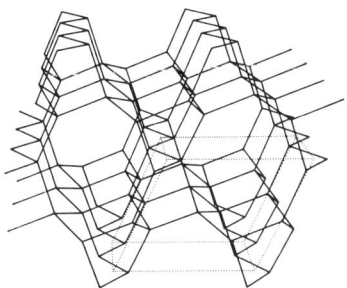

Figure 23. Bikitaite (1), $Li_2Al_2Si_4O_{12} \cdot 2\ H_2O$, viewed along [010]

monoclinic, P2₁, a = 8.61, b = 4.96, c = 7.61 A, β = 114.4°
CH: [010] **8** 3.2 × 4.9*

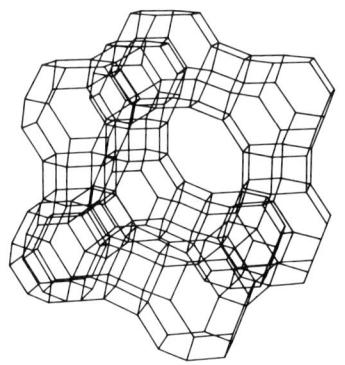 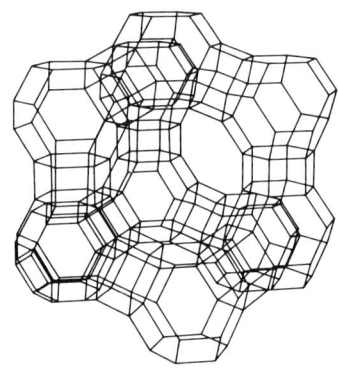

Figure 24. Faujasite (7, 9, 41), $(Na_2,Ca,Mg)_{29}Al_{58}Si_{134}O_{384} \cdot 240\ H_2O$, viewed along [111]

cubic, Fd3m, a = 24.7 A
CH: <111> **12** 7.4***
FP: (111). Isotypes: Linde X (11, 39), Linde Y (10)

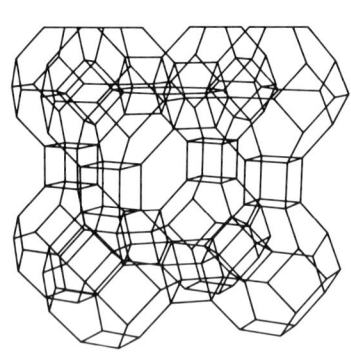 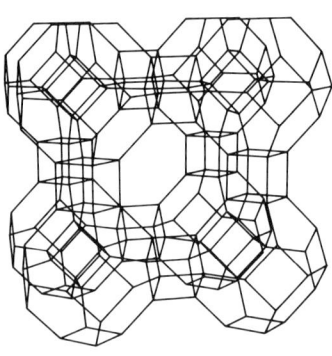

Figure 25. Linde A (11, 46), $Na_{12}Al_{12}Si_{12}O_{48} \cdot 27\ H_2O$, viewed along [100]

cubic, Pm3m, a = 12.32 A (pseudo-cell)
CH: <100> **8** 4.1***
Isotype: ZK-4 (26)

the chabazite group thus can be discussed using the concepts of polytypism (60). The chabazite and gmelinite frameworks, for instance, represent polytypes. Stacking faults, which are to be expected in the polytypic materials, frequently have been observed in these crystals, the fault plane being (001).

The phenomenon of polytypism is by no means limited to the chabazite group. Stacking faults also are likely to occur in most of the other

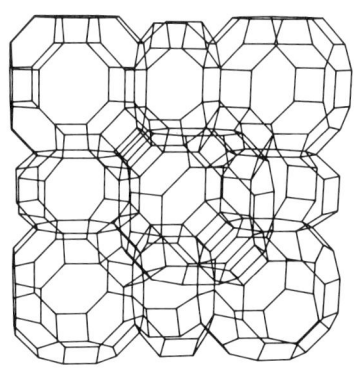 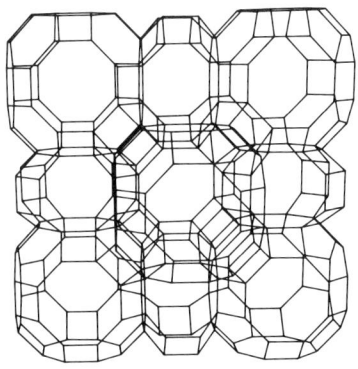

Figure 26. ZK-5 (36), $Na_{30}Al_{30}Si_{66}O_{192} \cdot 98\ H_2O$, viewed along [100]

cubic, Im3m, a = 18.7 Å
CH: <100> **8** 3.9*** | <100> **8** 3.9***

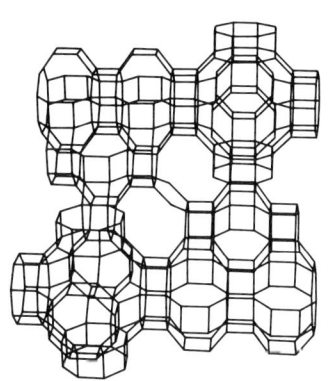 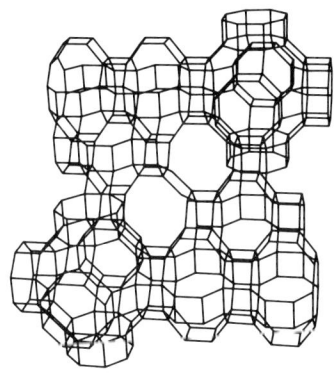

Figure 27. Paulingite (22), $(K_2,Ca,Na_2)_{76}Al_{152}Si_{520}O_{1344} \cdot \sim 700\ H_2O$, viewed along [100]

cubic, Im3m, a = 35.1 Å
CH: <100> **8** 3.9*** | <100> **8** 3.9***

zeolite frameworks. Possible fault planes (FP) have been listed in the figure captions in order to point this out. The existence of possible fault planes means that layer-like segments of the framework can be stacked in more than one way, giving rise to polytypism or "one-dimensional polymorphism." In all these cases, a number of similar structures (*i.e.*, polytypes) can be postulated quite readily.

Concluding Remarks

Isotypic species cannot always be recognized readily on the basis of powder patterns alone. Extensive structural investigations often are required in order to establish isotypism in cases of varying cell dimensions and symmetry. On the other hand, the x-ray patterns of polytypic species are frequently extremely similar and have led to the assignment of incorrect structures. A good example in this respect is provided by erionite and offretite (8). Because of this, careful reexamination of a number of discredited zeolitic species (such as stellerite, herschelite, etc.) would seem advisable.

Studies on the broader aspects of these framework structures are by no means exhausted. Further work is required on clinoptilolite, garronite, levynite, etc., to mention only a few natural zeolites.

Acknowledgment

The stereoplots shown in this paper have been generated with the aid of a computer program which was developed by Carroll K. Johnson (25). The work of one of the authors (W. M. M.) has been supported by the Schweiz. Nationalfonds.

Literature Cited

(1) Appleman, D. E., *Acta Cryst.* **1960**, 13, 1002, and personal communication.
(2) Baerlocher, Ch., Meier, W. M., *Helv. Chim. Acta* **1969**, 52, 1853.
(3) *Ibid.*, **1970**, in press.
(4) Baerlocher, Ch., Meier, W. M., in preparation.
(5) Barrer, R. M., Villiger, H., *Z. Krist.* **1969**, 128, 352.
(6) Bartl, H., Fischer, K. F., *Neues Jahrb. Mineral. Monatsch.* **1967**, 33.
(7) Baur, W. H., *Am. Mineralogist* **1964**, 49, 697.
(8) Bennett, J. M., Gard, J. A., *Nature* **1967**, 214, 1005, and personal communication.
(9) Bergerhoff, G., Baur, W. H., Nowacki, W., *Neues Jahrb. Mineral. Monatsch.* **1958**, 193.
(10) Breck, D. W., U. S. Patent **3,130,007** (1964).
(11) Broussard, L., Shoemaker, D. P., *J. Am. Chem. Soc.* **1960**, 82, 1041.
(12) Brown, G. E., Gibbs, G. V., *Am. Mineralogist* **1969**, 54, 101.
(13) Coombs, D. S., *Am. Mineralogist* **1952**, 37, 812.
(14) Coombs, D. S., *Mineral. Mag.* **1955**, 30, 699.
(15) Danoe, M., *Acta Cryst.* **1966**, 20, 812.
(16) Dent, L. S., Smith, J. V., *Nature* **1958**, 181, 1794.
(17) Erd, R. C., White, D. E., Fahey, J. J., Lee, D. E., *Am. Mineralogist* **1964**, 49, 831.
(18) Fischer, K., *Am. Mineralogist* **1963**, 48, 664, and personal communication.
(19) Fischer, K., *Neues Jahrb. Mineral. Monatsch.* **1966**, 1.
(20) Fischer, K., Meier, W. M., *Fortschr. Mineral.* **1965**, 42, 50.
(21) Galli, E., Gottardi, G., *Mineral. Petrog. Acta* **1966**, 12, 1.

(22) Gordon, E. K., Samson, S., Kamb, W. B., *Science* **1966**, 154, 1004.
(23) Gottardi, G., Meier, W. M., *Z. Krist.* **1963**, 119, 53.
(24) Jarchow, O., *Z. Krist.* **1965**, 122, 407.
(25) Johnson, C. K., "ORTEP: A Fortran Thermal-Ellipsoid Plot Program for Crystal Structure Illustrations," Oak Ridge National Laboratory, 1965.
(26) Kerr, G. T., Kokotailo, G. T., *J. Am. Chem. Soc.* **1961**, 83, 4675.
(27) Kerr, I. S., Williams, D. J., *Z. Krist.* **1967**, 125, 1.
(28) Kerr, I. S., Williams, D. J., *Acta Cryst.* **1969**, B25, 1183.
(29) Knowles, C. R., Rinaldi, F. F., Smith, J. V., *Indian Mineralogist* **1965**, 6, 127.
(30) Loens, J., Schulz, H., *Acta Cryst.* **1967**, 23, 434.
(31) McConnell, D., *Am. Mineralogist* **1952**, 37, 609.
(32) McConnell, D., *Mineral. Mag.* **1964**, 33, 799.
(33) Meier, W. M., *Z. Krist.* **1960**, 113, 430.
(34) *Ibid.*, **1961**, 115, 439.
(35) Meier, W. M., *S.C.I. Monograph Mol. Sieves* **1968**, 10.
(36) Meier, W. M., Kokotailo, G. T., *Z. Krist.* **1965**, 121, 211.
(37) Merkle, A. B., Slaughter, M., *Am. Mineralogist* **1967**, 52, 273.
(38) *Ibid.*, **1968**, 53, 1120.
(39) Milton, R. M., U. S. Patent 2,882,244 (1959).
(40) Náray-Szabó, St. V., *Z. Krist.* **1938**, 99, 277.
(41) Nowacki, W., Bergerhoff, G., *Schweiz. Mineral. Petrog. Mitt.* **1956**, 36, 621.
(42) Pauling, L., *Proc. Natl. Acad. Sci.* **1930**, 16, 453.
(43) Pauling, L., *Z. Krist.* **1930**, 74, 213.
(44) Perrotta, A. J., *Mineral. Mag.* **1967**, 36, 480.
(45) Perrotta, A. J., Smith, J. V., *Acta Cryst.* **1964**, 17, 857.
(46) Reed, T. B., Breck, D. W., *J. Am. Chem. Soc.* **1956**, 78, 5972.
(47) Sadanaga, R., Marumo, F., Takeuchi, Y., *Acta Cryst.* **1961**, 14, 1153.
(48) Sakavov, I. E., Shishakov, N. A., *Izv. Akad. Nauk SSSR, Ser. Khim.* **1963**, 1745.
(49) Shishakova, T. N., Dubinin, M. M., *Izv. Akad. Nauk SSSR, Ser. Khim.* **1965**, 1303.
(50) Smith, J. V., *Mineral. Soc. Am. Spec. Paper* **1963**, 1, 281.
(51) Smith, J. V., Rinaldi, R., Glasser, L. S. Dent, *Acta Cryst.* **1963**, 16, 45.
(52) Staples, L. W., Gard, J. A., *Mineral. Mag.* **1959**, 32, 261.
(53) Steinfink, H., *Acta Cryst.* **1962**, 15, 644.
(54) Strunz, H., "Mineralogische Tabellen," 5th ed., 1969.
(55) Taylor, W. H., *Z. Krist.* **1930**, 74, 1.
(56) Taylor, W. H., Jackson, R., *Z. Krist.* **1933**, 86, 53.
(57) Taylor, W. H., Meek, C. A., Jackson, W. W., *Z. Krist.* **1933**, 84, 373.
(58) Torrie, B. H., Brown, I. D., Petch, H. E., *Can. J. Phys.* **1964**, 42, 229.
(59) Vaughan, P. A., *Acta Cryst.* **1966**, 21, 983.
(60) Verma, A. R., Krishna, P., "Polymorphism and Polytypism in Crystals," Wiley, New York, 1966.
(61) Villiger, H., Ph.D. Thesis, University of London, 1969.
(62) Wyart, J., *Bull. Soc. Franc. Mineral. Crist.* **1940**, 63, 5.

RECEIVED February 16, 1970.

Discussion

R. M. Barrer (Imperial College, London): With the placing of Na-Pl as an isotype of gismondite, there appears to be only one structure having cubic secondary building units (8 linked tetrahedra), Linde A. Is this the case?

In the 1940's (*J. Chem. Soc.* **1948,** 127), I synthesized two phases, then termed P and Q, from analcite and $BaCl_2$ or $BaBr_2$. We have recently re-examined these and find that they have the ZK-5 framework. It thus appears that this framework was about the first of the purely synthetic zeolites to have been made.

D. H. Olson: Yes, zeolite A is the only known zeolite containing double four-membered rings as secondary building units.

15

Faujasite-Type Structures: Aluminosilicate Framework: Positions of Cations and Molecules: Nomenclature

J. V. SMITH

Department of the Geophysical Sciences, University of Chicago, Chicago, Ill. 60637

> *The framework distorts in response to cations and molecules. Type X zeolite has strong Al,Si long-range order, but the order–disorder in faujasite and Type Y zeolite is equivocal. Positions of framework hydroxyls in heated NH_4-exchanged faujasite were inferred from interatomic distances and infrared data. Exchangeable cations in strictly dehydrated specimens occupy sites offering minimum electrostatic energy; the center of the hexagonal prism is preferred. For incomplete dehydration (typical for most commercial catalytic processes), cations bond to residual molecules in the sodalite units. The location of exchangeable cations and water molecules in hydrated specimens is uncertain. Hydration complexes of cations occur in the supercage. In the Al-rich varieties, cations certainly enter the sodalite unit. In the Al-poor varieties, x-ray diffraction evidence on cation positions is equivocal.*

This paper reviews data available by January 1970 on the crystal structures of materials containing an aluminosilicate framework with the topology of the mineral faujasite. The nomenclature will be discussed in more detail at the end. Briefly, names will be assigned which specify the treatment applied to the 3 basic starting materials—*viz.*, faujasite, Linde Y, and Linde X zeolites. The latter 2 are synthetic materials prepared in hydrous sodium systems, the former being richer and the latter poorer in Si. The Si,Al content of Y overlaps that of faujasite.

Although x-ray diffraction techniques can yield data on atomic positions and occupation frequency which appear highly precise, these data are subject to personal interpretation. X-ray diffraction methods actually yield maps of electron density. Assignment of electron density to particular atoms involves chemical assumptions. No one will dispute assignment of atoms to electron density peaks in quartz, but in complex zeolites assignment of peaks can be most uncertain, especially when more than 1 kind of cation is present and when small molecules might occupy the same part of space as cations.

Another characteristic feature of x-ray diffraction is that each diffracted wave automatically sums the vector amplitude from all unit cells of each coherent crystal unit. Hence, random occupancy of 1 site by Al and Si atoms results in just 1 electron density concentration. Thermal vibration and random positional displacements of atoms from 1 unit cell to another yield a similar blurring-out of the electron density; furthermore, it is difficult experimentally to distinguish between reduction of height of an electron density peak resulting from the above 2 factors and that from a lower occupancy factor.

X-ray powder data yield much more uncertain results on atomic positions than data from single crystals. Large synthetic crystals have not been available until recently, and single-crystal data have been taken mostly on treated faujasite.

Although a zeolite is a single chemical system with mutual relationships between all parts, it is convenient to organize this paper in the sequence aluminosilicate framework, cations, and molecules.

Aluminosilicate Framework

The oxygen atoms lie at the corners of near-regular tetrahedra whose centers are occupied by either Al or Si atoms. Each corner is shared by 2 tetrahedra, and the linkage of the resulting framework is specified by placing tetrahedral centers at the corners of truncated octahedra whose centers lie at the positions of carbon atoms in the diamond structure. The truncated octahedra lie in such positions that they are joined by hexagonal prisms. Models built from cardboard polyhedra or from wire tetrahedra linked by spaghetti tubing are particularly easy to assemble. The construction key is to build a truncated octahedron first, assemble 4 hexagonal prisms onto any 4 hexagonal faces which are nonadjacent, and attach to the opposite face of each hexagonal prism a truncated octahedron such that its hexagonal faces are staggered with respect to those of the first truncated octahedron. Repeated application of this simple rule automatically generates the infinite framework.

Such abstract models of wire or cardboard are rather misleading when atomic interactions are considered, and a model assembled from balls—*e.g.*, cork or polystyrene—is very valuable, though not easy to make. Breck (*17*) figured several models, and Olson and coworkers gave valuable stereoscopic drawings in their papers. W. M. Meier and D. H. Olson (*44*) show stereoscopic drawings of the entire zeolite group.

The space inside a truncated octahedron is commonly called the sodalite cage. Each large interstice (commonly called a supercage) is linked to 4 other supercages through near-circular 12-membered rings and to 4 sodalite units through 6-rings. Each supercage shares 4-rings with 6 other sodalite units.

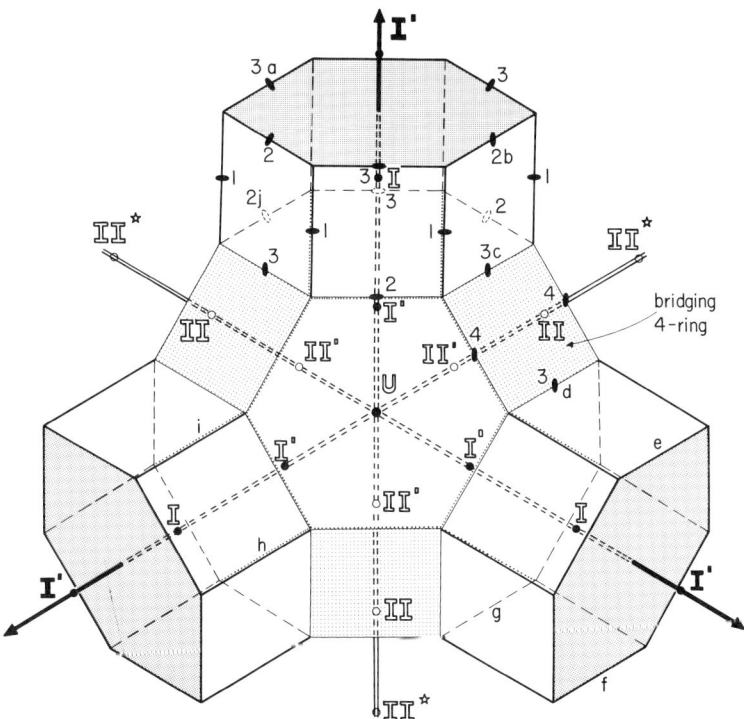

Figure 1. Idealized projection of sodalite unit with atom nomenclature

The intersections of the polyhedral edges show the positions of T atoms. Oxygen atoms are shown at the mid-points of the edges, but should be displaced to correspond to a tetrahedral environment for each T atom. Some of the positions for the 4 types of oxygens are shown. The center of the truncated octahedron is marked by U. Four axes of inverse 3-fold symmetry pass through this point: 3 are visible, and 1 is hidden because it lies perpendicular to the plane of the diagram. Four hexagonal prisms share a hexagonal face with the truncated octahedron; 1 is hidden at the back side. The cation sites are: I at the center of an hexagonal prism, I' displaced from a shared hexagonal face into the sodalite cage, II' displaced from an unshared hexagonal face into the sodalite cage, II slightly displaced into the supercage, and II* displaced considerably into the supercage. For abcdefghij, see Figure 2

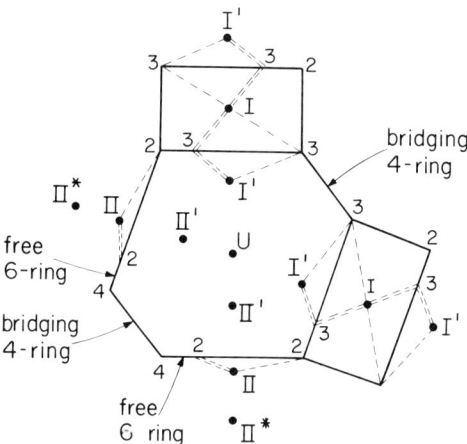

Figure 2. Section through abcdefghij in Figure 1 showing the relative positions of cation sites and framework oxygens. The broken lines show the principal cation-oxygen bonds in a dehydrated faujasite containing divalent cations.

Figures 1 and 2 show the generally accepted nomenclature for the atom sites. Readers should note that earlier papers use a variety of nomenclatures which must be checked very carefully. Figure 1 is a projection of a sodalite unit down a triad axis viewed from infinity. The silicon and aluminum atoms (collectively called T atoms) lie at the 4-fold intersections of the framework. Figure 2 is a section through Figure 1, as explained in the figure legend.

Oxygen atoms lie near the mid-points of the edges, but are displaced to attain the tetrahedral configuration around the T atoms. In addition, the atom positions in actual structures differ in detail from this idealized topologic pattern. The oxygen atoms form an interesting polyhedral arrangement in which tetrahedra are combined with truncated hexagonal prisms and polyhedra formed by truncating the truncated octahedra of T atoms. The interesting mathematical relations between the networks formed from T atoms and those formed from O atoms should be explored.

Each truncated octahedron defining the T atoms is composed of 4 hexagonal rings, each shared with a hexagonal prism (using O2 and O3 oxygen atoms), 4 free hexagonal rings shared with a supercage (using O2 and O4) and 6 bridging 4-rings (using O3 and O4). Each hexagonal prism contains 2 6-rings composed of O2 and O3 and 6 4-rings composed of O1, O2, and O3.

Site U is at the center of the truncated octahedron and lies at the intersection of 4 axes of inverse 3-fold rotation symmetry. Site I, at the center of each hexagonal prism, is a potential site for exchangeable cations. Site II projects slightly outwards from the free 6-ring into the supercage. Sites I' and II', respectively, project into the sodalite unit from the shared and free 6-rings. Figure 2 shows the distances between the positions more clearly than Figure 1. Site II* projects further into the supercage away from the free 6-ring than would be expected for a cation bonded to oxygen atoms of the 6-ring. Site V is near the center of the 12-membered ring between the supercages, and is not shown in the figures.

The early structure determinations of the framework topology (2, 14, 15, 16, 19) of faujasite have been confirmed by about 30 further determinations. Interest has shifted to 4 principal features of the aluminosilicate framework: shape, Si,Al order–disorder, attachment of protons to yield hydroxyl, and removal of hydroxyl, aluminum, or both from the framework.

Shape. The changes of shape of the tetrahedra are significant (Table I). The greatest distortion occurs in dehydrated forms, and the distortions can be explained quite satisfactorily by simple bonding theory. Thus, the more strongly an oxygen is bonded to the exchangeable cations, the more distant it tends to lie from the T atoms. Olson and Dempsey (48) listed quantitative data. Hydrated specimens have more uniform T–O distances and O–T–O angles than dehydrated ones, again readily

Table I. T–O Distances

Type	Reference	T–$O1^a$	T–$O2^a$	T–$O3^a$	T–$O4^a$
		Hydrated Forms			
Faujasite	(7)	1.643(3)	1.645(3)	1.657(3)	1.642(3)
Ca-faujasite	(10)	1.642(2)	1.640(2)	1 655(3)	1.644(3)
La-faujasite	(12)	1.643(3)	1.632(4)	1.647(4)	1.648(4)
Ce-faujasite	(49)	1.64 (1)	1.64 (1)	1.65 (1)	1.65 (1)
Na-X	(47)	1.627(7)	1.622(7)	1.616(7)	1.612(7)
		1.738(7)	1.719(7)	1.737(8)	1.723(7)
		Dehydrated Forms			
Ca-faujasite	(8)	1.633(5)	1.651(5)	1.671(5)	1.620(5)
Ni-faujasite	(46)	1.633(1)	1.641(1)	1.695(1)	1.613(1)
La-faujasite, 420°C	(11)	1.630(4)	1.624(3)	1.683(4)	1.599(3)
La-faujasite	(9)	1.632(3)	1.625(2)	1.689(3)	1.602(2)
Ce-faujasite	(49)	1.63 (1)	1.64 (1)	1.66 (1)	1.61 (1)
Ba-faujasite[b]	(52)	1.626(3)	1.638(3)	1.654(3)	1.637(4)
K-faujasite[b]	(52)	1.630(3)	1.649(2)	1.653(3)	1.647(3)
H-faujasite	(48)	1.653(2)	1.634(1)	1.662(2)	1.623(2)
Ca-X	(48)	1.653(7)	1.673(6)	1.678(8)	1.650(6)
Sr-X	(48)	1.655(6)	1.682(4)	1.680(5)	1.654(4)

[a] Standard errors in brackets to same significance level.
[b] Minor changes may occur during further refinement.

explainable by the more even distribution of the bonding in the former—
see, for example, quantitative data by Bennett and Smith (*10, 12*).

The $T–O–T$ angles fall mainly in the range 130°–155°C with the hydrated forms yielding on average a smaller spread. This range is typical of silicate frameworks (*40*).

Detailed explanation of the $T–O–T$ angles which link the tetrahedra is not so easy, though there is no reason to doubt that simple ionic bonding can yield a plausible model. Megaw, Kempster, and Radoslovich (*42*) interpreted the geometry of anorthite in terms of an engineering model with struts and springs representing rigid Si–O bonds and cation–cation repulsions. However, a more sophisticated treatment should consider covalent bonding [*see* Cruickshank for possible effect of π-bonding (*21*)].

Unfortunately, there is no reference structure free of cations, though perhaps a faujasite-type framework containing only SiO_2 ultimately will be synthesized. The dehydrated H-faujasite produced by heating ammonium-exchanged faujasite (*48*) is the closest to a cation-free faujasite, but even it has protons condensed with framework oxygens to form hydroxyls.

In faujasite, cations of types I', II', and II are permitted by the symmetry to adjust their distances from framework oxygens by sliding along the 3-fold axis. The Type I cations are fixed by symmetry, and the 6 O3 oxygens must adjust position in order to achieve a suitable bond length. Consequently, the primary factor in distortion of the framework should be the Type I cations.

Table II lists the 4 intertetrahedral angles for selected specimens. Table III shows the distances from site I to the O3 and O2 oxygens and the diameters of the 2 4-membered rings and the 2 6-membered rings. There is no obvious pattern to the angles listed in Table II. Faujasite and other hydrated forms (not shown) have angles which range over only a few degrees, but this small range also occurs in dehydrated Ce-faujasite. Hydrated Na–X has a wide range of angles, even when values averaged to symmetry $Fd3m$ are used.

Table III shows significant trends. In dehydrated H-faujasite, the 4-rings are approximately square. In dehydrated Ni-faujasite, occupation of I by the small Ni ion causes the I–O3 distance to fall from 2.68 in dehydrated H-faujasite to 2.29A. The I–O2 distance does not change significantly, but the bridging 4-ring (Figure 1) distorts tremendously with O3–O3 rising from 3.58 to 4.27, and O4–O4 falling from 3.52 to 3.03A. Reciprocally, occupation of I by the large cations K or Ba causes the I–O3 and O4–O4 distances to increase, and the O3–O3 distance to fall. In fact, there is an excellent correlation between the 3 distances for

Table II. Intertetrahedral Angles in Selected Specimens

Type	Reference	T–O1–T[a]	T–O2–T[a]	T–O3–T[a]	T–O4–T[a]
Faujasite	(7)	141	145	141	140
Dehydrated H-faujasite	(48)	139	147	140	145
Dehydrated Ni-faujasite	(46)	130	138	125	158
Dehydrated Ba-faujasite	(52)	137	156	147	145
Dehydrated K-faujasite	(52)	139	152	146	142
Dehydrated Ce-faujasite	(49)	142	148	143	144
Ce-faujasite	(49)	138	147	141	139
Na-X	(47)	132	141	136	147

[a] Angles rounded off to nearest degree. Standard deviations vary from 0.2° to 1.0°.

all structures. Qualitatively, the key effect is that cations induce distortions in the hexagonal prisms which twist the bridging 4-rings.

The distortion of the 4-rings of the hexagonal prisms is not so pronounced, and cannot be correlated simply to the I cations. The diameters of the 6-rings also vary only a small amount. Occupancy of I by small di- and tri-valent cations seems to increase the diameter of the free 6-ring at the expense of the diameter of the prism 6-ring. This is reasonable since the I cations pull the O3 oxygens towards the triad axis, whereas the O2 oxygens do not move an equal amount away from the triad.

Careful study of the above correlations may permit estimation of atomic positions to an accuracy of 0.1A or better.

Si,Al Order–Disorder. Extensive data on aluminosilicate framework structures with Si/Al ⩾ 1 have shown that not more than 1 Al atom is bonded to an oxygen atom, except in rare cases. Anorthite probably has 2 Al atoms attached to some oxygen atoms when crystallized dry at high temperature (38). Gehlenite has 2 Al atoms attached to oxygen atoms, but it is a sheet structure stable only at high temperature (37). Nepheline, with a framework structure, appears to have 2 Al atoms sharing an oxygen but the crystal structure should be re-examined (30). Since zeolites form at low temperature, the avoidance rule is probably applicable to them, except perhaps for very few atoms.

In faujasite-type structures, only the extreme Type X with Si/Al = 1 would be constrained mathematically by this avoidance rule to have complete Si,Al order. All faujasite-type materials with Si/Al > 1 mathematically may have either long-range or short-range disorder. However, from the viewpoint of thermodynamics, there should be a driving force at low temperature towards an ordered pattern with lower internal

Table III. Dimensions of the

	Reference
Ca-faujasite	(10)
La-faujasite	(12)
Dehydrated H-faujasite	(48)
Dehydrated K-faujasite	(52)
Dehydrated Ni-faujasite	(46)
Dehydrated Ca-faujasite	(8)
Dehydrated La-faujasite	(9)
Dehydrated Ba-faujasite	(52)
Na-X[c]	(47)

[a] Diagonals of bridging 4-ring.
[b] Diagonals of prism 4-ring.
[c] Diagonal of free 6-ring.

energy. At high temperature, the configurational entropy term may be more significant than the internal energy factor, thereby favoring disorder. Since synthetic zeolites are grown under conditions of high supersaturation, disorder is favored. These considerations have been treated exhaustively in papers on natural minerals, especially feldspars [see, for example, Ref. 6].

Olson (47) demonstrated from 3D single-crystal x-ray analysis that hydrated Na–X has symmetry $Fd3$ instead of $Fd3m$ for the ideal faujasite framework. The mean T–O distances of the 2 independent tetrahedra are 1.619(4) and 1.729(4)A. The latest compilation of T–O distances suggests reference values of 1.605 and 1.757A for Si–O and Al–O, respectively (55). Hence, the Olson data suggest occupancies of 0.91Si, 0.09Al and 0.18Si, 0.82Al in the tetrahedra. The cell content of 88 Al and 104 Si atoms prohibits complete order for $Fd3$ but the Olson data indicate very strong long-range order. Probably there is some residual disorder, since the first tetrahedron contains 0.09Al. Olson found that other cation forms of X also have symmetry $Fd3$, confirming the data from the Na form.

The data for the more silica-rich members are controversial. Smith (61) argued that electron density distributions reveal the long-range order, whereas cell dimensions depend on the short-range order. All x-ray determinations of electron density for faujasite and Y automatically indicate complete Si,Al disorder since there is only 1 T site when $Fd3m$ is used.

Several authors have discussed various aspects of order–disorder in these zeolites. The most complete discussion is by Dempsey (23) and Dempsey, Kühl, and Olson (25), who worked out mathematically the various ways of placing Si and Al atoms when $Si/Al \geq 1$. Unless there are restrictions on the cell edge and the symmetry, an infinite number of ordered structures is possible. Dempsey et al. restricted their treat-

Aluminosilicate Framework

SI-02	SI-03	04-04[a]	03-03[a]	01-01[b]	02-03[b]	02-04[c]	02-03[d]
3.52	2.76	3.73	3.41	3.70	3.75	5.15	5.14
3.51	2.78	3.79	3.42	3.69	3.72	5.15	5.12
3.52	2.68	3.52	3.58	3.61	3.64	5.20	5.10
3.45	2.86	3.97	3.36	3.73	3.63	5.14	5.20
3.51	2.29	3.03	4.27	3.54	3.43	5.30	4.88
3.46	2.49	3.27	4.10	3.83	3.37	5.19	5.01
3.34	2.54	3.41	4.02	3.86	3.25	5.27	4.96
3.36	2.84	3.98	3.42	3.79	3.50	5.19	5.14
3.62	2.62	3.54	3.66	3.68	3.71	5.25	5.15

[d] Diagonal of prism 6-ring.
[e] Averaged to pseudosymmetry $Fd3m$.

ment by considering identical units having zero net charge and dipole moment. For faujasite, the smallest unit is a double sodalite unit with 24 pairs of T atoms obeying the center of symmetry at site I of the shared hexagonal prism. The 4-fold multiplicity of the F-lattice permits ordering at the following contents of Al atoms per unit cell: 96, 88, 80, 72, 64, 56, 48, etc. For 12 Al atoms per sodalite unit, there is strict alternation of Si and Al atoms. The symmetry is $Fd3$ and all 6-rings have 3 Al alternating in the 1:3:5 positions. For 11 Al atoms, any 1 Si can be exchanged. Further exchange might produce 2 rings with Al at 1:3 positions. Dempsey et al. state that a lower electrostatic energy is obtained for 10 or fewer Al atoms if the Si and Al atoms redistribute, producing some rings with 2 Al atoms at 1:4 positions. Further discussion is too detailed to abstract here.

Similar considerations on Si,Al ordering were made by Niggli (44) for plagioclase feldspars. Whatever the details of ordering, it seems likely that as the Si,Al ratio changes in an ordered framework silicate, there will be readjustments of the topological distributions.

Dempsey, Kühl, and Olson (25) suggested that the cell dimensions supported the concept of Si,Al order in the faujasite-type zeolites. Figure 3 shows their data for cell dimensions of 21 specimens grown from hydrous sodium systems plotted against Al content derived from the bulk chemical compositions. Dempsey et al. inserted 4 straight lines with discontinuities close to 80 and 64 Al atoms per cell, and another one near 52 atoms per cell. The former 2 discontinuities were interpreted in terms of rearrangement of T atoms in accord with the theoretical analysis just described, while the latter was ascribed to possible formation of amorphous silica in the more siliceous bulk compositions.

Also in Figure 3 are the following data:

(a) Dotted line representing a least-squares linear fit to 37 data points with an experimental error of $\pm$ 0.005A in a and ± 0.5 wt % in

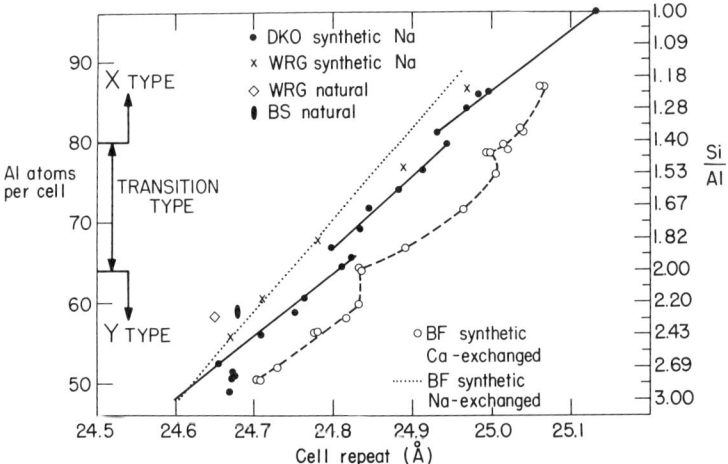

Figure 3. Cell dimensions of faujasite, Y, and X zeolites plotted against the number of Al atoms per unit cell. See text for explanation of symbols. The dotted line was obtained by Breck and Flanigen for zeolites grown from hydrous sodium systems.

SiO_2 and Al_2O_3. These data points obtained by Breck and Flanigen (*18*) were obtained for samples crystallized from hydrous sodium systems, as were those of Dempsey *et al.*

(b) Five data points obtained by Wright, Rupert, and Granquist (*69*), also for specimens obtained from hydrous sodium systems.

(c) Dashed line representing a subjective interpretation of 20 data points for dehydrated Ca-exchanged specimens (*18*). These might be interpreted in terms of discontinuities near 77 and 62 Al atoms per cell.

(d) A datum by Wright *et al.* (*69*) and one by Bennett and Smith (unpublished) for Na-exchanged faujasite from Sasbach. Both values are lower than the data for synthetic material of similar Al content. The Bennett-Smith specimen was checked by electron microprobe analysis for absence of all cations other than sodium.

The reviewer could not find definitive criteria to resolve these conflicting data. The likelihood of a systematic experimental bias between the techniques for cell-dimension measurement could only be tested by exchange of samples between the various workers. It is possible that there is an impurity in the zeolite causing the bulk chemical analysis to give a false estimate of the Al-content of the zeolite. Wright *et al.* suggested that amorphous silica occurred in their material, thereby indicating a falsely low Al-content of the zeolite. However, this should lead to a deviation opposite to that given in Figure 2 with respect to the data of Dempsey *et al.* Nevertheless, their implied suggestion of some impurity causing a difference between the bulk composition and that of the zeolite is a possible way out of the observed deviations. None of the authors give statistical criteria upon which the validity of the straight lines can be judged.

The cell dimensions of zeolites are determined by the total system, which includes cations and water molecules. Even if resolution of the differences in experimental data confirms the existence of discontinuities, the possibility of effects resulting from resiting of cations and molecules must be considered as well as Al,Si order–disorder. Certainly the X-zeolites studied by Olson (47) have long-range T order, but for faujasite and Y-zeolites the evidence from cell dimensions is equivocal. Ordered crystals and most natural minerals have undergone more annealing than synthetic specimens. The present data are consistent with the possibility that natural faujasite is more ordered than synthetic Y zeolite, but the evidence is inconclusive.

Obviously, a more direct method is needed to test the ordering of the silica-rich specimens. In feldspars, Laves and Hafner (39) successfully evaluated the Si,Al order–disorder using infrared absorption and nuclear quadrupole resonance techniques. The latter technique applied to Al and Na nuclei unfortunately requires single crystals at least several mm across. Infrared patterns of ordered feldspars showed multiple peaks for the T–O stretching and bending vibrations while disordered feldspars showed a single broad peak for each. Wright et al. (69) obtained patterns with single broad peaks for both faujasite and Types Y and X zeolites. The peaks changed position linearly with the Al content of the synthetic zeolites, but the natural specimen deviated significantly from the trend. At this time it is not clear what effect movement of cations and molecules (especially protons) has on the T–O vibrations, but such movement probably yields so many different crystal fields that the infrared peaks are broadened, thereby inhibiting separation of bands from Si and Al atoms.

Framework Hydroxyls. The structural nature of framework hydroxyls is particularly important because they probably act as Bronsted acid catalysts. The literature on catalysis is much too extensive to be reviewed here, and comments will be restricted to x-ray and infrared evidence on the location and nature of framework hydroxyls. Reviews of infrared data are given in this volume by Rabo and Poutsma and by Ward (see also 66, 67).

The clearest data are for H-forms of faujasite-type zeolites. The ammonium-exchanged form is heated, driving off ammonia and water, and thereby producing the H-form. It would be inconceivable chemically for the protons not to condense with framework oxygens. Olson and Dempsey (48) determined the crystal structure of H-faujasite held in the dehydrated form at room temperature. Unfortunately, there were no electron density peaks ascribable to the protons, probably because they do not have the same position in all unit cells. Since Si–OH bond lengths are about 0.08A longer than Si–O, the observed T–O bond lengths (Table I) of 1.653, 1.634, 1.663, and 1.623A indicated that most of the protons were attached to O3, and a smaller number to O1.

Olson and Dempsey predicted the position of the protons using electrostatic energy considerations. They postulated (their Figure 1) that in each of the shared 6-rings, a proton (denoted H2) is attached to only 1 of the 3 O3 atoms, and that the OH bond is directed towards the triad axis. A center of symmetry relates the protons of the shared 6-rings on opposite sides of the hexagonal prism. The other kind of proton (H1) is attached to an O1 atom and the OH bond is directed away from the triad axis. Olson and Dempsey related these positions to a preferred Si,Al ordering scheme mentioned in the preceding section.

The model is consistent with various infrared data. Many workers (*see* references in Ref. 48) found that H-Type Y yields infrared stretching frequencies at about 3750, 3650, and 3550 cm^{-1}. The highest frequency band has been found for all types of zeolites, and has been ascribed to hydroxyls completing the surface of individual crystallites or, perhaps more likely, to $Si(OH)_4$ occluded in the zeolite (1). The 3550 band is not perturbed by sorption of most molecules which cannot pass into the sodalite unit, and hence reasonably was ascribed to H2 protected inside the hexagonal prism. The 3650 band, which is perturbed by large sorbed molecules, was ascribed to H1. Other considerations supported this assignment (48, pp. 230–231).

Although the basic model looks very good, the actual situation must be more complex since faujasite and Y zeolites cannot have full long-range order. The actual positions of the H atoms must vary somewhat from 1 unit cell to another, and such variations may lead to particularly active catalytic sites in just a few of the unit cells. Hopefully, single-crystal infrared absorption studies, similar to those carried out for many silicates, will yield data on the orientation of the OH groups, thereby testing the Olson-Dempsey model.

Olson and Dempsey (48) briefly explained the disappearance of the lower frequency band upon absorption of large basic molecules like piperidine by mobility of the protons on the zeolite surface. Ward (66) presented some evidence for mobility of protons at high temperature.

X-ray data are much too insensitive to test for the location of framework OH groups in zeolites containing exchangeable cations. Nevertheless, the presence of similar OH stretching bands in many cation-bearing zeolites suggests that the protons are attached to the same oxygens. Another word of caution is desirable. The sequence of frequencies 3750, 3650, and 3550 corresponds qualitatively to the expected values for the crystal fields of environments of one-sided outer surface, one-sided large cage with diameter 11A, and small enclosed volume. Perhaps any hydroxyl projecting into the supercage will have a frequency near 3650 irrespective of which oxygen it is attached to, and similarly for the 3550 frequency and the sodalite unit.

Detailed interpretation of hydroxyl location in cation-bearing systems is given in Refs. 26, 48, 67, 68. The paper by Uytterhoeven et al. (65) reviews many of the presently available data, and lists many proposals concerning OH groups. Discussion of OH groups coordinated to cations is given later.

Oxygen Removal from the Framework. Progressive heating of ammonium-exchanged Type Y first gives dehydrated H-Type Y as just described, and then by removal of H_2O yields "decationated" material (32, 41, 63, 64, 65). McDaniel and Maher (41) produced an "ultrastable" form thermally stable to temperatures over 1000°C. Recent work by Kerr and coworkers (31, 32, 34, 35, 36) has shown that the reactions are complex, and depend critically on the method of de-ammoniation and dehydration.

The simplest explanation of dehydration of H-Y is that 2 hydroxyls from the framework condense to form water and an oxide ion. The oxide ion poses no problem since it can form part of the aluminosilicate framework, but the oxygen lost in the water leaves a vacancy in the framework. The vacancy can be accounted for if readjustment takes place yielding 3-coordinated cations in a framework which is still continuous. Oxygen attached to either 2 Al or 1 Al + 1 Si should be lost more easily than one attached to 2 Si. Hence, the simple model predicts that 3-coordinated Al and Si atoms are formed, since oxygens attached originally to 2 Al should be rare.

Structurally, there is considerable difficulty in devising an atomic pattern since removal of an oxygen leaves 2 highly charged cations facing each other without a shielding oxygen. An extreme alternative is that the framework recrystallizes into SiO_2 with elimination of the Al_2O_3 into an amorphous or poorly-crystalline phase.

$$2\ HAlSi_2O_6 \rightarrow H_2O + Al_2Si_4O_{11} \rightarrow Al_2O_3 + 4\ SiO_2$$

The final product could be a silica framework whose cell dimension should be approximately 24.0A when rehydrated (estimated from Si–O $\sim$ 1.605, Al–O $\sim$ 1.757A, starting material $\sim$ 59 Al per unit cell, $a \sim 24.7$A).

However, the actual reactions are not consistent with this simple scheme. Kerr (32) found that heating H-Y in a dry atmosphere led to a material with poor thermal stability, presumably because of many crystalline defects. Heating at 700°–800°C for 2–4 hours in an inert static atmosphere with water remaining near the zeolite produced a substance of unusually high thermal stability. About 25% of the Al is present as exchangeable cations, suggesting that about 12 Al per unit cell have left the framework. Kerr proposed a 2-step mechanism in which water reacts with the framework producing $(OH)_4$ groups (as occur in hydrogarnet) and $Al(OH)_3$. The latter reacts with the framework, removing a hydrogen and producing water and $Al(OH)_2^+$. The $(OH)_4$ groups break down, yielding 2 H_2O and 2 oxide ions which remain in the framework.

Kerr (33) removed Al from the framework of Na-Y by extraction with chelating agent H_4EDTA. He (34) found that the cell dimension of H-Y prepared by calcination of NH_4-exchanged Y was 24.74A but dropped to 24.51A after extraction by EDTA, reducing the Al content from 50 to 32.8 atoms per cell. The extracted material increased in stability both for thermal degradation at high temperature and hydrothermal degradation at low temperature. Kerr (31) found that thermal decomposition of NH_4-Y at 760 torr and 500°C yielded H-zeolite if ammonia was removed rapidly, and ultrastable zeolite if ammonia was removed slowly. Kerr and Shipman (36) described the reaction of H-Y with ammonia at high temperature producing "amido-zeolite Y."

The detailed structural mechanisms concerning the loss of ammonia and water from NH_4-exchanged zeolite have not been determined by x-ray diffraction work. It seems likely from a general body of knowledge on natural and synthetic zeolites that protons act as a catalyst in promoting a recrystallization of the framework.

J. M. Bennett and J. V. Smith (unpublished) made x-ray structural analyses of powdered samples of decationated Y ("ultrastable" type) kindly supplied by C. V. McDaniel. Although refinement was quite satisfactory from the technical viewpoint, the low resolution posed problems of interpretation. Details of the powder data obtained directly at temperatures of 25°, 400°, 700°, and 900°C may be obtained from the authors. Site I' was occupied at all temperatures, possibly resulting from some species containing aluminum. The data for the hydrated form at 25°C indicated occupancy of O1 and O4 by 69 ± 8 and 79 ± 9 oxygens (instead of ideal 96) when these sites were assigned the same B-values as O2 and O3. At 700° and 900°C, the occupancies were normal. These data might be explainable in terms of recrystallization, but must be tested by more accurate single-crystal data.

Kerr et al. (35) suggested that the term decationated not be used. They distinguished between: ammonium zeolite Y; hydrogen zeolite Y produced by controlled deammination and dehydration; dehydroxylated form produced by heating above 450°C at 10^{-6} torr or above 600°C at 760 torr, loss of 1 water per pair of Na^+-free AlO_2 giving negative 4-coordinated Al and positive 3-coordinated Si; ultrastable zeolite for material of McDaniel and Maher (41), with some framework Al turned into cationic Al.

Table IV. Cation Occupancy and Position

Type	Reference	I	I'	II'	II
		Strongly Dehydrated Forms			
Na-Y	(29)	7.8 Na	20.2 Na 0.069		31.2 Na 0.223
K-Y	(29)	12.0 K	14.6 K 0.075		31.0 K 0.250
Ag-Y	(29)	16 Ag	11.0 Ag 0.075		29.2 Ag 0.228
K-faujasite	(52)	8.6 K	12.9 K 0.072		31.7 K 0.250
Ba-faujasite	(52)	7.3 Ba	5.0 Ba 0.068		11.3 Ba 0.247
Ca-faujasite	(8)	14.2 Ca	2.6 Ca 0.061		11.4 Ca 0.229
Ni-faujasite	(46)	10.6 Ni	3.2 Ni 0.054 5.8 H_2O 0.081	1.9 Ni 0.207 1.9 H_2O 0.161	6.4 Ni 0.233
La-faujasite	(9)	11.8 La	2.5 La 0.067		1.5 La 0.234
La-faujasite (420°C)	(11)	11.7 La	2.5 La 0.066		1.4 La 0.233
		Partly Dehydrated Forms			
Na, Ce-faujasite	(49)	3.4 Na	11.5 Ce 0.067	16 H_2O 0.162	10.7 Na 0.236
La-Y 725°C	(62)	5.2 La	8.9 La 0.068		5.5 La 0.227
La-X	(49)		30 La 0.067	32 H_2O 0.165	
Ca-X	(48)	7.5 Ca	17.3 Ca	9.0 Ca 10.5 H_2O	17.3 Ca
Sr-X 16 hrs. 400°C	(48)	11.2 Sr	7.0 Sr	4.2 Sr 5.4 H_2O	19.5 Sr
Sr-X 16 hrs. 680°C	(48)	6.1 Sr	12.0 Sr	6.4 Sr 7.7 H_2O	20.3 Sr
La-X 425°C	(13)	5.0 La	15.2 La 0.065		4.9 La 0.230
La-X 735°C	(13)	5.2 La	14.1 La 0.065		6.3 La 0.231

Exchangeable Cations

The available data on the positions of exchangeable cations are very difficult to interpret since several specimens were not fully exchanged and most dehydrated specimens still contained significant amounts of water or other oxygen species. Room-temperature ion exchange often is incomplete. High-temperature exchange often is needed to obtain a mono-cationic form. The exchange process may result in entrance of protons instead of the chosen cation. Not all single crystals used for x-ray work have been checked by electron microprobe analysis or other techniques for the cation content. For convenience, the results will be discussed under 3 subheadings: dehydrated specimens, hydrated specimens, and samples containing sorbed molecules. Tables IV and V contain the indicated number and type of cations in the various sites together with the positional coordinate (x) specifying the location of type (xxx) on the 3-fold axis; see Refs. 20, 28 for additional data not listed in the tables.

Dehydrated Specimens. Ideally, the water molecules in a zeolite exist as discrete molecules and can be removed easily by heating. In

Table V. Site Occupancy

Type	Reference	I	I'
Faujasite	(7)		16 (Na, Ca) 0.070
Ca-faujasite	(10)		9.7 Ca 0.069
La-faujasite	(12)		3.3 La 0.069
Ce, Na, Ca-faujasite	(49)		18 Na 0.070
Na-X	(47)	9 Na	8 Na 0.060 11.H_2O 0.074
Sr_{12} Na-X	(51)	2.1 Sr	11.1 Sr 0.063
Sr_{30} Na_{24}-X	(51)	12 Na	7.3 Sr 0.065
La-X	(49)		12 La 0.067
La-X	(49)	5 La?	13.8 La 0.062

practice, the water molecules form hydration complexes with the exchangeable cations and interact electrostatically with framework oxygens. For monovalent cations, the hydration complexes are weakly bonded, and dehydration is relatively easy. For polyvalent cations, the hydration complexes are bonded more strongly. In addition, the cation may polarize the water molecule sufficiently that it splits into a hydroxyl group (which bonds to the cations) and a proton which attaches itself to a framework oxygen. Under conditions of severe hydration at high temperature, it is possible for oxide ions to be produced from the hydroxyls attached to the cations (53).

Table IV is split arbitrarily into sections covering strongly dehydrated and partly dehydrated forms. Dehydration of monovalent forms should be most complete. The data for K-Y (powder technique) and K-faujasite (single-crystal technique) are consistent within their respective errors showing occupancy of the I, I', and II sites. These sites are occupied also in Na- and Ag-Y (powder data) but with different occupancy levels. For all these forms, there should be about 58 cations per cell to balance the

in Hydrated Forms

II'	II	II*	V
32 H_2O 0.167		11 H_2O 0.272	
11.5 Ca 0.167		23 H_2O 0.274	2.2 Ca 0.484
28 H_2O 0.167		14 H_2O 0.272	10.3 La 0.491
32 H_2O 0.165		26 H_2O 0.264	5.8 Ce 0.191
26 H_2O 0.166	12 Na 0.230 12 Na 0.238 8 H_2O 0.245		
32 H_2O 0.173	15.0 Sr 0.246		
26 H_2O 0.170	11.5 Sr 0.246		
32 H_2O 0.165	17 La 0.235		4 La 0.500
24 H_2O 0.166	13.2 La 0.233		3.4 La 0.493

Al atoms, and the total occupancies estimated from the x-ray analyses (54, 57, 59, 60) agree fairly well.

Most thinking on cation distribution in silicates is based on Pauling's rules which are based on an electrostatic model. In applying them to zeolites, it should be noted that covalent bonding may occur, the Al-distribution in the framework may be important in causing exchangeable cations to associate with particular 6-rings containing high amounts of Al, and Pauling's rules were developed for crystals near chemical equilibrium, whereas dehydrated zeolites have been brought deliberately into a state unstable with respect to water. There are many comments in the literature on which the above remarks are based. The papers of Dempsey (23, 24) are particularly valuable since they provide detailed calculations of Madelung potentials.

Site I is the only site permitting a cation to be enclosed by framework oxygens. Sites I', II', and II permit only 1-sided or a waisted coordination to oxygens of the adjacent 6-ring. Hence, I is the preferred site unless other considerations such as electrostatic repulsion or cation–molecule attraction become important.

For monovalent cations, there are space problems. Site I can hold 16 cations while the other sites can hold 32 each. Cations simultaneously occupying I and I' are sharing the face of a coordination polyhedron, which is unfavored electrostatically. The sharing problem is particularly severe for sites II and II'. Electrostatically, the best way to distribute 58 monovalent cations in faujasite and Y is to put 32 in II, 6 in I, and 20 in I'. This distribution permits the I and I' cations to be distributed such that no I and I' cations share a 6-ring.

The observed occupancies of II agree well with the ideal value of 32, but the occupancies of the I and I' sites need detailed discussion. The single-crystal data for K-faujasite fit the prediction well, but the less accurate powder data for Na-, K-, and Ag-Y do not agree. The deviation for Na-Y is certainly within experimental error, and that for K-Y is probably within the error. Certainly, the good fit for the K-faujasite suggests that the K-Y data may be in error. The deviation for the Ag-Y form is rather high. Is it possible that it is meaningful and results from a stronger tendency for covalent bonding of Ag in site I than for the alkali metal forms? An accurate single-crystal analysis is desirable.

Currently there are no published data on the location of monovalent cations in dehydrated X zeolite. There are considerable problems in placing 86 cations per unit cell. Breck (17) placed 16 in I, 32 in II, and 38 in site III. In site III, the cation projects into the supercage and is bonded to the 4 oxygens of bridging 4-rings. Perhaps there is simultaneous occupancy of I and I', as suggested by the data for K- and Ag-Y.

In Y and faujasite, there should be about 29 divalent cations which could be entirely accommodated in sites I and II. The crystal of dehydrated Ca-exchanged faujasite (8) was shown free of other cations by microprobe analysis and was severely dehydrated at 475°C before being sealed in its capillary. The cation distribution is close to the theoretical suggestion, but there are 2.6 Ca atoms in site I'. Bennett and Smith pointed out that the 14.2 Ca in I and 2.6 Ca in I' are consistent with no sharing of a polyhedral face, since 14.2 + 2.6/2 yields 15.5, which is less than 16. Dempsey and Olson (27) suggested that presence of water molecules draws cations from I and II into I' such that n (I) + 0.5 n (I') = 16. There are insufficient data to test rigorously the detailed accuracy of this equation.

The total of Ba atoms in Ba-faujasite (23.6) is considerably lower than the expected value. Perhaps ion-exchange was incomplete, or perhaps protons were incorporated.

The data for Ni-faujasite are complicated by incomplete ion-exchange (27 Ni and 4 Ca indicated by chemical analysis) and possible presence of residual water resulting from dehydration at 400°C. Olson (46) observed double peaks at I' and II' ascribed to residual water molecules adjacent to the Ni ions. The situation is unclear but Olson suggested bonding between Ni ions and water molecules. The apparent deficiency of Ni ions (estimated total, 22.1) may result from neglect of other ions including Ca, or entrance of protons.

The data obtained by Bennett and Smith for La-faujasite dehydrated at 475°C show that the trivalent La ions prefer site I. A few ions prefer I' and II but these may be bonded to residual molecules. Perhaps high Al contents of a few 6-rings also may be a factor in the siting of these La atoms.

The details of the cation–oxygen bonding are obscured by the disordered cation distribution since a cation in site I may cause oxygen atoms in O3 to lie at different positions than when the cation is at site I'. Since x-rays average all atoms of a given type, the apparent position of O3 cannot be related directly to that of the cations. Taking this factor into account, the observed cation–oxygen distances in these dehydrated zeolites seem reasonable in relation to those of other silicates. In Table IV, the effect of cation radius is shown indirectly by the x-coordinates (*e.g.*, note the large x-coordinates for sites II of the Ba- and K-faujasites).

A surprising result of the study of dehydrated La-faujasite (9) was the large displacement of the La atom at S(I) along the triad axis. Thermal vibration seemed unlikely as an explanation, especially as data obtained directly at 420°C showed the same effect (11). Bennett and Smith suggested that the La atom might be bonded preferentially to only 3 of the O3 atoms, indicating covalent bonding. Such 1-sided bonding has been observed in other lanthanum–oxygen compounds (9).

Turning now to the incompletely-dehydrated varieties, the data are consistent with strong bonding between polyvalent cations and residual water molecules (45). Particular attention was paid to trivalent cations because of their importance in zeolite catalysts.

Olson, Kokotailo, and Charnell (49, 50) and Smith, Bennett, and Flanigen (62) independently found that trivalent ions were preferentially entering I' instead of the expected I. The latter authors heated $Na_{13}La_{16}$-Y in a powder diffractometer with the furnace containing flowing dry helium. At 725°C, 9 La atoms occupied I'. Olson, Kokotailo, and Charnell (49, 50) examined a crystal of $Na_8(Ca,Mg)_8Ce_{12}$-faujasite at room temperature after calcination at 350°C. The x-ray data indicated that all the Ce atoms occupied I', the other cations lying in I and II. Electron density in II' was ascribed to 16 water molecules, and the Ce atoms in I' were nicely interpreted as bonding to 3 O3 atoms of the adjacent 6-ring and to 1 or 2 water molecules in II'. The distances of 2.4–2.5A are quite consistent with this model. Smith et al. (62) suggested that residual water might occupy site U at the center of the sodalite unit (again 2.5A from site I') but their evidence from powder data is of marginal accuracy.

Following the above studies, Bennett and Smith (9, 11) showed that stricter dehydration caused La atoms in La-exchanged faujasite to move into I. Furthermore, the same dehydrated crystal of La-faujasite had essentially equal occupancy factors at 25° and 420°C, thereby ruling out any control of cation distribution from a pure temperature variation. Although not strictly proven from x-ray data, correlation of the above results with those obtained by many authors—see particularly Rabo et al. (53)—from infrared methods shows beyond reasonable doubt that the positions of cations depend strongly on even small quantities of residual molecules. It is possible that 1 water molecule is sufficient to bridge between 2 or more La atoms in I'. Hence, for 19 La atoms per unit cell of fully-exchanged Y-zeolite, only about 10 water molecules are needed, compared with 260 in the hydrated specimen.

Since most feed material for catalytic processes using zeolites contains a few parts per million of water, it is reasonable to expect that polyvalent ions in the zeolites will be occupying positions in the sodalite unit rather than in site I, and that the cations will be bonded to residual water or other oxygen species. Such bonding will reduce strongly the electrostatic field generated by the cation.

So far, water molecules have been assumed as the oxygen species. The x-ray data are not able to discriminate water molecules from hydroxyl or from oxide. Several authors interpreted infrared data in terms of metal–hydroxyl bonding as summarized in Ref. 64.

Table IV shows the cation distributions estimated for several forms of X dehydrated under various conditions. For trivalent cations (~30 per unit cell), sites I and II are most suitable electrostatically, but the sample of La-X calcined at unspecified temperature (probably 350°C) showed all the La atoms on I', while site II' was occupied by electron density explainable by 32 water molecules. This remarkable distribution yields a very stable chemical complex with each La bonded to 3 O3 and 3 H_2O at 2.5A and with each H_2O bonded to either 2 or 3 La atoms. The H_2O also is bonded weakly to 3 O2 atoms of a free 6-ring.

X-ray powder study of La-X in dry flowing helium at 425° and 735°C showed about 15 La atoms in I', with the others equally split between sites I and II. The simplest explanation is that partial dehydration has occurred, causing some La atoms to move into sites I and II.

The data for Ca- and Sr-varieties of X (Table IV) are interpretable in terms of partial dehydration.

Hydrated Species. Table V summarizes the assignments of cations and water molecules made by various authors for a variety of ion-exchanged forms. The data are highly unsatisfactory since no one has been able to find enough electron density peaks to account for all the ions and molecules or to find really reliable criteria to permit assignment to a particular species of every peak.

The first 4 structures of Table V show a remarkable identity of electron density peaks in the sodalite unit of polyvalent cation forms of faujasite. The authors have assigned the electron density to various atoms or water molecules, but if the data are transformed back into the number of electrons, they are consistent with 32 H_2O in II' and about 20–25 H_2O in I', irrespective of the type of exchangeable polyvalent cation. From the strict viewpoint of the crystal structure, the only obvious argument against such an assignment of water molecules is that the I'–O3 distances of 2.5A are rather short. However, it is possible that extremely strong electrostatic interactions might permit such short distances.

Sherry (56, 57) showed that the ion-exchange properties of Na-Y and Na-X zeolite were consistent with 16 Na atoms in the sodalite cage. Since Na scatters x-rays rather similarly to H_2O molecules, the presence of 16 Na atoms is consistent with the electron density measurements.

The problem arises with the atoms which scatter x-rays more strongly than water molecules. If La atoms account for the electron density peak at I' in La-faujasite, there are only 3.3 atoms in a 32-fold site. Such low occupancy is hard to believe for a crystal structure at low temperature in the presence of water molecules. Is it possible that in faujasite and Y, polyvalent cations prefer to be hydrated and occupy the supercage, while Na cations occupy both the sodalite unit and the supercage? Detailed x-ray studies of other cation forms than those of Table V would be valuable, especially in association with appropriate ion-exchange studies. In zeolite X, the data clearly indicated entrance of cations into the sodalite unit, but this is to be expected from the higher framework charge.

At least some cations in the supercage are surrounded entirely by water molecules. Olson, Kokotailo, and Charnell (49) located 6 Ce atoms of Ce-exchanged faujasite in site V displaced slightly from the center of a 12-membered ring. Bennett and Smith (12) located 10 La atoms at this site in a completely exchanged faujasite, and found density equivalent to 2 Ca atoms at the same site of a fully-exchanged Ca-faujasite (10). The 12-membered ring has a free diameter of about 7.5A, which is capable of accommodating a cation hydrated with about 6 water molecules. Probably the water molecules do not occupy the same position from 1 unit cell to the next, since no electron density peaks have been rigorously located and ascribed to them.

The site II* with $x \sim 0.27$ is occupied in all hydrated forms but its occupants are uncertain. Baur (7) assigned a water molecule to this site but the distances to the oxygens of the 6-ring—3.3A to 3 O2 and 3.5A to 3 O4—are rather large. Bennett and Smith (10) pointed out that the electron density peak is very broad, and suggested that a mixture of water molecules and hydrated cations occupies this site.

Detailed examination of the electron density in the supercage of Ca- and La-exchanged faujasites showed a nonzero electron density without any really significant peaks. Baur (7) correlated the absence of definite peaks with a variety of physical data, suggesting that the water molecules and cations act as a mobile electrolyte solution.

Olson's study of hydrated Na-X (47) has important implications for the positioning of ions and molecules. Although he was not able to find enough electron density to satisfy all the cations and water molecules, he found that 7 sites in the supercage were occupied at a low level, as well as the strongly-occupied sites in the sodalite unit listed in Table V. The X zeolite has strongly ordered Si and Al atoms such that the crystal field in each supercage will be fairly similar. In Y zeolite, the Si and Al atoms cannot have complete long-range order, and there must be a considerable variety of crystal fields. Consequently, it is possible that in the silica-rich varieties, the cations and water molecules of any single supercage tend to occupy specific sites but that the phase-amplitude integration of the x-ray beam gives the impression of a smeared-out fluid. Of course, there must be movement of ions and molecules to explain various physical data such as the high value of the self-diffusion coefficients, but a majority of the ions and molecules may be fixed for any chosen instant of time.

Olson (47) assigned 9 Na atoms to site I, 8 to I', and 24 to II. The first assignment should be correct because the interatomic distances are too small for a water molecule. The latter 2 assignments are consistent with the ion-exchange data of Sherry (56, 57) and with interatomic distances; however, the assignment is not completely unequivocal.

Olson made the important suggestion that in hydrated systems, cations only occupy site II when there are 3 Al atoms in the 6-ring to provide a favorable electrostatic environment. Since faujasite and Y zeolites probably do not have more than 2 Al atoms in almost all 6-rings, ions should only occupy site II in X zeolite. The available data are consistent with this suggestion.

Table V contains 2 sets of data for hydrated La-X zeolite, and data for X exchanged partly and then almost completed with Sr. The electron density of La is so high that at least 7 and perhaps as many as 14 La ions occupy I′. Site II′ surely is occupied by water molecules which bond to La ions in I′. Site II must be occupied strongly by La, and site V appears to be bonded weakly. The higher Al content of the 6-rings must favor occupancy of sites I′ and II, thereby depleting site V which was strongly occupied in faujasite and Y zeolites. The assignments of Na and Sr atoms given by Olson and Sherry seem reasonable.

Space precludes extensive discussion, but it is obvious that the structural basis of ion-exchange is very complex, involving an activation energy for formation of a hydration complex, and in some instances involving movement of cations through the 6-membered rings of the sodalite units. Interested readers are referred to papers by Olson and Sherry (51), Sherry (56, 57), and to 3 papers by Barrer and coworkers (3, 4, 5) which give data on various thermodynamic functions involved in ion exchange and cation hydration.

Sorption Complexes

Simpson and Steinfink (58, 59) determined the crystal structures of the 2 complexes m-dichlorobenzene–nickel faujasite and 1-chlorobutane–manganese faujasite. They were unable to locate the sorbed molecules from individual electron density peaks, but using liquid scattering functions (60), they were able to estimate the extent of occupancy of the supercage by the sorbed molecules. In both structures, the cations principally occupy I′ while electron density in II′ was ascribed to water molecules. This suggests that the crystals were only partially dehydrated. A few cations were ascribed to site II. In the Ni variety, some of the Ni ions were uniformly distributed through the sodalite unit for purposes of calculation. Simpson and Steinfink concluded that the sorbed organic molecules exist as a liquid in the supercages.

Nomenclature

Now that so many data have been accumulated on zeolites of the faujasite type, names must be used carefully to reduce confusion. It is highly desirable that the nomenclature be related to observable properties. There must be a loose broad name to encompass a group of materials possessing an important distinguishing quality, and there must be specific subnames to characterize subsidiary qualities.

In the general class of zeolites, the faujasite type is distinguished by the special topology of its aluminosilicate framework. To a mineralogist, the prototype faujasite is the mineral described by Damour (22) in 1842 from Sasbach, Kaiserstuhl, Germany. The single-crystal x-ray studies of Baur and coworkers (7) defined the topology of the aluminosilicate framework, but did not distinguish between the locations of Si and Al atoms, and did not fully define the positions of cations and molecules.

Synthetic zeolites produced by scientists at Union Carbide Corp. from sodium-bearing systems were shown by powder x-ray methods to have the same framework topology as that of the mineral faujasite (2, 19), and furthermore, the sorption characteristics were consistent with the geometry of the framework. Types Y and X zeolite were distinguished on the basis of various chemical and physical properties (18) such that the former ranges from 48 to 76 Al atoms per cell and the latter from 77 to 96. Most measurements have been made on Y with about 58 atoms per cell and X with about 88 Al atoms per cell. The range for Y encompasses the Al-content for natural faujasite which is near 59 atoms per cell, though some variation may occur in faujasite specimens.

So far, so good; the problems arise with the cations and sorbed molecules (especially water), both in amount and in structural position. Furthermore, the constituents of the framework cause difficulties—order–disorder of Al, Si atoms, attachment of protons to framework oxygens, and various kinds of defects.

To me, the most reliable nomenclature is based on actual physical and chemical operations. This is often clumsy, but should be replaced by a conceptual kind of nomenclature only when there is universal agreement that the concept is fully consistent with physical and chemical data. The topology of the faujasite framework is universally recognized, but the structural nature of chemical and physical variants is highly debatable.

There is no evidence to prove that the mineral faujasite has the same Si,Al arrangement as synthetic Y with the same Si/Al ratio. Furthermore, the mineral contains variable amounts of Na, Ca, Mg, and other cations. Hence, it should not be casually equated with synthetic Y grown from a pure Na-bearing system. Even when the mineral has been ion-exchanged to the fully Na-exchanged form it should not be equated with Y. Thus, I disagree with the nomenclature proposed by Wright, Rupert, and Granquist (*69, 70*), especially as their x-ray and infrared data indicate significant differences between the 2 types.

Dempsey *et al.* (*25*) restricted the use of the terms Y and X zeolite to the composition ranges 53–64 and 80–96 Al atoms per cell. The intermediate range was denoted the Transition type (Figure 3). The nomenclature proposed by Kerr *et al.* (*35*) for "decationated" zeolites was described earlier. Hopefully, further experimental data will reduce the uncertainty in the experimental data on the cell dimensions and the crystal structures of the zeolites, and provide a firmer basis for nomenclature.

At present, the most suitable nomenclature is one which specifies the treatment applied to a starting material—*e.g.*, dehydrated, Ca-exchanged Type Y zeolite. For brevity in this paper, terms such as La-Y have been used; such brevity should cause no confusion since La-exchanged Type Y zeolite is automatically implied. Readers are urged to read the earlier literature carefully since the term faujasite has sometimes been casually applied to Type Y zeolite.

Acknowledgment

R. M. Barrer, W. M. Meier, D. H. Olson, V. Schomaker, and H. Steinfink kindly provided material incorporated into the manuscript. Michael Bennett, Donald W. Breck, E. Dempsey, David Olson, Jules Rabo, and Hugo Steinfink kindly commented on the manuscript. I thank both the Petroleum Research Fund, administered by the American Chemical Society, and Union Carbide Corp. for grants-in-aid.

Literature Cited

(1) Angell, C. L., Schaffer, P. C., *J. Phys. Chem.* **1965**, 69, 3463.
(2) Barrer, R. M., Bultitude, F. W., Sutherland, J. W., *Trans. Faraday Soc.* **1957**, 53, 1111.
(3) Barrer, R. M., Davies, J. A., *J. Phys. Chem. Solids* **1969**, 30, 1921.
(4) Barrer, R. M., Davies, J. A., Rees, L. V. C., *J. Inorg. Nucl. Chem.* **1968**, 30, 3333.
(5) *Ibid.*, **1969**, 31, 2599.
(6) Barth, T. F. W., "Feldspars," Wiley, New York, 1969.
(7) Baur, W. H., *Am. Mineralogist* **1964**, 49, 697.
(8) Bennett, J. M., Smith, J. V., *Mater. Res. Bull.* **1968**, 3, 633.
(9) *Ibid.*, **1968**, 3, 865.
(10) *Ibid.*, **1968**, 3, 933.
(11) *Ibid.*, **1969**, 4, 7.
(12) *Ibid.*, **1969**, 4, 343.
(13) Bennett, J. M., Smith, J. V., Angell, C. L., *Mater. Res. Bull.* **1969**, 4, 77.
(14) Bergerhoff, G., Baur, W. H., Grutter, W. F., *Helv. Chim. Acta* **1956**, 39, 518.
(15) Bergerhoff, G., Baur, W. H., Nowacki, W., *Neues Jahrb. Mineral. Monatsh.* **1958**, 193.

(16) Bergerhoff, G., Koyama, H., Nowacki, W., *Experientia* **1956**, 12, 418.
(17) Breck, D. W., *J. Chem. Educ.* **1964**, 41, 678.
(18) Breck, D. W., Flanigen, E. M., "Molecular Sieves," p. 47, Society of the Chemical Industry (London), 1968.
(19) Breck, D. W., Flanigen, E. M., Milton, R. W., Reed, T. B., *Natl. Meeting, ACS, 134th, Chicago,* 1958.
(20) Broussard, L., Shoemaker, D. P., *J. Am. Chem. Soc.* **1960**, 82, 1041.
(21) Cruickshank, D. W. J., *J. Chem. Soc. (London)* **1961**, 5486.
(22) Damour, A., *Ann. Mines* **1842**, 1, 395.
(23) Dempsey, E., "Molecular Sieves," p. 293, Society of the Chemical Industry (London), 1968.
(24) Dempsey, E., *J. Phys. Chem.* **1969**, 73, 3660.
(25) Dempsey, E., Kuhl, G. H., Olson, D. H., *J. Phys. Chem.* **1969**, 73, 387.
(26) Dempsey, E., Olson, D. H., *J. Catalysis* **1969**, 15, 309.
(27) Dempsey, E., Olson, D. H., *J. Phys. Chem.* **1970**, 74, 305.
(28) Dodge, R. P., Union Carbide Corp., unpublished data.
(29) Eulenberger, G. R., Keil, J. G., Shoemaker, D. P., *J. Phys. Chem.* **1967**, 71, 1812.
(30) Hahn, T., Buerger, M. J., *Z. Krist.* **1955**, 106, 308.
(31) Kerr, G. T., *J. Catalysis* **1969**, 15, 200.
(32) Kerr, G. T., *J. Phys. Chem.* **1967**, 71, 4155.
(33) *Ibid.*, **1968**, 72, 2594.
(34) *Ibid.*, **1969**, 73, 2780.
(35) Kerr, G. T., Cattanach, J., Wu, E. L., *J. Catalysis* **1969**, 13, 114.
(36) Kerr, G. T., Shipman, G. F., *J. Phys. Chem.* **1968**, 72, 3071.
(37) Korczak, P., Raaz, F., *Ostterr. Akad. Wiss. Math. Naturw. Kl.* **1967**, 383.
(38) Laves, F., Goldsmith, J. R., *Z. Krist.* **1955**, 106, 227.
(39) Laves, F., Hafner, S., *Norsk. Geol. Tidsskr.* **1962**, 42, 57.
(40) Liebau, F., *Acta Cryst.* **1961**, 14, 1103.
(41) McDaniel, C. V., Maher, P. K., "Molecular Sieves," p. 186, Society of the Chemical Industry (London), 1968.
(42) Megaw, H. D., Kempster, C. J. E., Radoslovich, E. W., *Acta Cryst.* **1962**, 15, 1017.
(43) Meier, W. M., Olson, D. H., ADVAN. CHEM. SER. **1971**, 101, 155.
(44) Niggli, A., *Schweiz. Mineral. Petrog. Mitt.* **1967**, 47, 279.
(45) Olson, D. H., *J. Phys. Chem.* **1968**, 72, 1400.
(46) *Ibid.*, **1968**, 72, 4366.
(47) *Ibid.*, **1970**, in press.
(48) Olson, D. H., Dempsey, E., *J. Catalysis* **1969**, 13, 221.
(49) Olson, D. H., Kokotailo, G. T., Charnell, J. F., *J. Colloid Interface Sci.* **1968**, 28, 305.
(50) Olson, D. H., Kokotailo, G. T., Charnell, J. F., *Nature* **1967**, 215, 270.
(51) Olson, D. H., Sherry, H. S., *J. Phys. Chem.* **1968**, 72, 4095.
(52) Pluth, J., Schomaker, V., in preparation.
(53) Rabo, J. A., Angell, C. L., Schomaker, V., *Intern. Congr. Catalysis, 4th, Moscow, 1968,* Preprint 54.
(54) Rabo, J. A., Pickert, P. E., Boyle, J. E., U. S. Patent **3,130,006** (1964).
(55) Ribbe, P. H., Gibbs, G. V., *Am. Mineralogist* **1969**, 54, 85.
(56) Sherry, H. S., *J. Phys. Chem.* **1966**, 70, 1158.
(57) *Ibid.*, **1968**, 72, 4086.
(58) Simpson, H. D., Steinfink, H., *J. Am. Chem. Soc.* **1969**, 91, 6225.
(59) *Ibid.*, **1969**, 91, 6229.
(60) Simpson, H. D., Steinfink, H., *Acta Cryst.* **1970**, in press.
(61) Smith, J. V., *Lithos* **1970**, 3, 145.
(62) Smith, J. V., Bennett, J. M., Flanigen, E. M., *Nature* **1967**, 215, 241.
(63) Stamires, D. N., Turkevich, J., *J. Am. Chem. Soc.* **1964**, 86, 749.

(64) Uytterhoeven, J. B., Christner, L. G., Hall, W. K., *J. Phys. Chem.* **1965,** 69, 2117.
(65) Uytterhoeven, J. B., Schoonheydt, R., Liengme, B. V., Hall, W. K., *J. Catalysis* **1969,** 13, 425.
(66) Ward, J. W., *J. Phys. Chem.* **1969,** 73, 2086.
(67) Ward, J. W., Hansford, R. C., *J. Catalysis* **1969,** 13, 364.
(68) *Ibid.,* **1969,** 15, 311.
(69) Wright, A. C., Rupert, J. P., Granquist, W. T., *Am. Mineralogist* **1968,** 53, 1293.
(70) *Ibid.,* **1969,** 54, 1484.

RECEIVED February 13, 1970.

Discussion

D. H. Olson (Mobil Research & Development Corp., Princeton, N. J.): In view of the fact that the BF data for dehydrated Ca-exchanged faujasites show breaks agreeing with the position of the breaks shown by DKO for hydrated sodium faujasites, I think the suggestion that resiting of cations is responsible for the breaks can be ruled out.

J. V. Smith: The breaks in the DKO data of Figure 3 may result from a resiting of cations as a function of Si/Al ratio.

J. Paul Rupert (Carnegie-Mellon University, Pittsburgh, Pa. 15213): Our differing opinions with respect to designation may stem from differing points of view between mineralogists and physical chemists. We have suggested that when workers describe measurements made on faujasite structures they designate the material as "synthetic faujasite," specifying the number and type of exchange cations. Thus, "Zeolite Y" would be "synthetic faujasite, Na_{56}," clearly specifying the topology and composition. The physical property in question may be a function of both of these. Even though a synthetic faujasite may differ from the naturally-occurring material with respect to Si,Al ordering in the crystal, there is no assurance that different degrees of ordering do not occur in synthetic materials. Obviously, any nomenclature scheme must be a compromise between descriptiveness and convenience.

Alan C. Wright (Baroid Div., National Lead, Houston, Tex. 77001): With reference to your Figure 3, probably the apparent discrepancy between the DKO and WRG data sets can be ascribed to experimental technique. It is my understanding that the DKO x-ray measurements utilized only absolute 2-theta values since a relative scale was all that was sought. Our measurements, however, involved the use of an internal standard. Within error limits, the DKO and WRG data may be superimposed simply by shifting the Å scale. Obviously, our limited data are insufficient to confirm or deny the break points found by DKO.

Harry Robson (Esso Research Laboratory, Baton Rouge, La.): Since all the cation possibilities which we can locate are on the diagonal of the unit cell, I believe we could simplify the nomenclature by designating these sites by their position along this diagonal (numbers 00 to 50). This designation would be meaningful to crystallographers without reference to prior conventions.

N. G. Parsonage (Imperial College, London): With reference to Table V, I am surprised that some sites are good for both cations and water molecules (*e.g.*, Site I). After all, the cations should seek out positions of minimum coulombic potential energy (and consequently zero electric field). If, for example, we assumed that H_2O molecules were primarily held by field-induced dipole forces, they would seek positions of maximum field. They would not, then, go to the same sites as the cations.

J. V. Smith: I think one must be cautious of interpretation in terms of just one parameter such as coulombic potential energy. If coulombic energy were the sole term, Site I would always be occupied, whereas we know for sure that in some varieties of faujasite-type zeolites it is unoccupied. We must try to combine a variety of concepts in order to predict the positions of cations and molecules. I see no reason why one type of site cannot be occupied by cations as well as molecules. For example, if there are ten cations for a 16-fold site, the other six sites might be occupied by water molecules.

J. Turkevich (Princeton University, Princeton, N. J. 08540): In connection with Figure 3, I would like to know how the silicon–aluminum ratio is determined. The synthetic zeolites and maybe natural ones are obtained from silica-rich gels which increase in silica–alumina ratio in the mother liquid as the silicon–aluminum ratio increases in the crystals. Silica may be occluded and adsorbed as a colloidal gel on the zeolite and difficult to wash off. The silica ratio as measured will be higher and the lines asymetrically lower.

J. V. Smith: In my paper, I pointed out that there were difficulties in the interpretation of a bulk chemical analysis of a zeolite. Turkevich's remark supports my suggestion that the claim of DKO for breaks in the relation between cell dimension and Al content of synthetic faujasite-type zeolite must be treated cautiously.

E. M. Flanigen (Union Carbide Corp., Tarrytown, N. Y. 10591): In response to the question presented by Turkevich asking how the SiO_2/Al_2O_3 ratio was determined as presented by Smith in Figure 3.

Kühl responded that the data of DKO were obtained by wet chemical analysis. The data of Breck and Flanigen were also determined by chemical analysis. I would like to point out with respect to the BF curves in Figure 3 that the zeolite samples used were very carefully selected with respect to purity and homogeneity, as determined by x-ray analysis and oxygen adsorption characterization. Only preparations containing greater than 95% zeolite as determined from oxygen capacity were used to determine the curves. It was hoped that this procedure would minimize errors in bulk chemical analysis resulting from extraneous material.

G. H. Kühl (Mobil Research & Development Corp., Paulsboro, N. J. 08066): Referring to your Figure 3, statistical considerations show that there are three lines, although the center line is probably parallel to the

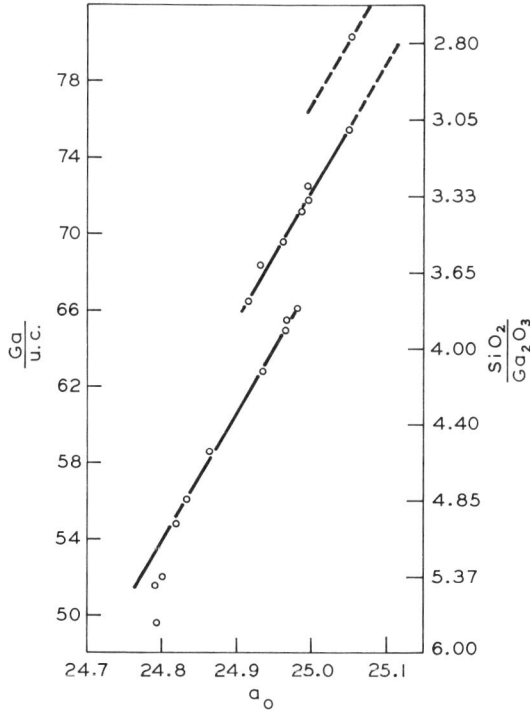

other two. The steps are expected to be more distinct in gallosilicate faujasite because of the larger gallium. Our measurements on available gallosilicate faujasite samples gave the expected results, proving that the breaks are real.

J. V. Smith: Statistical considerations are only as good as the assumptions. If one separates the data into three populations separated by the supposed discontinuities, the statistical reliability will appear much better than if one treats the data as a single population and searches for discontinuities. Undoubtedly, DKO use the first approach; *i.e.*, they are checking an assumed model rather than looking for breaks at any Al content. The data for the gallosilicate faujasite analogs certainly show much better evidence for breaks than do the aluminosilicate analogs. Hopefully, the data will be published shortly.

16

Infrared Structural Studies of Zeolite Frameworks

EDITH M. FLANIGEN and HASSAN KHATAMI

Union Carbide Corp., Linde Division Laboratory, Tarrytown Technical Center, Tarrytown, N. Y. 10591

HERMAN A. SZYMANSKI

Alliance College, Cambridge Springs, Pa. 16403

Mid-infrared spectroscopy has been applied to zeolite structural problems. The infrared spectrum in the region of 200 to 1300 cm^{-1} is a sensitive tool indicating structural features of zeolite frameworks. Preliminary interpretation suggests infrared specificity for zeolite structure type and group, and for structural subunits such as double rings and large pore openings. It is proposed that the major structural groups present in zeolites can be detected from their infrared pattern. This hypothesis is based on correlation of newly determined infrared spectra of synthetic zeolites with x-ray structure data for most of the known structural classes of zeolites. Other structural information obtained from infrared studies includes framework Si/Al composition, structural changes during thermal decomposition, and cation movement during dehydration and dehydroxylation.

The methods applied to the determination of zeolite structures have been the classical crystallographic techniques of x-ray diffraction and more recently electron diffraction. As a result of extensive x-ray structure studies on zeolites since about 1955, the framework structures of some 40 zeolites are known. A classification of zeolite structure types and groups has been proposed by Smith (43), Fischer and Meier (18), and Meier (28). The structure types and classes are based on a similarity in framework topology and common elements of secondary building units which comprise rings of tetrahedra, double rings, and larger symmetrical polyhedral units such as the 18-tetraherda "cancrinite" unit or the 24-tetrahedra truncated octahedron "sodalite" unit (28). It is difficult to determine atomic positions in the crystal structure of new zeolite species by x-ray techniques because of the large unit cells and the large number of possible ways of linking tetrahedra. The problem is magnified with synthetic zeolites where the structural investigator most frequently is limited to x-ray powder data owing to the unavailability of larger single crystals.

The object of this study was to apply mid-infrared spectroscopy to zeolite structural problems with the ultimate hope of using infrared, a relatively rapid and readily available analytical method, as a tool to characterize the framework structure and perhaps to detect the presence of the polyhedral building units present in zeolite frameworks. The mid-infrared region of the spectrum was used (1300 to 200 cm^{-1}) since that region contains the fundamental vibrations of the framework (Si,Al) O_4 tetrahedra and should reflect the framework structure. Infrared data in similar spectral regions have been published for many mineral zeolites (30) and a few synthetic zeolites (23, 49, 50). There is an extensive literature on infrared spectra of silica, silicates, and aluminosilicates (17). However, no systematic study of the infrared characteristics of zeolite frameworks as related to their crystal structure has appeared.

Figure 1. Infrared spectra of zeolites A, X, and Y and hydroxy sodalite (HS); Si/Al in X is 1.2, and in Y, 2.5

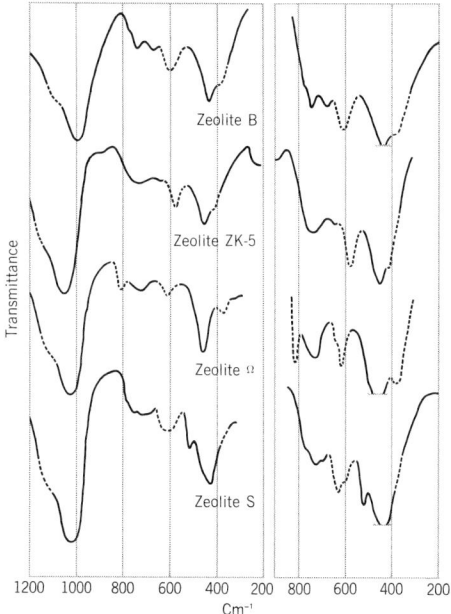

Figure 2. Infrared spectra of zeolites B(P1), ZK-5, Ω, and S; right hand portion of figure represents a higher zeolite concentration in the wafer than in the left portion for Figures 2, 5, and 7

Experimental

The majority of the infrared transmission spectra were obtained using the KBr wafer technique (37). Spectra of at least 2 and often more preparations of each zeolite were obtained before the spectrum was accepted as being characteristic of the zeolite species. Spectra also were obtained using CsI wafers and in the case of several zeolites with mineral oil mulls and pure zeolite self-supported wafers, to establish any matrix effect. Only minor spectral variations were observed among the several techniques and matrices. Except where otherwise noted, the spectra reported here are for hydrated zeolites in KBr or CsI wafers. A typical wafer concentration was 0.5 mg of zeolite in 300 mg of KBr or CsI; however, zeolite concentration sometimes was varied to obtain the desired absorbance or to increase the sensitivity to weak bands. Spectra were determined with a Perkin Elmer Model 621 double beam grating spectrometer. Essentially identical spectra were obtained using a Perkin Elmer Model 225 double beam grating spectrometer and a Beckman Model IR-12 double beam grating spectrometer as were obtained with the P-E 621 for zeolites A, X, and Y. A few spectra were measured for dehydrated zeolites by activating the zeolite powders in air at 350°C, rapidly quenching into dry mineral oil, and running the spectrum as a mineral oil mull. Dehydration studies reported for Ca-exchanged Y zeolite using self-supported wafers were carried out with a cell and technique essentially the same as that described by Angell and Schaffer (1). The spectral resolution is better than 5 cm^{-1} and the estimated accuracy ±5 cm^{-1} using the same technique and instrument, but can be as high as ±10 cm^{-1} with other measurement variations.

All of the synthetic zeolites investigated were prepared in this laboratory with the exception of the Zeolon product and ZK-5, and were fully characterized in terms of chemical composition, x-ray, and adsorption purity. All represent zeolite contents of greater than 90% and contained no crystalline impurities detectable by x-ray powder diffraction analysis. The Na "Zeolon" used was obtained from the Norton Co., and the experimental sample of ZK-5 was prepared by K. R. Muller at the Union Carbide European Research Associates Laboratory in Brussels, Belgium.

In the thermal decomposition studies of A, X, Y, and L zeolites, powder samples of each zeolite were heated in ambient air for 16 hours at increasing temperatures to yield a series of thermal decomposition products with successively lower residual zeolite x-ray crystallinity. The heated powders were hydrated at room temperature and run as KBr wafers to obtain their infrared patterns.

Results

Infrared spectra were obtained on the synthetic zeolites A, N-A, X, Y, B(Pl), KZ-5, omega (Ω), S, R, G, D, T, L, W, synthetic analogues of mordenite ("Zeolon"), and analcime (C), and for the related synthetic

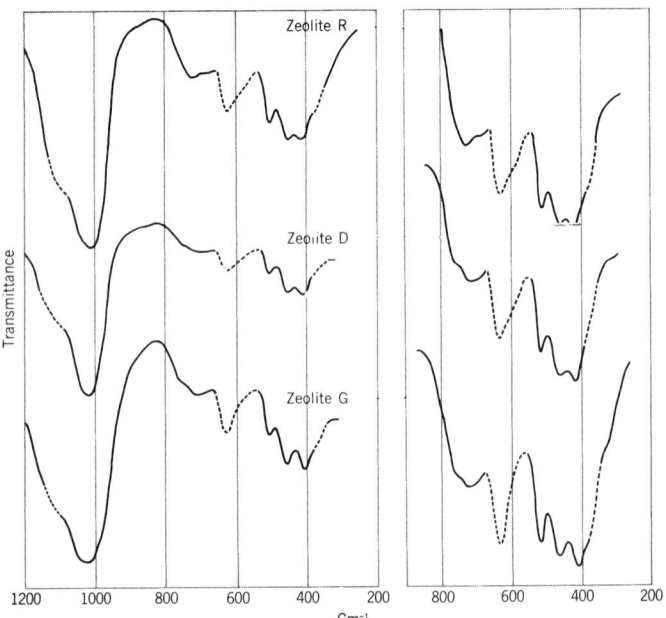

Figure 3. Infrared spectra of zeolites R, G, and D

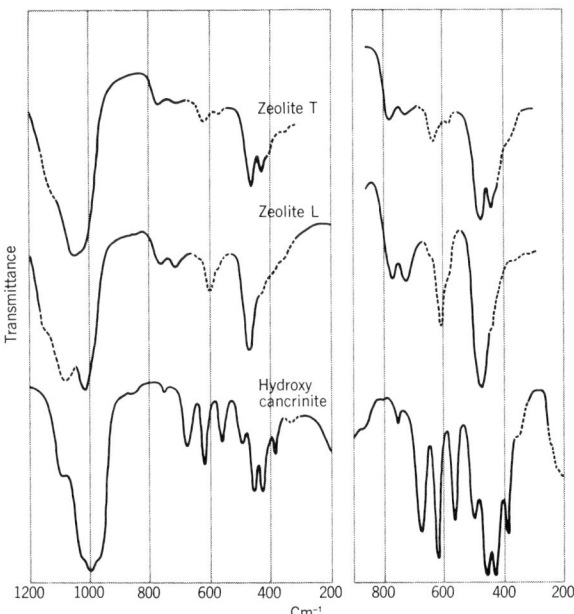

Figure 4. Infrared spectra of zeolites T and L, and hydroxy cancrinite (HC)

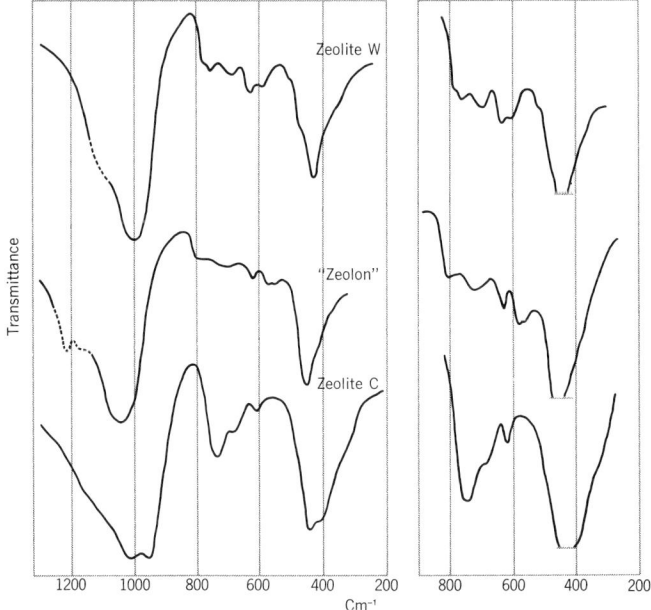

Figure 5. Infrared spectra of zeolites W, C, and "Zeolon"

Table I. Infrared Spectral

Zeolite	$\frac{SiO_2}{Al_2O_3}$	Asym. Stretch			Sym. Stretch			
A	1.88	1090vwsh	1050vwsh	995s			660vw	
$Ca^{ex}A$	1.9	1130vwsh	1055vwsh		742vwsh	705vwsh	665vw	
N-A	3.58	1131vwsh		1030s	750vwsh		675vw	
N-A	6.01	1151vwsh		1044s	750vwsh		698vw	
X	2.40		1060msh	971s	746m	690wsh	668m	
Y	3.42	1135msh		985s	760m		686m	
Y	4.87	1130msh		1005s	784m		714m	635vw
$La^{ex}Y$	5.0	1135msh		1006s	790m		705m	
Y	5.63	1130msh		1017s	789m		718m	645vw
B(P)	2.8	1105mwsh		995–1000s	772mwsh	738mw	670mw	
Hydroxy sodalite (HS)	2.0	1096vwsh		986s	729m	701mw	660m	
Ω	7.7	1130wsh		1024s	805mw	722mw		
ZK-5	6.0	1158wsh		1048s	890vwb	730mw		
R	3.25	1136mwsh		1007s	738w	678w		
G	5.44	1138mwsh		1027s	720w	696wsh		
D	4.62	1184mwsh		1018s	755wsh	711w		
S	2.5	1140wsh		1020s	770vwsh	722mw	690vwsh	
T	7.0	1156wsh	1059s	1010s	771w	718w		
Hydroxy cancrinite (HC)	2.0	1095mw	1035msh	1000s 965msh	755w	680m		
L	6.0	1160wsh	1080s	1015s	767mw	721mw		642vwsh
C	4.0	1162vwsh	1012s	952s	740m	686wb		
Zeolon	9.95	1216w	1180vwsh	1046s	795⎫ 772⎭wb	715⎫ 690⎭wb		
W	3.6	1128msh		1006s	786⎫ 756⎭mwb	691mwb		

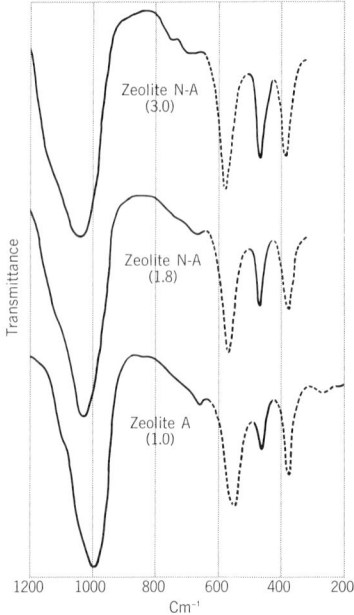

Figure 6. Infrared spectra of zeolites A and N-A; numbers in parenthesis are Si/Al values

Data for Synthetic Zeolites

Cm^{-1} [a]

Dbl. Rings			T-O Bend			Pore Opening?		
550ms			464m			378ms	260vwb?	
542ms			460m			376m		
572ms			474m			385m		
581ms			475m			393m		
560m			458ms		406w	365m	250vwb?	
564m		508vwsh	460ms			372m		
572m		500wsh	455ms			380m	260vwb?	
565m		500wsh	450ms			382m		
575m		504mwsh	456ms			383m	315vwsh	
600m				435ms		380mwsh		
			461ms	432ms			282vwb?	
610mw			451ms			372m		
	572m		445m		408wsh			
625m		508mw	452m	426m		370vwsh		
632m		515m	460m	408m		378vwsh		
631m		513m	459m	415m		376vwsh		
623, 595sh }mb		518mb	448m	424ms		370vwsh		
623mw	575w		467ms	433ms	410vwsh	366wsh		
624m	567m		498mw 458ms	429ms	390mw	353wb		
606m	580wsh		474ms	435wsh		375vwsh		
615vw			442ms	410msh				
621w	571, 555 }w		448ms			370vwsh		
637mw	590wb	512vwsh 483vwsh		432ns		375vwsh		

[a] s = strong; ms = medium strong; m = medium; mw = medium weak; w = weak; vw = very weak; sh = shoulder; b = broad.

felspathoid phases, hydroxy sodalite (HS) and hydroxy cancrinite (HC). The IR spectra are shown in Figures 1–6, and spectral frequencies listed in Table I. Cation and framework compositions and references describing their synthesis and properties are given for all of the zeolites in Table II. Structural characteristics based principally on Meier's (28) and Barrer's (2) reviews are compiled in Table III and a summary of structural elements and building units in zeolite frameworks given in Table IV. The definition of secondary building units (SBU) and building blocks used here is not as precise as that of Meier (28), and there is some minor variation from Meier's structural classification. The zeolites were chosen to represent a spectrum of structural types and SBU and polyhedral building units in the frameworks as well as a range of Si,Al framework compositions.

Correlation of the infrared spectra with zeolite structure has led us to propose the following interpretations and hypotheses. Each zeolite species has a typical infrared pattern. In addition, there are often general similarities among the spectra of zeolites with the same structural type and in the same structural group. The infrared spectra of zeolites in the 1300–200 cm^{-1} region appear to consist of 2 classes of vibrations: those caused by internal vibrations of the framework TO$_4$ tetrahedron, the primary building unit in all zeolite frameworks, which tend to be insensitive to variations in framework structure, and vibrations related to external linkages between tetrahedra which are sensitive to the framework structure and to the presence of some SBU and building block polyhedra such as double rings and the large pore openings. No vibrations specific to AlO$_4$ tetrahedra or Al–O bonds are assigned but rather vibrations of TO$_4$ groups and T–O bonds where the vibrational frequencies represent the average Si,Al composition and bond characteristics of the central T cation. The proposed infrared assignments are presented in detail in Table V and illustrated with the infrared spectrum of zeolite Y in Figure 7. The bands assigned to internal tetrahedral vibrations are shown in the

Table II. Compositions of Synthetic Zeolites

Zeolite	Cation[a]	SiO_2/Al_2O_3	Ref.
A	Na	2	(15, 31)
N-A	TMA[b], Na	2.5–6.0	(5)
X	Na	2.0–3.0	(16, 32)
Y	Na	>3–6	(11, 16)
HS (Hydroxy Sodalite)	Na	2–3	(4, 8)
ZK-5	Na, DDO[c]	4–6	(22)
B(Pl)	Na	2–5	(33)
Omega (Ω)	Na, TMA	5–12	(6, 19)
S	Na	4.6–5.9	(10)
R	Na	3.5–3.7	(35)
G	K	2–6	(3)
D	K, Na	4.6–5.0	(12)
T	K, Na	6.4–7.4	(13)
L	K; K, Na	5.2–6.9	(14, 16)
HC (Hydroxy cancrinite)	Na	2.0	(8)
W	K	3.3–4.9	(34)
Zeolon	Na	10–11	(41)
C	Na	2–6	(38, 39)

[a] Cation composition as synthesized.
[b] TMA = tetramethyl ammonium ion, (CH$_3$)$_4$N$^+$.
[c] DDO = (C$_8$H$_{18}$N$_2$)$^{2+}$ = [1,4-dimethyl-1,4-diazoniacyclo (2.2.2.) octane]$^{2+}$.

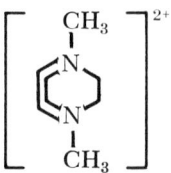

figures with a full line drawing, and those to external linkage modes with a broken line.

The first class of vibrations common to all zeolites and assigned to internal tetrahedron vibrations includes the 2 most intense bands in the spectrum, the strongest at 950–1250 cm^{-1} and the other of medium intensity at 420–500 cm^{-1}. We propose as a model for comparison and assignment of internal tetrahedral modes the comprehensive work of Lippincott et al. (26) on the infrared spectra of the polymorphs of silica. The primary building unit of a SiO$_4$ tetrahedron linked in a three-dimensional tetrahedral framework in the silicas is analogous to zeolite frameworks, and serves as a point of reference. Lippincott's Si–O vibrations become T–O vibrations in our assignments. After Lippincott, the strongest vibration in the 950–1250 cm^{-1} region is assigned to a T–O stretch involving motion primarily associated with oxygen atoms, or alternately described as an asymmetric stretching mode ← OT → ← O. The next strongest band (except for A zeolite) in the region of 420–500 cm^{-1} is assigned to a T–O bending mode. Stretching modes involving motions primarily associated with the T atoms, or alternately described as symmetric stretching modes ← OTO →, are assigned in the region of 650–820 cm^{-1}. The symmetric stretch modes are further classified into an internal tetrahedron stretch in the lower spectral region of 650–720 cm^{-1} and an external linkage symmetric stretch reflecting structure-sensitive external linkages in the higher region of 750–820 cm^{-1}. This distinction among symmetric stretch vibrations is somewhat tenuous and will be discussed further. All of the stretching modes discussed are sensitive to framework Si/Al composition and shift to lower frequency with increasing Al content. The T–O bending mode is not substantially affected by the substitution of Al for Si. The literature appears to be in general agreement in assigning fundamental tetrahedra modes in silica to Si–O stretches (asymmetric) from 800 to 1100 cm^{-1}, Si–Si stretching modes (symmetric) from 600 to 800 cm^{-1}, and bending and distortion modes under 600 cm^{-1} (36, 45, 46, 47).

The absence of any internal tetrahedral bands assigned to AlO$_4$ tetrahedra is in contrast to some previous literature assignments in aluminosilicate frameworks. The internal modes discussed are assigned by many authors to pure Si–O vibrations, and distinct Al–O vibrations assigned in the region of 715 to 815 cm^{-1} (17, 40). Stubican and Roy (45, 46, 47) point out that one should not expect vibrations of "pure" cation–anion (e.g., Si–O) bonds in aluminosilicate frameworks, discuss the influence of ions in second coordination spheres on Si–O vibrations, and introduce assignments of Si–O–Al combination frequencies. Kolesova (24) similarly assigns vibrations to the Si–O–Al unit. We prefer to describe the vibrational modes in terms of vibrating TO$_4$ units and not

Table III. Structural Characteristics

Zeolite	Structure Type/Group	Ideal Unit Cell Symmetry
A, N-A	A/faujasite	Cubic, $Pm3m$ (pseudo)
X, Y	Faujasite/faujasite	Cubic, $Fd3$ / $Fd3m$
Hydroxy Sodalite (HS)	Sodalite/chabazite	Cubic, $P\bar{4}3n$
ZK-5	ZK-5/faujasite	Cubic, $Im3m$
B(Pl)	B(Pl)/phillipsite	Cubic, $Im3m$
Ω	Ω/chabazite	Hex, $P6/mm^e$
S	Gmelinite/chabazite	Hex, $P6_3/mmc$
R, G, D	Chabazite/chabazite	Trigonal, $R\bar{3}m$
T	Offretite–erionite/chabazite	Offr.: Hex., $P\bar{6}m2$ Erion: Hex., $P6_3/mmc$
L	L/chabazite	Hex., $P6/mmm^f$
Hydroxy cancrinite	Cancrinite/chabazite	Hex., $P6_3$
W	Phillipsite–harmotome/phillipsite	Ortho., $B2mb$
Zeolon	Mordenite/mordenite	Ortho., $Cmcm$
C	Analcime/analcime	Cubic, $Ia3d$

Table IV. Building Units in Zeolite Structures[a]

Primary Building Unit—Tetrahedron (TO_4)
  Tetrahedron of 4 oxygen ions with a central tetrahedral ion (T) of Si^{4+} or Al^{3+}
  All oxygens shared between 2 tetrahedra, $(TO_2)_n$
Secondary Building Units, S.B.U.
  Rings, S-4, 6, 8
  Double rings, D-4, 6, 8, (14)
Larger Symmetrical Polyhedra[b]
  Truncated octahedron (T.O.) or sodalite unit
  11-Hedron[c] or Cancrinite (Cancr.) unit
  14-Hedron II[c] or gmelinite (gmel.) unit
Zeolite Structure
  Packing of S.B.U.'s and polyhedra in space

[a] See, for example, Meier (28) and Barrer (2) for a description and detailed discussion.

[b] Larger polyhedral units (>24 tetrahedra) as discussed by Ref. 2 and 28 are not considered here because they are not believed to be important in determining mid-infrared spectral characteristics.

[c] Barrer's (2) designations for building units.

of Synthetic Zeolites[a]

	Building Units[b]			Water Void Volume[c]	T/1000Å³
S-R	D-R	Pore Opening	Polyhedra		
4,6,8	D-4	S-8, planar	T.O.	0.47	12.8
4,6	D-6	S-12, nonplanar	T.O.	0.51	12.4
				0.50	12.8
4,6	—	—	T.O.	0.33[d]	17.2[d]
4,6,8	D-6	S-8, planar	—	0.44	14.7
4,8	D-4	S-8, nonplanar	—	0.41	15.8
4,6,8	D-12	D-12, planar	Gmel.	0.38	16.7
4,6,8	D-6	S-12, nonplanar	Gmel.	0.44	14.6
4,6,8	D-6	S-8, nonplanar	—	0.47	14.6
4,6,8	D-6	S-12, nonplanar	Cancr., Gmel.	0.40	15.7
4,6,8	D-6	S-8, nonplanar	Cancr.		
4,6,8	D-6	S-12, planar	Cancr.	0.32	16.3
4,6	—	S-12, nonplanar	Cancr.	0.21[d]	16.7[d]
4,8	—	S-8, nonplanar	—	0.31	15.8
4,5,8	—	S-12, nonplanar	—	0.28	17.2
4,6	—	—	—	0.18	18.6

[a] Adapted from Meier (28) and Barrer (2). Group classification that of Meier. Data shown are for the structure type.
[b] See Table IV for definition. Ideal size and symmetry of polyhedral units: D-4, 8 tetrahedra, cube, T_d; D-6, 12 tetrahedra, hexagonal prism, D_{6h} Cancr. 18 tetrahedra, D_{3h}; T.O., 24 tetrahedra, T_d; Gmel., 24 tetrahedra, D_{3h}.
[c] In cc/cc of crystal.
[d] Data for sodalite hydrate and cancrinite hydrate.
[e] From proposed tentative structure of Barrer (6).
[f] Structure of Barrer (7).

assign distinct SiO_4 vs. AlO_4 bands. The relative concentration of Si and Al in site T affects the frequency of the band, but not the number of bands. This is analogous to the crystallographic equivalency of Si and Al in the T site often encountered in x-ray diffraction analysis, where the T–O bond distances reflect the average Si,Al composition (28, 43).

The external linkage frequencies which are sensitive to topology and building units in the zeolite frameworks occur principally in 2 regions of the spectrum, 500–600 cm⁻¹ and 300–420 cm⁻¹. A medium intensity band in the former has been related to the presence of double ring polyhedra in the framework. Thus, all of the zeolite frameworks with D-4 and D-6 rings, X, Y, B, ZK-5, the chabazite group phases R, D, G, S, T, Ω, and L, have a medium (or in a few cases, mw or ms) band in the 550–630 cm⁻¹ region. The frequency of the band, as will be discussed later, shifts in

Table V. Zeolite Infrared Assignments, Cm⁻¹

Internal Tetrahedra
Asym. stretch   1250–950
Sym. stretch   720–650
T–O bend   420–500
External Linkages
Double ring   650–500
Pore opening   300–420
Sym. stretch   750–820
Asym. stretch   1050–1150 sh

some cases with Si/Al composition. The position of the band also appears to be related to structural type and class. In the faujasite group, A, X, Y, and ZK-5, the D-R frequency is found between 540 and 585 cm⁻¹ irrespective of Si/Al, and is symmetrical in shape. In B(Pl), the D-4 frequency is at 600 cm⁻¹, somewhat broad but symmetric. In all of the chabazite type phases (R, D, G) the D-6 band is at 625–630, with a common low-frequency asymmetry. In S, the band is broad, asymmetric, and centered near 600 cm⁻¹; it is resolved into 2 bands at 623 and 959 in higher concentration wafers. The D-6 band in Ω occurs as a relatively sharp but medium weak band at 610 cm⁻¹ with a high-frequency shoulder. Zeolites T and L, also in the chabazite group but containing 2 common building groups, D-6 and cancrinite units, show a medium (or mw) band at 623 and 606 cm⁻¹, respectively, with both a high and low frequency shoulder or satellite band. Thus, the chabazite group is characterized by a near medium band between 600 and 630 cm⁻¹ related to the D-6 unit, a frequency which is higher than the D-R region (540–585 cm⁻¹) in the faujasite group.

Zeolites devoid of D-R or larger symmetrical polyhedra are W, Zeolon, and C. The Zeolon and C phases show only weak absorption, and W medium weak, in the 540 to 630 cm⁻¹ portion of the spectrum. Zeolite W may be a borderline case since the medium weak band structure bears some resemblance to the D-R assignments in the chabazite family. The structure type assignment for W (phillipsite–harmotome) also may be questioned since it is based on a general resemblance of the x-ray powder diffraction pattern with that of the related mineral species. It is becoming increasingly apparent that such resemblance is insufficient to establish the framework topology unambiguously. This is especially true in the phillipsite and chabazite groups where a large number of stacking sequences and four- and eight-ring arrangements are possible, which would lead to similarities in x-ray powder patterns. [See, for example, Meier and Olson (29) and Beard (9)].

The second main extenal linkage frequency assigned is in the 300–420 cm⁻¹ portion of the spectrum and was initially assigned to a breathing motion of the isolated rings forming the pore opening in zeolites. It is a distinct medium to medium-weak band only in the spectrum of A, X, Y, and Ω zeolites, with a frequency of 360–395 cm^{-1}. It appears as a prominent shoulder or overlapping band at 350–420 cm^{-1} in the spectrum of B and ZK-5, as a weaker shoulder near or below 400 cm^{-1} in the chabazite group, and as a very weak shoulder in the remainder of the zeolites studied. All of the zeolites for which infrared data are presented contain pore openings of 8- to 12-R except C. Qualitatively, differences in the isolation, symmetry, and planarity of the ring which should affect ring distortion vibrations can be suggested to explain the observed variations in the prominence of the "pore opening" band, but the correlation is inconclusive. Interestingly, the band is prominent in those structures which have cubic unit cell symmetry, and decreases in prominence as the symmetry decreases. We prefer to maintain the suggested pore opening assignment, with reservation, for purposes of discussion.

Other infrared bands showing characteristics related to framework topology and assigned to external linkage modes (*see* Table V) are the shoulder near 1050–1150 cm^{-1} in the asymmetric stretch region on the high-frequency side of the principal T–O stretch band, and the higher frequency portion of the symmetric stretch region, 750–820 cm^{-1}. Variation in the features of the asymmetric shoulder, intensity, shape, and multiplicity of bands appears to be related to framework structure characteristics. The structural sensitivity of the 750–820 cm^{-1} region in the symmetric stretch region is based on variations observed among different structural types and classes, but principally on the thermal decomposition and cation movement studies for Y zeolite discussed subsequently. The spectral variation with framework structure appears to be much less in the lower frequency portion of the symmetric stretch region, 650–720 cm^{-1}.

In interpreting framework structure characteristics of zeolites from mid-infrared spectra, it is important that the over-all pattern in the spec-

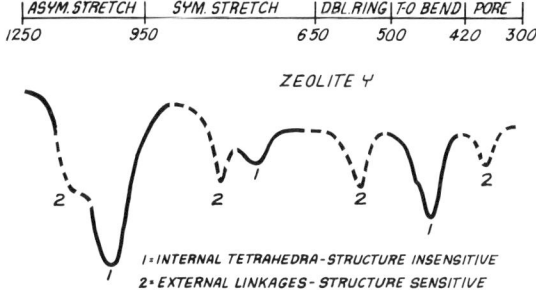

Figure 7. Infrared assignments illustrated with the spectrum of zeolite Y, Si/Al of 2.5

trum be considered, as well as those external linkage regions focused upon above. Slight structural differences often can be detected by the asymmetry of the bands. Common types and classes of zeolites have similar spectral characteristics. For example, although the T–O bend region, 420–500 cm^{-1}, is assigned to an internal tetrahedral mode less sensitive to framework structural variation, it nevertheless shows some structural specificity.

In the faujasite group of zeolites, the T–O bend occurs at 450–475 cm^{-1}; in Ω, also near 450; in the phillipsite group zeolites (B and W), at a lower frequency, near 430 cm^{-1}. A distinct two- to three-band T–O bend structure between 410 and 450 cm^{-1} is characteristic of the chabazite group of structures.

Structure types belonging to the same structure group also can be distinguished from each other in the T–O bend region. For example, in the chabazite group, the chabazite type phases (R, D, and G) have 3 prominent bands near 510, 455, and 420 cm^{-1}; in the gmelinite type (S), the 455 band becomes less distinct; in the offretite–erionite type (T), the 510 band disappears; and in the zeolite L, the 510 band is absent and the 430 band has become a weak shoulder. In N–A, the T–O bend occurs at 475 cm^{-1}, approximately 20 cm^{-1} higher than in A; in zeolite Y, the T–O bend near 455 cm^{-1} contains a high-frequency shoulder not present in X. Also, the 406 cm^{-1} pore opening band present in X zeolite is not found in Y zeolite.

The infrared spectrum of zeolite L according to our assignments is consistent with the recent structure proposed by Barrer and Villiger (7). There is less agreement between the same authors' structure tentatively proposed for Ω zeolite (6) and the infrared pattern characteristics. Their proposed structure based on x-ray powder data is hexagonal, and contains as polyhedral building units D-12 rings and gmelinite cages. Structurally, it is related to L, and classified with the extended chabazite group by the authors (6). Although the infrared pattern shows a medium weak band at 610 cm^{-1} which could be assigned to the D-12 ring, the over-all infrared spectrum of Ω shows the closest resemblance to the faujasite type zeolites, especially the distinct 370 cm^{-1} band in the Ω spectrum.

The uniqueness of the zeolite spectral features are further illustrated by comparing the infrared spectra of zeolites with those of felspathoid phases containing the same polyhedral building units. For example, the framework topology of the synthetic felspathoid phase hydroxy sodalite (HS) can be viewed as a cubic close packing of T.O. or sodalite units. The frameworks of A, X, and Y zeolites also contain cubic arrangements of T.O. units, but now with a second building unit interposed and linking the T.O. units, *i.e.*, a D-4 ring in A and a D-6 ring in X and Y. The effect of the combined D-R.–T.O. spatial packing is to build a much more open structure in the zeolites, with large water void volumes (~0.5 cc/cc of crystal; see Table III), and containing large three-dimensional pore openings (4–9 A in free diameter). The more open framework of the zeolites also results in the building units (here T.O. and D-4,6 rings) being more isolated, a point which will be developed further in the discussion. The infrared spectrum of HS is shown in Figure 1, together with the spectra for A, X, and Y zeolites. The difference in the spectral characteristics of HS compared with the zeolites is clear, especially the absence of the distinct external linkage bands assigned to the D-R (near 560 cm^{-1}) in the zeolite spectra. The sharper and more complex bands in the T–O bend (400–500 cm^{-1}) and symmetric stretch (650–800 cm^{-1}) regions observed for HS compared with the zeolites, should be noted also.

A second example is illustrated in Figure 4, which shows the infrared spectrum for the synthetic felspathoid hydroxy cancrinite (HC), and the zeolites T and L. The framework topology of HC is a hexagonal array of cancrinite units; the frameworks of zeolites T and L contain columns of alternating cancrinite—D-6 ring units along the hexagonal c direction. Perhaps 2 different structural groups are represented by the different bending vibrations. Again, the infrared spectrum of the felspathoid phase is sharper and more complex than those of the zeolites.

Spectral characteristics related to the presence of the larger symmetrical polyhedra building units in some zeolite frameworks, such as the T.O., cancrinite, and gmelinite units (*see* Tables III and IV), also were sought. Vibrations of such large groups (18 to 24 tetrahedra, ~ 8A in cross section) would not be expected to appear above 300 cm^{-1}. The infrared spectrum in the region below 300 cm^{-1} in KBr wafers is meaningless since that frequency is near the transmission cutoff for KBr and the lower frequency limit of the instrument (200 cm^{-1}). Careful measurements of the spectrum of some phases were made in CsI wafers where the resolution in the 200–300 cm^{-1} region is improved. A broad weak band near 250–300 cm^{-1} (listed in Table I and shown in Figures 1 and 6) which was observed in the spectra of the A, X, Y zeolites and in HS was initially considered as being indicative of the T.O. units present in these structures. Such an assignment was arbitrary and was not substantiated further. However, it still is suggested that vibrational bands characteristic of such symmetrical polyhedral building units may be found in the farther infrared region, below 300 cm^{-1}.

There is some indication that information of framework ordering and dislocation phenomena in zeolites may be obtained from mid-infrared spectroscopy. The multiplicity of peaks in the asymmetric stretch region in zeolite L, compared with the zeolites Ω and T with similar Si,Al framework composition and structure group characteristics, may indicate Si,Al ordering in zeolite L, or alternately 2 different Si,Al distributions in the

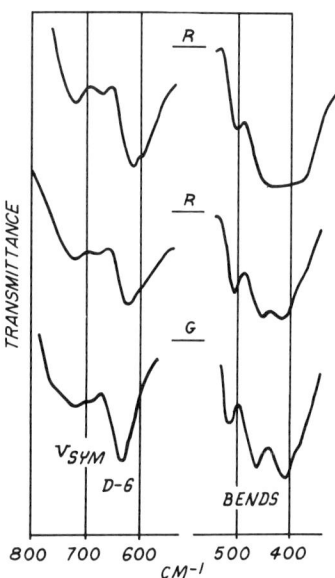

Figure 8. Infrared spectra for chabazite type zeolite phases, R (2 different synthesis preparations) and G

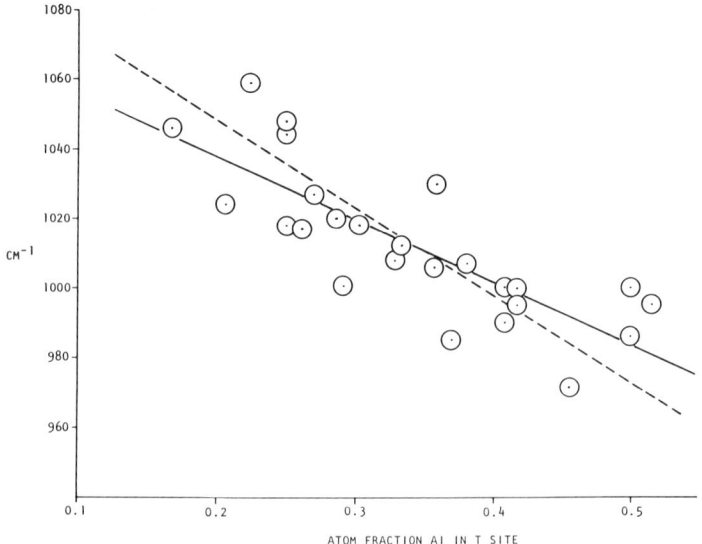

Figure 9. Frequency of the main asymmetric stretch band vs. the atom fraction of Al in the framework for all synthetic zeolites of this study

framework. Laves and Hafner (25) used infrared to evaluate Si,Al ordering in feldspars, where the infrared spectra of ordered feldspars show multiple sharp peaks in the asymmetric stretch and T–O bend regions, and disordered feldspars show a single broad peak in each. It is clear that the same interpretation cannot be generalized for synthetic zeolites, since A with the simplest single asymmetric stretch band has an ordered framework (43). Also, other parameters in addition to ordering can be related to number and shape of bands.

Evidence of dislocation phenomena could be suggested from spectral variations among members of the chabazite group. The latter grouping is based on various stacking sequences of parallel six-membered rings. Spectra for the synthetic chabazite type phases, R (2 different synthesis preparations) and G, are shown in Figure 8. Differences observed in the breadth and structure of bands in the symmetric stretch, D-6, and T–O bend regions may be indicative of some alteration in stacking sequence or variation in concentration and type of stacking faults.

Only short-range ordering and dislocation phenomena should be seen in the mid-infrared spectral region. Far infrared spectroscopy should yield interesting information on long-range order and dislocation phenomena in zeolite lattices.

The frequency shift of infrared stretch bands with substitution of tetrahedral Al for Si in aluminosilicate frameworks has been reported by many authors. Since the mass of Al and Si are nearly the same, the decrease in frequency with increasing Al concentration is related to variation in bond length and bond order. The longer bond length of Al–O and decreased electronegativity of Al results in a decrease in force constant (30, 45, 46, 47).

Milkey (30) reported a quantitative linear relationship between the main asymmetric stretch frequency (1000 1100 cm^{-1}) and atom fraction of Al in the tetrahedral site for a large number of tectosilicate minerals. He observed a decrease of 19 cm^{-1}/0.1 atom fraction of Al ion substitution. Stubican and Roy (45, 46, 47) also reported a linear relationship between Si–O stretch bands and tetrahedral Al substitution in layer lattice silicates. In synthetic saponites and beidellites, they observed a decrease of 15 cm^{-1}/0.1 atom fraction of Al in the main stretch frequency, corresponding to an increase in T–O bond length of 0.018 A. The magnitude of frequency shift is in good agreement with Milkey's. A decrease of 3 cm^{-1}/0.1 atom fraction of Al in the stretch frequency near 660 cm^{-1} was shown for saponites.

A similar treatment and result is shown in Figure 9 for the spectra of synthetic zeolites determined in this study. The frequency maximum of the main asymmetric stretch band is plotted against the atom fraction of Al in the framework tetrahedral site. Although a linear relationship is

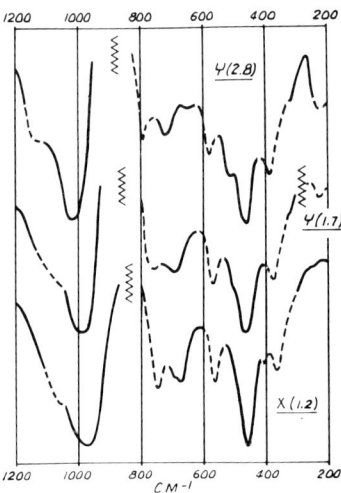

Figure 10. Infrared spectra for zeolites X and Y with different Si/Al contents; numbers in parenthesis refer to Si/Al in the zeolite

shown with a slope of 18.1 cm^{-1}/0.1 atom fraction of Al, also in good agreement with Milkey (*30*), there is considerable scatter in the data points. The dashed line in Figure 8 is the best line through the points when a y intercept of 1100 cm^{-1} (the average main stretch frequency observed for pure silica frameworks, Ref. *26*) is imposed. The slope of the dashed line is 25.3 cm^{-1}/0.1 Al. The scatter in Figure 9 appears to be related to some structure-sensitive characteristics in the main asymmetric stretch band in synthetic zeolites.

A shift in frequency with Si/Al content for several other classes of infrared bands also was found for the synthetic zeolites X and Y, both with a faujasite-type framework. The infrared spectra of zeolites X and Y are shown in Figure 10. Plots of frequency *vs.* fraction of Al in the framework are shown in Figure 11. A linear decrease in frequency with increase in fraction of Al in the framework was observed for the main asymmetric stretch band (970–1020 cm^{-1}), a symmetric stretch band (670–725 cm^{-1}), the D-6 ring band (565–580 cm^{-1}), and the 12-R pore opening band (360–385 cm^{-1}). The T–O bend frequencies for the same zeolites did not vary (*see* Table I). Kiselev *et al.* (*23, 50*) and Wright *et al.* (*49*) showed similar infrared spectral shifts with varying Si/Al for a series of X and Y zeolites. A linear decrease in frequency with increasing atom fraction of Al in the framework also was found in the type A zeolite phases, A and N-A (not shown), for the following bands: asymmetric stretch (990–1050 cm⁻¹, slope = 19 cm⁻¹/0.1 $\overline{Al}$), symmetric stretch (660–700, s = 14), D-4 ring (550–580, s = 12), and 8-R pore opening (375–395, s = 6).

The effect of dehydration on the infrared spectrum was studied for A, X, Y, L, and Ω zeolites. For these zeolites, the spectrum of the dehydrated form was essentially the same as that of the fully hydrated material. Minor changes in frequency maximum (usually an increase of 10–20 cm⁻¹) and shape of the main asymmetric stretch band were observed upon dehydration in some cases. The cation composition of all of the dehydrated zeolites mentioned was the as-synthesized alkali metal form (except Ω with Na and TMA cations; see Table II). The effect of dehydration on the infrared spectrum will depend on the degree of framework distortion affected by the dehydration reaction. For the zeolites mentioned above with alkali metal cations, the framework distortion is minimal on dehydration. The multivalent cation forms, M^{2+} and M^{3+}, of many zeolite frameworks, however, cause significant changes in the distortion of the framework elements and symmetry (42, 44) upon dehydration. Here one would expect changes in the infrared spectrum as found below. For the majority of the more open zeolite frameworks, ion exchange with multivalent cations would not be expected to change the infrared spectrum of the hydrated zeolite, since cation–framework interactions are usually weak in such cases. This is illustrated by infrared data given in Table I for (hydrated) Ca-exchanged A and (hydrated) La-exchanged Y, which do not differ significantly from those of the original Na forms.

The effect of cation movement and framework distortion on the infrared pattern of a Ca-exchanged Y zeolite (Si/Al = 2.5) is shown in Figure 12. Dehydration and complete dehydroxylation of zeolites with similar cation composition and framework topology (e.g., Ca-exchanged

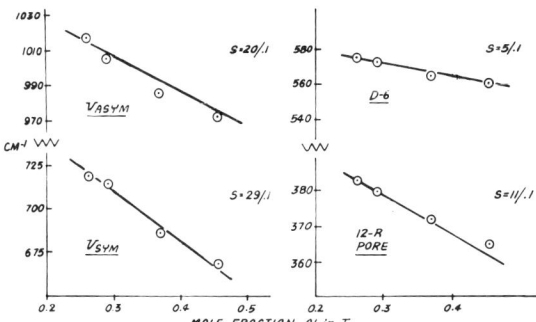

Figure 11. Frequency vs. atom fraction of Al in the framework for zeolites X and Y for several infrared bands

faujasite) causes (42) migration of Ca^{2+} cations from inside the sodalite (T. O.) cage into a position near the center of the D-6 ring (site I), with accompanying framework distortion and change of symmetry of the D-6 ring cage (42). In the infrared spectrum, this has been correlated with the following changes in the external linkage structure sensitive bands: the D-6 ring band near 570 cm^{-1} has shifted to 635; and the 390 "pore opening" band has shifted to 415 cm^{-1} (with shoulder at 400). In addition, changes in the character of the broad symmetric stretch band structure near 710 to 750 cm^{-1} are observed. The spectral changes are reversible upon rehydration. Kiselev et al. (4) showed similar results for dehydrated Ca, Sr, and NH_4-exchanged forms of X zeolites. Based on these results and analogous ones obtained for Ca-exchanged Y (Si/Al of 1.7) and La-exchanged Y (Si/Al 1.7 and 2.5), we propose that infrared is useful in determining the extent of cation movement in these and other zeolite frameworks where occupancy of cation sites causes distortion in framework elements.

The validity of assigning specific structure sensitive external linkage bands in the infrared spectra of some classes of zeolites is especially supported by the spectral changes observed during thermal decomposition of the zeolite crystal lattice. The infrared spectra for a series of thermally degraded zeolite Y (Si/Al of 2.5) phases with residual x-ray crystallinities of from 85% to 0% (no x-ray pattern detectable) are shown in Figure 13. The absorbance of the structure sensitive bands near 1130, 780, 570, and 380 cm^{-1} decreases proportional to the decrease in x-ray crystallinity. The internal tetrahedral vibrations near 1000, 710, and 460 cm^{-1} remain, with minor changes in breadth and position. Analogous data and results were

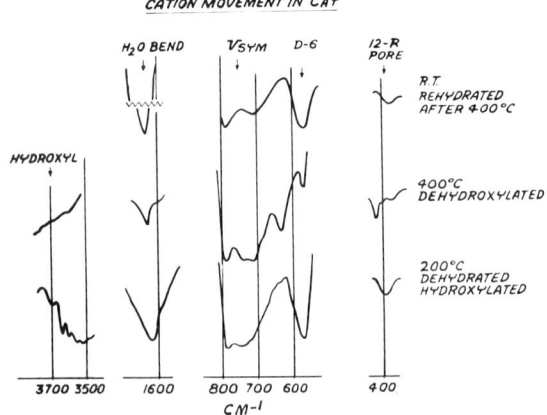

Figure 12. Infrared spectra for Ca-exchanged Y zeolite (Si/Al of 2.5) after dehydration, dehydroxylation, and rehydration

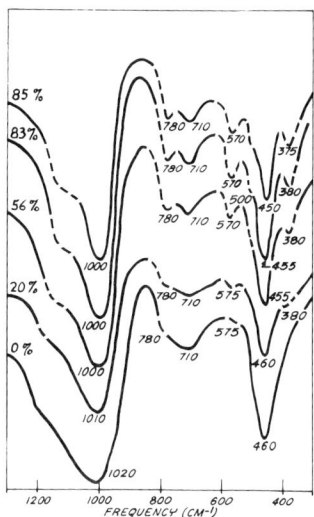

Figure 13. Infrared spectra of thermally decomposed Y zeolite (Si/Al of 2.5). The percentages are % residual x-ray crystallinity of the sample.

obtained for zeolites A, X, and L. The minimal frequency shifts in the internal tetrahedra bands during and after decomposition and the absence of any new bands in the amorphous product indicate a Si/Al content in the residue framework similar to that in the zeolite and no formation of octahedral Al or layer-type structure.

Discussion

There have been many literature reports of structure–infrared relationships for silica and aluminosilicate frameworks. [See, for example, White and Roy (48) where other work is cited, and the review by Dutz (17).] The present work extends such studies to the zeolite class of aluminosilicate structures. The assignment of specific frequencies to structural groups in the framework was developed from qualitative interpretation of structure–spectra correlations, and does not necessarily represent an assignment of a vibrational mode of the structural group. No attempt has been made to carry out theoretical calculations of the vibrational spectrum of the polyhedral building units to which infrared bands have been assigned, nor of the zeolite unit cells. Such calculations are possible, for example, by using a factor group analysis technique. However, because of the large number of atoms in the zeolite building units and unit cells, the number of calculated vibrations always would exceed the number of observed vibrations by many-fold, and coincidence of an observed infrared vibration with a calculated vibrational mode could not serve as a basis of assignment of spectral species, especially without supplemental Raman data. Attempts were made to determine the Raman spectra of zeolites A, X, Y, but were all unsuccessful. Raman instruments with both argon and neon laser radiation sources were used. No zeolite spectrum was observed. The absence of any detectable Raman spectrum for zeolites could be explained by experimental difficulties, excessive Rayleigh scattering of the small particle size synthetic zeolite crystals (1 to 50μ diameter) which obscures the Raman spectrum, or a real spectral characteristic of zeolites, *i.e.*, a very weak Raman spectrum.

Calculations of vibrational spectra have been reported for isolated silicate tetrahedra, rings, chains, and larger groups of tetrahedra in silicate structures, beginning with the pioneering work of Matossi (27), Saksena (40), and many others. Gaskell (20, 21) recently calculated the vibrational frequencies of typical silicate structures, framework, sheet, chain, pyrosilicate and island silicate, using an extended Wilson GF matrix method of calculation. Zijp (51) used the same method to calculate vibrational spectra of several $X_4Y_6Z_4$ molecular groups with T_d symmetry, such as As_4O_6, P_4O_6, and P_4O_{10}, polyhedral cage groups which show some

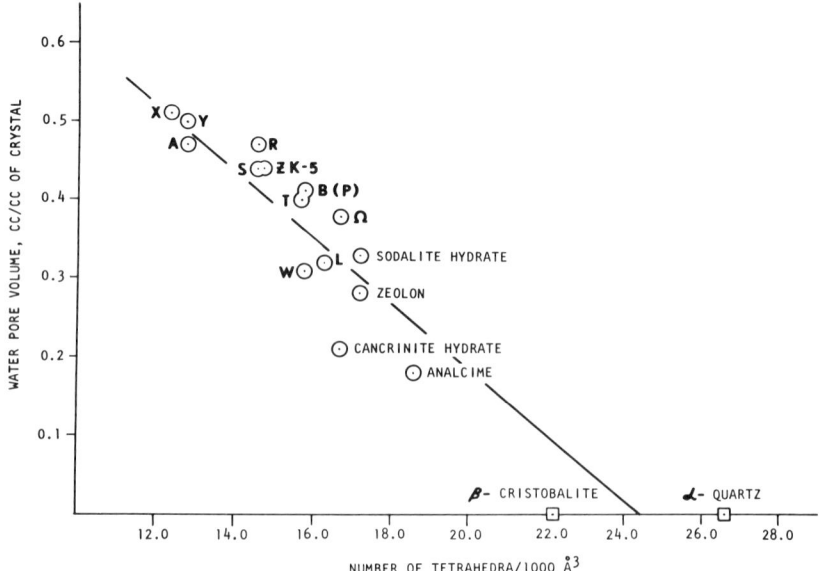

Figure 14. Water void volume in cc/cc of crystal vs. tetrahedra density in tetrahedra 1000/$Å^3$ for synthetic zeolites and related structures of this study; adapted from unpublished work of D. W. Breck

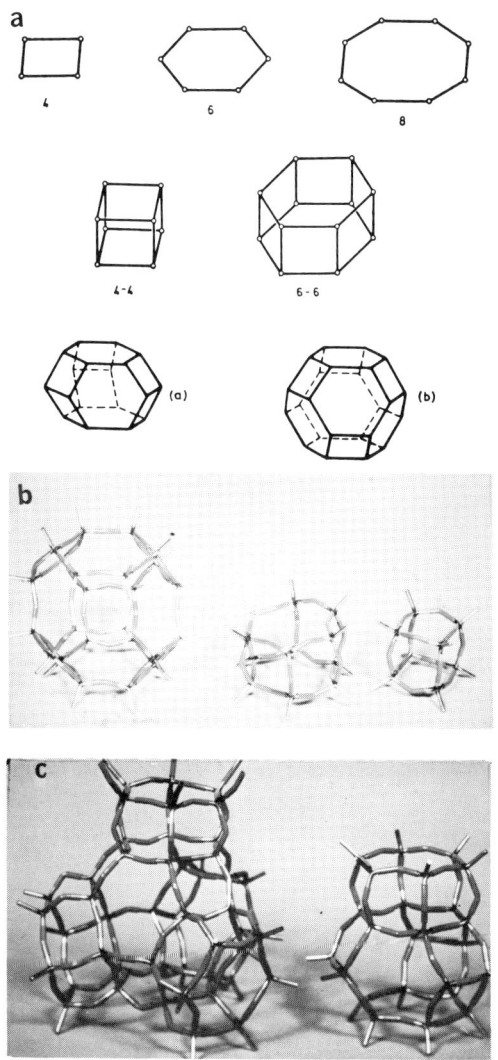

Figure 15. Building units in zeolite frameworks

(a) Geometric representation of (from top) S-4, 6, and 8 rings; D-4 and D-6 rings; and cancrinite (left) and truncated octahedron (sodalite) unit (right); from Meier (28)
(b) Wire tetrahedra models of T.O., D-6, and D-4 ring units (left to right)
(c) Groups of combined building units; left, tetrahedral arrangement of T.O. and D-6 rings in faujasite-type framework; right, cancrinite–D-6 ring group founds in columns || to c-axis in T and L zeolites

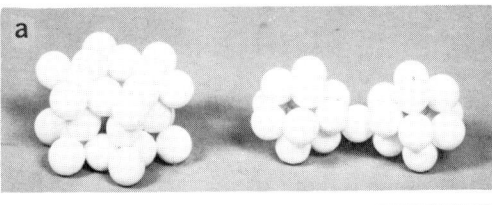
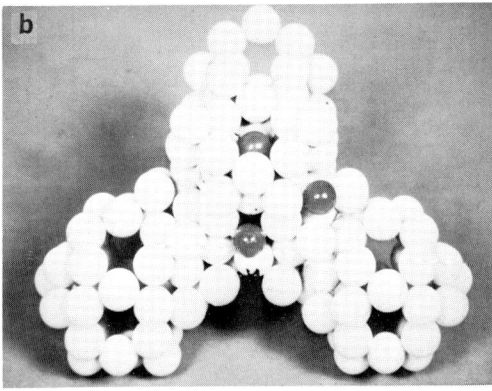

Figure 16. Oxygen packing models of (a) (left) D-6 ring, and (right) 2 linked D-4 rings as in the A-type framework, and (b) tetrahedral arrangement of T.O. units linked with D-6 ring found in faujasite type frameworks

resemblance to the polyhedral building units in zeolites. Both workers found some difficulty and ambiguity in frequency assignment.

Several aspects of the infrared spectra of zeolites should be emphasized. The spectra tend to be simple, with relatively few main vibrations which are somewhat broad and, in most cases, without fine structure. Indeed, many of their spectral features bear a strong resemblance to those of silicate glass structures. This indicates a high degree of lattice coupling of the tetrahedral vibrations in the framework. The uniqueness and structure specificity of their infrared spectrum increases with increasing openness of the framework and increasing unit cell symmetry. Values of framework densities (tetrahedra/1000 A^3) for the synthetic zeolites included in this study are plotted *vs.* their water void volumes in Figure 14. The zeolite frameworks which contain external linkage vibrations and the related double ring or larger polyhedral units have water void volumes greater than 0.3 cc/cc of crystal and framework densities of less than 17 tetrahedra/1000 A^3. In addition, their symmetry is either cubic or hexagonal. Other zeolite structures having similar large void volumes and low framework densities should show structure specific infrared spectral characteristics. From space filling considerations, any packing of a double ring with a larger symmetrical polyhedron unit in a three-dimensional lattice results in a framework with a density less than ∼ 17 tetrahedra/1000 A^3. Conversely, there is a high probability that zeolites with water void volumes greater than about 0.35 cc/cc contain such combinations of building units. Zeolites with the faujasite-type frameworks, A, X, and Y, have the most open structure (∼0.5 cc/cc and 12–13 tetrahedra/1000 A^3) and the most distinct spectral characteristics.

Detailed consideration has been given to the nature of the vibrational motion giving rise to the external linkage frequencies related to double rings and pore openings. The isolation of these building units in the most open zeolite frameworks may lead to uncoupling of their vibrations

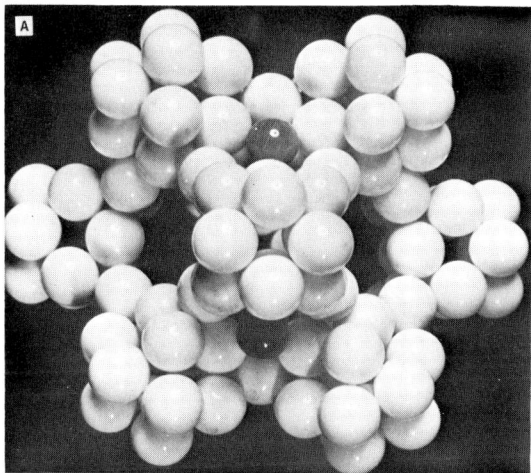

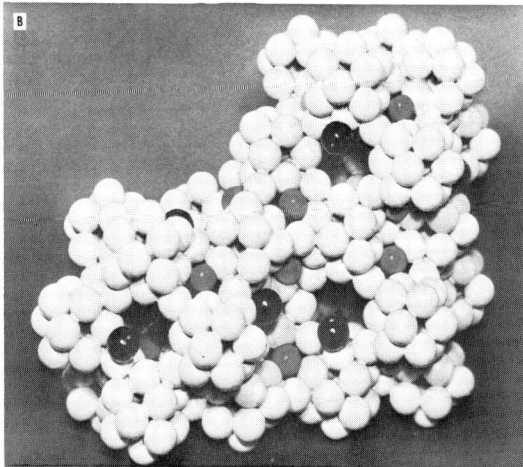

Figure 17. Oxygen packing model of (A) a unit cell, and (B) several unit cells of zeolite A. Dark balls represent sodium cations in structure.

compared with those of the internal tetrahedral modes, resulting in distinct group frequencies. The infrared bands associated with the polyhedral building units could arise from: vibrations of the group as a whole; unusual bond angles or symmetry characteristics of the group which affect infrared activity; unique oxygens or grouping of several oxygens; or over-all unit cell symmetry which is related to building unit symmetry. We suggest these vibrations involve primarily motion of oxygen ions and, based on their position in the spectrum, that the D-R band (500–650 cm^{-1}) is a stretching mode and the "pore opening" band (300–420 cm^{-1}), a bending or distortional mode. The "pore opening" assignment, as noted previously, is tenuous and that vibration may be a second one related to the double ring, or to some other structural element.

The geometry and oxygen packing characteristics of zeolite building units and unit cells are illustrated in Figures 15–18. Wire tetrahedra and oxygen-packing models are shown for unit cells of A-type and faujasite-type (zeolites X and Y) frameworks, and for several building units. The openness, high symmetry, and the unusual high-density packing of oxygen ions in the structural elements is apparent. Among the unique oxygens or groupings of oxygens are the bridging oxygen in the D-6 and D-4 rings, the oxygen linking 2 D-4 rings in zeolite A (Figure 16a, right), near-linear groups of 3 oxygens, and others.

A review of bond angles in zeolite structures, especially T–O–T angles which should be important for external linkage vibrations, showed

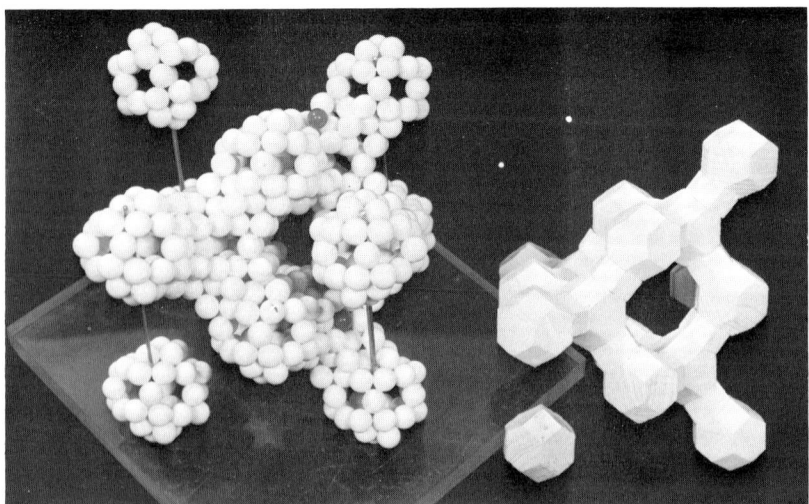

Figure 18. Oxygen packing model of unit cell of faujasite-type framework of zeolites X and Y, showing packing of T.O. and D-6 ring groups (left); and solid geometric representation of structure (right). Dark balls represent sodium cations in structure.

no obvious relationship with external linkage infrared bands. The majority of zeolite frameworks have T–O–T angles near 140° to 150°, and deviations from that value did not appear to correlate with infrared pattern.

The hypotheses and interpretations proposed here need further testing with other zeolite frameworks and additional spectral analysis. Infrared spectroscopy must be used as a supplemental tool to characterize zeolites, along with the more common methods of determining structure and properties; *e.g.*, x-ray crystallography, chemical analysis, adsorption, and other characterizations.

Acknowledgment

The authors thank R. L. Bujalski for his assistance in determining the infrared spectra and for his ingenuity and insight with respect to many of the problems encountered. We also thank J. T. Mullhaupt and D. W. Breck for their many helpful discussions throughout the course of this work and during the preparation of the manuscript.

Literature Cited

(1) Angell, C. L., Schaffer, P. C., *J. Phys. Chem.* **1965,** 69, 3464.
(2) Barrer, R. M., *Chem. Ind. (London)* **1968,** 1203.
(3) Barrer, R. M., Baynham, J. W., *J. Chem. Soc.* **1956,** 2894.
(4) Barrer, R. M., Baynham, J. W., Bultitude, F. W., Meier, W. M., *J. Chem. Soc.* **1959,** 195.
(5) Barrer, R. M., Denny, P. J., Flanigen, E. M., U. S. Patent **3,306,922** (1967).
(6) Barrer, R. M., Villiger, H., *Chem. Commun.* **1969,** 659.
(7) Barrer, R. M., Villiger, H., *Z. Krist.* **1969,** 128, 352–370.
(8) Barrer, R. M., White, E. A. D., *J. Chem. Soc.* **1952,** 1561.
(9) Beard, W. C., Advan. Chem. Ser. **1971,** 101, 237.
(10) Breck, D. W., U. S. Patent **3,054,657** (1962).
(11) *Ibid.,* **3,130,007** (1964).
(12) Breck, D. W., Acara, N. A., British Patent **868,846** (1961).
(13) Breck, D. W., Acara, N. A., U. S. Patent **2,950,952** (1960).
(14) *Ibid.,* **3,216,789** (1965).
(15) Breck, D. W., Eversole, W. G., Milton, R. M., Reed, T. B., Thomas, T. L., *J. Am. Chem. Soc.* **1956,** 78, 5963.
(16) Breck, D. W., Flanigen, E. M., "Molecular Sieves," p. 10–27, Society of the Chemical Industry, London, 1968.
(17) Dutz, Von H., *Ber. Deut. Keram. Ges.* **1969,** 46, 75.
(18) Fischer, K. R., Meier, W. M., *Fortschr. Mineral.* **1965,** 42, 50.
(19) Flanigen, E. M., Kellberg, E. R., Netherlands Patent **6,710,729** (1967).
(20) Gaskell, P. H., *Phys. Chem. Glasses* **1969,** 8, 69.
(21) Kerr, G. T., *J. Inorg. Chem.* **1966,** 5, 1539.
(22) Kerr, G. T., *Science* **1963,** 140, 1412.
(23) Kiselev, A. V., Lygin, V. I., "Infrared Spectra of Adsorbed Species," L. H. Little, Ed., p. 361–367, Academic, London, 1967.
(24) Kolesova, V. A., *Opt. Spectr.* **1959,** 6, 20.
(25) Laves, F., Hafner, S., *Norsk. Geol. Tidsskr.* **1962,** 42, 57.

(26) Lippincott, E. R., Van Valkenberg, A., Weir, C. E., Bunting, E. N., *J. Res. Natl. Bur. Std.* **1958**, A 61, 61.
(27) Matossi, F., *J. Chem. Phys.* **1949**, 17, 679.
(28) Meier, W. M., "Molecular Sieves," p. 10–27, Society of the Chemical Industry, London, 1968.
(29) Meier, W. M., Olson, D. H., ADVAN. CHEM. SER. **1971**, 101, 155.
(30) Milkey, R. G., *Am. Mineralogist* **1960**, 45, 990.
(31) Milton, R. M., U. S. Patent **2,882,243** (1959).
(32) *Ibid.*, **2,882,244** (1959).
(33) *Ibid.*, **3,008,803** (1961).
(34) *Ibid.*, **3,012,853** (1961).
(35) *Ibid.*, **3,030,181** (1962).
(36) Nakamoto, K., "Infrared Spectra of Inorganic and Coordination Compounds," p. 103–107, Wiley, New York, 1963.
(37) Rao, C. N. R., "Chemical Applications of Infrared Spectroscopy," Academic, New York, 1963.
(38) Saha, P., *Am. Mineralogist* **1959**, 44, 300.
(39) *Ibid.*, **1961**, 46, 859.
(40) Saksena, B. D., *Trans. Faraday Soc.* **1961**, 57, 242.
(41) Sand, L. B., "Molecular Sieves," p. 71–77, Society of the Chemical Industry, London, 1968.
(42) Smith, J. V., ADVAN. CHEM. SER. **1971**, 101, 171.
(43) Smith, J. V., *Mineralogical Soc. Am. Spec. Paper No. 1*, **1963**, 281.
(44) Smith, J. V., "Molecular Sieves," p. 28, Society of the Chemical Industry, London, 1968.
(45) Stubican, V., Roy, R., *Am. Mineralogist* **1961**, 46, 32.
(46) Stubican, V., Roy, R., *J. Am. Ceram. Soc.* **1961**, 44, 625.
(47) Stubican, V., Roy, R., *Z. Krist.* **1961**, 115, 200.
(48) White, W. B., Roy, R., *Am. Mineralogist* **1964**, 49, 1670.
(49) Wright, A. C., Rupert, J. P., Granquist, W. T., *Am. Mineralogist* **1968**, 53, 1293.
(50) Zhdanov, S. P., Kiselev, A. V., Lygin, V. I., Titova, T. I., *Russ. J. Phys. Chem.* **1964**, 38, 1299.
(51) Zijp, D. H., Ph.D. thesis, University of Amsterdam, Amsterdam, The Netherlands, 1960.

RECEIVED March 2, 1970.

Discussion

D. J. C. Yates (Esso Research Co., Linden, N. J. 07036): Do you have any further comments on your failure to observe bands in the Raman region of the spectrum?

E. M. Flanigen: I personally believe the problem in obtaining Raman spectra is the scattering from small particles, as suggested in the paper. From our reading of the Raman literature, most of the data reported on solids have been determined on particle sizes (if the size was given) above about 50 μ. In the case of synthetic zeolites for which we have reported infrared data, the particle size is of the order of 1 μ, and all we are observing in the Raman is a general Rayleigh scattering.

J. Turkevich (Princeton University, Princeton, N. J. 08540): Beuchler and I have also had difficulty in observing laser Raman spectra of zeolites and finely divided aluminas. It seems that the ease of obtaining Raman spectra increases markedly as one goes down the periodic table. Thus, finely divided molybdenum oxide or uranium oxide are easy to observe.

E. M. Flanigen: I would like to cite a recent reference on the Raman spectra of various silicates and aluminosilicates, W. P. Griffith, "Raman Studies on Rock-Forming Minerals. Part I. Orthosilicates and Cyclosilicates," *J. Chem. Soc.* (A) **1969,** 1372. This author apparently experienced no difficulty in obtaining the Raman spectra with a He–Ne laser Raman. The particle size of the samples was not specified, but described as "microcrystalline."

J. Turkevich: Is it not surprising that vibrations of such massive systems as rings are observed in the infrared region indicated? Could they be overtones of vibrations found in the far infrared?

E. M. Flanigen: You have raised a difficult point in our interpretation which we have considered and discussed in the paper. Our present feeling is that the internal tetrahedron vibrations are highly coupled, and that it is by reason of the large pore volume which isolates polyhedral units, such as the double four rings in zeolite A, that they can be seen as distinct entities. We propose that that is why, in the less open structures with higher densities and lower pore volumes, you do not see the same kind of general features in the infrared pattern as you do in the most open frameworks.

We do not believe they are overtone vibrations.

B. D. McNicol (Koninklijke/Shell Laboratorium, Amsterdam, Netherlands): Would it be possible to use this technique to look at the induction period of zeolite crystal growth—*e.g.*, identification of building units such as double six rings?

E. M. Flanigen: Yes, that is a possibility which we have considered.

17

Structural Studies on Erionite and Offretite

J. ALAN GARD and J. MERVYN TAIT

University of Aberdeen, Aberdeen AB9 2UE, Scotland

> *Structure analysis has confirmed the offretite model of Bennett and Gard and has located cation and water sites, including one K^+ in each cancrinite cage. Electron diffraction of many samples showed that the type locality is unique to date for offretite, which is partly disordered. All natural erionites examined were ordered. No ordered synthetic erionite was observed. One sample of ordered synthetic offretite was identified as "sausage-shaped" particles. The proportion of erionite in disordered intergrowths with offretite can, in theory, be estimated from intensities of reflections with 1 odd, but comparison of x-ray with electron diffraction data on the same samples suggests that x-ray estimates may be low if erionite domains are thin, due to gross broadening of lines with 1 odd.*

A structural scheme for the aluminosilicate frame of erionite was proposed by Staples and Gard (16) with the hexagonal space group $P6_3/mmc$, and $a = 13.26$, $c = 15.12$ A. Intensity data from x-ray fiber rotation gave a mean residual $R = 0.33$. Rings of 6 tetrahedra are stacked in AABAAC—sequence (Figure 1a), so that columns of alternating cancrinite cages and hexagonal prisms are cross-linked with single rings, to form cavities with a free diameter of 6.3 A, sharing windows 4.7 × 3.5 A. This scheme has been confirmed and refined by Kawahara and Curien (10) for erionite from Maze, Niigata, Japan [Shimazu and Kawakami (15)]. They also located one cation in each cancrinite cage and 2 peaks on the axis of each large cavity, which they interpreted as water molecules. Published cation exchange studies (4, 6, 11, 14) have shown that a residue of 2 K^+ ions in each unit cell of erionite or Zeolite T —a disordered synthetic erionite (1, 3)—is not exchangeable at temperatures below 300°C; loss of K^+ above 300° appears to cause partial disruption of the frame. Sherry (14) suggested that each cancrinite cage therefore contains a K^+ ion; this would agree with the structural work

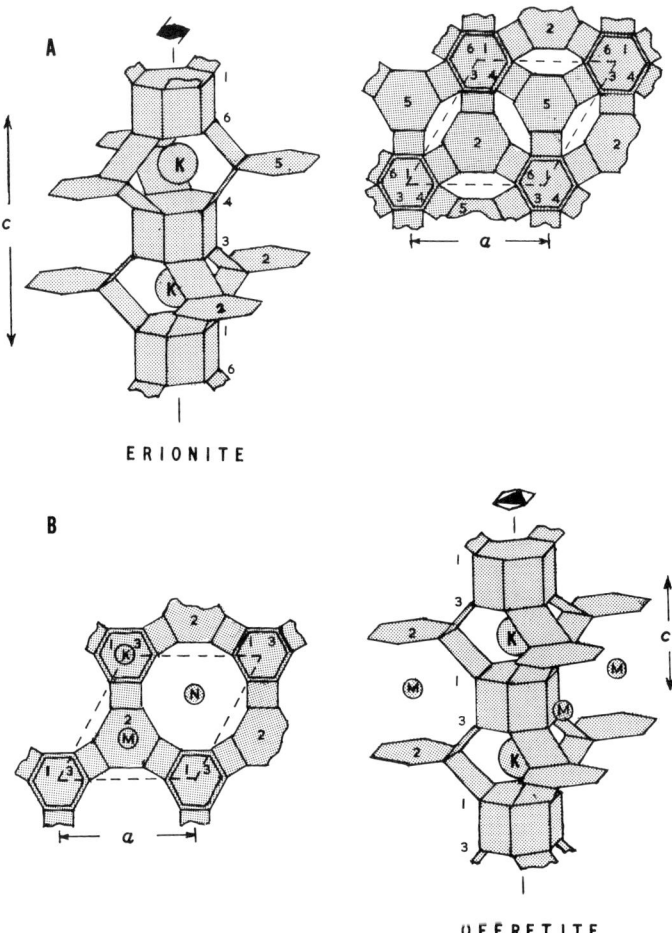

Figure 1. Projections of the frames of (A) erionite and (B) offretite. Relative heights of the rings of 6 tetrahedra are marked. Some cation sites are shown as K, M, and N.

mentioned above (*10*). Intensity data have been recorded with a Hilger Automatic Linear Diffractometer from another sample of erionite from Maze [Harada *et al.* (*8*)], and further refinement will be attempted.

Offretite was discovered by Gonnard (*7*) in 1890 on Mt. Simiouse, Montbrison, France. From x-ray powder data, Hey and Fejer (*9*) stated that it was identical with erionite. Bennett and Gard (*1*), however, found the *c*-spacing to be half that of erionite. Comparison of x-ray rotation photographs (Figure 2) and electron diffraction patterns (Figure 3) show clearly that all reflections with l odd for erionite are absent from the offretite patterns. Sheppard and Gude (*13*) have shown since

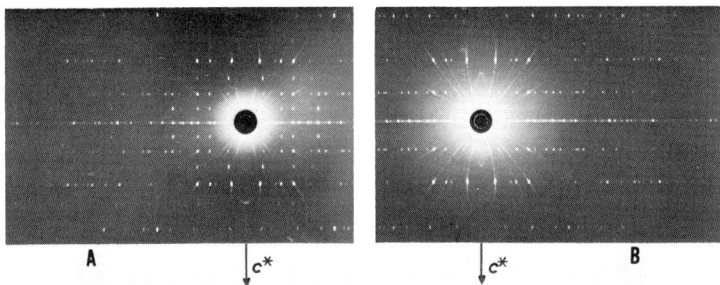

Figure 2. X-ray rotation photographs of (A) erionite and (B) offretite. Reflections with l odd are absent from (B), showing that c is halved for offretite.

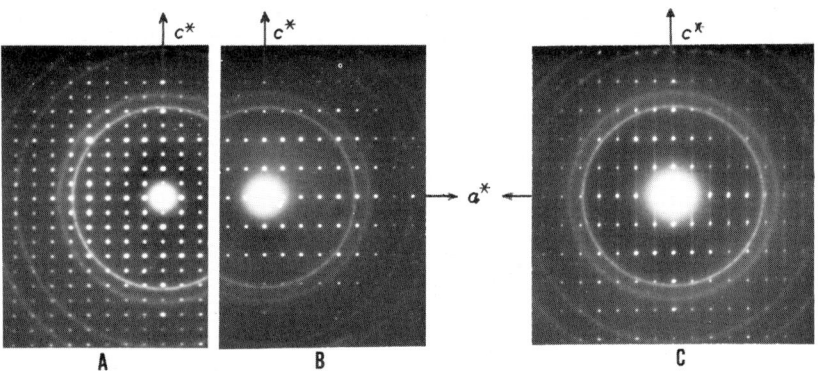

Figure 3. Electron diffraction of h0l zones: (A) erionite, (B) offretite, (C) offretite with streaks parallel to c^* indicating disorder

that the optic signs are different; offretite has negative, and erionite positive, elongation. They also gave accurate unit cell dimensions, $a = 13.291$, $c = 7.582$ A for offretite, and a chemical analysis corresponding to cell contents of $K_{1.1}Ca_{1.1}Mg_{0.7}Si_{12.9}Al_{5.2}O_{36.0}:15.3H_2O$. Bennett and Gard proposed a structure for the offretite frame in which the rings of 6 tetrahedra are stacked AABAAB— (Figure 1b). An ordered offretite should have channels with free diameter 6.3 A, and should be capable of sorbing much larger molecules or cations than would erionite, but a very small degree of stacking disorder would constrict the channels with windows identical with those in erionite. The resulting cavities, however, would be longer than those of erionite, with far-reaching effects on the diffusion rates and other properties [Robson *et al.* (*12*)]. Streaks parallel to c^* on electron diffraction patterns of some flakes (*e.g.*, Figure 3c) indicated disorder of this type which would restrict the channels in this way.

Intensity data have been collected with the Hilger A.L.D. for 363 independent reflections from a prism of the Mt. Simiouse specimen B.M. 68970. Using atomic coordinates derived from those of Staples and Gard for erionite, a residual $R = 0.34$ was obtained. Three-dimensional Fourier syntheses and least squares refinement improved R to 0.15. Displacements of the frame atoms have distorted both the double and single 6-rings to distinctly trigonal symmetry. One K^+ ion was identified in each cancrinite cage. Remaining cations were accounted for by partially occupied sites located on the axes parallel to c of the single 6-rings and the large channels (M and N in Figure 1b). Smaller peaks were interpreted as water molecules coordinated to these cations; some water molecules were octahedrally disposed around cations N to form columns packing the large channels. Refinement is proceeding.

Synthetic zeolites suitable for single-crystal x-ray analysis usually can be prepared only under special conditions [see e.g., Ciric (5)]. Electron diffraction patterns of crystals as small as 1 micron diameter, however, can be obtained readily. Given suitable means for orienting the crystals, electron diffraction is particularly useful for detecting disordered intergrowths of erionite and offretite, which were shown by Bennett and Gard (1) to be difficult to distinguish from offretite by x-ray powder techniques. Natural samples described as erionite from 12 localities and several synthetic samples have been examined with electron microscopy and diffraction to determine their structural types and the degree of disorder, if any. Synthetic samples included: Zeolite T from Linde Division of Union Carbide (3); 3 samples of "erionite" from Esso Research Laboratories, Baton Rouge, La. (12); Zeolite T from the batch prepared and assessed by Sherry (14). The following conclusions were drawn:

(a) Only one natural source of offretite, the type locality in France, was identified. Some fragments were disordered.

(b) All the other natural samples were fully ordered erionite. No streaking of reflections, parallel to c^* or otherwise, was detected.

(c) No fully ordered synthetic erionite has been observed yet.

(d) One sample of fully ordered synthetic offretite was identified among the Esso samples. It differs morphologically from any of the synthetic disordered erionites.

The synthetic offretite is particularly interesting, as it appears to have the ordered structure necessary for unrestricted channels. We do not have details of the method of preparation. All diffraction patterns (e.g., Figure 4a) were weak, but completely free from spots with 1 odd or streaks. The particles were "sausage-shaped" with c-elongation (Figure 4b). The rounded shape suggested a crystalline core enclosed in an amorphous layer, but this was discounted by taking dark-field micrographs using only diffracted beams, so that the crystalline regions ap-

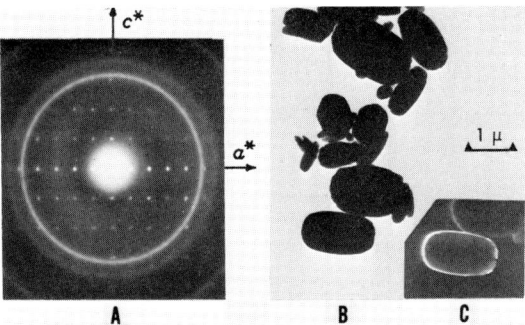

Figure 4. Synthetic offretite: (A) Electron diffraction of the h0l zone; odd-l reflections are absent. (B) Typical micrograph of "sausage-shaped" particles with c-elongation. (C) Dark-field image using electrons diffracted by the bottom particle, showing that the outer layer is crystalline.

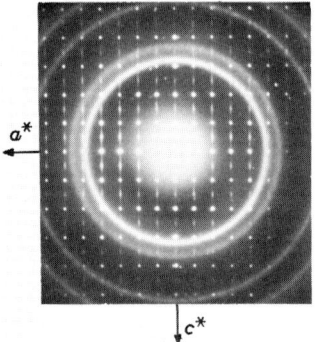

Figure 5. Electron diffraction of h0l zone of a crystal of disordered synthetic erionite; odd-l spots are diffuse and streaked parallel to c^*

peared bright. The bright outer layer in Figure 4c is therefore crystalline; the thicker center is dark owing to inelastic scatter.

The other synthetic samples comprised laths. In electron diffraction patterns, spots with l odd were always diffuse and often streaked along c^* (e.g., Figure 5). Assessment of the proportion of erionite in such intergrowths is technically important, as it affects diffusion rates and catalytic properties [Robson et al. (12)]. The 10.1, 20.1, and 21.1 x-ray powder lines are quite strong for ordered erionite, but they were either undetectable or very weak and diffuse for the other 2 Esso samples.

Intensities of odd-l spots on electron diffraction patterns indicated a considerable proportion of erionite, while streaking showed the presence of lamellae less than 50 A wide in the c-direction. Sherry (*14*) estimated 2–3% of erionite in his sample of Zeolite T. In electron diffraction patterns of this and the Linde sample, the odd-l spots were elongated, but not streaked, in the c^*-direction. Preliminary estimation of these intensities suggests a proportion of erionite significantly higher than 3%. The causes of these discrepancies must be investigated. The proportion of erionite controls the integrated intensity of each spot with l odd, but elongation parallel to c^* increases with "thinness" of ordered domains of erionite below a few hundred A. Where lamellae are extremely thin, x-ray lines become so diffuse that they are indistinguishable from background, although they are still visible as streaks with electron diffraction. Independent evidence for this was observed by Bhatty *et al.* (*2*), who found that several normally strong lines for anorthite, which were absent owing to twinning from x-ray photographs of certain synthetic samples, could be seen clearly with electron diffraction as pairs of spots joined by streaks. An x-ray estimate in this case would be zero instead of the true 100%, as the twins are structurally identical. Electron diffraction intensities also must be treated with caution, however, as multiple diffraction can enhance weak reflections.

Acknowledgment

The authors thank the British Museum (Natural History), K. Harada, H. E. Robson, and H. S. Sherry for specimens and advance information, L. Ingram and B. G. Cooksley for collecting the A.L.D. intensity data, and H. F. W. Taylor for his interest in this work and for advice on the structure analysis.

Literature Cited

(1) Bennett, J. M., Gard, J. A., *Nature, London* **1967**, 214, 1005.
(2) Bhatty, M. S. Y., Gard, J. A., Glasser, F. P., *Mineral. Mag.* **1970**, in press.
(3) Breck, D. W., Acara, N. A., U. S. Patent **2,950,952** (1960).
(4) Chen, N. Y., Rosinski, E. J., Wilson, J. R., Jr., private communication to H. S. Sherry, 1969.
(5) Ciric, J., *Science* **1967**, 55, 689.
(6) Eberley, P. E., Jr., *Am. Mineralogist* **1964**, 49, 30.
(7) Gonnard, F., *Compt. Rend.* **1890**, 111, 1002.
(8) Harada, K., Iwamoto, S., Kihara, K., *Am. Mineralogist* **1967**, 52, 1785.
(9) Hey, M. H., Fejer, E. E., *Mineral. Mag.* **1962**, 33, 66.
(10) Kawahara, A., Curien, H., *Bull. Soc. Franc. Mineral. Crist.* **1969**, 92, 250.
(11) Peterson, D. L., Helfferich, F., Blytas, G. C., *J. Phys. Chem. Solids* **1965**, 26, 835.
(12) Robson, H. E., Hamner, G. P., Arey, W. F., Jr., ADVAN. CHEM. SER. **1971**, 101, 607.

(13) Sheppard, R. A., Gude, A. J., III, Am. Mineralogist **1969**, 54, 875.
(14) Sherry, H. S., Proc. Intern. Conf. Ion Exchange, London, 1969, **1970**, in press.
(15) Shimazu, M., Kawakami, T., J. Japan. Assoc. Mineral. Petrol. Econ. Geol. **1967**, 57, 68.
(16) Staples, L. W., Gard, J. A., Mineral. Mag. **1959**, 32, 261.

RECEIVED February 10, 1970.

Discussion

W. H. Flank (Houdry Laboratories, Marcus Hook, Pa. 19061): Some micrographs of offretite crystallites have been observed in which the ends along the "c" axis appeared as sharp irregular terminations, like the fractured end of a bundle of prismatic needles. The over-all shape was sausage-like. This supports Gard's conclusion from the dark-field micrographs that the outer layer, as well as the crystallite core, is crystalline rather than amorphous.

W. Sieber and **W. M. Meier** (Eidgenössische Technische Hochschule, Zurich): A new member of the chabazite group, tentatively named LOSOD, has been synthesized in our laboratory. The synthesis mixture contains sodium and organic cations, but only sodium is built into the structure. The synthetic zeolite is hexagonal ($a = 12.91$ and $c = 10.54$ Å), and its framework is based on an ABAC stacking sequence of single 6-membered rings (W. Thöni and W. M. Meier, in preparation).

J. A. Gard: This suggests that two distinct mechanisms of synthesis are involved. Discrete cancrinite cages may form around K^+ ions during synthesis of erionite, offretite, and zeolite L and act as precursors which condense with similar units to form double 6-rings and single rings where the columns cross-link. K^+ ions are apparently not essential for synthesis of structures comprising only 6-rings, as they cannot be made by condensing whole cancrinite cages.

18

Linde Type B Zeolites and Related Mineral and Synthetic Phases

WILLIAM C. BEARD[1]

Union Carbide Corp., Tarrytown Technical Center, Tarrytown, New York 10591

> *The Linde Type B zeolites, synthesized in the system $Na_2O-SiO_2-Al_2O_3-H_2O$, have been correlated with synthetic phases produced by Barrer, and Taylor and Roy on the basis of powder x-ray diffraction patterns which show similarity with those of the mineral zeolites phillipsite, harmotome, and gismondine. The complex structural relationships among these zeolite phases are discussed, and the difficulties in identifying zeolite structures on the basis of a general similarity of x-ray powder diffraction patterns are illustrated. Structurally, the B zeolites may represent the following possibilities: (1) Displacive transformations due to variable cation composition and water content; (2) Twinning, such as is commonly encountered in phillipsite to yield lattice constants identical to the single crystal, but tetragonal diffraction pattern symmetry; (3) Intergrowths or stacking faults of several members within the phillipsite group.*

The Linde Type B zeolites are synthesized in the $Na_2O-SiO_2-Al_2O_3-H_2O$ system, and have an adsorption pore size of about 3.5A (6). The synthesis of various B-zeolites under a variety of conditions and their sequence of formation indicates that they are thermodynamically more stable than the more open structured zeolites, A, X, and Y. There exists a series of variants of synthetic B phases arbitrarily designated by "B" with subscripts 1 through 8. The designation of the zeolites in this series is based on differences in their respective x-ray powder diffraction patterns. These differences are of varying degree, and the major x-ray diffraction peaks are common to all phases; hence their tentative classifi-

---

[1] Present address: Department of Geology, The Cleveland State University, Cleveland, Ohio 44115.

cation collectively as "B" zeolites. The characterization and designation of the variants of the synthetic B phases was initially proposed by E. M. Flanigen and E. R. Kellberg of this laboratory; unpublished work.

The B zeolites have been called, at various times, phillipsite-like, harmotome-like, Na-P-like, and gismondine-like phases. This nomenclature has arisen by comparison with the x-ray diffraction patterns of mineral zeolite specimens. Since the B zeolites first were identified, however, the structures of phillipsite, harmotome, and gismondine have been determined, and a structure was proposed by Barrer (2), based on x-ray powder diffraction data, for Na-P1, the equivalent of cubic Linde B_1.

The following discussion attempts to explain the previous confusion of describing the B zeolite structures in terms of mineral zeolites by showing similarities among structures of the mineral phases and x-ray powder patterns of the mineral and synthetic phases.

Discussion

Harmotome. The structure of harmotome, $Ba_2Al_2Si_{12}O_{32} \cdot 12H_2O$, was determined by Sadanaga et al. (8). They give the space group as $P2_1/m$ and lattice constants $a_o = 9.87$, $b_o = 14.14$, $c_o = 8.72A$; $\beta = 124°50'$. That is, the structure is monoclinic, but the deviation from an orthorhombic cell is very slight. The pseudorhombic cell has $\beta = 90°23'$, $a_o = 9.87$, $b_o = 14.14$, and $c_o = 14.3A$, differing only slightly from a tetragonal cell. Figure 1 shows the relationship of the cells described above. The structure consists of double chains of $(Si,Al)O_4$ tetrahedra folded to an s-shaped configuration along the b-axis direction, offset in the direction normal to the a-b plane alternatingly by 1/2 the pseudorhombic c_o parameter, and connected by 4-rings tilted to the plane of the folded chains.

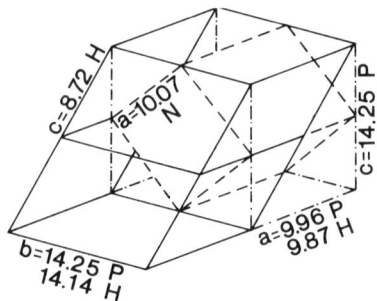

Figure 1. Relationship of the monoclinic harmotome cell (H) and the cubic Na-P1 cell (N) to the orthorhombic phillipsite cell (P)

Phillipsite. The crystal structure of phillipsite, probable composition $(K_xNa_{1-x})_5Si_{11}Al_5O_{32} \cdot 10H_2O$, was determined by Steinfink (12), space group, $B2mb$, $a_o = 9.96$, $b_o = 14.25$, $c_o = 14.25$A. From the lattice constants, one might question why the structure is not designated as tetragonal, but the geometrical arrangement of atoms about the c-axis does not permit an axis of 4-fold symmetry. The framework of this zeolite is essentially the same as that of harmotome.

Gismondine. The general arrangement of $(Si,Al)O_4$ tetrahedra in the framework of gismondine was proposed by Smith and Rinaldi (11) as one of the several possible arrangements of crosslinking the "double-crankshaft" tetrahedral chains common to feldspar structures. Fischer (5) confirmed that the N-arrangement of Smith and Rinaldi was indeed the gismondine structure. Like the other structures discussed above, gismondine has a similar x-ray diffraction pattern. The original designation of the B-zeolites was as gismondine-like phases (4). The structure model of gismondine shows double tetrahedral chains common to all of these structures.

Na-P1. The structure proposed for Na-P1 by Barrer (2) has the space group $Im3m$ with $a_o = 10.0$A. The framework is formed by joining double 4-rings of $(Al,Si)O_4$ tetrahedra (cubes) on their corners so that every cube is connected to 8 other cubes.

Barrer (2) also proposed a distortion of the cubic phase to a tetragonal structure (tetragonal Na-P2) by a displacive transformation.

The relationship of the cubic structure to harmotome/phillipsite, as suggested by Barrer before the structures of the latter 2 were known, is shown in Figure 1. The dimensions of the cell edge connecting the midpoints on the b and c unit cell lengths [$b_o = c_o = 14.25$A, Steinfink (12)] is 10.07A. This gives a cell with edges 10.07, 9.96A, which can be visualized as a distortion from an ideal cubic structure.

Table I. Typical Chemical Analyses of Linde Na-B Zeolites[a]

Zeolite	Composition, Moles/Al_2O_3			Range of $SiO_2:Al_2O_3$ Observed
	Na_2O	SiO_2	H_2O	
B_1	0.95	3.35	4.79	2.16–3.35
B_2	0.99	4.07	5.70	3.65–4.07
B_3	1.05	3.80	4.70	—
B_4	0.92	3.50	4.24	2.98–5.07
B_5	0.88	3.38	4.80	3.38–3.52
B_6	0.87	2.80	4.66	—
B_7	1.04	3.74	3.50	—
B_8	1.02	5.01	4.28	—[b]

[a] Unpublished data of E. M. Flanigen and E. R. Kellberg.
[b] The composition listed for B_8 is the starting composition of a cubic B zeolite before dehydration.

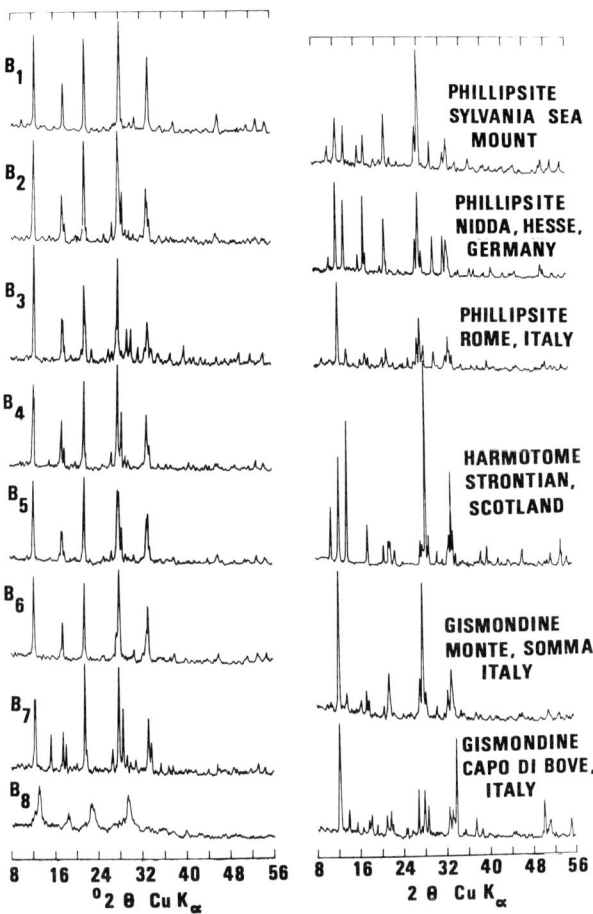

Figure 2. X-ray diffraction patterns of Linde B zeolites and phillipsite, harmotome, and gismondine

Although the cell dimensions for comparing the cubic Na-P1 with the harmotome/phillipsite framework are in good agreement, we now know from the structures described above that the arrangement of tetrahedra in the harmotome/phillipsite structure is definitely different from that proposed by Barrer (2) for the Na-P1 structure.

On the other hand, by looking at models of these 2 structures and superimposing 1 on the other, striking similarities are immediately apparent which are otherwise difficult to discern. First, the interplanar spacings essentially are identical, alone enough to give the suspicion that the x-ray patterns would probably be similar. Secondly, for a given volume there are the same number of tetrahedra, leading to similar densities and pore volumes.

Taylor and Roy (*13*) discussed ion exchanged derivatives of tetagonal Na-P structures. They define the "P zeolite group" as that group of zeolites composed of members having an aluminosilicate framework linked in a manner identical to that of the cubic Na-P1 zeolite, namely tetrahedral cube units joined by their corners. They state that the Na-P1 structure of Barrer (*2*) cannot be considered as a member of the harmotome/phillipsite group because the different linking of tetrahedra in the 2 structures would require a reconstructive type transformation. They studied the effect of ion exchange on the structure of the tetragonal Na-P zeolite and noted a maximum range of 7% in the c-dimension. On the basis of powder x-ray diffraction data, they distinguished 3 main structure divisions, depending on cations present: (1) Primitive cell, $a \geqslant c$: tetragonal Li, Na; cubic Mg, Ni, Cu; (2) Body-centered cell $a > c$: tetragonal K, Rb, Cs, Ag; (3) Body-centered cell, $c \geqslant a$: tetragonal Ca, Sr, Ba, Pb; cubic Cd.

Barrer's (*2*) Na-P1 cubic structure could be distorted to a body-centered tetragonal structure by displacive transformation, but distortion to a primitive tetragonal lattice without a reconstructive transformation seems impossible. Therefore, primitive cell tetragonal varieties of Na-P cannot be considered as belonging to the same group as the body-centered cubic and tetragonal Na-P's for the same reason given by Taylor and Roy (*13*) for excluding harmotome and phillipsite.

Linde Type B Zeolites. The Linde B zeolites are considered structurally the same as Barrer's Na-P phases. Table I lists the typical chemical compositions of the B zeolites. For some B zeolites, a range of SiO_2/Al_2O_3 is listed. The maximum observed variation in SiO_2/Al_2O_3 in the B series is from 2.2 to 5.1.

Synthetic phases apparently related to the phillipsite group occur in the K and K-Na systems. One such phase, designated Zeolite W by Linde (*7*), appears to be analogous to Barrer's K-M (*1*). Barrer also has reported a Ba-M phase in the barium system which is described as harmotome-like (*3*). Related phases which occur in systems other than pure Na will not be discussed further here.

X-ray diffraction patterns of Linde B zeolites are shown in Figure 2 and Table II. The similarity of the main diffraction peaks is obvious. The synthetic phases produced by various workers have been arranged in Table III to show their relationship to each other. Zeolite B_1 is correlated with the cubic body-centered phases of Barrer (*2*) (Na-P1) and Taylor and Roy's Na-P$_c$ (*13*). The Linde B_2, B_3, B_5, and B_6 phases are similar to the tetragonal body-centered phases of Barrer (*2*) (Na-P2) and

Table II. X-Ray Diffraction

B_1		B_2		B_3		B_4	
d, A	I/I_0	d, A	I/I_0	d, A	I/I_0	d, A	I/I_0
—	—	7.14	83	—	—	7.14	79
7.08	86	—	—	7.08	100	7.08	66
4.98	44	5.04	39	5.04	38	5.07	45
—	—	5.01	25	4.98	36	—	—
—	—	4.93	15	4.93	14	4.93	20
—	—	—	—	4.21	12	—	—
4.10	83	4.11	83	4.11	66	4.11	83
—	—	—	—	—	—	4.06	15
—	—	—	—	3.88	13	—	—
—	—	—	—	3.41	13	—	—
—	—	3.33	18	3.33	11	3.34	17
—	—	3.21	100	3.20	89	3.21	100
3.18	100	3.19	60	3.16	17	—	—
—	—	3.12	43	—	—	3.12	55
—	—	3.05	12	3.02	29	3.05	13
—	—	2.99	12	2.95	30	3.00	9
2.88	15	2.90	9	2.83	15	—	—
—	—	2.70	46	2.72	14	2.71	53
—	—	2.68	35	2.68	36	—	—
2.67	69	2.66	22	2.64	14	2.66	24
—	—	—	—	2.54	10	2.54	8
—	—	—	—	2.40	10	2.44	7
2.36	13	—	—	—	—	2.39	7
—	—	—	—	2.45	16	—	—
—	—	2.21	7	2.20	6	2.21	9
—	—	2.16	8	2.17	6	—	—
—	—	1.98	12	2.10	7	1.99	9
1.97	19	—	—	1.97	7	1.97	8
—	—	—	—	1.82	12	—	—
1.77	10	1.76	8	1.76	10	—	—
1.72	17	1.72	9	—	—	1.73	12
1.67	13	1.69	8	1.69	12	1.69	9

[a] Unpublished data of E. M. Flanigen and E. R. Kellberg.

Taylor and Roy's Na-P$_t$ (13), and possibly represent various degrees of distortion (displacive transformation) of the cubic body-centered structure. B_6 is quite similar to B_1 except for doubleting of peaks near 18°, 28°, and 34° $2\theta(\text{CuK}_\alpha)$. B_6 probably represents the least distortion from the cubic structure (B_1) of all the other B zeolites. Zeolites B_2 and B_5

Spacings of Linde Na-B Zeolites[a]

B_5		B_6		B_7		B_8	
d, A	I/I_0	d, A	I/I_0	d, A	I/I_0	d, A	I/I_0
7.14	96	7.14	91	—	—	—	—
—	—	—	—	7.08	68	—	—
—	—	—	—	7.03	68	—	—
—	—	—	—	—	—	6.56	100
—	—	—	—	5.75	38	—	—
—	—	5.13	13	—	—	—	—
5.07	39	5.04	42	5.04	40	—	—
5.01	40	—	—	—	—	—	—
4.93	14	—	—	4.90	29	—	—
—	—	—	—	—	—	4.79	50
—	—	—	—	4.42	10	—	—
4.11	100	4.11	87	4.11	100	—	—
—	—	—	—	4.04	22	—	—
—	—	—	—	—	—	3.86	67
3.33	17	—	—	3.33	23	—	—
3.21	88	3.23	18	3.20	100	—	—
3.18	86	3.19	100	—	—	—	—
3.12	44	—	—	3.11	50	—	—
3.04	13	—	—	3.04	19	—	—
3.00	8	—	—	2.98	12	3.00	88
2.90	11	2.90	16	2.89	14	—	—
2.71	25	2.71	35	2.70	53	—	—
2.68	58	2.69	61	—	—	—	—
2.67	20	—	—	2.65	28	—	—
—	—	—	—	2.53	12	—	—
—	—	—	—	2.44	9	—	—
—	—	2.37	12	2.39	9	—	—
—	—	—	—	2.05	7	—	—
—	—	—	—	1.98	12	—	—
1.97	14	1.97	16	—	—	—	—
—	—	—	—	1.87	9	—	—
—	—	1.78	12	1.76	9	—	—
1.72	13	1.72	14	1.72	13	—	—
—	—	1.68	14	1.69	10	—	—

are characterized by splitting of the lines in B_1 into doublets. B_3 is similar to B_2 and B_5 with 2 additional reflections at 22.9° and 40° 2θ. Zeolites B_4 and B_7 resemble the tetragonal primitive structure (Na-P_t) of Taylor and Roy (13) and show doublets in the first main peak of the pattern. Zeolite B_8 is a phase produced by partial dehydration of the cubic B_1

zeolite. The orthorhombic Na-P phase reported by Barrer (2) was not obtained by Taylor and Roy (13) or by Linde.

X-ray diffraction patterns of the mineral zeolites phillipsite, harmotome, and gismondine are shown in Figure 2 and Table IV. The Sylvania Sea Mount phillipsite is a deep-sea specimen on decomposed basalt obtained from the Scripps Institute, La Jolla, Calif. The Nidda, Germany, and Rome, Italy, phillipsites are from igneous rocks. The Nidda x-ray pattern checks in all major peaks with the ASTM card (13-455) for a phillipsite from the same locality, and is from the Harvard Museum collection (No. 102839). The Rome, Italy, specimen came from Ward's, Rochester, N. Y.

The harmotome specimen, from Strontian, Scotland (Harvard No. 86545), agrees with the ASTM pattern for a specimen from Strontian (13-494), but does not check with the pattern for one from North-West Ross-shire, Scotland (9-480).

Of the gismondine specimens Monte Somma (Ward's) and Capo di Bove (Harvard Museum), only the latter could be said to agree with the ASTM data (13-495) of a gismondine specimen from Fritz's Is., Pa.

From Table IV and Figure 2, the similarity in x-ray patterns for phillipsite, harmotome, and gismondine is apparent. All 3 zeolites have the following approximate interplanar spacings in common: 8.00, 7.15, 6.40, 5.35, 5.04, 4.12, 3.25, 3.20, 2.69. The variations among the patterns of 2 or more specimens identified as the same species are often as great as the variations between species. From the discussion of lattice parameter changes with cation composition and water content by Taylor and Roy (13, 14), changes in cell symmetry and size with accompanying diffraction pattern peak shifts and splitting are to be expected in natural zeolite specimens from different localities and exposed to different cation environments.

At least 6 of the diffraction peaks listed above as being common to the 3 mineral zeolites, namely the 7.15, 5.04, 4.12, 3.25, 3.20, and 2.69,

Table III. Relationship Between Linde B-Zeolites and Synthetic Phases of Barrer, and Tayor and Roy

Linde	Barrer (2)	Taylor and Roy (13)
B_1	Na-P1 (Cubic, body-centered)	Na-P_c (Cubic, body-centered)
B_2, B_3, B_5, B_6	Na-P2 (Tetragonal, body-centered)	Na-P_t (Tetragonal, body-centered)
B_4, B_7	—	Na-P_t (Tetragonal, primitive)
B_8	—	—
—	Na-P3 (Orthorhombic)	—

are common to the B zeolites. For this reason, they were originally identified as being phillipsite-, harmotome-, or gismondine-like phases.

One source of the difficulty may arise from the incorrect identification of mineral specimens whose x-ray patterns were used as standards for comparison. X-ray powder patterns of materials used in structure determinations would be of great help in clarifying the problem of identifying these zeolites. An alternative to the actual powder patterns is the calculation of powder patterns from structure data, as demonstrated by Smith (9).

From the foregoing discussion, it is seen that the mineral zeolites and Na-P1 possess similar lattice constants, d-values, and are derived from different arrangements of a common structural element, the double tetrahedral chain or "double-crankshaft." Added to this is the possibility of still further structures as yet unknown, based on the same structural units. This was pointed out by Smith and Rinaldi (11) in describing the series of structures derived by changing the tetrahedra in a 4-ring as either pointing upward (U) or downward (D). They said, "Because all types of structures based on the UUDD and related chains should give similar powder patterns, it is possible that some of the complexity may arise because of the existence of several unrecognized members of this structural family."

Conclusion

The Linde Type B zeolites have been correlated with synthetic phases produced by Barrer (2) and Taylor and Roy (13) on the basis of x-ray powder diffraction data. The powder patterns of the B zeolites also show similarity with those of the mineral zeolites phillipsite, harmotome, and gismondine.

Since the structures of the B zeolites have not been determined, classification is difficult, but from the x-ray powder diffraction data it seems that an assignment can be made to the phillipsite group as defined by Smith (10), *i.e.*, being a group of structures formed from parallel four- and eight-membered rings of $(Si,Al)O_4$ tetrahedra.

Assignment to this group does little to define the actual structure, however, for although Na-P1 and harmotome/phillipsite have similar lattice constants, etc., they have quite different structures.

The B series may represent displacive transformations from one or more basic structures. Structural changes due to cation composition and hydration state in the Na-P zeolites have been discussed by Taylor and Roy (13, 14). The B zeolite series, however, all contain sodium cations as synthesized, and the water content (Table II) is essentially constant, except for B_8 which is a dehydrated form produced from B_1. Another

Table IV. X-Ray Diffraction Spacings of

Phillipsite[a]		Phillipsite[b]		Phillipsite[c]	
d, Å	I/I_0	d, Å	I/I_0	d, Å	I/I_0
8.25	19	—	—	—	—
—	—	8.04	20	7.97	5
7.15	69	7.14	100	7.19	100
6.39	16	6.42	80	6.42	10
5.37	28	5.36	23	5.40	10
5.04	31	5.04	85	5.07	20
—	—	4.96	25	4.96	10
—	—	4.27	10	4.31	5
4.12	45	4.11	60	4.13	20
3.26	40	3.25	40	3.29	20
3.19	100	3.19	88	3.21	60
—	—	3.13	28	3.14	20
2.96	29	2.91	43	2.93	10
2.75	12	2.75	45	2.76	15
2.69	28	2.69	40	2.71	20
—	—	—	—	2.68	30
—	—	2.53	8	2.54	10
—	—	2.39	10	2.39	10
—	—	2.33	10	2.34	10
—	—	2.17	13	—	—
—	—	1.79	18	—	—

[a] Phillipsite, Sylvania Sea Mount, near Bikini Atoll.
[b] Phillipsite, Nidda, near Giessen, Germany.
[c] Phillipsite, Via Laurentia, Rome, Italy.
[d] Harmotome, Strontian, Scotland.
[e] Gismondine, Capo di Bove, Italy.
[f] Gismondine, Monte Somma, Italy.

Phillipsite, Harmotome, and Gismondine Zeolites

Harmotome[d]		Gismondine[e]		Gismondine[f]	
d, A	I/I$_0$	d, A	I/I$_0$	d, A	I/I$_0$
—	—	—	—	—	—
8.04	29	—	—	—	—
—	—	7.25	100	—	—
7.08	53	7.14	87	7.196	100
6.33	70	6.33	23	6.463	15
—	—	5.72	13	—	—
—	—	5.34	9	5.405	10
4.98	20	5.04	15	5.096	17
—	—	—	—	4.983	13
—	—	4.90	19	—	—
—	—	4.65	11	—	—
4.27	10	4.25	19	4.311	12
4.10	13	4.10	23	4.133	35
4.06	14	4.04	13	—	—
4.02	13	—	—	—	—
3.88	8	—	—	—	—
—	—	3.63	6	—	—
—	—	3.59	11	—	—
—	—	3.48	9	—	—
—	—	3.33	43	—	—
3.23	15	3.25	11	3.278	30
3.20	10	3.20	43	3.209	97
3.12	100	3.12	30	3.143	20
3.06	15	—	—	2.940	12
2.91	7	—	—	—	—
2.71	16	2.75	28	2.763	22
2.69	46	2.71	26	2.706	35
2.66	17	2.65	89	—	—
2.62	7	—	—	—	—
—	—	2.54	9	2.578	7
—	—	2.37	19	2.392	7
2.32	8	2.33	11	2.344	7
2.25	10	—	—	—	—
2.15	6	—	—	—	—
1.95	10	—	—	—	—
—	—	1.82	34	—	—
—	—	1.80	6	—	—
—	—	1.79	15	1.791	10
—	—	1.787	17	—	—
1.77	7	1.780	17	—	—
—	—	—	—	1.728	7
1.70	14	—	—	—	—
—	—	1.667	19	—	—
—	—	1.406	13	—	—
—	—	1.393	17	—	—

possibility for variation in the B series is twinning. Twinning in phillipsite is quite common, and indeed, Steinfink (12) noted that a twined phillipsite crystal gave a diffraction pattern displaying tetragonal symmetry with the same unit cell dimensions as the untwinned crystal.

Determination of the structural relationships between the Linde Type B zeolites and the related mineral zeolites by comparison of x-ray powder data would be greatly aided if powder data were available on the same specimen on which structure determinations were made.

The complex relationships among the family of zeolite structures discussed here aptly illustrate the difficulties in identifying zeolite framework structures on the basis of a general similarity in x-ray powder diffraction patterns.

Acknowledgment

The author is grateful to Union Carbide Corp. for permission to publish this work, and acknowledges the contributions of R. M. Milton, D. W. Breck, E. M. Flanigen, and E. R. Kellberg of the Linde Research Laboratory of Union Carbide Corp.

Literature Cited

(1) Barrer, R. M., Baynham, J. W., *J. Chem. Soc.* **1956**, 2882.
(2) Barrer, R. M., Bultitude, F. W., Kerr, I. S., *J. Chem. Soc.* **1959**, 1521–28.
(3) Barrer, R. M., Marshall, D. J., *J. Chem. Soc.* **1964**, 2296.
(4) Breck, D. W., Eversole, W. G., Milton, R. M., *J. Am. Chem. Soc.* **1956**, 78, 2338.
(5) Fischer, K., *Am. Mineralogist* **1963**, 48, 664–72.
(6) Milton, R. M., U. S. Patent **3,008.803** (1961).
(7) Milton, R. M., U. S. Patent **3,012,853** (1961).
(8) Sadanaga, R., Marumo, F., Takeuchi, Y., *Acta Cryst.* **1961**, 14, 1153–63.
(9) Smith, D. K., *Norelco Reptr.* **1968**, 15, 57–65.
(10) Smith, J. V., *Mineral. Soc. Am. Spec. Paper* **1963**, 1, 281–290.
(11) Smith, J. V., Rinaldi, F., *Mineral. Mag.* **1962**, 33 (258), 202–12.
(12) Steinfink, H., *Acta Cryst.* **1962**, 15, 644–51.
(13) Taylor, A. M., Roy, R., *Am. Mineralogist* **1964**, 49, 656–82.
(14) Taylor, A. M., Roy, R., *J. Chem. Soc.* **1965**, 4028–43.

RECEIVED February 13, 1970.

Discussion

W. M. Meier (Eidgenössische Technische Hochschule, Zurich): The structure of NaPl (B_1) has recently been solved using x-ray intensities obtained from a multiply-twinned crystal in addition to powder data. Refinement proceeded to an intensity R value of 0.077. The structure is based on a gismondine-type framework and is thus noncubic. The maximum possible symmetry of the framework is $14_1/amd$. The framework is remarkably flexible, and displacive changes are accompanied by marked changes of the lattice constants and symmetry. Since this type of framework can readily undergo displacive changes and twinning, it is no longer surprising that several apparently different P-phases have been recorded. Full details will be given in a forthcoming paper (C. Baerlocher and W. M. Meier, to be published in Z. Krist.).

G. H. Kühl (Mobil Research & Development Corp., Paulsboro, N. J. 08066): You have shown the x-ray diffraction patterns of three different "phillipsites" from different locations. There are numerous references in the literature to so-called phillipsites. The x-ray diffraction patterns are all different. I think it is time to agree on what phillipsite is. I suggest that the structure determined by Steinfink is that of a real phillipsite. All the other "phillipsites" reported are either mixtures or have different structures and should not be called phillipsite.

W. C. Beard: I agree that we should accept Steinfink's structure determination as that of a real phillipsite and that the practice be extended to cover the other zeolites for which structures have been determined. For practical identification of powder x-ray diffraction patterns, I would suggest using calculated powder diffraction patterns from the structure data as carried out by D. K. Smith (Ref. 9). I plan to do this for the three mineral zeolites discussed in this paper.

19

Crystal Structure of Gismondite, a Detailed Refinement

KARL F. FISCHER and VOLKER SCHRAMM

Lehrstuhl fuer Kristallographie, University of Saarbruecken, Germany

The crystal structure of gismondite, a zeolite with narrow channels, has been refined using about 2200 F_{obs} gathered with a single-crystal diffractometer. All H_2O molecules were located. The anisotropic refinement included determination of the Si/Al distribution and of site occupancies for the cation and 6 H_2O equipoints. Although the Si and Al atoms are nearly ordered, the results suggest different arrangements of H_2O in otherwise identical cages. The final R_w = 4.0% (conventional R = 5.5%).

The crystal structure of natural gismondite from Hohenberg near Buehne/Westfalia was deduced by Fischer (4). The intensity data as well as the structure appeared to be suitable for a detailed refinement. Redetermination of the cell constants gave a = 9.843 ± 0.015; b = 10.023 ± 0.003; c = 10.616 ± 0.005 A; γ = 92°25′ ± 15′ (first setting, space group $P2_1/a$). To obtain the Si/Al distribution as well as physically meaningful r.m.s. vibrational amplitudes, the set of Cu-K_α intensities used for solving the structure had to be extended into the range of Mo-K_α radiation (see Fischer, 5). Of the ca. 16,000 asymmetric reflections outside the Cu-K_α but within the Mo-K_α range, about 250 were selected for measurement. After an intermediate stage of the refinement with copper data alone, structure factors and their contributions from the Si and Al atoms were computed for the Mo-K_α range. Reflections with both large F_{cal} and high contributions from the tetrahedral atoms were measured. Additional measurements were done with Mo-K_α radiation for about 50 reflections in the copper sphere for precise scaling. Furthermore, some 200 reflections were remeasured with Cu-K_α radiation. All intensity data were gathered with a single-crystal diffractometer of the equi-inclination type (2, 6). The same crystal was used as in Ref. 4 (see also 9). The absorption correction employed a method patterned after Weber

(11). This appeared necessary, as the "spherical" crystal was actually an ellipsoid with its rotation axis parallel to the two-fold crystallographic axis c and about 10% shorter than in the a and b direction. No correction for extinction was applied. After proper scaling of the 3 sets of data, multiple (or symmetrical equivalent) reflections were averaged. Thus, the final set of 2205 F_{obs} required only one scale factor. Individual standard deviations were assigned to the data. They are based mainly on counting statistics, peak-to-background ratio, smoothness of the background, and quality of the peak profile.

Refinement

Conventional least-squares refinement with a full-matrix program was carried out starting with the parameters of the crystal structure published earlier (4). Generally, after one cycle with variation of the positional parameters and the scale factor, 2 cycles followed with refinement of the coordinates and the temperature factors only. Thus, correlation effects between scale factor and temperature parameters were minimized. Scattering factors for fully ionized atoms were used and introduced according to the procedure described by Onken and Fischer (8). Using anisotropic vibration amplitudes, convergence was reached with $R = 6.3$ and $R_w = 4.7\%$. At this stage, a three-dimensional Fourier and difference Fourier synthesis was computed. It revealed splitting of $H_2O(4)$ into 2 sites close to each other, each having a population density of about 0.5. In order to test this new model, refinement was continued as above. After a few cycles, R_w reached 4.3%. A significance test patterned after Hamilton (7) showed that the model with 6 H_2O sites provides a superior description of the structure. A second difference Fourier map contained no important maxima.

After convergence, the set of data was divided into 3 parts. Set I consisted of reflections with $\sin \theta / \lambda < 0.3178$ A^{-1} of them. A few were omitted since they were apparently in error owing to extinction or incorrect measurements. With 238 F_{obs}, the over-all scale factor and site occupancies c_i of Ca plus the 6 H_2O equipoints were determined almost independently of the temperature factors. All the other parameters were kept constant. After 3 cycles, the results listed in the upper part of Table I were obtained.

Data set II included 1845 F_{obs} with 0.3178 $A^{-1} < \sin \theta / \lambda < 0.75$; data set III contained 68 reflections with $\sin \theta / \lambda > 0.75$ A^{-1}. They were used in the computation for determining the Si/Al distribution after Fischer (5), to be discussed in detail elsewhere. The results of these computations are listed in the lower part of Table I.

Table I. Site Occupancies and Si/Al Distributions

Equipoint	Composition[a]	
Ca^{2+}	91.8%	(0.5)
$H_2O(1)$	88%	(1)
$H_2O(2)$	100%	(1)
$H_2O(3)$	95%	(1)
$H_2O(4)$	61%	(1)
$H_2O(5)$	48%	(1)
$H_2O(6)$	40%	(1)
Si(1)	ca. 100%	Si^{4+}
Si(2)	ca. 100%	Si^{4+}
Al(1)	ca. 80%	Al^{3+} + 20% Si^{4+}
Al(2)	ca. 100%	Al^{3+}

[a] Est. standard deviation in parentheses.

Table II. Final

Atom	x	y	z	β_{11}
Si(1)	0.18142(9)	0.41470(10)	0.11288(10)	0.00123(8)
Si(2)	0.16038(9)	0.90820(10)	0.87012(10)	0.00114(8)
Al(1)	0.16921(10)	0.09656(11)	0.11329(11)	0.00094(9)
Al(2)	0.14900(10)	0.59053(11)	0.86690(11)	0.00077(9)
O(1)	0.9996(3)	0.0787(4)	0.1562(3)	0.0012(2)
O(2)	0.2125(3)	0.2624(3)	0.0763(3)	0.0038(3)
O(3)	0.0254(3)	0.4361(3)	0.1485(3)	0.0024(3)
O(4)	0.3037(3)	0.2450(3)	0.4027(3)	0.0039(3)
O(5)	0.2136(3)	0.9994(3)	0.9861(3)	0.0048(3)
O(6)	0.2595(3)	0.0449(3)	0.2437(3)	0.0028(3)
O(7)	0.2777(3)	0.4645(3)	0.2288(3)	0.0025(3)
O(8)	0.2253(3)	0.5107(3)	0.9944(3)	0.0023(3)
Ca	0.3537(1)	0.7192(1)	0.0764(1)	0.0033(1)
$H_2O(1)$	0.5023(4)	0.2596(4)	0.1048(4)	0.0036(3)
$H_2O(2)$	0.5410(4)	0.5914(4)	0.1262(3)	0.0060(4)
$H_2O(3)$	0.5020(4)	0.9113(4)	0.1174(4)	0.0056(4)
$H_2O(4)$	0.2306(11)	0.7732(8)	0.2369(9)	0.0243(17)
$H_2O(5)$	0.4028(11)	0.7407(10)	0.3180(9)	0.0157(15)
$H_2O(6)$	0.1326(16)	0.7620(17)	0.1794(20)	0.0163(21)

[a] Standard deviations of the last digit(s) in parentheses.

After all refinements, a final least-squares cycle using all 2205 F_{obs} and varying all parameters except those listed in Table I yielded the following discrepancy factors:

$$\text{conventional } R = 5.5\%$$
$$\text{weighted } R_w = 4.0\%$$
$$\text{error of fit} = 3.5$$

General Description of the Structure

The published crystal structure (4) was essentially confirmed, apart from the splitting of one H_2O equipoint. The atomic coordinates and temperature parameters obtained in the final cycle are listed in Table II. The composition found by the detailed structure investigation can be derived from Table I as $Ca_{0.92}[Al_{1.8}Si_{2.2}O_8] \cdot 4.3_2\ H_2O$. Pieper (9) reports 4.3 H_2O, determined analytically from the same specimen. The framework of the structure is pseudotetragonal, if all tetrahedral sites are considered equivalent. Channels parallel to the a- and b-axes are limited with respect to their cross-section by 8-membered rings of $(Si,Al)O_4$ tetrahedra.

Atomic Parameters[a]

β_{22}	β_{33}	β_{12}	β_{13}	β_{23}
0.00151(9)	0.00065(8)	0.00020(7)	−0.00023(7)	0.00035(7)
0.00118(9)	0.00083(8)	0.00017(7)	0.00008(7)	−0.00007(7)
0.00163(10)	0.00074(9)	0.00010(8)	−0.00022(7)	−0.00032(8)
0.00128(10)	0.00055(9)	0.00023(8)	0.00011(8)	0.00028(8)
0.0063(4)	0.0031(3)	0.0006(2)	−0.0004(2)	0.0008(2)
0.0019(3)	0.0035(3)	0.0004(2)	0.0014(2)	0.0003(2)
0.0047(3)	0.0015(2)	0.0002(2)	−0.0004(2)	−0.0004(2)
0.0022(3)	0.0022(2)	0.0007(2)	0.0008(2)	−0.0002(2)
0.0038(3)	0.0024(2)	−0.0006(3)	0.0002(2)	−0.0019(2)
0.0036(3)	0.0019(2)	0.0001(2)	−0.0010(2)	−0.0002(2)
0.0039(3)	0.0017(2)	−0.0001(2)	−0.0007(2)	0.0001(2)
0.0021(3)	0.0018(2)	−0.0000(2)	0.0004(2)	0.0009(2)
0.0087(1)	0.0061(1)	−0.0008(1)	−0.0009(1)	−0.0023(1)
0.0087(5)	0.0033(4)	0.0008(3)	0.0002(3)	−0.0006(4)
0.0082(4)	0.0034(3)	0.0014(3)	0.0012(3)	0.0019(3)
0.0080(4)	0.0040(4)	0.0001(3)	0.0014(3)	−0.0024(3)
0.0066(9)	0.0117(13)	−0.0041(12)	0.0133(13)	−0.0024(8)
0.0124(13)	0.0040(8)	−0.0036(11)	0.0047(9)	−0.0019(8)
0.0147(22)	0.0228(30)	0.0107(21)	0.0168(22)	0.0080(20)

Details of the Framework

A list of selected interatomic distances and bond angles is given in Tables III and IV. Neglecting the influence of Ca and Si–O–Al angles on the T–O bond length—which has been pointed out by Brown *et al.* (1)—and averaging all Si–O and Al–O distances, one obtains 1.61_7 and 1.73_8 Å, respectively. They agree well with the Smith-Bailey plot (Figure 2 of

Table III. Selected Interatomic Distances[a] in Å

Si_1–O_2	1.617(2)	Al_1–O_1	1.733(4)	Ca–O_4	2.445(3)
–O_3	1.605(4)	–O_2	1.744(2)	–O_8	2.549(6)
–O_7	1.619(3)	–O_5	1.731(3)	–$H_2O(1)$	2.394(4)
–O_8	1.631(3)	–O_6	1.736(3)	–$H_2O(2)$	2.239(6)
Average	1.618	Average	1.736	–$H_2O(3)$	2.406(7)
				–$H_2O(4)$	2.173(10)
Si_2–O_1	1.610(4)	Al_2–O_3	1.734(4)	–$H_2O(5)$	2.617(3)
–O_4	1.627(3)	–O_4	1.736(4)	–$H_2O(6)$	2.489(12)
–O_5	1.608(3)	–O_7	1.734(3)		
–O_6	1.617(3)	–O_8	1.757(3)		
Average	1.616	Average	1.740		

Tetrahedron Around Si_1		Tetrahedron Around Al_1		Some O–H_2O and H_2O–H_2O Distances		
O_2–O_3	2.699(7)	O_1–O_2	2.860(7)	O_1	–$H_2O(3)$	2.932(3)
–O_7	2.651(4)	–O_5	2.910(5)	O_2	–$H_2O(2)$	2.869(6)
–O_8	2.634(3)	–O_6	2.756(5)	O_2	–$H_2O(5)$	2.968(6)
O_3–O_7	2.630(5)	O_2–O_5	2.805(3)	O_3	–$H_2O(2)$	2.854(5)
–O_8	2.643(5)	–O_6	2.865(4)	O_4	–$H_2O(1)$	2.978(6)
O_7–O_8	2.587(4)	O_5–O_6	2.805(4)	O_6	–$H_2O(4)$	2.727(3)
Average	2.64	Average	2.83	O_8	–$H_2O(2)$	2.860(6)
				$H_2O(1)$–$H_2O(2)$		2.912(6)
Tetrahedron Around Si_2		Tetrahedron Around Al_2		–$H_2O(3)$		2.915(5)
O_1–O_4	2.675(7)	O_3–O_4	2.893(8)	$H_2O(2)$–$H_2O(4)$		2.938(11)
–O_5	2.675(5)	–O_7	2.789(5)	–$H_2O(5)$		2.899(7)
–O_6	2.600(5)	–O_8	2.964(5)	–$H_2O(6)$		2.805(14)
O_4–O_5	2.604(4)	O_4–O_7	2.892(4)	$H_2O(3)$–$H_2O(5)$		2.876(5)
–O_6	2.644(4)	–O_8	2.661(4)	–$H_2O(6)$		2.951(14)
O_5–O_6	2.627(4)	O_7–O_8	2.830(4)	$\{H_2O(4)$–$H_2O(5)$		1.941(13)$\}$
Average	2.64	Average	2.84	$\{$ –$H_2O(6)$		1.142(16)$\}$
				$H_2O(5)$–$H_2O(6)$		2.264(16)

[a] Standard deviations of the last digit(s) are quoted in parentheses.

Ref. 10) if pure Si tetrahedra are assumed and if the over-all Si content of all Al tetrahedra is approximately 10%, according to the chemical composition derived above. The mean Al–O distances of each separate Al tetrahedron (1.73_6 and 1.74_0), however, are not consistent with our result on Si/Al distribution in these sites. This observation suggests that surrounding cations and T–O–T angles (with corresponding changes in the bond status of the oxygen—cf. Ref. 1 and 3) cannot always be neglected for an estimate of Si/Al distribution from T–O distances. More Si/Al site analyses independent of bond length arguments seem, therefore, desirable.

The TO_4 tetrahedra are, in general, of rather regular shape. One substantial deviation from the ideal tetrahedral angle (O_4–Al_2–$O_8 = 99°$), and accordingly a shorter distance O_4–O_8 (2.66 Å), can be explained by

Table IV. Selected Bond Angles[a] in Degrees

Angles O–T–O

O_2–Si_1–O_3	113.8(3)		O_1–Si_2–O_4	111.5(3)
–O_7	110.0(2)		–O_5	112.5(2)
–O_8	108.4(2)		–O_6	107.4(2)
O_3–Si_1–O_7	109.3(2)		O_4–Si_2–O_5	107.2(2)
–O_8	109.5(2)		–O_6	109.2(2)
O_7–Si_1–O_8	105.5(2)		O_5–Si_2–O_6	109.1(2)
O_1–Al_1–O_2	110.7(3)		O_3–Si_2–O_4	113.0(3)
–O_5	114.3(2)		–O_7	107.1(2)
–O_6	105.3(1)		–O_8	116.2(2)
O_2–Al_1–O_5	107.6(2)		O_4–Si_2–O_7	112.9(2)
–O_6	110.9(2)		–O_8	99.3(2)
O_5–Al_1–O_6	108.0(2)		O_7–Si_2–O_8	108.3(2)

Angles T–O–T

Si_2–O_1–Al_1	152.6(1)
Si_1–O_2–Al_1	142.9(2)
Si_1–O_3–Al_2	154.7(2)
Si_2–O_4–Al_2	142.1(1)
Si_2–O_5–Al_1	146.4(2)
Si_2–O_6–Al_1	145.6(2)
Si_1–O_7–Al_2	142.2(2)
Si_1–O_8–Al_2	139.0(2)

Additional Angles for O_4 and O_8

Ca–O_4–Si_2	117.6(2)
–Al_2	99.9(2)
Ca–O_8–Si_1	109.3(2)
–Al_2	95.5(1)

[a] Standard deviations as in Table III.

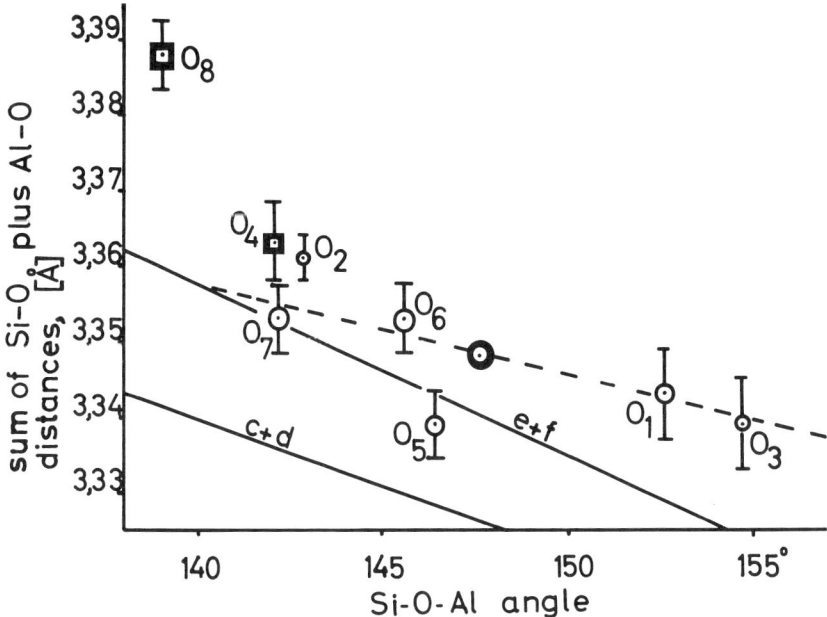

Figure 1. Correlation between bond distance and angle

Al_2–Ca repulsion (edge O_4–O_8 shared by the Al_2 tetrahedron and the coordination figure of Ca).

In general agreement with Brown et al. (1), we observe a correlation between T–O distance and T–O–T angle. As only 6 of the oxygens have no close contact to Ca, this structure offers too few data for quantitative conclusions. For the same reason, the sum of Si–O and Al–O distances for each of these oxygens (with estimated standard deviations) were plotted vs. the Si–O–Al angle; Figure 1 reveals a general decrease of bond length with increasing Si–O–Al angle (broken line). The regression lines observed by (1) and given in Table 2 of their paper were added up for Si–O–Al types $c + d$ (frameworks without anorthite) and $e + f$ (anorthite), and are shown as solid lines in Figure 1. Our results thus confirm the influence of d-p π-bonding (3) on the Si–O and Al–O bonds in aluminosilicate framework structures, as was first demonstrated by Brown et al. (1). Our data appear to agree better with the anorthite regression line. The smaller slope found by us may not be significant because of too few points in the plot and/or may be influenced by the surrounding H_2O molecules which offer the possibility of forming hydrogen bridges. The heavy circle in Figure 1 represents the average Si–O–Al angle and distance sum for the 6 oxygens in question and matches approximately with the average T–O–T angle reported in Table 3 of (1) (difference 0.9°) and with the observed bond lengths (difference 0.01_2 A) given there. For comparison, the 2 Ca-coordinated oxygen atoms O_4 and O_8 are represented by open squares in Figure 1. The mean Si–O and Al–O bond lengths are 1.61_3 and 1.73_5 for those oxygens with bonds to Si and Al alone, and 1.62_9 and 1.74_7 A, respectively, for O_4 and O_8 averaged.

The vibration amplitudes of the framework atoms appear to be quite normal (Table V). The average for the r.m.s. radial thermal displacement of the T and O atoms is 0.12_7 and 0.21_3 A, respectively. Judged from the vibration ellipsoids of the T atoms, the framework appears to exhibit some anisotropic vibrations. Details of the interpretation of the anisotropic temperature coefficients will be discussed in a separate paper.

Table V. R.M.S. Radial Thermal Displacements[a] in A

Si_1	0.132(2)	$H_2O(1)$	0.285(5)	O_1	0.236(6)
Si_2	0.128(2)	$H_2O(2)$	0.300(5)	O_2	0.218(5)
Al_1	0.131(2)	$H_2O(3)$	0.301(6)	O_3	0.211(4)
Al_2	0.115(3)			O_4	0.207(4)
		$H_2O(4)$	0.470(11)	O_5	0.239(4)
Ca	0.309(1)	$H_2O(5)$	0.406	O_6	0.208(4)
		$H_2O(6)$	0.530(8)	O_7	0.205(4)
				O_8	0.178(5)

[a] Standard deviations as in Table III.

Ca and H_2O Molecules

At each intersection of the 2 channels parallel to the a and b axes, a small cavity is formed. It is occupied by Ca bound to O_4 and O_8 of the framework and coordinated to an average of 4.3 H_2O. The distance Ca–$H_2O(4)$ appears too short. If one takes the "thermal" vibration into account, one obtains a revised distance of 2.23 A which may be tolerable. The occupancy of the Ca site is roughly in accordance with the over-all Al content of the framework, which is not precisely determined. The population densities of the Ca as well as of the H_2O molecules apparently are well established, as can be seen from Table VI. Here, correlation coefficients are given between the over-all scale factor s and the occupancies c for Ca and the 6 H_2O sites. There exist only 2 strong correlations: $c(Ca)$ vs. s and $c(H_2O(4))$ vs. $c(H_2O(6))$. The first is caused by the high scattering power of Ca; the second can be explained by the close "neighborhood" of the sites (see below).

Table VI. Correlation Matrix Obtained from Last Cycle Computed with Data Set I

	s	Ca	$H_2O(1)$	$H_2O(2)$	$H_2O(3)$	$H_2O(4)$	$H_2O(5)$	$H_2O(6)$
s	1							
c for Ca	−0.47	1						
$H_2O(1)$	−0.18	0.12	1					
$H_2O(2)$	−0.14	0.07	0.10	1				
$H_2O(3)$	−0.16	0.01	0.07	0.12	1			
$H_2O(4)$	−0.03	−0.01	−0.14	−0.01	−0	1		
$H_2O(5)$	−0.01	−0.06	0.07	0.07	0.01	−0.03	1	
$H_2O(6)$	0.01	−0.04	0.10	0.03	0.07	−0.36	0.05	1

At least 2 of the distances between H_2O molecules are too short to be tolerated (in brackets in Table III). However, the distance $H_2O(5)$–$H_2O(6)$ of 2.26 A becomes 2.33 A if independent movement of the atomic thermal vibrations is taken into account. We conclude that at least 2 (or possibly 3) different types of cavity fillings exist: if $H_2O(4)$ is present, the site for neither $H_2O(5)$ nor $H_2O(6)$ can be occupied. This is surprising for a framework which is almost fully ordered.

The ellipsoids of thermal vibrations of Ca and $H_2O(1)$ to (3) are of normal size and shape. For the partially occupied H_2O sites (no. 4 to

Table VII. Strong "Vibrational" Anisotropy of Some H_2O Molecules, R.M.S. Components in A

Axis	r_1	r_2	r_3
$H_2O(4)$	0.13	0.17	0.42
$H_2O(5)$	0.12	0.22	0.32
$H_2O(6)$	0.08	0.23	0.47

6), they appear strongly elongated (*cf.* Table VII). One can, therefore, suspect that the long axes partially represent positional disorder as well.

Acknowledgment

The authors express sincere thanks to the Deutsche Forschungsgemeinschaft for support. The computations were carried out at the Deutsches Rechenzentrum, Darmstadt. We thank G. Pieper, Frankfurt, for additional intensity measurements and B. Klar for programming the absorption correction.

Literature Cited

(1) Brown, G. E., Gibbs, G. V., Ribbe, P. H., *Am. Mineralogist* **1969**, 54, 1044–1061.
(2) Buerger, M. J., *Acta Cryst.* **1956**, 9, 834.
(3) Cruickshank, D. W. J., *J. Chem. Soc.* **1961**, 5486–5504.
(4) Fischer, K., *Am. Mineralogist* **1963**, 48, 664–672.
(5) Fischer, K., *Tschermaks Mineral. Petrog. Mitt.* **1965**, 10, 203–208.
(6) Fischer, K., Pieper, G., Weber, F., *Fortschr. Mineral.* **1966**, 43, 211–229.
(7) Hamilton, W., "Statistics in Physical Science," pp. 157–162, Table V, Ronald Press, New York, 1964.
(8) Onken, H., Fischer, K., *Z. Krist.* **1968**, 127, 188–199.
(9) Pieper, G., Ph.D. thesis, p. 48, Frankfurt/Main, 1967.
(10) Smith, J. V., Bailey, S. W., *Acta Cryst.* **1963**, 16, 801–811.
(11) Weber, K., *Acta Cryst.* **1963**, 16, 535–542.

RECEIVED February 13, 1970.

20

Refinement of the Crystal Structure of Laumontite

VOLKER SCHRAMM and KARL F. FISCHER

Lehrstuhl fuer Kristallographie, University of Saarbruecken, Germany

> *A published crystal structure has been refined by least-squares based on about 8600 reflections measured with an automated three-circle single-crystal diffractometer. A distance least-squares computation (DLS) revealed the space group Am (No. 8) to be the most probable one. The published structure was found to be essentially correct. The weighted R-factor for all reflections used is 9.4%. Further detailed work is in progress.*

A prismatic specimen of laumontite from the island of Mull, Great Britain, was investigated. Its size is $0.22 \times 0.09 \times 0.08$ mm^3, and its lattice parameters are $a_o = 7.549 \pm 0.008$ A; $b_o = 14.740 \pm 0.015$ A; $c_o = 13.072 \pm 0.012$ A; $\gamma = 111.9 \pm 0.1°$. The unit cell contains 4 formula units of Ca[Al$_2$Si$_4$O$_{12}$] · ($\leqslant$ 4H$_2$O). An automated Siemens single-crystal diffractometer was used for measuring 8584 reflections with Nb-filtered Mo-K$_\alpha$ radiation. No absorption correction was necessary because of its size. During data reduction, individual standard deviations were assigned to the $|F_{obs}|$. The purpose of this study was to examine the details of the laumontite structure first established by Bartl and Fischer (2). Based on 1196 Weissenberg film reflections, they refined to $R_{hkl} = 12.6\%$. The average standard deviation of their atomic coordinates was reported as 0.03 A.

Symmetry

Laue-group and extinction rules suggest A2, Am, and A2/m as the probable space groups. Because of the pyroelectic behavior of the crystals noted by Coombs (4), only an acentric space group is possible. In order to deduce the space group, a least-squares distance refinement using the method of Meier and Villiger (9) was performed for all 3 space groups.

Table I. Atomic Parameters

Atom	x/a_0	y/b_0	z/c_0	B
Ca(1)[a]	0.7263(0)	0.2216(0)	0.5	1.69(4)
Ca(2)	0.2419(3)	0.7656(2)	0.5	0.50(2)
Si(I)	0.3310(6)	0.2594(3)	0.3814(2)	0.75(3)
Al(II)	0.7565(6)	0.3700(3)	0.3082(2)	0.72(3)
Si(III)	0.1638(6)	0.4169(3)	0.3833(2)	0.77(3)
Si(IV)	0.8152(6)	0.5779(3)	0.3828(2)	0.73(3)
Al(V)	0.2256(7)	0.6282(3)	0.3098(2)	0.62(3)
Si(VI)	0.6465(6)	0.7361(3)	0.3823(2)	0.61(3)
O(1)	0.273(1)	0.2352(6)	0.5	1.10(10)
O(2)	0.556(1)	0.2874(6)	0.3760(6)	1.00(10)
O(3)	0.933(1)	0.3505(5)	0.3809(6)	0.92(8)
O(4)	0.281(1)	0.3511(5)	0.3417(5)	0.92(7)
O(5)	0.752(1)	0.3358(5)	0.1838(4)	1.13(7)
O(6)	0.230(1)	0.4405(5)	0.5	0.96(8)
O(7)	0.752(1)	0.4788(5)	0.3131(4)	1.30(7)
O(8)	0.201(1)	0.5040(5)	0.3063(4)	0.91(6)
O(9)	0.752(1)	0.5389(5)	0.5	0.82(8)
O(10)	0.042(1)	0.6391(6)	0.3824(7)	1.84(12)
O(11)	0.705(1)	0.6469(7)	0.3392(7)	1.81(12)
O(12)	0.215(1)	0.6656(5)	0.1845(4)	0.88(7)
O(13)	0.419(1)	0.7067(6)	0.3753(6)	1.50(9)
O(14)	0.727(1)	0.7613(6)	0.5	0.72(8)
O(15)	0.249(2)	0.3785(9)	0.1280(8)	5.64(23)
O(16)	0.835(1)	0.6194(5)	0.1079(5)	3.24(11)

Site Occupancies

| Ca(1) | 0.90(5) | Ca(2) | 0.88(4) | O(15) | 0.88(10) |

[a] The x- and y-coordinates of Ca(1) were held constant (see text).

Table II. Interatomic Distances in Angstroms

Ca(1)	–O(2)	2.479(9)	Ca(2)	–O(10)	2.454(10)
	–O(3)	2.500(9)		–O(13)	2.465(9)
	–O(16)	2.421(8)		–O(15)	2.348(12)
Si(I)	–O(1)	1.616(5)	Al(II)	–O(2)	1.783(9)
	–O(2)	1.595(9)		–O(3)	1.742(8)
	–O(4)	1.618(8)		–O(5)	1.699(7)
	–O(12)	1.587(7)		–O(7)	1.617(7)
Si(III)	–O(3)	1.653(8)	Si(IV)	–O(7)	1.635(7)
	–O(4)	1.630(8)		–O(9)	1.644(5)
	–O(6)	1.603(5)		–O(10)	1.607(9)
	–O(8)	1.573(7)		–O(11)	1.639(11)
Al(V)	–O(8)	1.770(7)	Si(VI)	–O(5)	1.634(8)
	–O(10)	1.739(10)		–O(11)	1.634(10)
	–O(12)	1.741(7)		–O(13)	1.607(9)
	–O(13)	1.717(9)		–O(14)	1.646(5)

Table III. Bond Angles in Degrees; Central Atom is Vertex; Average Standard Deviations Ranging from 0.3 to 0.6 Degrees

O(1)–Si(I)	–O(2)	106.0	O(7)–Si(IV)	–O(9)	105.0
	–O(4)	111.2		–O(10)	113.2
	–O(12)	108.1		–O(11)	108.4
O(2)–	–O(4)	109.2	O(9)–	–O(10)	108.1
	–O(12)	113.1		–O(11)	113.0
O(4)–	–O(12)	109.2	O(10)–	–O(11)	109.3
O(2)–Al(II)	–O(3)	97.0	O(8)–Al(V)	–O(10)	109.2
	–O(5)	110.8		–O(12)	107.7
	–O(7)	109.4		–O(13)	116.5
O(3)–	–O(5)	114.0	O(10)–	–O(12)	110.5
	–O(7)	116.1		–O(13)	100.1
O(5)–	–O(7)	109.0	O(12)–	–O(13)	112.6
O(3)–Si(III)	–O(4)	109.2	O(5)–Si(VI)	–O(11)	110.8
	–O(6)	108.9		–O(13)	108.9
	–O(8)	107.2		–O(14)	106.0
O(4)	–O(6)	104.6	O(11)–	–O(13)	110.1
	–O(8)	107.6		–O(14)	108.4
O(6)–	–O(8)	119.1	O(13)–	–O(14)	112.6
Si(I) –O(1)–Si(I)		147.5	Si(III)–O(8) –Al(V)		138.2
Si(I) –O(2)–Al(II)		138.5	Si(IV) –O(9) –Si(IV)		137.5
Al(II) –O(3)–Si(III)		128.0	Si(IV) –O(10)–Al(V)		131.9
Si(I) –O(4)–Si(III)		137.0	Si(IV) –O(11)–Si(VI)		137.0
Al(II) –O(5)–Si(VI)		136.2	Si(I) –O(12)–Al(V)		135.2
Si(III)–O(6)–Si(III)		144.2	Al(V) –O(13)–Si(VI)		139.9
Al(II) –O(7)–Si(IV)		143.3	Si(VI) –O(14)–Si(VI)		138.4

The distance least-squares (DLS) computations (*11*) were initiated on the atomic parameters of Bartl and Fischer (*2*). The result of this computation favors the space group *Am*. This contradicts the assumption of Amirov *et al.* (*1*) who adopted the space group *A2*. Further details of DLS-computations will be published elsewhere.

Refinement of the Structure

For refinement, we assumed completely ionized atoms. The full-matrix least-squares program LQJL developed from SBLQ utilizes an exponential parameter-representation of *f*-curves as follows (*6*)

$$F(s) = \exp\left\{\sum_{n=1}^{7} a_n \cdot s^{n-1}\right\}, \text{ with } s = \sin\theta/\lambda$$

The large set of data allowed a proper choice of partial data sets for various groups of parameters to be refined, starting with the parameters given by Bartl and Fischer (*2*). For refining the scale factor and site

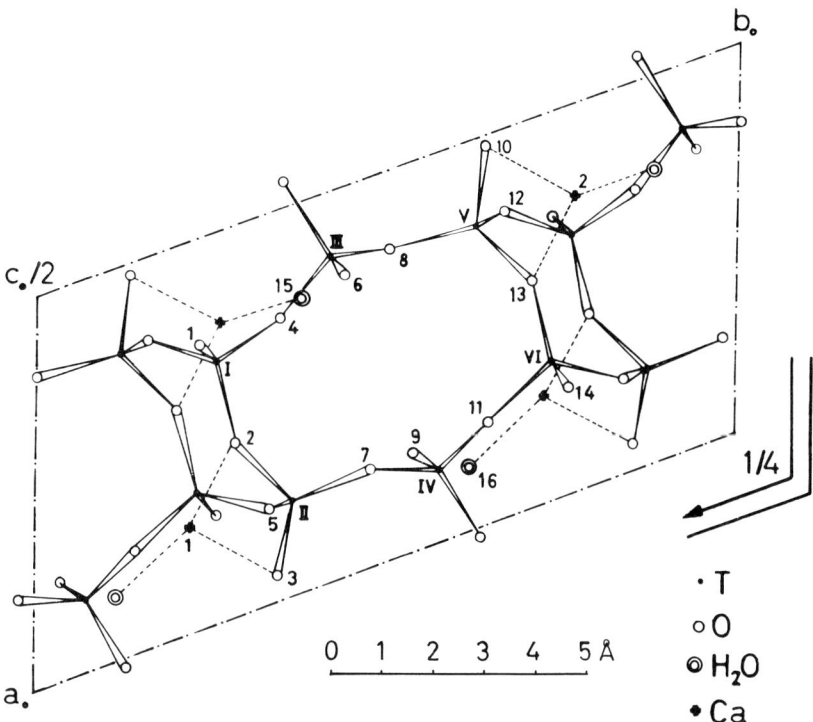

Figure 1. Laumontite Am, first setting, lower part of the unit cell, c-coordinates from 0.0 to 0.5

occupancies of Ca and H_2O (A-cycles), about 130 data with $\sin \theta/\lambda = 0.35$ Å^{-1} were employed. For varying coordinates and temperature factors (B-cycles), we used data sets with increasing numbers of $|F_{obs}|$ starting with the strongest ones and proceeding to all measured reflections. During the progress of refinement, a sequence of one A-cycle and 2 B-cycles was followed. Within the B-cycles, the "heavy" atoms (Ca, Si, Al) were refined, keeping the other coordinates fixed. For variation of coordinates and temperature factors of all atoms, the x- and y-coordinates of Ca 1 were held constant in order to define the origin. Similarly, only the 2 Ca occupancies were varied together with the scale factor at the beginning of the A-cycles. In the subsequent A-cycles, H_2O population densities were added to the variables. The final weighted R-factor including all 8584 F_{obs} was 9.4%.

Description of the Structure

Tables I, II, and III present the refined atomic coordinates as well as the bond distances and angles. The setting used (*see also* Figure 1) is the same as given by Bartl and Fischer (2). The aluminosilicate framework consists of bent 6-rings of (Si/Al)O_4-tetrahedra, which form the asymmetric unit. The concentration of the Al-atoms in the equipoints II and V is probably incomplete as judged from the Al–O-distances. Two "pure" Si_4O_{12} rings—all parallel to (110)—are linked together by 4 isolated Al atoms forming Al_4Si_8-cages. Each cage is connected with its neighbor by 4 linking oxygen atoms, resulting in a chain along the *a*-axis. Four of these chains, which are linked together by only a few Si–O–Al bonds, surround a channel outlined by substantially distorted 10-membered rings. The narrowest cross-section of this channel is 4.6 × 6.3 $Å^2$, based on 1.32 Å ionic radius of oxygen. The Ca atoms are surrounded by 4 O and 2 H_2O in nearly prismatic, six-fold coordination.

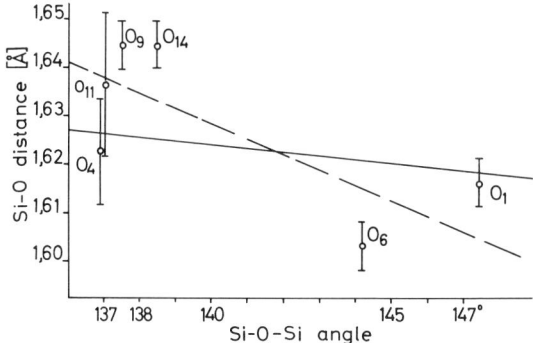

Figure 2. Comparison of present data with that of Brown et al. (3)

Based on the Smith and Bailey diagram (Ref. *10*, Figure 2), and neglecting all other influences on bond lengths, one would expect the following approximate Al contents of the tetrahedral sites: 68% in Al(II), 15% in Si(IV), 94% in Al(V), and 15% in Si(VI). Assuming these values, one would obtain about 1.9 Al atoms per 6 tetrahedral sites, corresponding to about 94% average occupancy of the 2 Ca-equipoints. The Ca occupancies found in this refinement correspond to this value within their standard deviations.

The T–O–T angles are smaller than in similar structures (8). Three of the 4 oxygens bound to Ca atoms also have T–O–T angles below average, in agreement with some other structures (*e.g.*, Ref. 7).

Table IV represents averaged T–O–T angles and T–O distances for different oxygen coordinations according to Table III of Brown *et al.* (3).

Table IV. Observed Mean T–O Bond Lengths and T–O–T Angles

Coordination About Oxygen			No. of Oxygens	T–O–T	Si–O	Al–O
Si	Al	Ca				
2	0	0	6	140.0	1.63	—
1	1	0	4	138.0	1.61	1.71
1	1	1	4	134.6	1.62	1.75

We are in general qualitative agreement with the results of these authors. In detail:

(1) T–O–T angle decreases with an increase of the Al- and Ca-coordination of oxygen.

(2) T–O bond length increases owing to an additional coordination of Ca to an oxygen.

(3) A plot of Si–O bond lengths vs. Si–O–Si bond angle similar to Brown et al. (3) is shown in Figure 2. The distances are drawn with their respective standard deviations. The full line is the regression line of type b (see Ref. 3, Table II). Our data suggest a steeper slope (broken line). This must not be significant because the Al content of the T atoms is not yet well known and because we have too few data.

Judging these observations, one has to keep in mind that the refinement of the structure is presently at an intermediate stage. Our preliminary results appear to support the ideas of Brown et al. (3) on d-p π bonding based on Cruickshank's theory (5).

Anisotropic refinement, investigation of the H_2O sites, and determination of the Si/Al distribution is in progress.

Acknowledgment

Thanks are due to the Deutsche Forschungsgemeinschaft, who supported the project. We also thank the Institut fuer Neutronenphysik I (T. Springer, director) of the Kernforschungsanlage, Juelich, for providing computer time. Some preliminary computations were done at the Deutsches Rechenzentrum, Darmstadt.

Literature Cited

(1) Amirov, S. T., Ilyuchin, W. W., Belov, N. W., *Dokl. Akad. Nauk SSSR* **1967**, 174, 667–670.
(2) Bartl, H., Fischer, K. F., *Neues Jahrb. Mineral. Monatsch.* **1967**, 2/3, 33–42.
(3) Brown, G. E., Gibbs, G. V., Ribbe, P. H., *Am. Mineralogist* **1969**, 54, 1044–1061.
(4) Coombs, D. S., *Am. Mineralogist* **1952**, 37, 812–830.
(5) Cruickshank, D. W. J., *J. Chem. Soc.* **1961**, 5486–5504.

(6) Fischer, K. F., Z. *Krist.* **1968**, 127, 110–120.
(7) Fischer, K. F., Schramm, V., ADVAN. CHEM. SER. **1971**, 101, 250.
(8) Liebau, F., *Acta Cryst.* **1961**, 14, 1103–1109.
(9) Meier, W. M., Villiger, M., Z. *Krist.* **1969**, 129, 411–423.
(10) Smith, J. V., Bailey, S. W., *Acta Cryst.* **1963**, 16, 801–811.
(11) Villiger, H., "DLS—A FORTRAN Computer Program for the Least-Squares Refinement of Interatomic Distances," Institut fuer Kristallographie und Petrographie, Eidg. Technische Hochschule, Zuerich, 1969.

RECEIVED February 13, 1970.

21

Crystal Structures of Ultrastable Faujasites

P. K. MAHER, F. D. HUNTER, and J. SCHERZER

W. R. Grace & Co., Washington Research Center, Clarksville, Md.

The crystal structures of 4 ammonium exchanged, heat-treated faujasites were determined from x-ray powder data. Structure I, often called "decationated Y," has lost 15 framework aluminum atoms and 21 framework $O_{(1)}$ atoms (bridging oxygen atoms) per unit cell, and 15 $Al(OH)_2^+$ ions are present in the sodalite cages. Structure II, called ammonium–aluminum Y hydrate, shows a complete rehydroxylation of the vacant $O_{(1)}$ positions. Structure III, called ultrastable Y, shows the same 15 framework aluminum atoms absent, and the removal of 25 $O_{(3)}$ and 13 $O_{(4)}$ framework oxygen atoms. Structure IV, which is a repetitive exchanged and heat-treated version of Structure III, has a mean Si–O bond length of 1.610 A, which indicates that little framework aluminum is present.

McDaniel and Maher (9) have reported a new form of the zeolite faujasite that is thermally stable at temperatures in excess of 1000°C. This stability is in contrast with other reported hydrogen or "decationated" faujasites, which are less stable than the cation forms. According to these authors, the conversion of faujasite to the hydrogen form leads to a partial destruction of the cation exchange sites which has been explained only by an alteration in the silica–alumina anionic framework.

Ambs and Flank (2) have concluded on the basis of limited data that the thermal stability of synthetic faujasite is dependent only on the level of sodium present. They further assert that no difference exists between decationated Y and ultrastable materials.

Kerr (7) reported that hydrogen zeolite Y, heated 2–4 hours at 700°–800°C in an inert atmosphere where the chemical water remains in the environment of the hydrogen zeolite, yields a substance of unusually high thermal stability (1000°C). He showed that in this zeolite approximately 25% of the aluminum is present in the cationic form.

Kerr (8) also noted the removal of aluminum from sodium Y zeolite through the use of dilute solutions of ethylenediaminetetraacetic acid. He observed that the removal of up to 50% of the aluminum yields highly crystalline products of improved thermal stability and increased sorptive capacity. In a more recent article, Kerr (6) has shown that 2 distinct products can be obtained by the thermal decomposition of ammonium Y at 760 torr and 500°C. According to this article, the geometry of the zeolite bed during calcination determines the nature of the product. Bed geometries that impede the removal of ammonia from the bed yield an ultrastable product. He found that upon cation exchange of this material using 0.1 N NaOH solution, the Si/Al ratio increased from 2.85 for the original material to 3.58 for the exchanged material.

This x-ray structural investigation was undertaken in an effort to clarify the structure of the ultrastable Y zeolite. It was felt that the loss of framework aluminum, the existence of cationic aluminum, and alterations in the framework structure could be revealed by this technique even though powder data, which are inherently less accurate than single-crystal data, had to be used.

Experimental

In order to understand the structural changes which occur, 4 different samples, each from different stages in the preparation of the ultrastable material, were studied. The first sample was selected from step 2 of Procedure "A" outlined by McDaniel and Maher (9). This sample was obtained from NaY which had undergone ammonium sulfate exchange to reduce the sodium oxide content to 2.5% and then had been calcined at 540°C for 3 hours in a muffle furnace. Analyses of this sample showed that its unit cell composition was $Na_9(AlO_2)_{53}(SiO_2)_{139}$ (Structure I).

The second sample was chosen from step 3 of the procedure and consisted of a portion of the first sample which had been treated twice with ammonium sulfate solution at 100°C and then oven-dried. The unit cell composition of this sample is $(NH_4,H)_{8.5}Na_{0.5}(AlO_2)_{53}(SiO_2)_{139}nH_2O$ (Structure II). Hydrogen and/or ammonium ions were assumed to be present in this sample to balance the charges owing to uncompensated aluminum tetrahedra.

The third sample (Structure III) was the ultrastable Y material itself, $Na_{0.5}(AlO_2)_{53}(SiO_2)_{139}$. It was obtained by calcination of the second sample at 870°C for 5 hours in a muffle furnace. The fourth sample studied (Structure IV) was a portion of the ultrastable material which was subjected to 2 additional cycles of 100°C ammonium sulfate treatment followed by calcination for 5 hours at 870°C. By chemical analysis, this material had the same silica and alumina content as the second and third samples.

The unit cell values were 24.56, 24.59, 24.31, and 24.24 ± 0.02 A for structures I, II, III, and IV, respectively. The space group was assumed

Table I. Atomic Coordinates

Atom	Occupancy Parameter	X	Y	Z	T
Structure I					
Si,Al	0.92(3)[a]	−0.0529(7)	0.1228(12)	0.0364(9)	3.37(0.50)
O_1	0.78(4)	0.0000	0.1136(16)	−0.1136(16)	5.64(2.07)
O_2	1.00	0.0012(11)	0.0012(11)	0.1447(17)	4.19(2.28)
O_3	1.00	0.1743(11)	0.1743(11)	−0.0288(15)	0.08(1.53)
O_4	1.00	0.1816(13)	0.1816(13)	0.3216(16)	2.42(1.88)
Al_1 (S_{II})	0.46(4)	0.0800(20)	0.0800(20)	0.0800(20)	4.02(2.51)
$OH_1(S_{III})$	1.00	0.1618(15)	0.1618(15)	0.1618(15)	6.59(2.70)
Na_1 (S_V)	0.32(8)	0.5000	0.5000	0.5000	5.38(11.78)
Structure II					
Si,Al	1.00	−0.0530(8)	0.1256(10)	0.0371(7)	1.39(0.38)
O_1	1.00	0.0000	0.1107(14)	−0.1107(14)	5.04(1.51)
O_2	1.00	−0.0017(14)	−0.0017(14)	0.1404(17)	4.78(1.56)
O_3	1.00	0.1732(12)	0.1732(12)	−0.0315(15)	0.08(2.20)
O_4	1.00	0.1766(12)	0.1766(12)	0.3230(17)	0.58(1.89)
Al_1 (S_{II})	0.42(5)	0.0899(15)	0.0899(15)	0.0899(15)	1.89(3.18)
OH_1 (S_{III})	0.47(7)	0.1617(43)	0.1617(43)	0.1617(43)	4.34(8.92)
N_1 (S_V)	0.37(12)	0.5000	0.5000	0.5000	12.88(7.26)
Structure III					
Si,Al	0.92(8)	−0.0540(5)	0.1261(6)	0.0358(6)	2.95(0.22)
O_1	1.00	0.0000	0.1058(7)	−0.1058(7)	2.76(0.81)
O_2	1.00	−0.0044(8)	−0.0044(8)	0.1476(10)	3.50(0.79)
O_3	0.74(7)	0.1752(9)	0.1752(9)	−0.0416(15)	0.60(1.3)
O_4	0.86(8)	0.1861(11)	0.1861(11)	0.3195(14)	5.51(1.2)
Al_1 (S_{II})	0.46(4)	0.0913(18)	0.0913(18)	0.0913(18)	20.74(4.0)
Structure IV					
Si,Al	1.00	−0.0532(4)	0.1262(4)	0.0359(4)	1.90(0.15)
O_1	1.00	0.0000	0.1061(6)	−0.1061(6)	3.44(0.89)
O_2	1.00	−0.0041(7)	−0.0041(7)	0.1464(8)	3.14(0.96)
O_3	0.87(9)	0.1745(6)	0.1745(6)	−0.0366(11)	2.50(0.87)
O_4	1.00	0.1809(6)	0.1809(6)	0.3200(8)	0.84(0.83)
Al_1	0.23(3)	0.0806(22)	0.0806(22)	0.0806(22)	8.08(3.6)

[a] The values in parentheses are the estimated standard deviations.

to be $Fd3m$ in all cases. The x-ray intensity data were collected on a Norelco powder diffractometer equipped with a Hamner solid state detection system utilizing a scintillation detector and pulse height analyzer. The data for the first 2 structures were collected under a nitrogen atmosphere, and that for the latter 2 were collected under a partial vacuum of 88 mm of Hg. The areas of the x-ray peaks were obtained by the use of a planimeter. Three-dimensional Fourier techniques and least-squares refinement procedures similar to those outlined by Shoemaker *et al.* (*3*, *12*) were employed in the structure determinations.

Results

The positional and thermal parameters for all 4 structures are given in Table I. The standard deviations given in this table were calculated

Table II. Summary of Framework and Nonframework Positions

Position	Structure I	Structure II	Structure III	Structure IV
Si,Al	0.92, 176.6	1.00, 192.0	0.92, 176.6	1.00, 192
$O_{(1)}$	0.78, 74.9	1.00, 96	1.00, 96	1.00, 96
$O_{(2)}$	1.00, 96	1.00, 96	1.00, 96	1.00, 96
$O_{(3)}$	1.00, 96	1.00, 96	0.74, 71.0	0.87, 83.5
$O_{(4)}$	1.00, 96	1.00, 96	0.86, 82.6	1.00, 96
S_{II}	0.46, 14.7 Al^{3+}	0.42, 13.4 Al^{3+}	0.46, 4.9 AlO(OH)	0.23, 2.4 AlO(OH)
S_{III}	1.00, 32 OH^-	0.47, 15.0 OH^-		
S_V	0.32, 5.1 Na^+	0.37, 5.9 NH_4^+		

in the usual manner from the residuals and are probably unrealistically small. Tables of observed and calculated intensities are available from the authors upon request. In Table II, a summary of the site occupancy numbers for all the structures studied is presented. The $R(I)$ values for the 4 structures are 0.168, 0.172, 0.128, and 0.089 for Structures I, II, III, and IV, respectively. $R(I)$ is defined as $[\Sigma w(I_{obs.} - I_{calc.})^2/\Sigma w(I_{obs.})^2]^{1/2}$, where w is the weight, and $I_{obs.}$ and $I_{calc.}$ are the observed and calculated intensities.

For Structure I, which will be termed sodium–aluminum Y by us and has been termed "decationated" Y by Rabo and others (*11*, *13*), an occupancy factor of 0.92 for the framework Si,Al sites indicates the loss

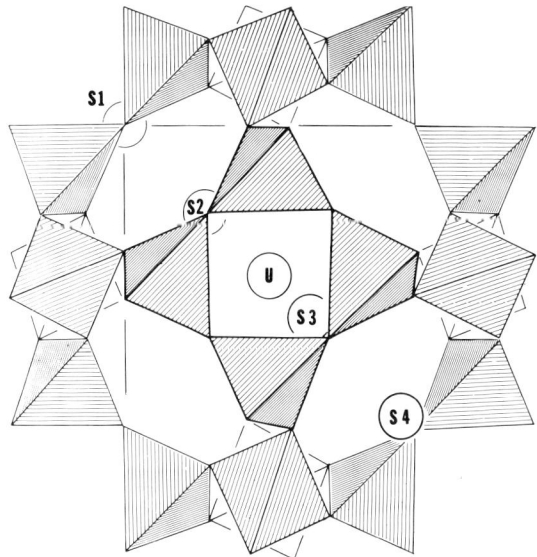

Figure 1. Projection of the sodalite cage for faujasite showing the available nonframework sites. Site 5 is in the center of the 12-membered ring and is not shown here.

of 15.4 aluminum ions from the framework. This number agrees well with the 14.7 aluminum cations per unit cell found to occupy S_{II} sites (*see* Figure 1 for a definition of the sites). Also, 21 $O_{(1)}$ oxygen atoms per unit cell have been removed from the framework by the 540°C calcination. Site S_{III} is occupied completely by hydroxyl groups. The 5.1 atoms per unit cell residing in S_V sites were assumed to be sodium ions. The aluminum cations located in S_{II} sites are coordinated to 3 hydroxyls at 2.03 A and 3 framework $O_{(3)}$ oxygens at 2.68 A to form a distorted octahedral complex. Of the 3 hydroxyl groups coordinated to each aluminum ion in S_{II} sites, 2 are shared with another S_{II} aluminum ion and the other is not shared.

Table III. Metal to Oxygen Bond Distances, A

Atom Set	Structure I	Structure II	Structure III	Structure IV
Si,Al–$O_{(1)}$	1.75(4)	1.73(3)	1.61(3)	1.62(2)
Si,Al–$O_{(2)}$	1.67(4)	1.62(4)	1.64(2)	1.61(2)
Si,Al–$O_{(3)}$	1.62(4)	1.63(4)	1.60(3)	1.61(2)
Si,Al–$O_{(4)}$	1.71(4)	1.62(4)	1.65(3)	1.60(2)
Mean Value	1.69	1.65	1.62	1.61
$Al_{(1)}$–$O_{(3)}$	2.68(6)	3.02(5)	3.28(6)	2.85(3)
$Al_{(1)}$–$OH_{(1)}$	2.03(6)	1.77(11)		

As seen in Table III, the mean Si,Al–O bond distance of 1.69 A is relatively long compared to the values of 1.635 and 1.644 A obtained by Eulenberger *et al.* (*3*) for NaY and Olson and Dempsey (*10*) for hydrogen faujasite. Both of these materials had somewhat lower silica-to-alumina ratios than our sodium–aluminum Y sieve. A projection of the sodalite cage for this structure (Figure 2) shows that the cage is distorted as compared with the projections obtained for the other structures (Figure 3). The relatively long Si,Al–O bond lengths, the distortion of the sodalite cage, and the loss of 22% of the bridging $O_{(1)}$ oxygen atoms explain the instability of this material at higher temperatures.

In Structure II, which we call ammonium–aluminum Y hydrate, the occupancy factors for all the framework positions were unity. Site S_{II} is occupied by 13.4 aluminum ions, and S_{III} sites are filled by 15.0 hydroxyl groups per unit cell. The 5.9 ions located in S_V sites are thought to be ammonium ions. The (approximately) 2 aluminum ions present per sodalite cage are coordinated to the same 2 hydroxyl groups from S_{III} sites at a distance of 1.77 A. The nearest framework oxygens are 3 $O_{(3)}$ oxygens at 3.02 A. The mean Si,Al–O bond length is 1.65 A.

The occupancy parameters for ultrastable Y, Structure III, show the absence of the same 15.4 aluminum ions as in Structure I, but 25.0 $O_{(3)}$

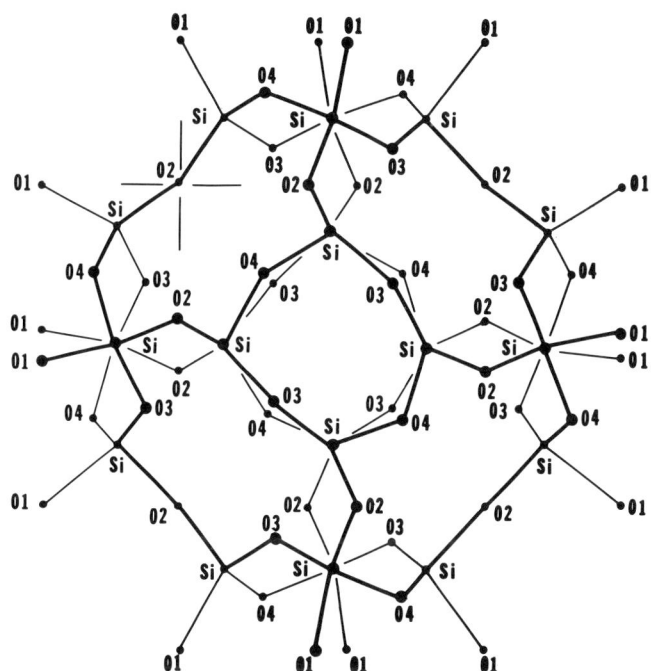

Figure 2. Projection of the sodalite cage for sodium–aluminum Y (Structure I)

and 13.4 $O_{(4)}$ oxygen atoms rather than the 21 $O_{(1)}$ atoms are missing from the framework. The loss of a total of 38.4 oxygen atoms per unit cell represents the removal of 10% of the framework oxygen atoms present in the faujasite structure. Sites S_{II} are filled by 14.7 atoms per unit cell which, as displayed by the high thermal parameter obtained and the elongated peak in the electron density map, seem to be displaced from the three-fold axis. When a position off this axis was assumed for these atoms, a more reasonable temperature parameter and a distance of 1.91 Å corresponding to an aluminum–oxygen species are obtained.

In Structure IV, which is termed high-silica Y, the framework occupancy factors show that the repeated calcination and boiling ammonium sulfate treatment has restored all the framework atoms except for 12.5 of the $O_{(3)}$ oxygen atoms. Sites S_{II} are occupied by 7.4 ions per unit cell. These ions, whose identity is not certain, have 3 framework $O_{(3)}$ oxygens as nearest neighbors at a distance of 2.85 Å. The mean Si,Al–O bond distance of 1.610 (8) Å is only 1 σ greater than the value of 1.603 Å reported by Jones for a pure Si–O bond length (5).

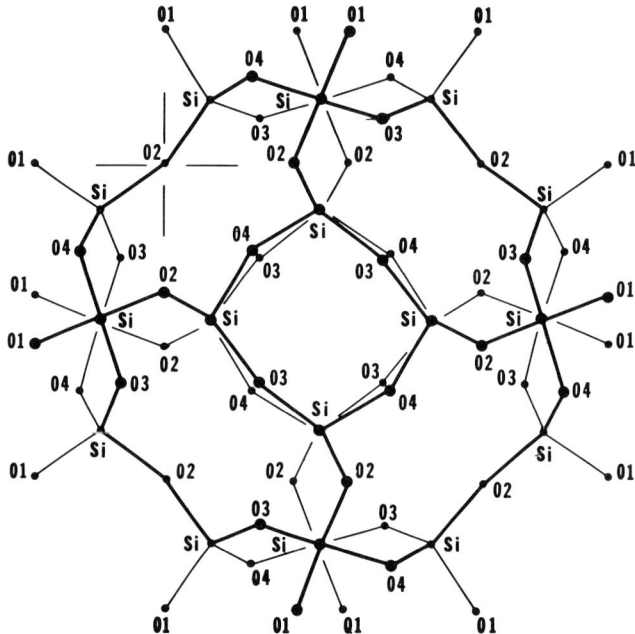

Figure 3. Projection of the sodalite cage for high-silica Y (Structure IV)

Discussion

The structure obtained for the sodium–aluminum Y sieve is in good agreement with other experimental results which have been reported. This material was made under the prescribed conditions for producing "decationated" Y. However, as our data have shown, this material contains aluminum cations which were derived from the framework aluminum. Hence, it is apparent why we prefer the name sodium–aluminum Y to "decationated" Y. McDaniel and Maher (9) reported a decrease in the ion exchange capacity of "decationated" Y sieves. In view of our results, this decrease probably results from the removal of framework aluminum, which was subsequently ion-exchanged by the sodium or silver ions used in their experiments. The lowering of the amount of framework aluminum would result in fewer aluminum tetrahedra whose negative charges must be balanced and hence a decrease in the ion exchange capacity.

Kerr (6) has shown that during the thermal decomposition of ammonium Y zeolite at 500°C there is a loss of aluminum if a "deep bed" geometry is employed. In the particular material that Kerr studied, sodium hydroxide treatment of the "deep bed" sample led to the increase of the Si/Al ratio from 2.85 in the original ammonium zeolite to 3.58. Sodium hydroxide treatment of our sodium–aluminum Y sample resulted in a new Si/Al ratio of 3.19 according to chemical analyses. This number is somewhat smaller than the ratio of 3.62 which we get from the x-ray results if we take into account the removal of 15 framework aluminum atoms, but is considerably higher than the ratio of 2.62 in the original material.

Kerr's postulation of the formation of $Al(OH)_2^+$ species as part of the mechanism for aluminum removal agrees with our x-ray findings.

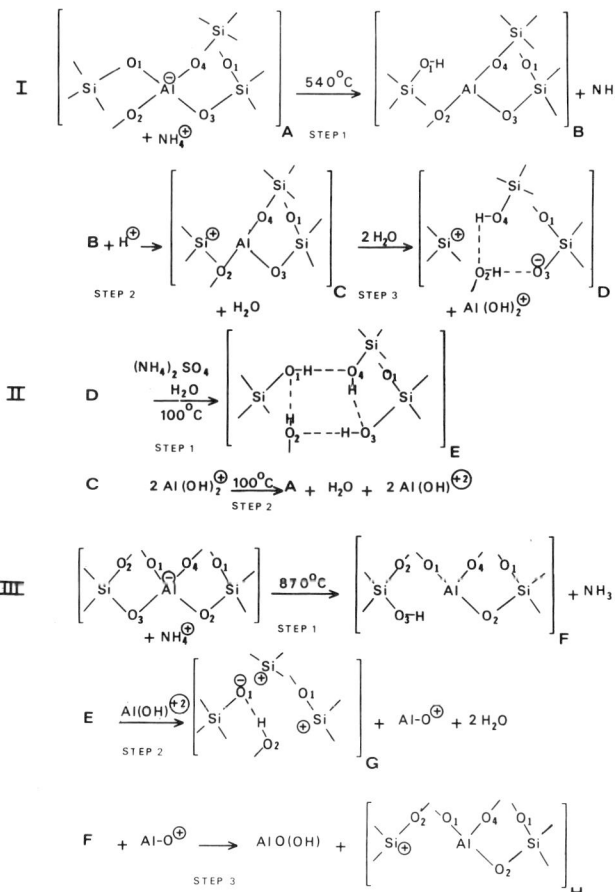

Figure 4. Proposed mechanism for the structures studied; percentages are based on the total aluminum content

Structure I   ≈ 60% A + 11% C + 28% D
Structure II ≈ 72% A + 28% E
Structure III ≈ 49% A + 28% G + 23% H

However, in view of the loss of 22% of the framework $O_{(1)}$ oxygen atoms which we observed, we feel that the mechanism for aluminum removal may be somewhat different from that given by Kerr. The mechanism as we envision it is given in Reaction I of Figure 4 and depends upon the reaction of ammonium ions with the framework $O_{(1)}$ oxygens which are conventionally the bridging oxygens. Upon heating to higher temperatures, protonation and the subsequent removal of these oxygen atoms as water occur. These steps leave the aluminum ions accessible to attack by steam. Hence, according to this reaction and our x-ray results, the structure of sodium–aluminum Y contains 21 SiO_3^+, 15 $Al(OH)_2^+$, and 9 Na^+ (total of 45) species or ions per unit cell to balance the 15 SiO_4^- and 32 AlO_4^- (total 47) groups. The remaining 6 aluminum atoms per unit cell would be involved in neutral AlO_3 groups.

McDaniel and Maher have found that Step 3 of their procedure for making ultrastable Y is extremely important. They noted that treatment of the 540°C calcined, low-sodium oxide material (3% Na_2O) with a solution of ammonium sulfate at 100°C for a prolonged period was essential to the stabilization step. In repeated experiments in which this material was subjected to only 2 cycles of 15 to 20 minutes each of boiling ammonium sulfate treatment, the molecular sieve collapsed upon high-temperature (810° to 927°C) calcination (9).

Our structural results on the ammonium–aluminum Y (Structure II) material are in complete agreement with the above findings. The complete occupancy of all $O_{(1)}$ positions indicates that the ammonium sulfate treatment serves not only to remove the remaining sodium ions but also to rehydroxylate (Figure 4, Reaction II, Step 1) the tri-coordinated silicon atoms. Evidently this reaction is difficult to affect, and the 15-minute treatments are not sufficient to rehydroxylate the material. Since the $O_{(1)}$ oxygens are the prism or bridging oxygens, it is understandable that high-temperature calcination would result in the loss of the structure when 22% of them are absent. We have repeated McDaniel and Maher's experiment and found that the sodium–aluminum Y (Structure I) sieve which has undergone only two 15–20 minute treatments in boiling ammonium sulfate has a surface area of 300 m^2/gram at 810°C and less than 10 m^2/gram at 927°C in spite of almost complete removal of sodium oxide. These values are in striking contrast to the 700 m^2/gram which is typical of the ultrastable Y sieve. We hope that future x-ray studies of the sieve, which has undergone only two 15-minute ammonium sulfate treatments, will verify that the vacant $O_{(1)}$ sites have not been filled in this material.

Since the structure of ultrastable Y (Structure III) shows the loss of 15 aluminum atoms, it is thought that the lack of vacancies in the Si,Al sites in the ammonium-aluminum Y (Structure II) sieve is owing to water molecules or ammonium ions occupying positions near these holes. In the structure of ammonium–aluminum Y, there are 13 $Al(OH)^{2+}$ ions per unit cell. Based on the 38 framework aluminum atoms found in Structure III, a charge balance calculation for Structure II shows that there are only 12 negative charges which must be compensated. These remaining 12 negative charges would be balanced by 12 ammonium ions. Six of these ammonium ions are located in S_V sites. Upon heating this sample, the loss of ammonia would result in the eventual protonation of 12 more $O_{(3)}$ atoms (Figure 4, Step I of Reaction III). Protonation of $O_{(3)}$ rather than any of the other oxygen atoms would be preferred since Olson and Dempsey have concluded that in hydrogen faujasite (*10*) 32 protons are located on $O_{(3)}$ atoms and the remainder on $O_{(1)}$ atoms.

Such protonation easily would explain the results obtained for ultrastable Y (Structure III). High-temperature calcination of the ammonium–aluminum sieve results in the dehydroxylation of the silanol groups to form water and hydroxyl groups which could react with the $Al(OH)^{2+}$ species to form amorphous alumina as suggested by Reaction III, Steps 2 and 3, of Figure 4. Thus, approximately 27 $O_{(3)}$ atoms and 15 $O_{(4)}$ atoms per unit cell would be lost on the basis of the proposed mechanisms. The experimental results show the removal of 25 $O_{(3)}$ and 13 $O_{(4)}$ atoms.

Interpretation of these results shows that for the ultrastable Y sieve, little or no cationic species are needed to balance the framework charges. The above reactions (Figure 4, Steps 2 and 3 of Reaction III) would result in 15 SiO_4^- and 26 AlO_4^- groups in the framework which would be balanced by the 42 SiO_3^+ groups per unit cell created by the loss of the 42 oxygen atoms. This material would have more of a Lewis acid surface rather than a Bronsted acid type and rehydroxylation is not achieved easily. The formation of predominantly Lewis acid and base sites would explain the low ion exchangeability observed by McDaniel and Maher in this material.

By repeated ammonium sulfate treatments at 100°C and high-temperature calcination at 870°C, it is conceivable that eventually a majority of the aluminum can be removed. However, to give stability to the structure, it is logical that transport of silica from another portion of the sample should occur. The results of these 2 phenomena are depicted by the structure found for the high-silica Y sample studied (Structure IV). The absence of approximately 12 $O_{(3)}$ atoms means that 12 SiO_3^+ groups must be present. Hence, it is possible that there are still approximately 12 aluminum atoms per unit cell present in the framework. The close agreement of the average (Si,Al)–O bond of 1.610 (8) Å to the value of 1.603 Å given by Jones for a pure Si–O bond supports the removal of a majority of framework aluminum. The occupancy factor of unity for the framework tetrahedral sites indicates that these sites must have been refilled by silicon atoms. Since these reactions were performed under hydrothermal conditions which facilitate the transport of silica (*14*), these suppositions are quite reasonable. Chemical analyses show that the ultrastable Y material, high-silica Y, and ammonium–aluminum Y all have essentially the same silica and alumina content. This indicates that the thermal treatments have caused the removed aluminum to form some type of insoluble, polymeric alumina compound which would not be removed by the ammonium sulfate treatments at 100°C.

Acknowledgment

We thank R. M. Barrer for his valuable comments and P. H. Emmett for his helpful discussions. We are indebted to the Analytical Research Department for chemical analyses. Computer programs used were NRC-12 Fourier program (*1*) and POWOW (*4*) least squares program. Computations were done on the Univac 1108 computer at the University of Maryland.

Literature Cited

(1) Ahmed, F. G., "NCR Crystallographic Programs for the IBM/360 System," National Research Council of Canada, Ottawa, 1968.
(2) Ambs, W. J., Flank, W. H., *J. Catalysis* **1969**, 14, 118.
(3) Eulenberger, G. R., Shoemaker, D. P., Keil, J. G., *J. Phys. Chem.* **1967**, 71, 1812.
(4) Hamilton, W. C., POWOW, Brookhaven National Laboratory, Brookhaven, New York, 1962.
(5) Jones, J. B., *Acta Cryst.* **1968**, B 24, 355.
(6) Kerr, G. T., *J. Catalysis* **1969**, 15, 200.
(7) Kerr, G. T., *J. Phys. Chem.* **1967**, 71, 4155.
(8) *Ibid.*, **1968**, 72, 2594.
(9) McDaniel, C. V., Maher, P. K., *Conf. Mol. Sieves*, Society of Chemical Industry, London, 1967.
(10) Olson, D. H., Dempsey, E., *J. Catalysis* **1969**, 13, 221.
(11) Rabo, J. A., Pickert, P. E., Boyle, J. E., U. S. Patent **3,130,006** (1964).
(12) Seff, K., Shoemaker, D. P., *Acta Cryst.* **1967**, 22, 162.
(13) Uytterhoeven, J. B., Christner, L. G., Hall, W. K., *J. Phys. Chem.* **1965**, 69, 2117.
(14) Wyart, J., Sabatier, G., *Compt. Rend.* **1954**, 238, 702.

RECEIVED March 24, 1970.

Discussion

Hans Villiger (Martinswerk GmbH, Bergheim, Germany): I am rather puzzled by the small temperature factor of O_3 which is persistent throughout Structures I–III. Could you comment on this situation?

F. D. Hunter: I don't recall that it was always the O_3 temperature factor which was low. I do remember that in each structure there was one oxygen whose temperature factor was unusually low. Generally, in the structures, the first five observations had measured values which were lower than the calculated ones. This is probably owing to extinction and absorption. If the data were corrected for these two factors, all the temperature factors probably would have refined into the expected values.

D. H. Olson (Mobil Research & Development Corp., Princeton, N. J.): The results reported here are the best structural data to date pertaining to the structural transformations occurring during the preparation of ultrastable faujasite. However, I feel the warning presented earlier by J. V. Smith should be kept in mind. The large standard deviations of the population parameters which give site occupancies and vacancies should be considered while deriving a mechanistic picture from the structural data.

D. J. C. Yates (Esso Research Co., Linden, N. J. 07036): The term "calcination" is usually taken to mean heating in air but it could be important to know if static air or moving air is used. For instance, if static air is used, then one is heating the zeolite in a mixture of ammonia and water which were introduced into the zeolite by the ammonium sulfate exchange.

F. D. Hunter: The heating was done in a muffle furnace. We think that we are simulating the deep-bed conditions described by Kerr.

P. K. Maher: We have examined the calcination treatment under controlled atmospheres and have found that this static muffle treatment is the same as treatment in low partial pressure steam.

J. B. Uytterhoeven (University Leuven, 3030 Heverlee, Belgium): In Structures I and II, the presence of hydroxyls on S-III positions is postulated; in Structures III and IV, hydroxyls belonging to AlO(OH) on Site II are shown. Do you have infrared or other data which demonstrate a different nature for these hydroxyls?

J. Scherzer: Since we can expect that OH groups attached to non-framework aluminum will absorb in the same general region of the IR spectrum where absorption bands of structural hydroxyl groups of faujasite-type zeolites occur, their identification would be difficult because of possible band overlapping. This is true especially for materials with Structure III and IV since we have found that their IR spectra show a rather broad band in the 3600 cm^{-1} region.

G. T. Kerr (Mobil Research & Development Corp., Princeton, N. J.): In your paper you said, "McDaniel and Maher reported a decrease in the ion exchange capacity of 'decationated' Y sieves. In view of our results, this decrease probably results from the removal of framework aluminum, which was subsequently ion-exchanged by the sodium or silver ions used in their experiments." I can find nothing in the McDaniel and

Maher paper where they showed removal of aluminum from the zeolite by ion exchange. Please clarify this point.

F. D. Hunter: I think we said "probably" this occurred. However, we think that our Structure III provides the answer to the low ion-exchangeability which McDaniel and Maher observed.

W. H. Flank (Houdry Laboratories, Marcus Hook, Pa. 19061): Ambs and Flank were somewhat misquoted in your paper when it stated that faujasite stability was claimed by us to be dependent only on sodium level. We claimed that a continuous distribution of materials existed whose relative properties are a function of sodium level. This view was more clearly defined and discussed in an exchange of letters in a recent issue of *J. Catalysis*. We recognize the existence of differences in degree rather than differences in kind.

Structure I (Na,Al Y) is claimed unstable at higher temperatures. We prepared a number of samples of this type and found them to be generally quite stable, with collapse temperatures higher than 950°C. We didn't use "Step 3" at all. Perhaps you get degradation at 810°–927° because you used a partial "Step 3" comprising brief exchange.

Silica transport under hydrothermal conditions is cited to support some mechanistic postulations regarding Structure IV. Why should the explanation for the occupancy factor in Structure IV be qualitatively different than for Structure II, since Structure IV only received an intensification of the treatment given Structure II? The type of hydrothermal environment present can have an important bearing on this silica transport phenomenon. In view of your use of ammonium sulfate, did you in fact achieve the conditions for transport cited by Wyart and Sabatier? Residual sulfate would be expected to be quite persistent in such a system. A further point regarding Structure IV is that we have found that the lattice parameter is a smooth function of the degree of calcination severity, so that it might be expected to be smaller than the lattice parameter for Structure III on that basis alone.

F. D. Hunter: You have several questions which I will try to answer. I think you are comparing Structure II and Structure IV. These structures are quite different. Structure II resulted from a 540°C calcination and then a rehydration owing to the ammonium sulfate treatment. Structure IV resulted from repeated 100°C ammonium sulfate exchanges and calcination at 870°C. Hence, you are trying to compare a hydrated sieve to a calcined or dehydrated one. Also, since the framework aluminums are still missing in Structure III, they should be missing in II.

22

Zeolites in Sedimentary Deposits of the United States—A Review

RICHARD A. SHEPPARD

U. S. Geological Survey, Federal Center, Denver, Colo. 80225

> *Zeolites are among the most common authigenic silicate minerals in sedimentary deposits, occurring in rocks that are diverse in age, lithology, and depositional environment. Zeolites are particularly common in those sedimentary rocks that originally contained abundant silicic vitric material. Of the more than 30 naturally occurring zeolites, only 6 commonly occur in bedded deposits: analcime, chabazite, clinoptilolite, erionite, mordenite, and phillipsite. Most zeolites in sedimentary rocks formed during diagenesis by reaction of vitric material with interstitial water. The formation of zeolites is favored by a relatively high pH and high activities of alkali ions in the interstitial water. The zeolites, except analcime, formed directly from the silicic glass by a solution–precipitation mechanism. Most analcime formed during later diagenesis from alkalic, silicic zeolite precursors.*

Zeolites are among the most common authigenic silicate minerals that occur in sedimentary rocks. They have formed in rocks that are diverse in lithology, age, and depositional environment, as summarized by Hay (48). Authigenic zeolites are especially common in those Cenozoic sedimentary rocks that originally contained silicic vitric material. Nearly monomineralic beds of zeolite are known from many areas of the United States, but most zeolitic sedimentary rocks consist of 2 or more zeolites as well as clay minerals, silica minerals, feldspars, and searlesite of authigenic origin. Zeolitic rocks also commonly contain relict vitric material and pyrogenic or detrital grains.

Zeolites have been recognized in sedimentary deposits since 1891, when Murray and Renard (89) described phillipsite in deep-sea deposits. However, prior to the early 1950's, most zeolite occurrences were reported from fracture- and vesicle-fillings in igneous rocks, particularly

Table I. Occurrences of

Locality No., Figure 1	Locality
1	Near Vaughn, Cascade County, Mont.
2	Near Twin Creek, Bear Lake County, Idaho
3	Near Gros Ventre River, Teton County, Wyo.
4	Near Dubois, Fremont County, Wyo.
5	Near Lander, Fremont County, Wyo.
6	Near Thermopolis, Hot Springs County, Wyo.
7	Near Hyattville, Big Horn County, Wyo.
8	South Fork of the Powder River, Natrona County, Wyo.
9	Near Casper, Natrona County, Wyo.
10	Near Lysite Mountain, Hot Springs County, Wyo.
11	Beaver Rim, Fremont County, Wyo.
12	Near Green River, Sweetwater County, Wyo.
13	Near Ludlow, Harding County, S. D.
14	Cathedral Bluffs, Rio Blanco County, Colo.
15	Near Piceance Creek, Rio Blanco County, Colo.
16	Anvil Points, Garfield County, Colo.
17	Along Piceance Creek, about 20 miles west of Meeker, Rio Blanco County, Colo.
18	Lone Tree Mesa, Montrose County, Colo.
19	Near Slick Rock, San Miguel County, Colo.
20	Near Vernal, Uintah County, Utah

Analcime in Sedimentary Rocks

Occurrence	References
Siltstone and sandstone in the Taft Hill Member of the Blackleaf Formation of Cretaceous age and tuff in the Bootlegger Member of the Blackleaf Formation of Cretaceous age	(146)
Tuff in the Twin Creek Limestone of Jurassic age	(41)
Ocher oolitic beds in the Popo Agie Member of the Chugwater Formation of Triassic age	(66)
Ocher oolitic beds in the Popo Agie Member of the Chugwater Formation of Triassic age	(57, 66)
Ocher oolitic beds in the Popo Agie Member of the Chugwater Formation of Triassic age	(57, 66)
Purple and ocher units of the Popo Agie Member of the Chugwater Formation of Triassic age	(57)
Bentonite in the Mowry Shale of Cretaceous age	(128)
Bentonite in the Mowry Formation of Cretaceous age	(128)
Bentonite in the Mowry Formation of Cretaceous age	(128)
Tuff in the Tepee Trail Formation of Eocene age	(140)
Tuff in the Wagon Bed Formation of Eocene age	(7)
Tuff in the Green River Formation of Eocene age	(10, 61)
Lignite in the upper member of the Tongue River Formation of Paleocene age	(104)
Tuff in the Green River Formation of Eocene age	(9)
Oil shale in the Green River Formation of Eocene age	(131, 132)
Oil shale in the Green River Formation of Eocene age	(50)
Tuff in the Parachute Creek Member of the Green River Formation of Eocene age	(12)
Tuffaceous mudstone in the Brushy Basin Member of the Morrison Formation of Jurassic age	(68)
Tuffaceous mudstone in the Brushy Basin Member of the Morrison Formation of Jurassic age	(113)
Ocher oolitic beds in the Chinle Formation of Triassic age	(67)

Table I.

Locality No., Figure 1	Locality
21	White River Canyon, Uintah County, Utah
22	Near Two Water Creek, Uintah County, Utah
23	Near Duchesne, Duchesne County, Utah
24	Near Currant, Nye County, Nev.
25	Nevada Test Site, Nye County, Nev.
26	Teels Marsh, Mineral County, Nev.
27	Near Silver Peak, Esmeralda County, Nev.
28	Deep Springs Lake, Inyo County, Calif.
29	Saline Valley, Inyo County, Calif.
30	Owens Lake, Inyo County, Calif.
31	Lake Tecopa, Inyo County, Calif.
32	Searles Lake, San Bernardino County, Calif.
33	Mojave Desert, eastern Kern County and San Bernardino County, Calif.
34	Near Delano, Kern County, Calif.
35	Near Wikieup, Mohave County, Ariz.
36	Maggie Canyon, Mohave County, Ariz.
37	Near Horseshoe Reservoir, Maricopa County, Ariz.
38	Near Eloy, Pinal County, Ariz.
39	Willcox Playa, Cochise County, Ariz.
40	Along San Simon Creek, Cochise and Graham Counties, Ariz.

Continued

Occurrence	References
Tuff in the Green River Formation of Eocene age	(9)
Tuff in the Parachute Creek Member of the Green River Formation of Eocene age	(13)
Dolomitic oil shale of the Green River Formation of Eocene age	(80)
Tuff in Horse Camp Formation of Miocene and Pliocene age	(86)
Tuff and lapilli tuff of Tertiary age	(58, 59, 60)
Tuff in lacustrine deposit of Quaternary age	(16, 47, 48)
Tuff in the Esmeralda Formation of Miocene and Pliocene age	(82, 100)
Saline crusts of Holocene age	(62)
Mud of Holocene age	(44)
Tuff and tuffaceous sediments of Pleistocene age	(47, 48)
Tuff in lacustrine rocks of Pleistocene age	(120)
Tuff and mudstone of Quaternary age	(49, 130)
Tuff and mudstone of late Tertiary and Quaternary age	(2, 27, 90, 118, 129)
Pond series soil	(4)
Tuff in unnamed lacustrine formation of Pliocene age	(102, 103, 116)
Sandstone of the Chapin Wash Formation of Pliocene(?) age	(73)
Tuff in the Verde Formation of Pliocene(?) or Pleistocene age	(117)
Silty claystone of late Tertiary age	(6)
Mudstone of Pleistocene age	(92)
Tuff in unnamed lacustrine formation of late Cenozoic age	(96, 107)

Table I.

Locality No., Figure 1	Locality
41	Near Nutrioso, Apache County, Ariz.
42	Near Red Wash, San Juan County, N. M.
43	About 2.5 miles southeast of Senorito, Sandoval County, N. M.
44	Wichita Mountains, Kiowa County, Okla.
45	Near Terlingua, Brewster County, Tex.
46	Near Yardley, Bucks County, Pa.
47	Near Frenchtown, Hunterdon County, N. J.
48	Near Pursglove, Monongalia County, W. Va.

basaltic rocks. Of the more than 350 published reports that describe zeolites in sedimentary rocks throughout the world, about 75% were published in the 1960's. The factors chiefly responsible for this recent surge of reports are: (1) the widespread use of x-ray powder diffraction techniques in the study of fine-grained sedimentary rocks, (2) the exploration for zeolite deposits suitable for commercial use, and (3) the review papers by Coombs and others (*18*) and Deffeyes (*22*), both of which emphasized the widespread and relatively common occurrences of zeolites in sedimentary rocks.

This report summarizes the chemistry and physical properties of those zeolites from sedimentary deposits of the conterminous United States and briefly describes their occurrence and origin. Excluded from this discussion are those zeolites in sedimentary rocks that resulted from low-grade metamorphism (*17*) or hydrothermal activity (*29, 136*).

Description and Occurrence of Authigenic Zeolites

Of the more than 30 natural zeolites, only 6 commonly occur in sedimentary deposits. These are analcime, chabazite, clinoptilolite, erionite, mordenite, and phillipsite. The zeolites are very finely crystalline and have similar optical and physical properties; therefore, x-ray powder diffraction techniques generally are used for their identification.

Continued

Occurrence	References
Sandstone in unnamed formation of Tertiary age	(155)
Tuffaceous mudstone of the Brushy Basin Member of the Morrison Formation of Jurassic age	(68)
Siliceous tuff in the Brushy Basin Member of the Morrison Formation of Jurassic age	(109)
Arkose in the Tepee Creek Formation of Permian age	(5, 77, 78)
Black tarry shale of late Mesozoic or early Tertiary age	(79)
Argillite in the Lockatong Formation of Triassic age	(142, 143, 145)
Argillite in the Lockatong Formation of Triassic age	(141, 142, 143, 145)
Concretion in the Pittsburgh coal bed of the Monongahela Formation of Pennsylvanian age	(33)

Analcime. Analcime is one of the more abundant zeolites occurring in sedimentary rocks. Analcime has an ideal formula of $NaAlSi_2O_6 \cdot H_2O$, but the analcime of sedimentary rocks is generally more siliceous. Most analcime in sedimentary rocks is unsuitable for chemical analysis because of abundant inclusions of clay minerals, feldspar, or opal. The composition of these analcimes can, however, be inferred from their index of refraction or cell dimension, utilizing the data of Saha (105, 106) for synthetic analcimes. Both the index of refraction and the a cell dimension decrease with increasing Si/Al ratio. Coombs and Whetten (19) studied analcimes from various sedimentary rock units throughout the world and determined a range in Si/Al ratio of about 2.0–2.8. Subsequent studies have shown a similar range in composition for analcimes in certain lacustrine formations of the western United States. For example, analcime in tuffs of the Miocene Barstow Formation of southeastern California shows a range in Si/Al ratio of about 2.2–2.8 (123), and analcime in tuffs of the Eocene Green River Formation of southwestern Wyoming shows a range of about 2.0–2.9 (61). The meager chemical data suggest that analcime from sedimentary rocks contains only minor amounts of cations other than sodium.

Since the discovery of analcime in the Green River Formation (8) and in lacustrine tuffs near Wikieup, Ariz. (102), analcime has been reported in sedimentary rocks that are diverse in age, lithology, and sedimentary environment (Table I, Figure 1). Analcime occurs in rocks that range in age from Pennsylvanian to Holocene, but it is especially common relative to other zeolites in rocks of Mesozoic age, particularly those older than Cretaceous. Saline lacustrine deposits, regardless of age, very commonly contain analcime. Except for occurrences in the Triassic Lockatong Formation of New Jersey and Pennsylvania (*145*) and the Pennsylvanian Monongahela Formation of West Virginia (*33*), analcime is restricted to sedimentary rocks of the western United States. Analcime, unlike other zeolites in sedimentary rocks, does occur in rocks that lack evidence of volcanic material. Analcime is apparently a common constituent of saline, alkaline soils such as those of southern California (*4*).

Chabazite. Chabazite has an ideal formula of $Ca_2Al_4Si_8O_{24} \cdot 12H_2O$, but natural chabazites show considerable variation in cation content and $Si/Al + Fe^{3+}$ ratio (Figure 2). Ideal chabazite has a $Si/Al + Fe^{3+}$ ratio of 2, but chabazite from sedimentary rocks has a $Si/Al + Fe^{3+}$ ratio of about 3.2–3.8. Chabazite and herschelite (a sodic variety of chabazite) from mafic volcanic rocks generally have a $Si/Al + Fe^{3+}$ ratio near 2. The sedimentary chabazites generally have alkalis in excess of alkaline earths and sodium greatly in excess of potassium. Regis and Sand (*96*), however, have described a calcic sedimentary chabazite from southeastern Arizona. Although some workers have termed these sodic and siliceous sedimentary chabazites "herschelite," the name herschelite should probably be restricted to sodic chabazites that have a $Si/Al + Fe^{3+}$ ratio near 2. What distinguishes the sedimentary chabazites from chaba-

Figure 1. Map showing the occurrences of analcime in sedimentary rocks in the United States. Data for localities are given in Table I.

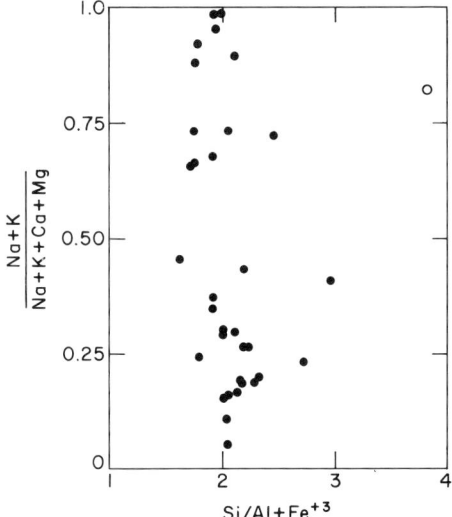

Figure 2. Plot showing the compositional variation of chabazite
● Chabazite (and herschelite) from mafic igneous rocks
○ Chabazite from tuff in the lacustrine Barstow Formation.

zite and herschelite of igneous rocks is not the cation content but the high $Si/Al + Fe^{3+}$ ratio (40).

Indices of refraction and cell dimensions of chabazite from sedimentary rocks are lower than those for chabazite from nonsedimentary rocks. Sedimentary chabazite commonly shows a range in the mean index of refraction of about 1.46–1.47, whereas chabazite and herschelite from igneous rocks show a range of about 1.47–1.49. A siliceous chabazite from a lacustrine tuff in the Barstow Formation of California (40) has relatively small cell dimensions that give a cell volume about 2–3% smaller than typical aluminous chabazite. Thus, the low indices of refraction and small cell dimensions of sedimentary chabazite seem to correlate with its high silicon content.

Chabazite was unknown from sedimentary deposits prior to its discovery by Hay (47) in tuffs and tuffaceous clays at Olduvai Gorge, Tanzania. Since then, authigenic chabazite has been recognized in silicic tuffs from Arizona, California, Nevada, and Wyoming (Table II, Figure 3). Most of the occurrences are in lacustrine rocks of late Cenozoic age. There are no reported occurrences of chabazite in rocks older than Eocene in the United States. Monomineralic beds of chabazite are rare, but extensive and nearly pure beds have been reported from lacustrine

Table II. Occurrences of Chabazite, Erionite,

Locality No., Figure 3	Locality	Zeolites
1	Near Bearbones Mountain, Lane County, Ore.	Mordenite
2	Vicinity of Stein's Pillar, Crook County, Ore.	Mordenite
3	Near Durkee, Baker County, Ore.	Erionite
4	Near Rome, Malheur County, Ore.	Erionite, mordenite, phillipsite
5	West face of Hart Mountain, Lake County, Ore.	Mordenite, phillipsite
6	Near Harney Lake, Harney County, Ore.	Erionite, phillipsite
7	Beaver Rim, Fremont County, Wyo.	Chabazite, erionite, phillipsite
8	Near Split Rock, Natrona County, Wyo.	Phillipsite
9	Near Green River, Sweetwater County, Wyo.	Mordenite
10	Near Mud Buttes, Butte County, S. D.	Phillipsite
11	Sheep Mountain Table, Shannon County, S. D.	Erionite
12	Near Creede, Mineral County, Colo.	Mordenite
13	Pine Valley, Eureka County, Nev.	Erionite, phillipsite
14	West flank of the Shoshone Range, Lander County, Nev.	Erionite
15	Reese River, Lander County, Nev.	Erionite
16	Jersey Valley, Pershing County, Nev.	Erionite, phillipsite
17	Near Lovelock, Pershing County, Nev.	Mordenite

Mordenite, and Phillipsite in Sedimentary Rocks

Occurrence	References
Tuff and lapilli tuff in the Little Butte Volcanic Series of Oligocene and Miocene age	(*85, 91*)
Tuff in the John Day Formation of Oligocene and Miocene age	(*150*)
Tuff of Tertiary age	(*24, 135*)
Tuff and tuffaceous sandstone in an unnamed lacustrine formation of Pliocene age	(*26, 95, 121, 137*)
Tuff and tuffaceous sedimentary rocks of late Oligocene or early Miocene age	(*148*)
Tuff and tuffaceous sedimentary rocks in the Danforth Formation of Pliocene age	(*149*)
Tuff in the Wagon Bed Formation of Eocene age	(*7, 144*)
Tuff in the Moonstone Formation of Pliocene age	(*74*)
Tuff in the Tipton Shale Member of the Green River Formation of Eocene age	(*38*)
Bentonite in the Gammon Ferruginous Member of the Pierre Shale of Cretaceous age	(*112*)
Tuff in the Arikaree Formation of Miocene age	(*22*)
Tuff in the Windy Gulch Member of the Bachelor Mountain Rhyolite of Oligocene age	(*94*)
Tuff in the Hay Ranch Formation of Pliocene and Pleistocene age	(*22, 97*)
Tuff in unnamed lacustrine formation of Pliocene age	(*23*)
Tuff in unnamed lacustrine formation of Pliocene age	(*23*)
Tuff in unnamed lacustrine formation of Pliocene age	(*23*)
Tuff in unnamed lacustrine formation of late Tertiary age	(*108*)

Table II.

Locality No., Figure 3	Locality	Zeolites
18	Near Copper Valley, Churchill County, Nev.	Mordenite
19	Near Eastgate, Churchill County, Nev.	Erionite
20	Teels Marsh, Mineral County, Nev.	Phillipsite
21	Near Silver Peak, Esmeralda County, Nev.	Mordenite, phillipsite
22	Nevada Test Site, Nye County, Nev.	Chabazite, mordenite
23	Owens Lake, Inyo County, Calif.	Erionite, phillipsite
24	Lake Tecopa, Inyo County, Calif.	Chabazite, erionite, phillipsite
25	Searles Lake, San Bernardino County, Calif.	Phillipsite
26	Mojave Desert, eastern Kern County and San Bernardino County, Calif.	Chabazite, erionite, mordenite, phillipsite
27	Near Nipomo, San Luis Obispo County, Calif.	Mordenite
28	Union Pass, Mohave County, Ariz.	Mordenite
29	Near Wikieup, Mohave County, Ariz.	Chabazite, erionite, phillipsite
30	Near Horseshoe Reservoir, Maricopa County, Ariz.	Erionite, phillipsite
31	Near Morenci, Greenlee County, Ariz.	Mordenite
32	Near Bear Springs, Graham County, Ariz.	Chabazite, erionite, phillipsite
33	Along San Simon Creek, Cochise and Graham Counties, Ariz.	Chabazite, erionite

Continued

Occurrence	References
Tuff in unnamed lacustrine formation of late Tertiary age	(*117*)
Tuff in unnamed lacustrine formation of late Tertiary age	(*122*)
Tuff in lacustrine deposit of Quarternary age	(*16, 47, 48*)
Tuff in the Esmeralda Formation of Miocene and Pliocene age	(*82, 83, 100*)
Tuff and lapilli tuff of Tertiary age	(*58, 59, 60*)
Tuff and tuffaceous sediments of Pleistocene age	(*47, 48*)
Tuff and tuffaceous rocks of Pleistocene age	(*120*)
Tuff of Quaternary age	(*49, 130*)
Tuff and tuffaceous rocks of late Tertiary and Quaternary age	(*40, 118, 123*)
Tuff in the Obispo Formation of Miocene age	(*42, 138*)
Tuff and lapilli tuff in the Golden Door Volcanics of Tertiary age	(*39*)
Tuff in unnamed lacustrine formation of Pliocene age	(*116*)
Tuff in the Verde Formation of Pliocene(?) or Pleistocene age	(*117*)
Tuff and lapilli tuff in unnamed formation of Tertiary age	(*116*)
Tuff in unnamed lacustrine formation of late Cenozoic age	(*117*)
Tuff in unnamed lacustrine formation of late Cenozoic age	(*96, 107*)

Figure 3. Map showing occurrences of chabazite, erionite, mordenite, and phillipsite in sedimentary rocks in the United States. Data for localities are given in Table II.

tuffs along the San Simon Creek in southeastern Arizona (96) and near Wikieup in northwestern Arizona (116).

Clinoptilolite. Clinoptilolite is a member of the heulandite structural group. Although there is still some disagreement on the distinction between these closely related zeolites, most workers agree that clinoptilolite is the Si-rich (56, 88) and alkali-rich (76) member. The compositions of clinoptilolite and heulandite from various rock types are represented in Figure 4. Except for slight overlap in the Si/Al + Fe^{3+} and Na + K/ Na + K + Ca + Mg ratios, plots of the compositions cluster into 2 groups. Heulandite characteristically has a Na + K/Na + K + Ca + Mg ratio less than 0.5 and a Si/Al + Fe^{3+} ratio near 3. Most clinoptilolites have a Na + K/Na + K + Ca + Mg ratio greater than 0.6 and range in Si/Al + Fe^{3+} ratio from about 4.0 to 5.0. The clinoptilolites from sedimentary rocks show a range in Si/Al + Fe^{3+} ratio of about 4.1–5.6. Sodium is the predominant cation in most clinoptilolites; however, potassic clinoptilolites are known from California (124) and Japan (81). The few clinoptilolites that have a Na + K/Na + K+ Ca + Mg ratio less than 0.6 are calcic specimens from volcanic rocks in Bulgaria (71) and Italy (1).

Indices of refraction and thermal treatment have been used to distinguish clinoptilolite from heulandite. Mason and Sand (76) suggested that clinoptilolite can be identified by a β index of refraction of 1.485 or lower or that heulandite can be identified by a β index of 1.488 or higher. Heulandite is thermally unstable above about 250°C, whereas clinoptilolite is stable to 750°C or higher (88, 115). However, some members of the heulandite structural group from sedimentary rocks display anomalous optical properties and thermal behavior and cannot be classified conveniently as clinoptilolite or heulandite (46, 114).

The original description of clinoptilolite is of material from amygdales in a basaltic rock from Wyoming (93, 110). Subsequent occurrences of clinoptilolite have been reported chiefly from sedimentary rocks, especially those originally rich in silicic vitric material. Clinoptilolite is the zeolite most often reported from sedimentary rocks in recent years, and it occurs in many rock types from lacustrine, fluviatile, and marine environments (Table III, Figure 5). Although clinoptilolite is most abundant in rocks of Cenozoic age, it has been reported from rocks as old as Cretaceous in Montana, South Dakota, and Wyoming. Occurrences of clinoptilolite are especially common in the western United States. Clinoptilolite is also a common and, locally, abundant constituent in the Tertiary sedimentary rocks of the Coastal Plain from southeastern Texas to North Carolina and northward to western Kentucky (Figure 5).

Erionite. Erionite is generally alkalic and has a $Si/Al + Fe^{3+}$ ratio of about 2.9–3.7. Most analyzed specimens show a relatively high content of potassium, although erionites with sodium in excess of potassium are reported (122). The only calcic erionite reported is a specimen from

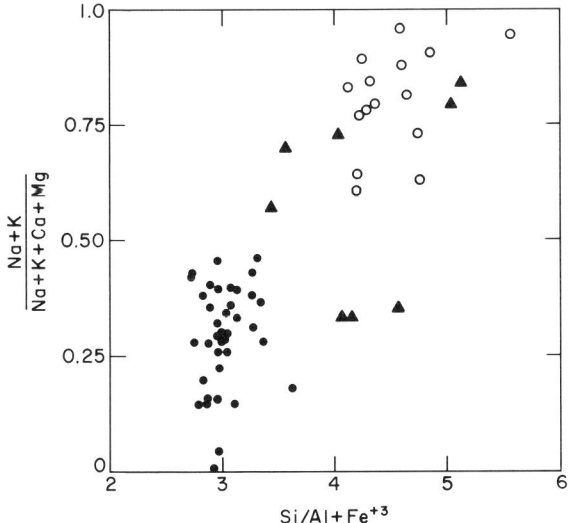

Figure 4. Plot showing the compositional variation of clinoptilolite and heulandite

○ *Clinoptilolite from sedimentary rocks*
▲ *Clinoptilolite from volcanic rocks*
● *Heulandite from igneous rocks*

Table III. Occurrences of

Locality No., Figure 5	Locality
1	Near Vaughn, Cascade County, Mont.
2[a]	Near Livingston, Park County, Mont.
3	Near Preston, Franklin County, Idaho
4	Near Renton, King County, Wash.
5	Near Bearbones Mountain, Lane County, Ore.
6	Near Stein's Pillar, Crook County, Ore.
7	Near Deep Creek, Wheeler County, Ore.
8	Near the Painted Hills, Wheeler County, Ore.
9	Sucker Creek, Malheur County, Ore.
10	Near Sheaville, Malheur County, Ore.
11	Near Rome, Malheur County, Ore.
12	East face of Steens Mountain, Harney County, Ore.
13	Near Harney Lake, Harney County, Ore.
14	West face of Hart Mountain, Lake County, Ore.
15	Near Pedro, Weston County, Wyo.
16	Near Lysite Mountain, Hot Springs County, Wyo.
17[a]	Snake River Canyon, Lincoln County, Wyo.
18	Beaver Rim, Fremont County, Wyo.
19	Near Cameron Spring on Beaver Rim, Fremont County, Wyo.
20	Near Split Rock, Natrona County, Wyo.

Clinoptilolite in Sedimentary Rocks

Occurrence	References
Tuff and tuffaceous siltstone and sandstone in the Taft Hill, Vaughn, and Bootlegger Members of the Blackleaf Formation of Cretaceous age	(*15, 34, 146*)
Tuffaceous mudstone, siltstone, and sandstone in the Livingston Group of Cretaceous age	(*99, 127*)
Tuff in the Salt Lake Group of late Tertiary age	(*22*)
Sandstone and conglomerate in unnamed marine formation of Oligocene age	(*87*)
Tuff and lapilli tuff in the Little Butte Volcanic Series of Oligocene and Miocene age	(*85, 91*)
Tuff in the John Day Formation of Oligocene and Miocene age	(*150*)
Tuff in the lower part of the John Day Formation of Oligocene and Miocene age	(*31, 32, 154*)
Tuff and claystone in the lower part of the John Day Formation of Oligocene and Miocene age	(*45, 46*)
Tuff and tuffaceous sandstone in the Sucker Creek Formation of Miocene age	(*72*)
Tuff probably equivalent to part of the Sucker Creek Formation of Miocene age	(*125*)
Tuff and tuffaceous sandstone in unnamed lacustrine formation of Pliocene age	(*95, 121, 137*)
Tuff in the Pike Creek Formation of Oligocene(?) and Miocene age	(*147*)
Tuff and tuffaceous sedimentary rocks in the Danforth Formation of Pliocene age	(*148*)
Tuff and tuffaceous sedimentary rocks of late Oligocene or early Miocene age	(*149*)
Bentonite in the Pierre Shale of Cretaceous age	(*11*)
Tuff in the Tepee Trail Formation of Eocene age	(*74*)
Shale in the Aspen Formation of Cretaceous age	(*51*)
Tuff in the Wagon Bed Formation of Eocene age	(*7, 144*)
Tuffaceous sandstone in the White River Formation of Oligocene age	(*144*)
Tuff in the Moonstone Formation of Pliocene age	(*74*)

Table III.

Locality No., Figure 5	Locality
21	Near Green River, Sweetwater County, Wyo.
22	Near Twin Buttes, Sweetwater County, Wyo.
23	Near Chamberlain, Buffalo County, S. D.
24	Sheep Mountain Table, Shannon County, S. D.
25	Near Vermillion Cliffs, Moffat County, Colo.
26	Near Creede, Mineral County, Colo.
27	Near Mountain Green, Morgan County, Utah
28	Northern part of the Markagunt Plateau, Iron County, Utah
29	Near Elko, Elko County, Nev.
30	Near Carlin, Eureka County, Nev.
31	West flank of the Shoshone Range, Lander County, Nev.
32	Reese River, Lander County, Nev.
33	Jersey Valley, Pershing County, Nev.
34	Near Lovelock, Pershing County, Nev.
35	Near Eastgate, Churchill County, Nev.
36	Teels Marsh, Mineral County, Nev.
37	Near Silver Peak, Esmeralda County, Nev.
38	Near Goldfield, Esmeralda County, Nev.
39	Nevada Test Site, Nye County, Nev.

Continued

Occurrence	References
Tuff in the Tipton Shale Member of the Green River Formation of Eocene age	(*38*)
Tuff and tuffaceous sandstone in the Bridger Formation of Eocene age	(*117*)
Bentonite in the Sharon Springs Member of the Pierre Shale of Cretaceous age	(*112*)
Tuff in the Arikaree Formation of Miocene age	(*111*)
Tuff in the Bridger Formation of Eocene age	(*39*)
Tuff in the Windy Gulch Member of the Bachelor Mountain Rhyolite of Oligocene age	(*94*)
Tuff in the Salt Lake Group of Tertiary age	(*117*)
Tuffaceous sandstone of Oligocene and Miocene(?) age	(*3*)
Oil shale in unnamed formation of Oligocene age	(*22*)
Tuff in the Safford Canyon Formation of Oligocene(?) or Miocene(?) age and the Carlin Formation of Pliocene age	(*22, 97*)
Tuff in unnamed lacustrine formation of Pliocene age	(*23*)
Tuff in unnamed lacustrine formation of Pliocene age	(*23*)
Tuff in unnamed lacustrine formation of Pliocene age	(*23*)
Tuff in unnamed lacustrine formation of late Tertiary age	(*117*)
Tuff in unnamed lacustrine formation of late Tertiary age	(*117*)
Tuff in lacustrine deposit of Quaternary age	(*16, 47, 48*)
Tuff in the Esmeralda Formation of Miocene and Pliocene age	(*82, 83, 100*)
Tuffaceous sandstone in the Siebert Formation of Miocene(?) age	(*84*)
Tuff and lapilli tuff of Tertiary age	(*20, 36, 58, 59, 60*)

Table III.

Locality No., Figure 5	Locality
40	Near Bullfrog Hills, Nye County, Nev.
41	Death Valley, Inyo County, Calif.
42	Lake Tecopa, Inyo County, Calif.
43	Owens Lake, Inyo County, Calif.
44	Mojave Desert, eastern Kern County and San Bernardino County, Calif.
45	Near Branciforte Creek, Santa Cruz County, Calif.
46	Near Nipomo, San Luis Obispo County, Calif.
47[a]	Near Oakview, Ventura County, Calif.
48	Near San Pedro, Los Angeles County, Calif.
49	Near Wikieup, Mohave County, Ariz.
50	Near Dome, Yuma County, Ariz.
51	Near Horseshoe Reservoir, Maricopa County, Ariz.
52	Near Nutrioso, Apache County, Ariz.
53	Near Morenci, Greenlee County, Ariz.
54	Along San Simon Creek, Cochise and Graham Counties, Ariz.
55	Near Bayard, Grant County, N. M.
56	Near Coy City, Karnes County, Tex.
57	Near Tilden, McMullen County, Tex.
58	Near Meridian, Lauderdale County, Miss.
59	Near Nettleboro, Clarke County, Ala.

Continued

Occurrence	References
Tuff of Tertiary age	(*20*)
Tuff in the Furnace Creek Formation of Pliocene age	(*75*)
Tuff in lacustrine rocks of Pleistocene age	(*120*)
Tuff and tuffaceous sediments of Pleistocene age	(*47, 48*)
Tuff and tuffaceous rocks in numerous formations of late Tertiary and Quaternary age	(*2, 70, 118, 119, 123*)
Tuffaceous sandstone in the Santa Margarita Formation of Miocene age	(*37*)
Tuff in the Obispo Formation of Miocene age	(*11, 138*)
Bentonite in the Modelo Formation of Miocene age	(*69*)
Dolomitic sandstone in the Monterey Formation of Miocene age	(*134*)
Tuff in unnamed lacustrine formation of Pliocene age	(*116*)
Bentonite of Tertiary(?) age	(*11*)
Tuff in the Verde Formation of Pliocene(?) or Pleistocene age	(*117*)
Tuff and sandstone in unnamed formation of Tertiary age	(*155*)
Tuff and lapilli tuff in unnamed formation of Tertiary age	(*116*)
Tuff in unnamed lacustrine formation of late Cenozoic age	(*96, 107*)
Tuff in the Sugarlump Tuff of Oligocene age	(*65*)
Tuff and tuffaceous sandstone in the Jackson Group of Eocene age	(*151, 152*)
Tuff in the Jackson Group of Eocene age	(*25*)
Tuffaceous sandstone in the Meridian Sand of Eocene age	(*153*)
Tuffaceous sandstone in the Meridian Sand of Eocene age	(*153*)

Table III.

Locality No., Figure 5	Locality
60	Near McKenzie, Butler County, Ala.
61	Near Paducah, McCracken County, Ky.
62[a]	Near Jackson, Madison County, Tenn.
63[a]	Near Caryville, Washington County, Fla.
64	Near Coosawhatchie, Jasper County, S. C.
65	Central S. C.
66	Near Eward, Beaufort County, N. C.

Figure 5. Map showing the occurrences of clinoptilolite in sedimentary rocks in the United States. Data for localities are given in Table III.

basalt near Maze, Japan (*43*). Ferric iron seems to substitute for aluminum in some sedimentary erionites, as well as in other zeolites from sedimentary rocks. An analysis of erionite from lacustrine tuff near Rome, Oregon (*26*), suggests that ferric iron can substitute for as much as 15% of the aluminum.

Continued

Occurrence	References
Tuff and tuffaceous claystone in the Tallahatta Formation of Eocene age	(98)
Claystone in the Clayton(?) Formation of Paleocene age and clay in the Porters Creek Clay of Paleocene age	(30, 133)
Fossiliferous rock of Paleocene age	(139)
Suwanee Limestone of Oligocene age	(139)
Clay in the Hawthorn Formation of Miocene age and the Santee Limestone of Eocene age	(55)
Mudstone in the Black Mingo Formation of Paleocene and Eocene age	(54)
Phosphorite in the Pungo River Formation of Miocene age	(14, 101)

a Zeolite was identified as heulandite.

A decrease in the indices of refraction and cell dimensions of erionite can be correlated with an increase in the $Si/Al + Fe^{3+}$ ratio (122). Indices of refraction show a range of about 1.46–1.48, but most erionites from sedimentary deposits have indices in the lower part of the range. Erionite shows about a 1% decrease in cell volume from the most aluminous analyzed specimen to the most siliceous specimen.

Erionite was considered an extremely rare mineral prior to the work of Deffeyes (22, 23) and Regnier (97), who showed it to be a common authigenic zeolite in the altered silicic tuffs of lacustrine deposits in north-central Nevada. Since then, erionite has been recognized in silicic bedded tuffs from many western states (Table II, Figure 3). Erionite, like chabazite, has not been reported from sedimentary rocks older than Eocene. Most occurrences of erionite are in upper Cenozoic lacustrine deposits. Extensive and relatively pure beds of erionite occur in southeastern Oregon, southeastern California, and north-central Nevada.

Mordenite. Mordenite has been confused with clinoptilolite or heulandite in sedimentary rocks because of the similarity in chemistry and indices of refraction (18, 126). X-ray diffractometer techniques, fortunately, are adequate for positive identification. Mordenites from non-sedimentary rocks show a range in $Si/Al + Fe^{3+}$ ratio of about 4.3–5.3 and generally show an excess of alkalis over alkaline earths. Sodium is generally greatly in excess of potassium. The relatively low potassium content is about the only chemical parameter that distinguishes mordenite from clinoptilolite. The only reported analyzed mordenite from a sedimentary rock is from a silicic tuff in the Barstow Formation of California (123). This mordenite is typically sodic and has a $Si/Al + Fe^{3+}$ ratio of about 4.7.

Indices of refraction for nonsedimentary mordenites range from about 1.47 to 1.49 (21). The mordenites from sedimentary rocks commonly have indices in the lower part of this range.

Mordenite was recognized only recently as a rock-forming constituent of sedimentary deposits in the United States. In 1964, mordenite was reported from tuffaceous rocks of California (118) and Nevada (83). Since then, occurrences of mordenite have been recorded from Cenozoic tuffs in many of the western states (Table II, Figure 3). Although mordenite occurs in lacustrine rocks, most occurrences are in rocks from other depositional environments. Clinoptilolite and opal are commonly associated with mordenite in the sedimentary deposits.

Phillipsite. Phillipsite shows a wide variation in $Si/Al + Fe^{3+}$ ratio and cation content, although it is consistently high in potassium content. The compositions of phillipsite from various rock types are represented in Figure 6. On the basis of $Si/Al + Fe^{3+}$ ratio, the phillipsites can be classed into 3 groups that show some overlap. The least siliceous group is from nonsedimentary rocks and is characterized by a $Si/Al + Fe^{3+}$ ratio of 1.3–2.4. Phillipsites of this group generally contain a higher percentage of alkaline earths than those of the other 2 groups; some specimens have alkaline earths (chiefly calcium) in excess of alkalis. An intermediate group has a $Si/Al + Fe^{3+}$ ratio of 1.9–2.8, but most specimens are in the range of 2.4–2.8. These phillipsites are from deep-sea sediments and are characteristically rich in alkalis. The most siliceous group has a $Si/Al + Fe^{3+}$ ratio of 2.6–3.4, but most specimens have a $Si/Al + Fe^{3+}$ ratio greater than 3.0. These phillipsites are from tuffs in saline lacustrine deposits. Like the marine phillipsites, these lacustrine phillipsites are rich in alkalis; however, the 2 groups differ in the predominant alkali. Marine phillipsites generally have potassium in excess of sodium, whereas lacustrine phillipsites have sodium in excess of potassium. Hay (47) has suggested that ferric iron may substitute for about 5% of the aluminum in phillipsite from lacustrine deposits.

Indices of refraction for phillipsite range from about 1.44 to 1.51 and seem to vary inversely with the $Si/Al + Fe^{3+}$ ratio (47). Indices of the relatively aluminous nonsedimentary phillipsites are 1.48–1.51 (21); indices of the intermediate marine phillipsites are 1.48–1.49; and indices of the siliceous lacustrine phillipsites are 1.44–1.48. Phillipsite from lacustrine sedimentary rocks is commonly zoned and shows a difference

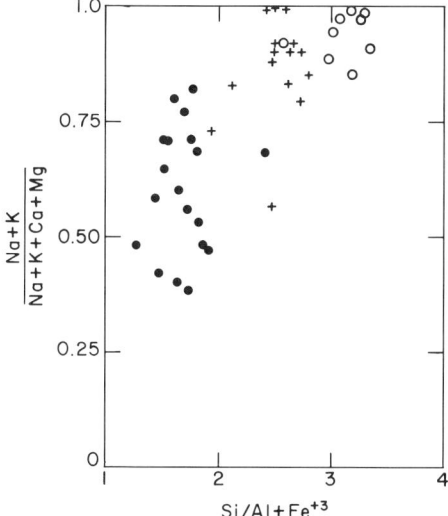

Figure 6. Plot showing the compositional variation of phillipsite

○ Phillipsite from saline lacustrine deposits
+ Phillipsite from deep-sea deposits
● Phillipsite from mafic igneous rocks

in index of refraction of as much as 0.02 between the interior and marginal parts of crystals.

Phillipsite was found long ago in deposits on the sea floor (89). Since the discovery by Deffeyes (22) of phillipsite in lacustrine tuffs of Nevada, this zeolite has been reported commonly as a rock-forming constituent in tuffaceous rocks of the western United States (Table II, Figure 3). Phillipsite occurs in sedimentary rocks that range in age from Cretaceous to Holocene, but it is especially common in lacustrine deposits of late Cenozoic age, particularly those deposits of saline, alkaline lakes. Extensive and relatively pure beds of phillipsite are reported from southeastern Oregon, southeastern California, and Nevada.

Genesis of Zeolites

Most zeolites in sedimentary rocks formed during diagenesis by the reaction of aluminosilicate materials with the pore water. Silicic volcanic glass is the aluminosilicate material that most commonly served as a precursor for the zeolites, although materials such as clay minerals, plagioclase, leucite, and nepheline also have reacted locally to form zeolites (48). Solution of silicic glass by the pore water provided the constituents necessary for the formation of the zeolites. Deffeyes

(22) emphasized that zeolites formed during diagenesis—not by devitrification of the glass in the solid state but by solution of the glass and subsequent precipitation of zeolites from the solution. Except for analcime, all the zeolites in sedimentary deposits that have not been deeply buried or exposed to hydrothermal solutions seem to have formed directly from vitric matrial. The genesis of analcime may involve an intermediate step, as described in the following discussion.

Experimental work by others indicates that the activity ratio of alkali ions to hydrogen ions and the activity of silica are the major chemical parameters of the pore water that control whether clay minerals, zeolites, or feldspars will form at conditions that approximate surface temperatures and pressures (35, 52, 53). The formation of zeolites and feldspars is favored over clay minerals by relatively high alkali ion to hydrogen ion activity ratios and by relatively high silica activities.

The high alkali ion to hydrogen ion activity ratio necessary for the formation of zeolites in a silicic vitric tuff or tuffaceous sediment can be simply characteristic of the original water trapped during sedimentation in a saline, alkaline lake. Brine of sodium carbonate–bicarbonate composition seems to have been particularly effective in the alteration of silicic glass to zeolites. These brines commonly have a pH of 9–10 (63, 64), which probably accounts for the relatively high solubility of the glass as well as the relatively fast rate of solution of the glass (47). From a study of vitric tuffs in Teels Marsh, Nev., Hay (48) concluded that zeolites in the uppermost tuff formed in less than 1000 years.

Studies of tuffs deposited in relatively young saline lakes where water analyses are available have shown a strong correlation between salinity and the authigenic silicate mineralogy (47, 48). Tuffaceous sediments deposited in fresh lakes contain unaltered glass or glass altered to clay minerals, chiefly montmorillonite. Those tuffaceous sediments deposited in saline lakes are altered and now contain zeolites, potassium feldspar, and searlesite. The occurrence of authigenic feldspar and searlesite correlates with waters of the highest salinities.

Older lacustrine deposits that contain interbedded saline minerals also show a correlation between the inferred salinity of the depositional environment and the authigenic mineralogy of tuffaceous sediments. In the Pleistocene deposits of Lake Tecopa, Calif. (120), glass is unaltered in tuff deposited in fresh water near the lake shore and inlets; however, the tuffs consist chiefly of phillipsite, clinoptilolite, and erionite where deposited in moderately saline water and of potassium feldspar and searlesite where deposited in the highly saline water of the central part of the basin. Individual tuffs show a lateral gradation in a basinward direction of unaltered glass to zeolites and then to potassium feldspar with searlesite. A similar correlation between salinity and authigenic silicate mineralogy has been demonstrated for Miocene Barstow Formation of California (*123*) and the Eocene Green River Formation of Wyoming (*38, 48*).

Pore water in tuffs and tuffaceous sediments of depositional environments other than saline, alkaline lakes can attain a high alkali ion to hydrogen ion activity ratio after burial through solution and hydrolysis of the vitric material by subsurface water. Hay (*46*) proposed solution and hydrolysis of rhyolitic and dacitic glass by subsurface water to account for the formation of clinoptilolite in tuff and tuffaceous claystone in the lower part of the John Day Formation in central Oregon. The upper part of the formation contains unaltered glass or montmorillonite. The subsurface water, which originated from meteoric water, increased in pH and concentration of alkalis as it moved downward through the formation. Thus, clinoptilolite formed at the depth in the formation where the alkali ion to hydrogen ion activity ratio was highest. In zeolitic rocks of the John Day Formation and in similar zeolite deposits, montmorillonite probably crystallized before the zeolite. The early alteration of glass to montmorillonite would probably increase the pH and concentration of alkalis in the pore water, thereby providing a chemical environment more favorable for the formation of zeolites. Hay (*48*) suggested that this early alteration of glass to montmorillonite is an important factor for the subsequent crystallization of zeolites in tuffs deposited in marine and fresh-water environments.

Zeolite deposits that formed by the above mechanism commonly show a vertical zonation of authigenic silicate minerals similar to that in the John Day Formation. Tertiary tuffs at the Nevada Test Site in southern Nevada were altered after burial by subsurface water (*59*), but the authigenic mineral zonation is more complex than that in the John Day Formation. The upper zone consists of unaltered glass with local concentrations of chabazite or clay minerals. Zeolitic tuff continues downward for as much as 6000 feet. A zone rich in clinoptilolite underlies the zone of unaltered glass and is succeeded downward by zones rich in mordenite and analcime, respectively.

Ever since the discovery of analcime in tuffaceous rocks, most workers have assumed that the analcime formed directly from vitric material. The presence of vitroclastic texture and pyrogenic crystals in some analcimic tuffs seemed sufficient evidence; however, these criteria do not necessarily prove that the glass altered directly to analcime. Hay (*48*) and Sheppard and Gude (*123*) concluded from a study of tuffs in saline-lake deposits that analcime commonly formed from alkalic, silicic zeolite precursors. Formation of analcime from clinoptilolite and phillipsite was documented in tuffs of the Miocene Barstow Formation. Relict fresh glass has not been confirmed in analcimic tuffs from any area; thus, there is doubt that analcime ever has formed directly from glass.

The alkalic, silicic zeolites that occur in tuffaceous sedimentary rocks would seem to be particularly susceptible to alteration in the diagenetic environment because of their open structure. Chemical factors that favor the reaction of early-formed alkalic, silicic zeolites to analcime are a high Na^+/H^+ ratio, relatively low activity of H_2O, relatively low activity of SiO_2, and high pH. Studies of zeolitic tuffs in saline-lake deposits such as the Eocene Green River Formation (38, 48) and the Miocene Barstow Formation (123) suggest that a moderately to highly saline pore water facilitates the conversion of alkalic, silicic zeolites to analcime during later diagenesis. A relatively high salinity would reduce the activity of H_2O and favor the formation of a mineral less hydrous than clinoptilolite or phillipsite (48, 123).

Analcime in some nontuffaceous saline-lake deposits probably formed by direct precipitation from the lake water. The analcime in the Triassic Lockatong Formation either precipitated directly or formed at an early stage of diagenesis from a colloidal precursor or aluminosilicate mineral (141, 145). At Lake Natron, Kenya, analcime in Quaternary nontuffaceous clays was precipitated from a sodium carbonate brine (48). A sodium aluminosilicate gel was recently found at Lake Magadi, Kenya (28). Analcime could form during diagenesis by crystallization of such a gel. Analcime in other nontuffaceous lacustrine rocks apparently formed during diagenesis by reaction of plagioclase, montmorillonite, or kaolinite with the pore water (49, 92).

Acknowledgment

Grateful appreciation is expressed to those colleagues in the U.S. Geological Survey who provided unpublished data. B. M. Madsen and J. D. Vine critically read the manuscript and made helpful suggestions.

Literature Cited

(1) Alietti, Andrea, *Petrog. Acta* **1967**, 13, 119–138.
(2) Ames, L. L., Jr., Sand, L. B., Goldich, S. S., *Econ. Geol.* **1958**, 53, 22–37.
(3) Anderson, J. J., *Geol. Soc. Am. Spec. Papers* **1969**, 121, 586.
(4) Baldar, N. A., Whittig, L. D., *Soil Sci. Soc. Am. Proc.* **1968**, 32, No. 2, 235–238.
(5) Bellis, W. H., Mankin, C. J., "Clays and Clay Minerals—Proc. Clay Minerals Conf., 15th, Pittsburgh, Pa., 1966," S. W. Bailey, Ed., p. 191, Pergamon, New York, 1967.
(6) Blackmon, P. D., Oral communication, 1966.
(7) Boles, J. R., M.S. thesis, University of Wyoming, Laramie, Wyo., 1968.
(8) Bradley, W. H., *Science* **1928**, 67, 73–74.
(9) Bradley, W. H., *U.S. Geol. Surv. Profess. Papers* **1929**, 158-A, 1–7.

(10) Bradley, W. H., *U.S. Geol. Surv. Profess. Papers* **1964**, 496-A, 86 pp.
(11) Bramlette, M. N., Posnjak, Eugen, *Am. Mineralogist* **1933**, 18, 167–171.
(12) Brobst, D. A., Oral communication, 1969.
(13) Cashion, W. B., *U.S. Geol. Surv. Profess. Papers* **1967**, 548, 48 pp.
(14) Cathcart, J. B., *Proc. Forum Geol. Ind. Minerals, 4th, Austin, Tex.*, *1968*, L. F. Brown, Jr., Ed., p. 23–34.
(15) Cobban, W. A., *Billings Geol. Soc. Guidebook Ann. Field Conf., 6th*, *1955*, p. 107–119.
(16) Cook, H. E., Hay, R. L., *Geol. Soc. Am. Spec. Papers* **1965**, 82, 31–32.
(17) Coombs, D. S., *Trans. Roy. Soc. New Zealand* **1954**, 82, Pt. 1, 65–109.
(18) Coombs, D. S., Ellis, A. J., Fyfe, W. S., Taylor, A. M., *Geochim. Cosmochim. Acta* **1959**, 17, 53–107.
(19) Coombs, D. S., Whetten, J. T., *Geol. Soc. Am. Bull.* **1967**, 78, 269–282.
(20) Cornwall, H. R., "Petrologic Studies A Volume in Honor of A. F. Buddington," A. E. J. Engel, H. L. James, and B. F. Leonard, Eds., p. 357–371, Geological Society of America, New York, 1962.
(21) Deer, W. A., Howie, R. A., Zussman, J., "Rock-Forming Minerals," Vol. 4, Wiley, New York, 1963.
(22) Deffeyes, K. S., *J. Sediment. Petrol.* **1959**, 29, 602–609.
(23) Deffeyes, K. S., *Am. Mineralogist* **1959**, 44, 501–509.
(24) Eakle, A. S., *Am. J. Sci.* **1898**, 6, 66–68.
(25) Eargle, D. H., *U.S. Geol. Surv. Bull. 1251-D*, **1968**, 25pp.
(26) Eberly, P. E., Jr., *Am. Mineralogist* **1964**, 49, p. 30–40.
(27) Erd, R. C., Morgan, Vincent, Clark, J. R., *U.S. Geol. Surv. Profess. Papers* **1961**, 424-C, C294–C297.
(28) Eugster, H. P., Jones, B. F., *Science* **1968**, 161, 160–163.
(29) Fenner, C. N., *J. Geol.* **1936**, 44, 225–315.
(30) Finch, W. I., U.S. Geol. Surv. Geol. Quad. Map GQ-557, 1966.
(31) Fisher, R. V., *Ore Bin* **1962**, 24, 197–203.
(32) Fisher, R. V., *Ore Bin* **1963**, 25, 185–197.
(33) Foster, W. D., Feicht, F. L., *Am. Mineralogist* **1946**, 31, 357–364.
(34) Fox, R. D., *Montana Bur. Mines Geol. Bull.* **1966**, 52, 64 pp.
(35) Garrels, R. M., Christ, C. L., "Solutions, Minerals, and Equilibria," Harper and Row, New York, 450 pp., 1965.
(36) Gibbons, A. B., Hinrichs, E. N., Botinelly, Theodore, *U.S. Geol. Surv. Profess. Papers* **1960**, 400-B, B473–B475.
(37) Gilbert, C. M., McAndrews, M. G., *J. Sediment. Petrol.* **1948**, 18, 91–99.
(38) Goodwin, J. H., Surdam, R. C., *Science* **1967**, 157, 307–308.
(39) Gude, A. J., III, Oral communication, 1966.
(40) Gude, A. J., III, Sheppard, R. A., *Am. Mineralogist* **1966**, 51, 909–915.
(41) Gulbrandsen, R. A., Cressman, E. R., *J. Geol.* **1960**, 68, 458–464.
(42) Hall, C. A., Turner, D. L., Surdam, R. C., *Geol. Soc. Am. Bull.* **1966**, 77, 443–445.
(43) Harada, Kazuo, Iwamoto, Shigeki, Kihara, Kuniaki, *Am. Mineralogist* **1967**, 52, 1785–1794.
(44) Hardie, L. A., *Geochim. Cosmochim. Acta* **1968**, 32, 1279–1301.
(45) Hay, R. L., "Petrologic Studies—A Volume in Honor of A. F. Buddington," A. E. J. Engel, H. L. James, B. F. Leonard, Eds., p. 191–216, Geological Society of America, New York, 1962.
(46) Hay, R. L., *Calif. Univ. Pubs. Geol. Sci.* **1963**, 42, No. 5, 199–262.
(47) Hay, R. L., *Am. Mineralogist* **1964**, 49, 1366–1387.
(48) Hay, R. L., *Geol. Soc. Am. Spec. Papers* **1966**, 85, 130 pp.
(49) Hay, R. L., Moiola, R. J., *Sedimentology* **1963**, 2, No. 4, 312–332.
(50) Heady, H. H., *Am. Mineralogist* **1952**, 37, 804–811.
(51) Heinrich, E. W., *Am. Mineralogist* **1963**, 48, 1172–1174.
(52) Hemley, J. J., *Am. J. Sci.* **1959**, 257, 241–270.
(53) Hemley, J. J., *Geol. Soc. Am. Spec. Papers* **1962**, 68, 196.

(54) Heron, S. D., Jr., S. C. Develop. Board Div. Geology Geol. Notes, **1969**, 13, No. 1, 27–41.
(55) Heron, S. D., Jr., Johnson, H. S., Jr., Southeastern Geol. **1966**, 7, No. 2, 51–63.
(56) Hey, M. H., Bannister, F. A., Mineral. Mag. **1934**, 23, 556–559.
(57) High, L. R., Jr., Picard, M. D., J. Sediment. Petrol. **1965**, 35, 49–70.
(58) Hoover, D. L., Geol. Soc. Am. Spec. Papers **1966**, 87, 286–287.
(59) Hoover, D. L., Geol. Soc. Am. Mem. **1968**, 110, 275–284.
(60) Hoover, D. L., Shepard, A. O., Am. Mineralogist **1965**, 50, 287.
(61) Iijima, Azuma, Hay, R. L., Am. Mineralogist **1968**, 53, 184–200.
(62) Jones, B. F., U.S. Geol. Surv. Profess. Papers **1965**, 502-A, 56 pp. [1966].
(63) Jones, B. F., "Geology, Geochemistry, Mining," J. L. Rau, Ed., p. 181–200, Northern Ohio Geological Society, Cleveland, 1966.
(64) Jones, B. F., Rettig, S. L., Eugster, H. P., Science **1967**, 158, 1310–1314.
(65) Jones, W. R., Hernon, R. M., Moore, S. L., U.S. Geol. Surv. Profess. Papers **1967**, 555, 144 pp.
(66) Keller, W. D., J. Sediment. Petrol. **1952**, 22, 70–82.
(67) Keller, W. D., J. Sediment. Petrol. **1953**, 23, 10–12.
(68) Keller, W. D., U.S. Geol. Surv. Bull. 1150, **1962**, 90 pp.
(69) Kerr, P. F., Econ. Geol. **1931**, 26, 153–168.
(70) Kerr, P. F., Cameron, E. N., Am. Mineralogist **1936**, 21, 230–237.
(71) Kirov, G. N., Sofia Univ. Annuaire, Fac. Geologie et Geographie **1965**, 60, 193–200.
(72) Kittleman, L. R., Green, A. R., Hagood, A. R., Johnson, A. M., McMurray, J. M., Russell, R. G., Weeden, D. A., Oregon Univ. Mus. Nat. Hist. Bull. **1965**, 1, 45 pp.
(73) Lasky, S. G., Webber, B. N., U.S. Geol. Surv. Bull. 961, **1949**, 86 pp.
(74) Love, J. D., Written communication, 1964.
(75) McAllister, J. F., Written communication, 1965.
(76) Mason, B. H., Sand, L. B., Am. Mineralogist **1960**, 45, 341–350.
(77) Merritt, C. A., Okla. Geol. Surv. Bull. **1958**, 76, 70 pp.
(78) Merritt, C. A., Ham, W. E., Am. Assoc. Petrol. Geologists Bull. **1941**, 25, 287–299.
(79) Milton, Charles, Wash. Acad. Sci. J. **1936**, 26, No. 9, 386.
(80) Milton, Charles, Chao, E. C-T., Fahey, J. J., Mrose, M. E., Rept. Intern. Geol. Congr. 21st **1960**, Pt. 21, 171–184.
(81) Minato, Hideo, Takano, Yukio, Nendo Kagaku **1964**, 4, 12–22.
(82) Moiola, R. J., Geol. Soc. Am. Spec. Papers **1964**, 76, 116–117.
(83) Moiola, R. J., Am. Mineralogist **1964**, 49, 1472–1474.
(84) Moiola, R. J., Written communication, 1964.
(85) Moore, J. G., Peck, D. L., J. Geol. **1962**, 70, 182–193.
(86) Moores, E. M., J. Geol. **1968**, 76, 88–98.
(87) Mullineaux, D. R., Oral communication, 1969.
(88) Mumpton, F. A., Am. Mineralogist **1960**, 45, 351–369.
(89) Murray, John, Renard, A. F., "Report on the Scientific Results of the Voyage of H.M.S. Challenger During the Years 1873–76," Edinburgh, Neill and Co., 520 pp., 1891.
(90) Neal, J. T., Langer, A. M., Kerr, P. F., Geol. Soc. Am. Bull. **1968**, 79, 69–90.
(91) Peck, D. L., Griggs, A. B., Schlicker, H. G., Wells, F. G., Dole, H. M., U.S. Geol. Surv. Profess. Papers **1964**, 449, 56 pp.
(92) Pipkin, B. W., Am. Assoc. Petrol. Geologists Bull. **1967**, 51, 478.
(93) Pirsson, L. V., Am. J. Sci. **1890**, 40, 232–237.
(94) Ratte, J. C., Steven, T. A., U.S. Geol. Surv. Profess. Papers **1967**, 524-H, 58 pp.
(95) Regis, A. J., Sand, L. B., Am. Mineralogist **1966**, 51, 270.

(96) Regis, A. J., Sand, L. B., "Clays and Clay Minerals—Proc. Clay Minerals Conf., 15th, Pittsburgh, Pa., 1966," S. W. Bailey, Ed., p. 193, Pergamon, New York, 1967.
(97) Regnier, J. P. M., *Geol. Soc. Am. Bull.* **1960**, 71, 1189–1210.
(98) Reynolds, W. R., "Facies Changes in the Alabama Tertiary—Alabama Geol. Soc. Guidebook, 4th Ann. Field Trip," C. W. Copeland, Ed., p. 26–37, Alabama Geological Society, 1966.
(99) Roberts, A. E., *U.S. Geol. Surv. Profess. Papers* **1963**, 475-B, B86–B92.
(100) Robinson, P. T., *J. Sediment. Petrol.* **1966**, 36, 1007–1015.
(101) Rooney, T. P., Kerr, P. F., *Science* **1964**, 144, 1453.
(102) Ross, C. S., *Am. Mineralogist* **1928**, 13, 195–197.
(103) Ross, C. S., *Am. Mineralogist* **1941**, 26, 627–629.
(104) Rozendal, Roger, *Proc. S. D. Acad. Sci., 1956*, **1957**, 35, 39–41.
(105) Saha, Prasenjit, *Am. Mineralogist* **1959**, 44, 300–313.
(106) Saha, Prasenjit, *Am. Mineralogist* **1961**, 46, 859–884.
(107) Sand, L. B., Regis, A. J., *Geol. Soc. Am. Spec. Papers* **1966**, 87, 145–146.
(108) Sand, L. B., Regis, A. J., *Geol. Soc. Am. Spec. Papers* **1968**, 101, 189.
(109) Santos, E. S., Oral communication, 1969.
(110) Schaller, W. T., *Am. Mineralogist* **1932**, 17, 128–134.
(111) Schultz, L. G., *U.S. Geol. Surv. Open-File Rept.* **1961**, 60 pp.
(112) Schultz, L. G., "Clays and Clay Minerals—Proc. Natl. Conf. Clays and Clay Minerals, 11th, Ottawa, 1962, W. F. Bradley, Ed., p. 169–177, Macmillan, New York, 1963.
(113) Shawe, D. R., *U.S. Geol. Surv. Profess. Papers* **1968**, 576-B, 34 pp.
(114) Shepard, A. O., *U.S. Geol. Surv. Profess. Papers* **1961**, 424-C, C320–C323.
(115) Shepard, A. O., Starkey, H. C., *U.S. Geol. Surv. Profess. Papers* **1964**, 475-D, D89–D92.
(116) Sheppard, R. A., "Mineral and Water Resources of Arizona," p. 464–467, U.S. 90th Congr., 2d sess., Comm. on Interior and Insular Affairs, Comm. Print, 1969.
(117) Sheppard, R. A., Unpublished data.
(118) Sheppard, R. A., Gude, A. J., III, *U.S. Geol. Surv. Profess Papers* **1964**, 501-C, C114–C116.
(119) Sheppard, R. A., Gude, A. J., III, *U.S. Geol. Surv. Profess. Papers* **1965**, 525-D, D44–D47.
(120) Sheppard, R. A., Gude, A. J., III, *U.S. Geol. Surv. Profess. Papers* **1968**, 597, 38 pp.
(121) Sheppard, R. A., Gude, A. J, III, *U.S. Geol. Surv. Profess. Papers* **1969**, 650-D, D69–D74.
(122) Sheppard, R. A., Gude, A. J., III, *Am. Mineralogist* **1969**, 54, 875–886.
(123) Sheppard, R. A., Gude, A. J., III, *U.S. Geol. Surv. Profess. Papers* **1969**, 634, 35 pp.
(124) Sheppard, R. A., Gude, A. J., III, Munson, E. L., *Am. Mineralogist* **1965**, 50, 244–249.
(125) Sheppard, R. A., Walker, G. W., "Mineral and Water Resources of Oregon," p. 268–271, U.S. 90th Congr., 2d sess., Comm. on Interior and Insular Affairs, Comm. Print, 1969.
(126) Shumenko, S. I., *Dokl. Akad. Nauk SSSR* **1962**, 144, 1347–1350.
(127) Sims, J. D., *Geol. Soc. Am. Spec. Papers* **1969**, 121, 636–637.
(128) Slaughter, Maynard, Earley, J. W., *Geol. Soc. Am. Spec. Papers* **1965**, 83, 116 pp.
(129) Smith, G. I., Almond, Hy, Sawyer, D. L., Jr., *Am. Mineralogist* **1958**, 43, 1068–1078.
(130) Smith, G. I., Haines, D. V., *U.S. Geol. Surv. Bull. 1181-P*, **1964**, 58 pp.
(131) Smith, J. W., *Am. Assoc. Petrol. Geologists Bull.* **1963**, 47, 804–813.
(132) Smith, J. W., Milton, Charles, *Econ. Geol.* **1966**, 61, 1029–1042.

(133) Sohn, I. G., Herrick, S. M., Lambert, T. W., *U.S. Geol. Surv. Profess. Papers* **1961**, 424-B, B227–B228.
(134) Spotts, J. H., Silverman, S. R., *Am. Mineralogist* **1966**, 51, 1144–1155.
(135) Staples, L. W., Gard, J. A., *Mineral. Mag.* **1959**, 32, 261–281.
(136) Steiner, Alfred, *Econ. Geol.* **1953**, 48, 1–13.
(137) Studer, H. P., "Clays and Clay Minerals—Proc. Clay Minerals Conf., 15th, Pittsburgh, Pa., 1966," S. W. Bailey, Ed., p. 187, Pergamon, New York, 1967.
(138) Surdam, R. C., Hall, C. A., *Geol. Soc. Am. Spec. Papers* **1968**, 101, 338.
(139) Switzer, G. S., Boucot, A. J., *J. Paleontol.* **1955**, 29, No. 3, p. 525–533.
(140) Tourtelot, H. A., *U.S. Geol. Surv. Oil Gas Inv. Prelim. Chart 22*, 1946.
(141) Van Houten, F. B., *J. Geol.* **1960**, 68, 666–669.
(142) Van Houten, F. B., *Am. J. Sci.* **1962**, 260, 561–576.
(143) Van Houten, F. B., *Kan. Geol. Surv. Bull. 169*, **1964**, 2, 497–531.
(144) Van Houten, F. B., *U.S. Geol. Surv. Bull. 1164*, **1964**, 99 pp.
(145) Van Houten, F. B., *Am. J. Sci.* **1965**, 263, 825–863.
(146) Van Loenen, R. E., Written communication, 1968.
(147) Walker, G. W., Repenning, C. A., U.S. Geol. Surv. Misc. Geol. Inv. Map I-446, 1965.
(148) Walker, G. W., Swanson, D. A., *U.S. Geol. Surv. Bull. 1260-M*, **1968**, 16 pp.
(149) Walker, G. W., Swanson, D. A., *U.S. Geol. Surv. Bull. 1260-L*, **1968**, 17 pp.
(150) Waters, A. C., *Ore Bin* **1966**, 28, 137–144.
(151) Weeks, A. D., Eargle, D. H., "Clays and Clay Minerals—Proc. Natl. Conf. Clays and Clay Minerals, 10th, 1961, Ada Swineford, Ed., p. 23–41, Macmillan, New York, 1963.
(152) Weeks, A. D., Levin, Betsy, Bowen, R. J., *Geol. Soc. Am. Bull.* **1958**, 69, 1659.
(153) Wermund, E. G., Moiola, R. J., *J. Sediment. Petrol.* **1966**, 36, 248–253.
(154) Wilcox, R. E., Fisher, R. V., U.S. Geol. Surv. Geol. Quad. Map GQ-541, 1966.
(155) Wrucke, C. T., *U.S. Geol. Surv. Bull. 1121-H*, **1961**, 26 pp.

RECEIVED February 4, 1970. Publication authorized by the Director, U. S. Geological Survey.

Discussion

L. B. Sand (Worcester Polytechnic Institute, Worcester, Mass. 01609): Please define authigenic as used relevant to the origin of zeolites in these deposits.

R. A. Sheppard: Authigenic refers to those minerals that crystallized in sedimentary rocks after deposition of the original detrital grains. Generally, the zeolites crystallized after burial of the enclosing rock. The depth of burial probably ranged from millimeters to several thousand feet.

23

Clinoptilolite from Japan

HIDEO MINATO and MINORU UTADA

Institute of Earth Science and Astronomy, College of General Education, University of Tokyo, Komaba, Megro-ku, Tokyo, Japan

> *In Japan, clinoptilolite is the commonest zeolite formed from altered pyroclastics. Four modes of occurrence have been found, replacement of vitric materials and precipitation in interstitial voids being predominant. By chemical analyses, clinoptilolite is classified into 3 types, Ca-, Na-, and K-type. In a ternary diagram of Ca(Mg), Na, and K, the field of clinoptilolite is not overlapped by that of heulandite. The x-ray powder profiles of clinoptilolites resemble that of heulandite, but their thermal behavior differs. When heated to 250°C, heulandite changes to heulandite-B; this transition is not observed in clinoptilolite. Furthermore, the thermal behavior of Ca-clinoptilolite differs from alkali-clinoptilolite. This may be attributed to the difference of dehydration between Ca-clinoptilolite and alkali-clinoptilolite, which seems to depend on the atomic ratio of Ca and alkalies.*

Clinoptilolite was named by Schaller (6) as a new mineral of the mordenite group, but Hey and Bannister (1) concluded that "clinoptilolite" was merely high-silica heulandite. Recently, Mumpton (5) redefined clinoptilolite as a high-silica member of the heulandite group. Mason and Sand (2), however, contend that the differences between clinoptilolite and heulandite do not lie in the content of Si, but of Na and K. We describe the mode of occurrence of clinoptilolite in Tertiary acidic tuffs in Japan and discuss the difference between clinoptilolite and heulandite on the basis of several mineralogical studies.

As shown in Figure 1, clinoptilolite seems to be concentrated in the Green Tuff Region, which is so named because of the green-color altered pyroclastics. In the Paleo-Setouchi Region, clinoptilolite is recognized commonly. Apart from these Neogene systems, some occurrences of this zeolite have been reported from the Paleogene through Cretaceous systems.

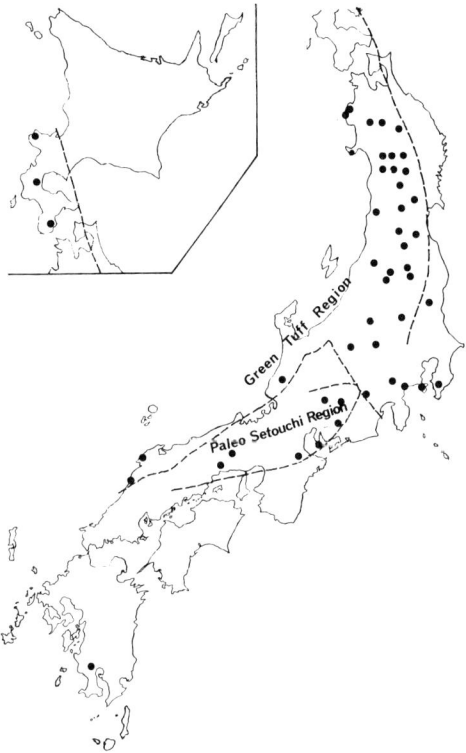

Figure 1. Distribution of clinoptilolite in Japan

The following 4 modes of occurrence have been found, with (a) being predominant.

(a) Replacement of vitric materials and precipitation in interstitial spaces. Vitric materials replaced by clinoptilolite are most frequently rhyolitic through dacitic, although sometimes they are andesitic. There are cases of complete replacement, and other cases in which unaltered vitric materials remain, with many intermediate stages. On the other hand, as is observed typically in the Paleo-Setouchi sediments, some occurrences show layers of zeolitized tuff and fresh tuff alternately overlapped in the vertical direction. The original texture is preserved in large measure, but the extent of preservation depends on the degree of zeolitization. Where clinoptilolite replaces vitric materials, microcrystalline or cryptocrystalline aggregates generally are found, with well-preserved texture.

(b) Cementation of clastics. This does not appear to be different from precipitation in interstitial spaces in terms of origin. Even if no vitric materials can be detected, there is no denying, as far as the geology of Japan is concerned, the relation to the original vitric materials.

Table I. Chemical Analyses of Clinoptilolites

	Shizuma (S)		Itaya (I)		Futatsui (F)	
	Wt. %	Mol. prop.	Wt. %	Mol. prop.	Wt. %	Mol. prop.
SiO_2	65.17	1.085	66.68	1.110	67.08	1.116
TiO_2	0.16	0.002	0.16	0.002	0.26	0.003
Al_2O_3	13.38	0.131	11.30	0.110	12.00	0.118
Fe_2O_3	1.06	0.007	0.89	0.006	0.68	0.004
MnO	none	—	trace	—	none	—
MgO	0.53	0.013	1.14	0.028	0.80	0.020
CaO	3.22	0.057	1.86	0.033	0.80	0.014
K_2O	2.82	0.030	4.25	0.045	3.21	0.033
Na_2O	1.62	0.026	0.43	0.007	2.14	0.035
$H_2O(+)$	6.48	0.360	9.48	0.526	8.21	0.456
$H_2O(-)$	4.95	0.275	4.53	0.251	5.60	0.311
Total	99.39		100.72		100.74	

(c) Replacement of plagioclase. Generally speaking, it is rare that plagioclase and other phenocrystal minerals contained in zeolite-bearing rocks suffer alteration. Zeolitization is seen, however, along the crystal margins, or cleavage of plagioclase only in the case of advanced zeolitization.

(d) Segregation veins. Though in relatively rare cases, "segregation veins" consisting of clinoptilolite are found, frequently accompanied by no other coexisting minerals. No hydrothermal veins have been reported.

The chemical composition of clinoptilolite resembles that of heulandite, but it may be characterized as compared with that of heulandite by high alkali and Si content and low Ca and Al content. Table I shows

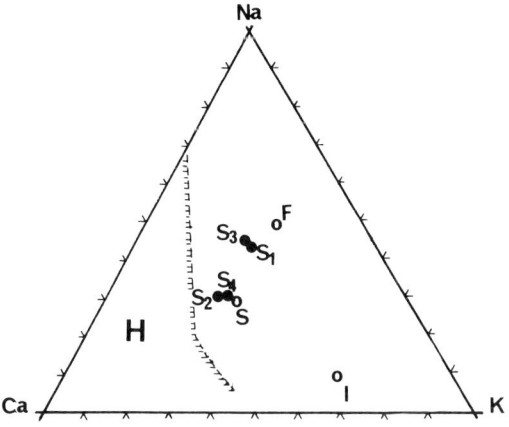

Figure 2. Ca(+Mg)-, Na-, and K-ratios of clinoptilolite. H = field of heulandite

the chemical composition of specimens of clinoptilolites from Japan which are assumed to be of high purity. In a ternary diagram of the atomic proportions among Ca($+$Mg), Na, and K, the field of clinoptilolite is not overlapped by that of heulandite, as shown in Figure 2.

Furthermore, x-ray powder profiles of clinoptilolite resemble those of heulandite, but their thermal behavior differs. Heulandite changes to heulandite-B by heating 4 hours at 250°C, and this transition is easily detected by the changes of the position of the (020) reflection pattern. The change appears as a shrinkage of this reflection from about 8.9 to 8.6A, a transition not observed in any clinoptilolite.

By chemical analyses, as shown in Table I, clinoptilolite may be classified into 3 types, Ca-, Na-, and K-type; Na-type clinoptilolite is the commonest. K-type clinoptilolite was first described by Minato and

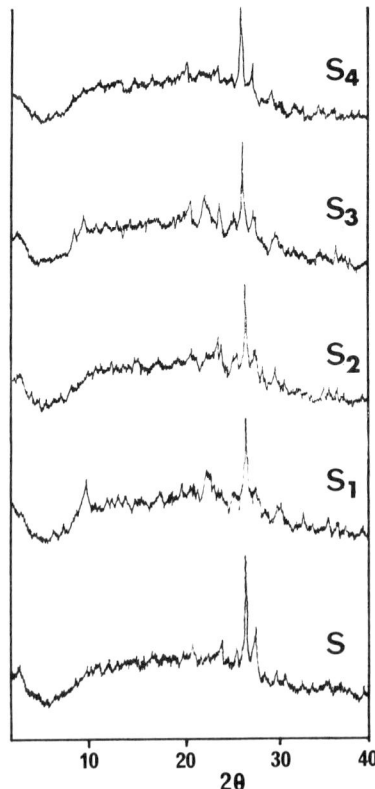

Figure 3. X-ray powder profiles (Cu Kα radiation) of untreated clinoptilolite from Shizuma (S) and treated materials (S_1–S_4)

Takano (3) (1964), and Ca-type clinoptilolite by Minato and Utada (4) (1968). A Na-type specimen from Futatsui (F), a K-type one from Itaya (I) and a Ca-type one from Shizuma (S) are plotted in the ternary diagram of Ca(+Mg), Na, and K (Figure 2).

A difference in thermal behavior is observed between the Ca-type clinoptilolite and the Na- and K-types. The temperature for the destruction of crystal structure by heating is lower in the Ca-type than in the Na- and K-types. The intensity of the x-ray powder patterns of the Ca-type is reduced almost to nil by 4-hour heating at 400°–450°C, while with the Na- and K-type the pattern is reduced to nil by 4-hour heating at 700°C.

By immersion in NaCl solution, Ca-type clinoptilolite is easily changed to Na-type. Powdered Ca-type clinoptilolite with small amounts of quartz and plagioclase from Shizuma (S) was treated for 24 hours at room temperature with a 5% NaCl solution. The product (S_1) was a Na-type clinoptilolite. S_1 can be reverted to the Ca-type by similar treatment with a 5% $CaCl_2$ solution. This product (S_2) was a Ca-type clinoptilolite. The same treatments were repeated and 2 more products—S_3 (Na-type) and S_4 (Ca-type)—were obtained. The atomic proportions of Ca(+Mg), Na, and K in these products S_1, S_2, S_3, and S_4 are plotted in Figure 2. No change takes place in the x-ray powder patterns because of these treatments. Untreated material (S) and 4 treated materials (S_1–S_4) were tested by the heating procedures mentioned above. X-ray powder profiles of S, S_1, S_2, S_3, and S_4 after 4-hour heating at 450°C are shown in Figure 3. The profiles of S, S_2, and S_4 are those of Ca-type clinoptilolite and S_1 and S_3 are Na-type.

Furthermore, endothermic peaks in differential thermal analyses of clinoptilolites show that the dehydration of Ca-type clinoptilolite is completed at lower temperature than that of Na- and K-types. The difference in the destruction of crystals by heating may reflect the difference in the behavior on dehydration of the clinoptilolites. From these facts, the atomic ratio of Ca and alkalies may be related to behavior on dehydration and to the destruction of the crystal structure.

Literature Cited

(1) Hey, M. H., Bannister, F. M., *Mineral. Mag.* **1934**, 23, 556-559.
(2) Mason, B., Sand, L. B., *Am. Mineralogist*, **1960**, 45, 341-350.
(3) Minato, H., Takano, Y., *J. Clay Sci. Soc. Japan* **1964**, 4, 12-22.
(4) Minato, H., Utada, M., *J. Clay Sci. Soc. Japan* **1968**, 7, 25-32.
(5) Mumpton, F. A., *Am. Mineralogist* **1960**, 45, 351-369.
(6) Schaller, W. T., *Am. Mineralogist* **1932**, 17, 128-134.

RECEIVED January 12, 1970.

Discussion

F. A. Mumpton (State University College at Brockport, N. Y.): I would like to compliment our Japanese colleagues on the fine nature and amount of research which they have carried out in recent years on sedimentary zeolites, especially in the area of utilization for which they are well known. I hope that our industrial friends in Union Carbide, W. R. Grace, Mobil, and Norton Companies take note of the size and scale of the zeolite mining operations which you showed in your slide.

I believe you stated that when you sodium-exchanged heulandite, it showed the same thermal behavior as the normal mineral; and when you calcium exchanged clinoptilolite, it did not show a heulandite–heulandite B transformation but merely a somewhat lower stability than normal clinoptilolite. Is this correct?

Hideo Minato: Zeolite production in Japan is 5000–6000 t/month, 1000–1500 t from Futatsui and 4000–4500 t from Itaya. A few other workers produce 100–500 t/month.

Chemical treatment of heulandite is the same as that of Ca-clinoptilolite, and the stability of heulandite in heating is lower than that of clinoptilolite.

D. B. Hawkins (University of Alaska, College, Alaska): My observation of the hydrothermal behavior of exchanged clinoptilolite is perhaps pertinent to yours on thermal behavior. I find that clinoptilolite can be transformed to a Ba form by exchange at 80°C. The hydrothermal behavior of Ba-clinoptilolite differs from that of the natural clinoptilolite in that the latter transforms to mordenite at $\sim$320°C and 15,000 psi, whereas the Ba form does not. Thus, the hydrothermal and thermal behavior of clinoptilolite is profoundly affected by exchangeable cations.

H. Minato: I agree with you. Furthermore, I think there is some relationship between the behavior of exchangeable cation and H_2O and (OH).

24

Present Status of the Zeolite Facies

DOUGLAS S. COOMBS

Geology Department, University of Otago, Dunedin, N. Z.

Zeolite sequences originating under diagenetic, burial metamorphic, and hydrothermal conditions are reviewed briefly. Some of the assemblages appear to represent equilibrium. The zeolite facies concept is discussed, and some possible mineral assemblages are analyzed. Mordenite, heulandite, laumontite, and wairakite facies of Seki are treated as subfacies of a more comprehensive zeolite facies. Probable maximum temperatures for these and contiguous low-grade mineral facies when $P_{H_2O} = P_{total}$ are shown on a P–T grid. Under other conditions, the boundaries are displaced, sometimes drastically, to lower temperatures. Distinct facies series, depending on P/T ratios, are recognizable. Zeolites and other calcium aluminosilicates can be suppressed and replaced by clay–carbonate assemblages when μ_{CO_2} is sufficiently high.

The general pattern of distribution of zeolites in several types of geologic situation is now reasonably well known and occurrences in sedimentary rocks have been reviewed recently with a large bibliography (21). Occurrences in joints and other miscellaneous situations will not be discussed here. Important cases are:

(1) In sediments from the deep ocean basins.

(2) As products of diagenesis in nonmarine beds, in part the product of reaction between glassy tuffs and alkalic groundwater (22, 25, 40) and in part of reaction between highly alkaline lake waters and sediments, both tuffaceous and nontuffaceous. Diagenetic zeolites include erionite, chabazite, phillipsite, natrolite, mordenite, clinoptilolite, and analcime. In many cases, the mineral assemblages do not appear to represent thermodynamic equilibrium, and analcime tends to replace the other phases during aging (21). In some cases, laumontite is reported (4, 48).

(3) In burial metamorphic sequences, including products of "epigenesis" of Russian writers, often very thick, and usually marine (6, 8, 13, 26, 30, 34, 49). In the upper members of such sequences, analcime may

co-exist with quartz, and members of the clinoptilolite–heulandite group are common. Mordenite is found occasionally. At greater depth, the characteristic assemblage is quartz–albite–laumontite–chlorite, laumontite eventually giving way to the less hydrous nonzeolitic calcium aluminosilicates, prehnite, pumpellyite, epidote, and sometimes lawsonite. The zeolites in part replace glass of vitric tuffs, and in part they occur as cement and as products of metamorphic reactions. Laumontite and albite replace detrital calcic plagioclase. Laumontite also partially replaces fossil shell material (8, 27) or inorganic calcite, presumably by reaction with aluminous clay minerals. In regions inferred to have had a high geothermal gradient, a zone of wairakite may intervene between, or overlap, zones characterized by laumontite and by prehnite–pumpellyite (19, 37, 43). Gross overlapping of zones is characteristic. Thus, in Taringatura, southern New Zealand, relict heulandite persists in tightly cemented, impermeable beds deep in the laumontite zone; alteration of detrital plagioclase to laumontite plus albite, producing a characteristic color mottling on a scale of a few millimeters (as in the "facies moucheté" of the Alps, Ref. 31, 32; cf. 24) occurs spasmodically in coarser-grained sandstones at least 7 km higher in the section than the lowest heulandite.

(4) As products of hydrothermal metamorphism in geothermal fields, such as Wairakei (10, 42), Pauzhetsk (39), and Onikobe (38). In these, a generalized downward sequence with increasing temperature is mordenite with or without analcime, rare heulandite, laumontite, wairakite, less hydrous phases.

(5) Perhaps least well systematized are the occurrences of zeolites in volcanic rocks and breccias, the source of so many handsome show-case specimens. Walker (44, 45) has described a low-temperature depth zonation transecting volcanic stratigraphy in the lava piles of Northern Ireland and Iceland; crude correlation between zeolite species and silica-saturation of host lavas can also be detected (10).

Definition of the Zeolite Facies

According to Fyfe and Turner (17), "A metamorphic facies is a set of mineral assemblages, repeatedly associated in space and time, such that there is a constant and therefore predictable relation between mineral composition and chemical composition." In his original definitions, Eskola (14, 16) stipulated that such assemblages represent chemical equilibrium. Later (15), he abandoned the requirement of equilibrium, demonstration of this in a rigorous sense being impracticable. Nevertheless, if the physical conditions of metamorphism are to be inferred from laboratory studies of mineral equilibria, the assumption that mineral facies represent chemical equilibria is normally required.

On the basis of observations in southern New Zealand (9), Turner (18) defined a zeolitic facies to include only regionally developed assemblages, including laumontite–albite–quartz, "that largely replace the pre-existing rocks and conform to the mineralogical and chemical requirements of a metamorphic facies . . ." Analcime and heulandite assemblages were excluded as being diagenetic rather than metamorphic.

Largely to avoid involvement in definitions of metamorphism, and in the belief that assemblages of minerals reflect the physico-chemical conditions of their formation whether they be metamorphic, hydrothermal, or diagenetic, Coombs et al. (8, 10) treated zeolitic assemblages in terms of Eskola's mineral facies concept (15) and broadened Turner's definition of the zeolite facies "to include at least all those assemblages" produced under conditions in which quartz–analcime, quartz–heulandite, and quartz–laumontite commonly are formed.

Winkler (46) restricted the zeolitic facies, his "laumontite–prehnite–quartz facies" (cf. Figure 1d), to supposedly metamorphic, nondiagenetic assemblages containing laumontite as the critical phase. Hay (21) questioned the validity of the concept on the grounds that zeolitic mineral assemblages vary as a function of age, and may represent incomplete replacement of one mineral by another, neither of these feaures being compatible with standard definitions of metamorphic facies. This may be true of the very low-temperature assemblages referred to under (2) above. However, laumontite–albite–quartz, analcime–quartz, and less certainly heulandite–quartz are now well known in rocks from many regions and environments, and of ages extending well back into the Paleozoic. There is now reasonably good experimental evidence (2, 5, 23), as well as textural evidence, that analcime–quartz has a low-temperature field of stability relative to albite. Liou and Ernst (28, 29) have successfully reversed reactions between laumontite, wairakite, and anorthite, showing that each has its own stability field relative to the others in the presence of water and quartz. Textural evidence is as unequivocal as it can be that in the laumontite zone, laumontite–albite is stable relative both to heulandite and to calcic plagioclase. Whether heulandite has a true stability field relative to laumontite is not clear yet. The reservations of Hay can now be put aside, at least in the case of certain quartz–analcime and laumontite assemblages.

Wairakite, Mordenite, Heulandite, and Laumontite Facies

In 1966, Seki (36) proposed a new facies, the wairakite facies, characterized by the occurrence of wairakite–quartz with clay minerals, in the absence of prehnite or pumpellyite. On field, experimental, and thermodynamic grounds, this proposed facies evidently represents, at least for conditions where $P_{H_2O} = P_{fluid}$, higher temperatures than the laumontite assemblages, and relatively low pressure. Subsequently, Seki (35) introduced separate mordenite, heulandite, and laumontite facies. Our earlier definition of the zeolite facies was intended to be broad enough to allow extension to accommodate such assemblages. If Seki's 4 facies are given independent status, the zeolite facies should be dropped. Alternatively, and accepting for the moment that the heulandite and mordenite assemblages are sufficiently reproducible in place and time to accord with definitions of facies, all 4 become subfacies of a broadened zeolite facies. The choice between these alternatives is one of personal preference, but the second probably best serves the needs of the greater number of geologists, and is adopted here.

Analysis of Zeolite Facies Assemblages

Some possible suites of assemblages in the zeolite facies are represented by diagrams a to g in Figure 1. All these suites are represented, for example, in Triassic sediments of the Murihiku Supergroup of New Zealand (*cf.* 6). The arrows represent dehydration or decarbonation reactions, and hence conventionally increasing metamorphic grade. A possible earlier reaction is:

clinoptilolite = heulandite + albite and/or adularia + quartz + water

It is to be noted:

(1) Heulandite (Figure 1b) can co-exist, apparently stably, with prehnite (3, 10).

(2) There are alternative routes from the suite of assemblages represented by Figure 1a to those represented by Figures 1d and 1g. The route followed will be controlled by μ_{CO_2}.

(3) Before the stability limit of laumontite is reached, a reaction such as

laumontite + chlorite = pumpellyite + clay mineral + H_2O (Figure 1 g → h)

will greatly restrict the range of whole-rock composition in which laumontite will occur. Typical volcanogenic sediments will be represented by assemblages such as prehnite–pumpellyite–chlorite–quartz–albite. At this stage, it may not be practicable to distinguish the assemblages observed in most rocks from those of the prehnite–pumpellyite group of facies (7, 8, 11, 20) into which Figure 1h represents a passage.

(4) A change of mineral facies cannot be established with certainty unless mineral aggregates of similar bulk composition with respect to nonmobile components are compared. In practice, boundaries between the lower-grade mineral facies are defined for bulk compositions projecting onto the Ca-zeolite–chlorite join in the Al_2O_3–CaO–(Mg,Fe)O diagram (*cf.* Figure 1) or somewhat to the Al_2O_3-poor side of this join. Many volcanic rocks, volcanogenic sediments, and greywackes project into this region.

(5) Calcic zeolites, prehnite, pumpellyite, and pistacitic epidote are not mutually exclusive, but are related by a series of reactions such as

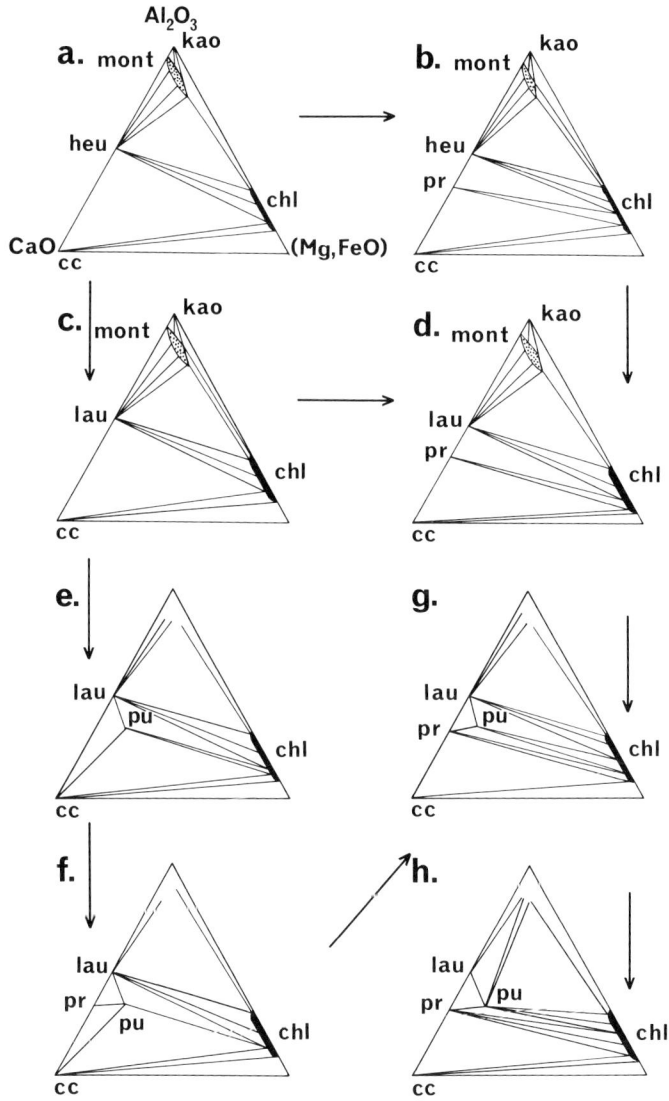

Figure 1. Some zeolite facies assemblages

cc: calcite; chl: chlorite; heu: heulandite; kao: kaolinite; lau: laumontite; mont: montmorillonite; pr: prehnite; pu: pumpellyite. Further study of the aluminous clay minerals appropriate to Figures e to h is required. The joins calcite–dolomite and dolomite–chlorite presumably will be stable, but are not observed commonly in zeolite facies assemblages.

those implied by Figure 1. It may be possible locally to recognize subfacies corresponding to the individual assemblage diagrams of Figure 1.

(6) Calcite–chlorite, calcite–prehnite–chlorite, and calcite–prehnite are not fully diagnostic assemblages; they are stable both in the zeolite facies and in the prehnite–pumpellyite facies. Calcite–chlorite is stable over a still wider range of facies. Parts of the same bed of greywacke or feldspathic sandstone commonly contain assemblages such as laumontite–chlorite, calcite–chlorite, and prehnite–chlorite (*12, 31*). Such occurrences do not imply a mixture of facies; they may result from subtle differences in bulk chemistry, either primary or the result of the minor metasomatic changes that are so characteristic of low metamorphic grades (*7, 9, 45*).

(7) Wairakite-bearing assemblages similarly often contain prehnite and/or pumpellyite (*37, 43*), implying the existence of reactions analogous to those involved in Figure 1. The zeolite facies in the broad sense, as well as the heulandite, laumontite, and wairakite subfacies, are evidently broad groupings of suites of assemblages, containing key zeolite species where the bulk chemical composition is appropriate, but highly dependent in detail on subtle physico-chemical controls.

The Role of CO_2 and H_2O

Zen (*50*) pointed out that zeolite and clay–carbonate or "greenschist" assemblages can be related under isothermal, isobaric conditions as a function of differing values of μ_{H_2O}, μ_{CO_2}. A more complete chemographic analysis (Figure 2, cf. *1, 11, 47*) illustrates a possible progression through the zeolite facies, prehnite–pumpellyite facies, and greenschist facies, as μ_{H_2O} is reduced at (T, P) constant, μ_{CO_2} being low. At higher values of μ_{CO_2}, each of these facies gives way to clay–carbonate or pyrophyllite–carbonate assemblages. A clay–carbonate assemblage hardly can be regarded as belonging to the zeolite, prehnite–pumpellyite, or greenschist facies, and perhaps should be regarded as defining a further mineral facies (*11*) stable at low temperatures, μ_{CO_2} being relatively high. The diagram applies to isobaric, isothermal conditions, but the sequences of assemblages shown from top to bottom and right to left are also sequences that will be encountered with progressively increasing temperature. An excellent natural example of low-temperature hydrothermal metamorphism at relatively high μ_{CO_2} is provided by the Salton Sea field, where clay–carbonate assemblages pass directly into epidote-bearing greenschist facies assemblages at temperatures of about 300°, without the prior formation of zeolites, prehnite, or pumpellyite (*33*).

The chemical potential of water in fluid phases is reduced as salinity, or CO_2 content, is increased. Increasing salinity of intrastratal waters at depth thus will have the same effect as increasing temperature in promoting a downwards zonation from more hydrous to less hydrous phases (*22*). Furthermore, differences in pressure, salinity, CO_2 content of

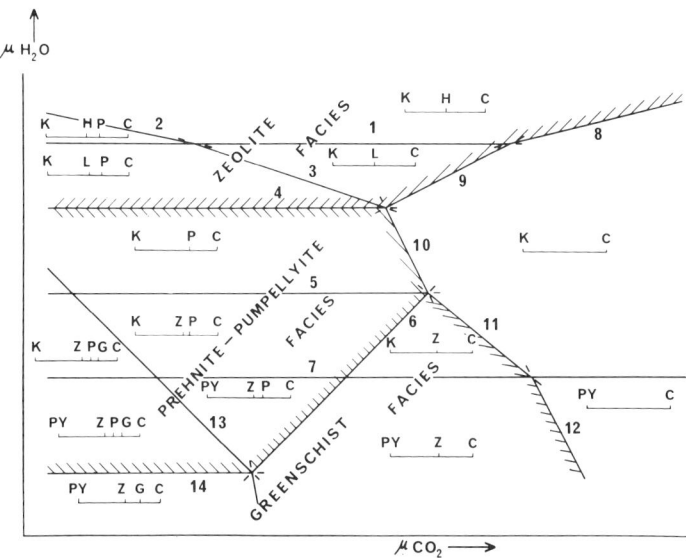

Figure 2. Schematic $\mu_{H_2O}-\mu_{CO_2}$ diagram for part of the system Al_2O_3–CaO in the presence of excess quartz, at some arbitrary value of T and total P. The precise position of the kaolinite–pyrophyllite reaction, relative to the other reactions shown, is uncertain.

C: calcite; G: grossular; H: heulandite; K: kaolinite; L: laumontite; P: prehnite; PY: pyrophyllite; Z: zoisite.

In any divariant assemblage, 2 of these phases can co-exist as indicated by adjacent symbols in the phase compatibility diagrams. For example, in the uppermost field, i.e., above the univariant lines 2, 1, and 8, the phases kaolinite, heulandite, and calcite are stable where the ratio of Al_2O_3:CaO is 1:0, 1:1, and 0:1, respectively. Kaolinite and heulandite co-exist for finite ratios of Al_2O_3:CaO > 1:1; heulandite and calcite co-exist for finite ratios Al_2O_3:CaO < 1:1.

waters within different beds of the one sedimentary sequence, and osmotic effects can account for the seemingly bewildering overlap of zeolite subfacies. Thus, it has been calculated (5) that the reaction analcime + quartz = albite + water, at equilibrium for pure water at about 190°C, or somewhat lower, may be displaced to about 100° in saturated NaCl solution, or even lower in other salt solutions.

Throughout its discontinuous length of outcrop, the Taveyanne sandstone of the western Alps varies in grade from laumontite, to prehnite–pumpellyite, to pumpellyite–actinolite schist facies (12, 32). A feature of the laumontite rocks is the occurrence of dehydration zones along joints and minor faults, in which laumontite has been replaced by prehnite and pumpellyite, or epidote. An attractive explanation of this phenomenon is that whereas P_{H_2O} in the country rock approximated lithostatic pressure (controlled by overriding nappes), solutions in open joints and faults may have been able to communicate with much higher structural levels, thus approaching hydrostatic pressure. In such a case, equilibrium is likely to be displaced towards less hydrous assemblages.

A Problem in Redefinition of the Zeolite Facies

In spite of statements sometimes made to the contrary, prehnite, pumpellyite, and lawsonite emphatically are not zeolites, and are not diagnostic of the zeolite facies. The zeolite facies might be defined as that set of mineral assemblages that is characterized by the association calcium zeolite–chlorite–quartz in rocks of favorable bulk composition. In this way, attention would be concentrated on bulk compositions typical of many volcanogenic sediments. A corollary would be the separate recognition of a clay–carbonate facies, as indicated above. Under such a restricted definition, the assemblage sodium–zeolite–quartz (normally

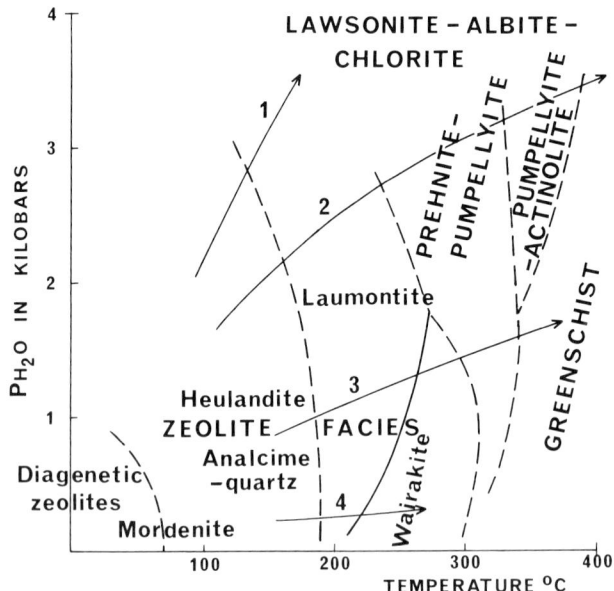

Figure 3. Possible P-T fields for low-grade mineral facies and subfacies, calibrated for the laumontite–wairakite and analcime + quartz–albite reactions (21, 3). Where P_{H_2O} < P_{total}, boundaries are displaced to the left. Arrows represent facies series as follows, slightly modified from Seki (35).

(1) High pressure, low temperature
(2) Intermediate
(3) Low pressure intermediate
(4) Lowest pressure

analcime–quartz) would no longer be restricted to the zeolite facies since, in some sedimentary environments, it appears to be stable along with aluminous clay minerals, dolomite, and calcite, an assemblage that would be diagnostic of the clay–carbonate facies.

Many workers probably will prefer a loosely defined zeolite facies which includes all mineral assemblages characterized in rocks of appropriate composition by zeolites other than the analcime of silica-deficient environments. The latter can persist to magmatic or near magmatic temperatures.

Zeolite Facies Petrogenetic Grid and Facies Series

Figure 3 (*cf.* 28) shows possible *P-T* fields for zeolite subfacies and contiguous facies. The figure is calibrated for the laumontite–wairakite (29) and analcime + quartz–albite (5) equilibria; the other fields are drawn to be compatible with field observations. As has been pointed out by Seki (35), a spread of facies series can be recognized ranging from high-pressure, low-temperature types (curve 1) typified by parts of the Kanto Mountains, Japan, the Franciscan of California, and the Bryneira Group in southern New Zealand, where locally there is a direct passage from laumontite to lawsonite assemblages (27), through intermediate types (curve 2) such as in much of the southern and western sections of the New Zealand Geosyncline and Kii Penninsula, Japan, to low-pressure intermediate (curve 3) as in Tanzawa Mountains, Japan, and lowest pressure (curve 4) as in active geothermal fields.

The temperatures shown are estimated maxima for the facies indicated; whenever P_{H_2O} is less than P_{total}, or whenever the chemical potential of water is lowered by the presence of solutes, the boundaries will be displaced, perhaps drastically, towards lower temperatures.

Acknowledgment

This paper has been made possible by receipt of research grants from the New Zealand University Grants Committee and the U.S. National Science Foundation, and by the hospitality of the Institute of Mineralogy, University of Geneva. It has benefitted from critical reading by E-an Zen.

Literature Cited

(1) Albee, A. L., Zen, E-an, *Akad. Nauk SSSR, Inst. Fiz. Tverdogo Tela* **1969**, 1, 249–260.
(2) Althaus, E., personal communication, 1968.
(3) Boles, J. R., personal communication, 1969.
(4) Bur'yanova, Ye. Z., Bogdanov, V. V., *Lithology Mineral Resources* **1967**, 195–202, transl.
(5) Campbell, A. S., Fyfe, W. S., *Am. J. Sci.* **1965**, 263, 807–816.

(6) Coombs, D. S., *Australian J. Sci.* **1962**, 24, 203–215.
(7) Coombs, D. S., "On the Mineral Facies of Spilitic Rocks and Their Genesis," **1970**, in press.
(8) Coombs, D. S., *Rept. Intern. Geol. Congr., 21st, 1960, Part XIII*, 339–351.
(9) Coombs, D. S., *Trans. Roy. Soc. N. Z.* **1954**, 82, 65–109.
(10) Coombs, D. S., Ellis, A. J., Fyfe, W. S., Taylor, A. M., *Geochim. Cosmochim. Acta* **1959**, 17, 53–107.
(11) Coombs, D. S., Horodyski, R. J., Naylor, R. S., *Am. J. Sci.* **1970**, 268, 142–156.
(12) Coombs, D. S., Martini, J., Vuagnat, M., "Low Grade Mineral Facies in the Taveyanne Sandstone, French and Swiss Alps," in preparation.
(13) Dickinson, W. R., Ojakangas, R. W., Stewart, R. J., *Bull. Geol. Soc. Am.* **1969**, 80, 519–526.
(14) Eskola, P., *Commiss. Geol. Finlande Bull.* **1915**, 44, 1–145.
(15) Eskola, P., "Die Enstehung der Gesteine," T. F. W. Barth, C. W. Correns, P. Eskola, Eds., pp. 263–407, Springer, Berlin, 1939.
(16) Eskola, P., *Norsk. Geol. Tidsskr.* **1920**, 6, 143–194.
(17) Fyfe, W. S., Turner, F. J., *Contrib. Mineral. Petrol.* **1966**, 12, 354–364.
(18) Fyfe, W. S., Turner, F. J., Verhoogen, J., *Geol. Soc. Am. Mem.* **1958**, 73.
(19) Harada, K., *J. Geol. Soc. Japan* **1968**, 74, 239–244.
(20) Hashimoto, M., *J. Geol. Soc. Japan* **1965**, 72, 253–265.
(21) Hay, R. L., *Geol. Soc. Am. Spec. Paper* **1966**, 85.
(22) Hay, R. L., *Univ. Calif. Publ. Geol. Sci.* **1963**, 42, 199–262.
(23) Hemley, J. J., *Geol. Soc. Am. Spec. Paper* **1965**, 87, 78.
(24) Hoare, J. M., Condon, W. H., Patton, W. W., *U. S. Geol. Surv. Profess. Papers* **1964**, 501-C, 74–78.
(25) Hoover, D. L., *Geol. Soc. Am. Mem.* **1968**, 110, 275–284.
(26) Kaplan, M. Ye, *Dokl. Akad. Nauk SSSR*, 163, 976–979; *Earth Sci. Sect. Transl.* **1965**, 163, 166–168.
(27) Landis, C. A., Ph.D. thesis, University of Otago, 1969.
(28) Liou, J. G., *J. Petrol.* **1970**, in press.
(29) Liou, J. G., Ernst, W. G., *Izv. Akad. Nauk, SSSR Earth Sci. Sect.*, in press.
(30) McKelvey, B. C., Ph.D. thesis, University of New England, Armidale, New South Wales, 1967.
(31) Martini, J., *Bull. Suisse Mineral. Petrog.* **1968**, 48, 539–654.
(32) Martini, J., Vuagnat, M., *Bull. Suisse Mineral. Petrog.* **1965**, 45, 281–293.
(33) Muffler, L. J. P., White, D. E., *Geol. Soc. Am. Bull.* **1969**, 80, 157–182.
(34) Packham, G. H., Crook, K. A. W., *J. Geol.* **1960**, 68, 392–407.
(35) Seki, Y., *J. Geol. Soc. Japan* **1969**, 75, 255–266.
(36) Seki, Y., *J. Japan Assoc. Mineral. Petrol. Econ. Geol.* **1966**, 55, 254–261, 56, 30–39.
(37) Seki, Y., Oki, Y., Matsuda, T., Mikami, K., Okumura, K., *J. Japan. Assoc. Mineral. Petrol. Econ. Geol.* **1969**, 61, 1–75.
(38) Seki, Y., Onuki, H., Okumura, K., Takashima, I., *Japan. J. Geol. Geography* **1969**, 40, 63–79.
(39) Senderov, E. E., *Geochem. Intern.* **1965**, 3, 1143–1155 (transl. from Russian).
(40) Sheppard, R. A., Gude, A. J., *U. S. Geol. Surv. Profess. Papers* **1969**, 634.
(41) Smith, R. E., *J. Petrol.* **1968**, 9, 191–219.
(42) Steiner, A., *Econ. Geol.* **1953**, 48, 1–13.
(43) Surdam, R. C., Ph.D. dissertation, University of California, Los Angeles, 1967.
(44) Walker, G. P. L., *J. Geol.* **1960**, 68, 515–528.
(45) Walker, G. P. L., *Mineral. Mag.* **1951**, 29, 773–791.
(46) Winkler, H. G. F., "Petrogenesis of Metamorphic Rocks," Springer-Verlag, Berlin-Heidelberg, 1965.
(47) Wise, W. S., Eugster, H. P., *Am. Mineralogist* **1964**, 49, 1031–1083.

(48) Zaporozhtseva, A. S., *Izv. Akad. Nauk SSSR Geol. Ser.* **1960**, 61–69; transl. **1961,** 52–59.
(49) Zaporozhtseva, A. S., Vishnevskaya, T. N., Dubar, G. D., *Dokl. Akad. Nauk SSSR* 141, 448–450, *Earth Sci. Sect., Transl.* **1963,** 141, 1264–1266.
(50) Zen, E-an, *Am. J. Sci.* **1961,** 50, 259, 401–409.

RECEIVED March 4, 1970.

Discussion

L. B. Sand (Worcester Polytechnic Institute, Worcester, Mass. 10609): Why can't we extend your facies concept to include the sediments, particularly pyroclastics, altered to zeolites in the lowest grade metamorphism category following the diagenic phase?

D. S. Coombs: I have tried to demonstrate that the criticisms levelled against the zeolite facies are rather certainly invalid in the case of quartz–laumontite and certain other assemblages. Some writers would restrict the zeolite facies to rocks containing these. Personally, I would extend the facies in a broad way to include all the low-temperature zeolitic assemblages, though I would recognize that some of the low-temperature assemblages probably do not represent strict equilibrium. When opal or cristobalite is present, for example, this fact is obvious. Nevertheless, we may still have a metastable equilibrium between such cristobalite and a silica-rich hydrous zeolite.

V. C. Juan (National Taiwan University, Taipei, Taiwan): Since it is difficult to differentiate zeolite facies by diagenesis from those by metamorphism, an understanding of zeolite facies in the sense of mineral facies would be appropriate. However, prehnite and pumpellyite are closely associated with hydrothermally formed zeolites and when μ_{CO_2} increases in the fluid phase, in many cases we found no zeolite facies under the greenschist facies. Therefore, it might be possible to put the prehnite–pumpellyite as a subfacies under the zeolite facies.

D. S. Coombs: We both emphasize that the prehnite and pumpellyite fields overlap those of zeolites. However, rocks mapped in the prehnite–pumpellyite facies (lacking zeolites) occupy an intermediate position between zeolite facies and pumpellyite–actinolite or greenschist facies in many terrains. In many parts of the circum-Pacific region, rocks of the prehnite–pumpellyite facies occupy a larger area than those of contiguous facies. In geological terms, the concept of a prehnite–pumpellyite facies is useful. Furthermore, as the prehnite–pumpellyite facies is defined on the absence of zeolites and the presence of these other phases in rocks of appropriate composition, it would be quite inappropriate mineralogically to classify it as a subfacies of the zeolite facies.

25

Graphical Analysis of Zeolite Mineral Assemblages from the Bay of Fundy Area, Nova Scotia

ALDEN B. CARPENTER

Department of Geology, University of Missouri, Columbia, Mo. 65201

> *Zeolites occur in consistent combinations throughout a wide area in the Bay of Fundy region. Three-phase zeolite-bearing assemblages observed include quartz–analcite–heulandite, analcite–mesolite–thomsonite, heulandite–mesolite–thomsonite, analcite–chabazite–heulandite, and analcite–natrolite–stilbite. The stability fields of these minerals can be represented conveniently on activity–activity diagrams. These facts indicate that the appropriate experimental work can provide data which may be used for making quantitative estimates of the conditions under which these zeolites form.*

The chemical compositions of the naturally occurring sodium–calcium zeolites can be expressed in terms of the four-component system $NaAlSiO_4$–$CaAl_2Si_2O_8$–SiO_2–H_2O. This system may be represented graphically on triangular coordinate paper, using the first 3 components and projecting all compositions from H_2O to this plane. The general range of variation in the chemical composition of the common sodium calcium zeolites is shown by this projection in Figure 1.

Zeolite occurrences in the Bay of Fundy area, Nova Scotia, are reported here for 2 localities. The mineral assemblages in the altered basalts at Cape Blomidon include (1) heulandite–laumontite, (2) analcite–heulandite, (3) quartz–analcite–heulandite, (4) analcite–mesolite–thomsonite, and (5) heulandite–mesolite–thomsonite. These data are summarized in Figure 2.

The spatial relationships of the minerals in assemblages 2, 4, and 5 indicate that all of the zeolites in each of these assemblages were growing

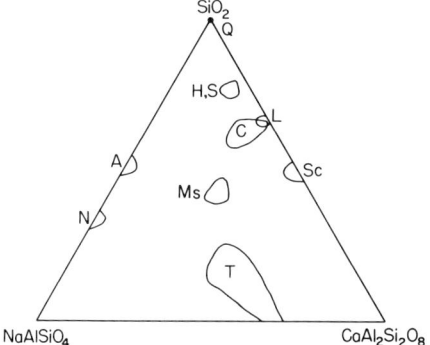

Figure 1. General range of variation in the chemical composition of the common sodium–calcium zeolites

Q = quartz
H = heulandite
S = stilbite
A = analcite
C = chabazite
L = laumontite
N = natrolite
Ms = mesolite
T = thomsonite
Sc = scolecite

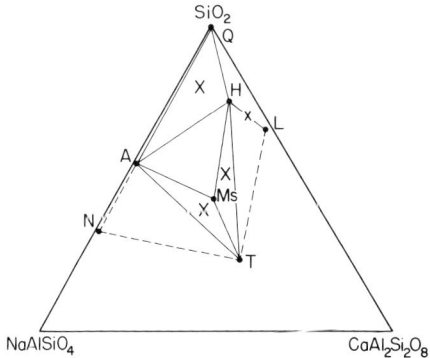

Figure 2. Common zeolite mineral assemblages at Cape Blomidon, Nova Scotia

X = observed three-phase assemblage
x = observed two-phase assemblage
Other abbreviations as in Figure 1

contemporaneously at some point in time. In assemblage 3, the quartz is present as euhedral crystals but appears to be older than the heulandite and analcite. Since the quartz does not show evidence of corrosion, the quartz–analcite–heulandite is regarded tentatively as an equilibrium assemblage. The assemblages analcite–natrolite, natrolite–thomsonite, and thomsonite–laumontite are inferred from the geometry of the phase diagram.

The mineral assemblages at Wasson's Bluff are (1) heulandite–stilbite, (2) chabazite–stilbite, (3) chabazite–natrolite (3), (4) analcite–chabazite–heulandite, (5) analcite–natrolite–stilbite, and (6) gmelinite–stilbite–quartz. These data are summarized in Figures 3 and 4.

The spatial relationships of the minerals in assemblages 3, 4, and 5 indicate that all of the zeolites in each of these assemblages were growing contemporaneously at some point in time. In all of the heulandite–stilbite specimens, the stilbite appears to be younger than the heulandite and there does not appear to be any good indication of a period of con-

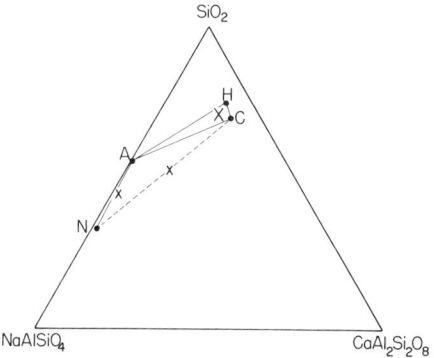

Figure 3. Early minerals and mineral assemblages at Wasson's Bluff, Nova Scotia

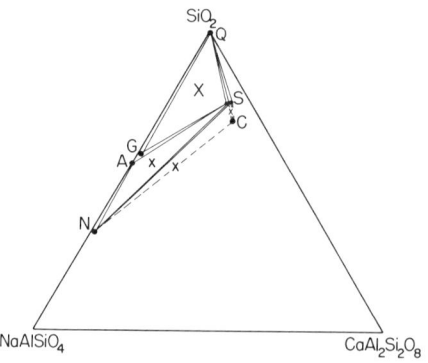

Figure 4. Stilbite-bearing mineral assemblages at Wasson's Bluff, Nova Scotia

temporaneous growth. Heulandite and stilbite from throughout the Bay of Fundy area have very similar or identical Na/Ca and Al/Si ratios. These minerals are thus related, in this particular region, by the reaction

$$Na_{0.33}K_{0.08}Ca_{0.97}Al_{2.35}Si_{6.65}O_{18} \cdot 6H_2O + H_2O \rightarrow$$
$$Na_{0.33}K_{0.08}Ca_{0.97}Al_{2.35}Si_{6.65}O_{18} \cdot 7H_2O$$

heulandite   + water →   stilbite

From this relationship and assuming that the zeolites formed during a period of falling temperatures, one would expect the 2 minerals to coexist in equilibrium for only a short time and that stilbite should be the younger mineral. For this reason and because stilbite in fact does appear to be younger than heulandite, it is assumed that stilbite-bearing assemblages formed later and at lower temperatures than heulandite-bearing assemblages.

The occurrence of gmelinite (assemblage 6) presents some problems in interpretation. In the one specimen available for study, the gmelinite crystals were deposited on top of fine-grained euhedral quartz, and both of these minerals are partially coated by stilbite. The gmelinite crystals are somewhat corroded, but this may be the result of exposure to rainwater. Gmelinite from Five Islands (3), near Wasson's Bluff, is very nearly a purely sodic zeolite and the refractive index of the Wasson's Bluff sample ($n_0 = 1.463$) indicates that it is also very sodic. Sodium gmelinite is chemically related to analcite by the reaction

$$NaAlSi_2O_6 \cdot H_2O + 2H_2O \rightarrow NaAlSi_2O_6 \cdot 3H_2O$$

analcite   + water→   gmelinite

The abundance of analcite in certain types of sedimentary rocks and the general rarity of gmelinite suggests that sodic gmelinite is generally unstable with respect to analcite plus water.

Coombs et al. (1) have pointed out that the sequence of minerals at a locality may be a function of the changing chemical composition of the depositing solutions or changing P-T conditions or both. The construction of qualitative activity–activity diagrams, similar in appearance to the chemical potential diagrams used by Korzhinskii (2), is useful for obtaining some insight into the relationship between mineralogy and solution composition. The diagram in Figure 5 is based on the mineral compositions and mineral assemblages at Cape Blomidon summarized in Figure 2 and should be valid for the temperature range implied by that set of data. This diagram also indicates that an equilibrium assemblage of 3 zeolites in the system $NaAlSiO_4$–$CaAl_2Si_2O_8$–SiO_2–H_2O coexists with an aqueous solution of unique chemical composition.

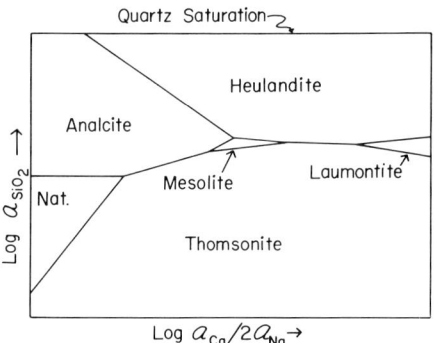

Figure 5. Qualitative activity–activity diagram for a portion of the system $NaAlSiO_4$–$CaAl_2Si_2O_8$–SiO_2–H_2O for temperatures and pressures, mineral compositions implied by Figure 2

Phase equilibrium studies on the stabilities of the three-phase mineral assemblages listed above and determinations of the chemical composition of aqueous solutions in equilibrium with these assemblages will provide data which can be used for a quantitative assessment of the conditions under which these and other comparable zeolite occurrences have formed.

Acknowledgment

I thank P. Blair Hostetler for reading the manuscript and offering many useful suggestions. D. Herman Jackson and Glen W. Heckman assisted in preparing a compilation of zeolite chemical analyses. This investigation was supported by a Grant-In-Aid-Of-Research from the Society of the Sigma Xi.

Literature Cited

(1) Coombs, D. S., Ellis, A. J., Fyfe, W. S., Taylor, A. M., *Geochim. Cosmochim. Acta* **1959**, 17, 53.
(2) Korzhinskii, D. S., "Physicochemical Basis of the Analysis of the Paragenesis of Minerals," Consultants Bureau, New York, 1959.
(3) Walker, T. L., Parsons, A. L., *Univ. of Toronto Studies, Geol. Ser.* **1922**, 14, 13.

RECEIVED February 4, 1970.

Discussion

V. C. Juan (National Taiwan University, Taipei, Taiwan): Zeolites and low-temperature plagioclase can be closely correlated. We have done work with thomsonite dehydrated into anorthite, analcime into albite, and in between with intermediate plagioclase into an association of analcime, garronite, and thomsonite. The divisions between different zeolites are also correlated with the structural division of plagioclase.

A. B. Carpenter: Thank you for your comments. Your remarks again illustrate the value of a graphical method which can show the relationships between zeolites and plagioclase feldspars.

26

Composition and Origin of Clinoptilolite in the Nakanosawa Tuff of Rumoi, Hokkaido

AZUMA IIJIMA

Geological Institute, University of Tokyo, Hongo, Tokyo, Japan

> *Clinoptilolite occurs as an alteration product of rhyolitic glass in the Upper Cretaceous Nakanosawa tuff, forming the uppermost zeolitic zone of diagenetic alteration of the Cretaceous geosynclinal deposits in north-central Hokkaido. The clinoptilolite contains abundant alkalies in which K predominates over Na. Two modes of occurrence were distinguished; clinoptilolite replaces glass shards and it fills vesicles within the zeolitized glass. The vesicles probably were filled after the shards were replaced. From the electron probe analyses, the 2 clinoptilolites have small but significantly different chemical compositions. Crystals filling the vesicles are more aluminous than those replacing glass, whereas the Ca:Na:K ratios are almost the same. This suggests substitution of the type $Ca \rightleftharpoons 2(Na,K)$, instead of $CaAl \rightleftharpoons (Na,K)Si$.*

Migration of some chemical elements takes place extensively during zeolitic alteration of volcanic glass. Abundant Si and Al as well as alkalies and alkali-earths in nephelinite and alkali basalt glass are dissolved by percolating groundwater in palogonite tuffs on Oahu, Hawaii. Most of these elements, especially Si, Al, and K, precipitate as authigenic zeolites on the palagonitized glass particles (6). Felsic glass in vitric tuffs frequently reacts with interstitial solution to form zeolites. Clinoptilolite is characteristically an alteration product of volcanic, especially felsic, glass at rather low temperature. It is found in the Cenozoic deep-sea sediments (7, 16) and in altered vitric tuffs in alkali-saline lake deposits (5, 18). In the thick sequence of zeolitized marine and non-marine deposits, clinoptilolite forms the uppermost zeolitic zone of alteration (9, 10). Chemical change of felsic glass to clinoptilolite, however, has not necessarily been clarified.

Table I. Mineralogical Composition of the Nakanosawa Tuff Collected From a Railroad Cut on the North Side of the Washing Plant of the Chikubetsu Coal Mine

Mineral	Vol %
Montmorillonite and opal[a]	60.0
Clinoptilolite[a]	37.2
Celadonite[a]	0.2
Leucoxene[a]	0.1
Plagioclase	1.2
Quartz	1.0
Biotite	0.1
Silicified wood fragment	0.2
	100.0

[a] Authigenic mineral.

Relatively coarse crystals of authigenic clinoptilolite were found in the Upper Cretaceous Nakanosawa tuff, 40 meters thick, which is distributed in the Chikubetsu Anticline of Rumoi, north-central Hokkaido (3, 4). Though it has undergone zeolitic and argillaceous alteration, the vitric tuff is rather friable and porous. Rhyolitic glass alters completely to an aggregate of montmorillonite, clinoptilolite, opal, and much less celadonite. Primary crystals of plagioclase(An_{20-43}), quartz, and biotite remain unaltered, and the vitroclastic texture is well preserved. The mineralogical composition of the tuff is shown in Table I.

The Nakanosawa clinoptilolite forms the uppermost zone of zeolitic diagenesis of the 6500-meter-thick Cretaceous geosynclinal marine sediments in the Rumoi district. There was overburden about 3500 meters thick of the Younger Tertiary deposits at the end of the Tertiary period (13). The geothermal gradient of the area at that time is estimated to have been about 20°C/km from the gibbsite–diaspore (11) and the analcime–albite (9) transition temperatures. Consequently, the clinoptilolite formed at temperatures below 80°C, provided that the surface temperature averaged 10°C and that the gradient had not increased since the Late Cretaceous period. In this area, no specific source of heat, such as intrusive bodies of granitic rocks, volcanism, etc., has been noted.

There are 2 modes of occurrence of clinoptilolite: replacing glass shards and filling small vesicles within the zeolitized glass (Figures 1, 2). Slender crystals, as much as 0.1 mm in length, fill the pores which were formed by complete dissolution of glass shards. They grow on, and perpendicular to, rims of opal and montmorillonite which line the pseudomorphs of glass shards. Some void space frequently remains, as seen in Figure 2, strongly suggesting that the clinoptilolite precipitated from an interstitial solution which had filled the pores.

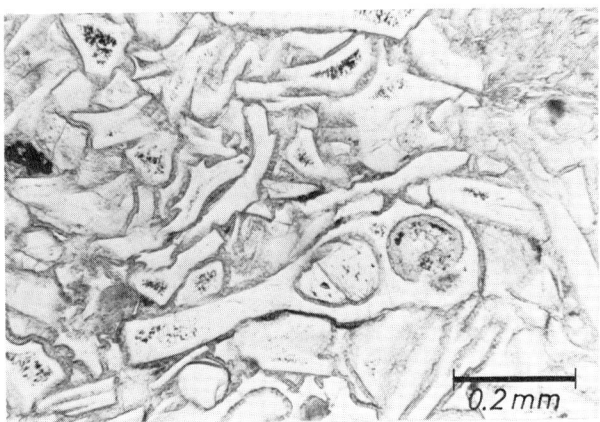

Figure 1. Microscopic photograph of zeolitized Nakanosawa rhyolitic vitric tuff. Clinoptilolite replaces glass shards and fills vesicles within the glass.

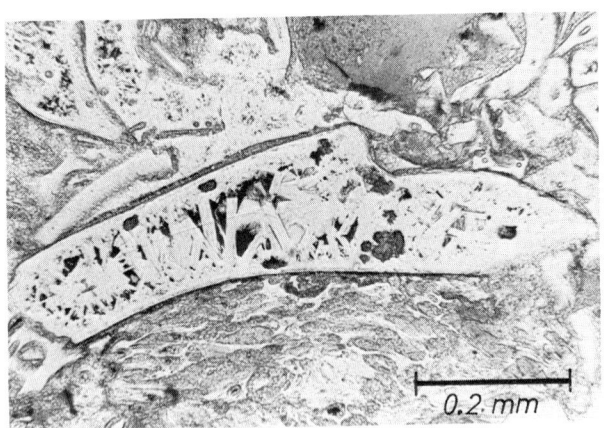

Figure 2. Slender crystals of clinoptilolite in pseudomorphic pore of a glass shard which is fringed by montmorillonite and opal. Some void (grey-black) remains.

The Nakanosawa clinoptilolite has very low refractive indices and birefringence: $\alpha = 1.447 \pm 0.002$, $\gamma = 1.479 \pm 0.002$, and $\gamma\text{-}\alpha = 0.002$. It shows parallel extinction and is length-fast. No difference in optical characters was found between the 2 types, and zonal structure is not observed microscopically. The x-ray powder diffraction pattern agrees with that of the Patagonia clinoptilolite (12). The Nakanosawa clinoptilolite was unchanged up to 700°C, thus fitting Mumpton's criterion (15). It was not destroyed by hot 6N HCl in 2 hours.

Chemical compositions of the Nakanosawa tuff and clinoptilolite from it are given in Table II. The clinoptilolite is characterized by a high content of alkalies in which K exceeds Na. It contains an appreciable amount of Sr. The bulk analysis of tuff indicates that Sr is concentrated selectively by the clinoptilolite.

The contents of Si, Al, Ca, Na, and K of clinoptilolite were measured on the polished thin section by the electron microprobe analyzer, using

Table II. Chemical Composition[a] of the Nakanosawa Tuff and Clinoptilolite

	Bulk %[b]	Clinoptilolite %[c]	Corrected %[d]	Mole Ratio	Atoms per Unit Cell	
SiO_2	68.20	67.55	66.16	1.1016	Si	29.83
TiO_2	0.09	0.04	0.04	0.0005	Ti	0.01
Al_2O_3	11.77	11.13	11.70	0.2296	Al	6.22
Fe_2O_3	0.75	0.15	0.16	0.0020	Fe^{3+}	0.05
FeO	0.22	0.12	0.13	0.0018	Fe^{2+}	0.05
MnO	0.08	0.08	0.08	0.0011	Mn	0.03
MgO	0.52	0.45	0.47	0.0117	Mg	0.32
CaO	1.44	1.70	1.79	0.0319	Ca	0.86
SrO	0.05	0.16	0.17	0.0016	Sr	0.04
Na_2O	1.75	1.47	1.55	0.0500	Na	1.35
K_2O	2.57	2.94	3.09	0.0656	K	1.78
$H_2O(+)$	7.28	8.55	8.98	0.4985	O	72.00
$H_2O(-)$	4.88	5.38	5.65	0.3136	H_2O	22.23
P_2O_5	0.05	0.03	0.03	0.0004		
	99.65	99.75	100.00			

[a] Analyst: H. Haramura.
[b] Bulk of zeolitized rhyolite vitric tuff of the Nakanosawa tuff; the same sample as listed in Table I.
[c] Clinoptilolite from tuff, including 4.6% excess silica as opal, detected by x-ray and calculated from the Si:Al ratio by electron microprobe analyses.
[d] Clinoptilolite recalculated with corrections.

alkali feldspar and plagioclase as comparison standards. Vaporization of such lighter elements as Na was protected by using a weak sample current, 0.01 mA, and a large beam. It is difficult to recognize any systematic change of the composition within single crystals. There is, however, a small but significant difference in the Si:Al ratio between the 2 clinoptilolites. The crystals replacing glass are slightly more siliceous than those filling vesicles (Figure 3). The Si:Al ratio averages 4.86 for 10 samples in the shard pseudomorphs and 4.74 for 10 samples in the vesicles. No significant difference in the Ca:Na:K ratio could be distinguished between the 2 clinoptilolites (Figure 4).

The significance of the difference in the Si:Al ratio between the 2 clinoptilolites is not clear. The Si:Al ratio of analcime in the presence of quartz depends not only on the temperature at which the analcime forms (17) but also on the salinity of the interstitial solution (8). It is difficult to imagine that the 2 clinoptilolites formed at significantly different temperatures. The distinction in the Si:Al ratio of clinoptilolites probably reflects the change of chemical composition of interstitial solution, which was originally sea water, as reacted with rhyolite glass. At the initial stage, glass particles reacted with trapped sea water to form opal and montmorillonite, which fringed the particles. As glass was progressively dissolved, the pH and salinity of the interstitial water became sufficiently high to precipitate clinoptilolite. In the clinoptilolite–mordenite zone of the Miocene Sainokami formation of Akita, large amounts of K, Na, and Si were dissolved out of rhyolitic glass, as Fe, Mn, Mg, and H_2O were adsorbed (Table III). Even the relatively fresh glass that is optically isotropic probably lost some Si and Na, as inferred from the H_2O content (1). The content of alumina is, however, little or not changed. The Si:Al ratio in rhyolitic glass of the Nakanosawa tuff should decrease as the alteration progressed. It is probable, therefore, that relatively high-silica crystals of clinoptilolite precipitated earlier in the pseudomorphous pores of glass shards, and relatively low-silica ones filled the vesicles later. Two large-zoned clinoptilolites in the Barstow Formation of Southern California showed that the rim contains 2.9–3.5 wt % less SiO_2 than the core (19). This mode of occurrence is consistent with the above discussion.

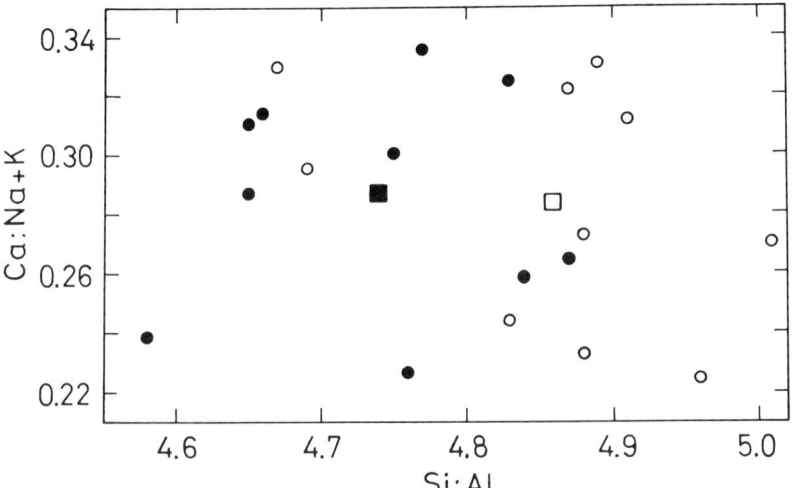

Figure 3. The Si:Al vs. the Ca:Na + K ratios in the Nakanosawa clinoptilolite grains. Crystals replacing glass (○) and the mean (□); filling vesicles (●) and the mean (■). Data from electron probe analyses.

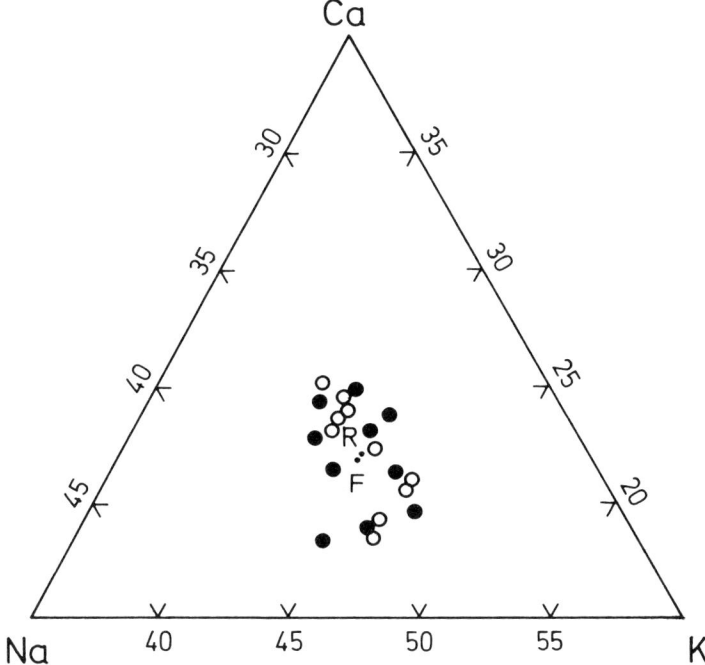

Figure 4. The Ca:Na:K ratio in the Nakanosawa clinoptilolites which replace glass (○) and fill vesicles (●). R: mean of the replacing, F: mean of the filling. Data from electron probe analyses.

In general, low-silica zeolites have a higher exchange capacity and contain more bivalent cations than high-silica varieties which contain abundant univalent cations. This assumes that the crystal lattice can accommodate both univalent and bivalent cations. Clinoptilolite shows a large cation-exchange capacity. Ca $\rightleftharpoons$ 2(Na,K) and CaAl $\rightleftharpoons$ (Na,K)Si substitution were observed on treating with 5% NaCl and $CaCl_2$ solutions at room temperature for 24 hours (*14*). It is unknown what types of substitution occur in such complex multi-ion solutions as sea water and interstitial water, however. In the Nakanosawa clinoptilolites, the substitution (Na,K)Si $\rightleftharpoons$ CaAl is not evident, for both the low- and high-silica varieties have nearly the same Ca:Na:K ratio. The early-formed high-silica crystals originally might have had more abundant univalent cations, and only the substitution (Ca,(Mg),(Sr)) $\rightleftharpoons$ 2(Na,K) occurred at the last stage of zeolitization in which the low-silica crystals precipitated. Alternatively, the Ca:Na:K ratio may depend on the chemical composition of interstitial water of any post-zeolitization time.

The chemical composition of clinoptilolite is variable in different crystals in different parts of the same specimen. This suggests that the

Table III. Chemical Compositions of Fresh and Altered Rhyolitic Obsidian, and the Gain and Loss of Chemical Components During Zeolitic–Argillaceous Alteration[a]

	Fresh Obsidian[b]		Altered Obsidian[c]		Gain and Loss	
	%	Grams/Liter	%	Grams/Liter	%[d]	%[e]
SiO_2	71.3	1675	58.2	1210	− 28	− 37
TiO_2	0.25	6	0.28	6	0	− 17
Al_2O_3	12.1	284	15.7	327	+ 15	0
Fe_2O_3[f]	1.90	45	5.38	112	+ 149	+ 115
MnO	0.08	2	0.18	4	+ 100	+ 50
MgO	0.20	5	3.40	71	+1320	+1140
CaO	1.37	32	1.49	31	− 3	− 16
Na_2O	3.60	85	1.77	37	− 57	− 62
K_2O	2.56	60	0.88	18	− 70	− 73
H_2O	6.64[g]	156	12.72[g]	264	+ 69	+ 47
	100.0	2350	100.0	2080		
S.G.[h]	2.35		2.08			

[a] Electron microprobe analyses, Analyst: Y. Nakamura. The corrections were made after Bence and Albee's procedure(2).
[b] Fresh rhyolitic obsidian in autobrecciated lava of the Miocene Sainokami formation, Odate, Akita(PA5-12a).
[c] Altered part of the obsidian(PA5-12b).
[d] Gain and loss when volume of glass was unchanged.
[e] Gain and loss when alumina was unchanged and the volume increased 1.15 times.
[f] Total iron.
[g] Subtracting the sum of other components from 100%.
[h] Measured by the Berman balance using toluene.

zeolitic reactions did not occur under constant conditions throughout space and time, and that the whole rock was not in an equilibrium state.

Acknowledgment

R. L. Hay and F. A. Mumpton read the manuscript.

Literature Cited

(1) Aramaki, S., Haramura, H., *J. Geol. Soc. Japan* **1966**, 72, 69–73.
(2) Bence, A. E., Albee, A. L., *J. Geol.* **1968**, 76, 382–403.
(3) Hariya, Y., *Rept. Geol. Surv. Hokkaido* **1964**, 32, 84–85.
(4) Hattori, Y., *J. Geol. Soc. Japan* **1965**, 71, 149–151.
(5) Hay, R. L., *Geol. Soc. Am. Spec. Papers* **1966**, 85, 1–130.
(6) Hay, R. L., Iijima, A., *Geol. Soc. Am. Mem.* **1968**, 116, 331–376.
(7) Heath, G. R., *Geol. Soc. Am. Bull.* **1969**, 80, 1997–2018.
(8) Iijima, A., Hay, R. L., *Am. Mineralogist* **1968**, 53, 184–200.
(9) Iijima, A., Utada, M., ADVAN. CHEM. SER. **1971**, 101, 342.
(10) Iijima, A., Utada, M., *Sedimentology* **1966**, 7, 327–357.
(11) Kennedy, G. C., *Am. J. Sci.* **1959**, 257, 563–573.
(12) Mason, B., Sand, L. B., *Am. Mineralogist* **1960**, 45, 341–350.
(13) Matsuno, K., Kino, Y., *Geol. Surv. Japan* **1960**, 1–43.

(14) Minato, H., Utada, M., *J. Japan. Assoc. Mineral. Petrol. Econ. Geol.* **1968**, 60, 213–221.
(15) Mumpton, F. A., *Am. Mineralogist* **1960**, 45, 351–367.
(16) Murata, K. J., Erd, R. C., *J. Sediment. Petrol.* **1964**, 34, 633–655.
(17) Saha, P., *Am. Mineralogist* **1959**, 44, 300–313.
(18) Sheppard, R. A., Gude, A. J., III, *U. S. Geol. Surv. Profess. Papers* **1968**, 597, 1–38.
(19) Sheppard, R. A., Gude, A. J., III, *U. S. Geol. Surv. Profess. Papers* **1969**, 634, 1–35.

RECEIVED January 12, 1970.

27

Present-Day Zeolitic Diagenesis of the Neogene Geosynclinal Deposits in the Niigata Oil Field, Japan

AZUMA IIJIMA

Geological Institute, University of Tokyo, Hongo, Tokyo, Japan

MINORU UTADA

Institute of Earth Science, University of Tokyo, Komaba, Tokyo, Japan

Three deep drillings recently have penetrated through nearly horizontal strata of thick sedimentary piles beneath the Niigata Plain. Authigenic zeolites, associated with opal or quartz, were found in felsic tuffs of the drillings, and are distributed in a vertically zonal arrangement which is divided into 5 zones: from the surface, (I) fresh glass, (II) alkali clinoptilolite, (III) mordenite, (IV) analcime, and (V) albite. The zones are forming in response to temperature increases with increasing depth of burial. The clinoptilolite or mordenite–analcime and the analcime–albite transitions occur at $84°$–$91°C$ and $120°$–$124°C$, respectively, in the presence of quartz and interstitial water. The zeolitic transitions may be a geothermometer in marine deposits which have not been subjected to local hydrothermal alteration.

More than 8000 meters of the Upper Cenozoic marine deposits fill the Uetsu geosyncline along the coast of the Sea of Japan in Northern Honshu. Three drillings, each about 5000 meters deep, promoted by the Japanese Petroleum Development Corp., recently have penetrated the deposits beneath the Niigata Plain, one of the important oil and gas fields in Japan. They are Obuchi, Shimoigarashi, and Masugata, which are located geologically in a broad synclinorium (Figure 1). The penetrated

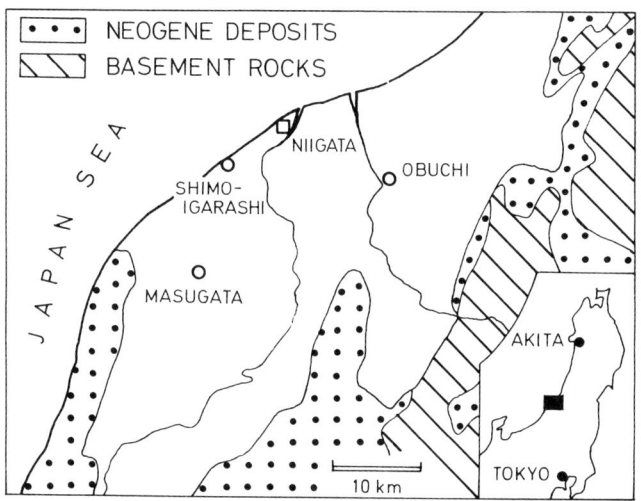

Figure 1. Location and geological map

strata are horizontal or gently folded, and maximum dips are 10°. All were deposited in marine environments since the middle Miocene. Numerous beds of pyroclastic rock are found throughout the deposits, intercalated by normal clastic sediments. Most are thin and fine- to medium-grained vitric tuffs of rhyolite and dacite. The tuffs alter, to varying degrees, to an aggregate of authigenic minerals.

Authigenic minerals in 224 samples were identified by x-ray powder diffraction and microscope. They are as follows:

zeolites: clinoptilolite, mordenite, analcime, laumontite.
clay minerals: montmorillonite, corrensite, chlorite, celadonite, illite.
silica minerals: opal, quartz.
alkali feldspars: albite, monoclinic K-feldspar.
carbonate: calcite.

These minerals, especially zeolites, are distributed in vertically zonal arrangement, which we divide into 5 zones—i.e., from the surface, (I) fresh glass, (II) alkali clinoptilolite, (III) mordenite, (IV) analcime, and (V) albite zones (Figure 2).

Zone I is characterized by the general occurrence of fragments of fresh rhyolite and dacite glass which are coated with montmorillonite and opal. Zone II is characterized by alkali clinoptilolite that is stable to heating at 750°C for 12 hours and treating with hot 6N hydrochloric acid for 1 hour. The minute crystals of alkali clinoptilolite fill the voids resulting from dissolved glass shards, and are associated with opal, montmorillonite, and sometimes celadonite. The mode of occurrence strongly suggests that it was formed by reaction of felsic glass with interstitial

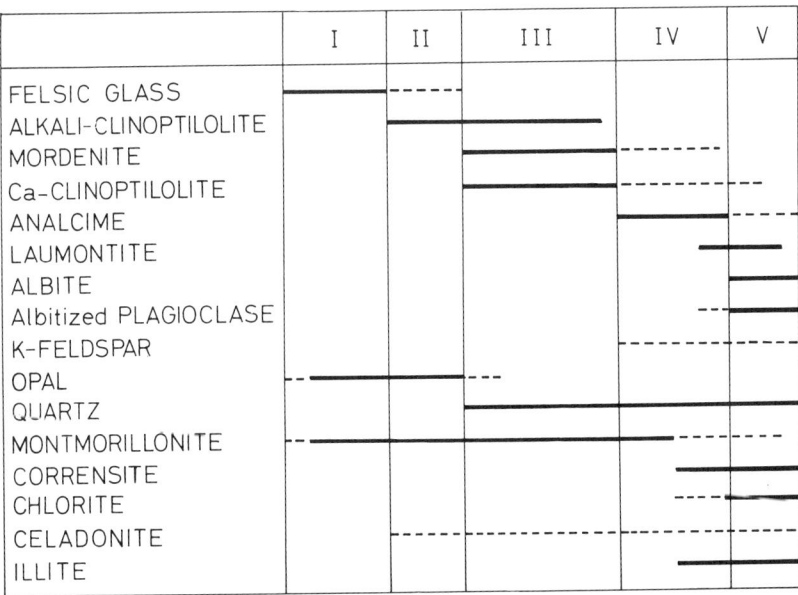

Figure 2. Diagenetic zones of authigenic minerals in tuffs and tuffaceous sediments. The length of the zone is roughly proportional to the depth range.

water—originally sea water. Primary fragments of quartz, feldspars, and biotite are unaltered. Fragments of relic glass are not uncommon.

Zone III is characterized by mordenite, though its occurrence is rather sporadic. In tuffs of the Masugata drilling, Ca-clinoptilolite is common, which is stable to heating up to 750°C but destroyed by hot hydrochloric acid. Alkali clinoptilolite is common in tuffs of the other 2 drillings. Mordenite and clinoptilolite sometimes coexist, but also occur separately. They fill the voids resulting from dissolved glass shards and cement the interstitial pores of tuffs where they are associated with opal or chalcedonic quartz, montmorillonite, and sometimes celadonite. Mordenite and alkali clinoptilolite were probably formed by reaction of felsic glass with interstitial solutions, whereas Ca-clinoptilolite may be produced by the cation exchange reaction of alkali clinoptilolite. The primary minerals of the tuffs are unaltered, and the original vitroclastic texture is well preserved. No fresh glass shards remain in this zone.

Analcime is specific in Zone IV. Under the microscope, it occurs commonly as the pseudomorphs of clinoptilolite and mordenite which replaced glass shards or cemented the interstitial pores of tuffs. Mordenite and Ca-clinoptilolite frequently coexist with analcime, however. Authigenic quartz appears as microcrystalline aggregates in the cement and is associated with montmorillonite, corrensite, or chlorite. Montmorillonite gradually changes to chlorite through corrensite or swelling chlorite in the lower part of this zone. Primary crystals of quartz are almost unaltered except where replaced by calcite. Plagioclase is slightly albitized in the lower part of the zone. The original vitroclastic texture is recognized easily. Tuffs in this zone retain 15 to 20% porosity (8, 9).

Zone V is characterized by the coexistence of authigenic albite with analcime. The albite is formed by reaction of analcime with quartz in the presence of interstitial water. Albitization of primary plagioclase (oligoclase to andesine) occurs extensively. The original vitroclastic texture becomes vague through crystallization of quartz and albite. Laumontite precipitates in cavities, lined with quartz, of augite dacite lava and volcanic breccia of the Masugata drilling and dolerite sheet of the Obuchi drilling.

The study area has been the site of deposition since the middle Miocene, and the deposits have been little, if any, disturbed by tectonic

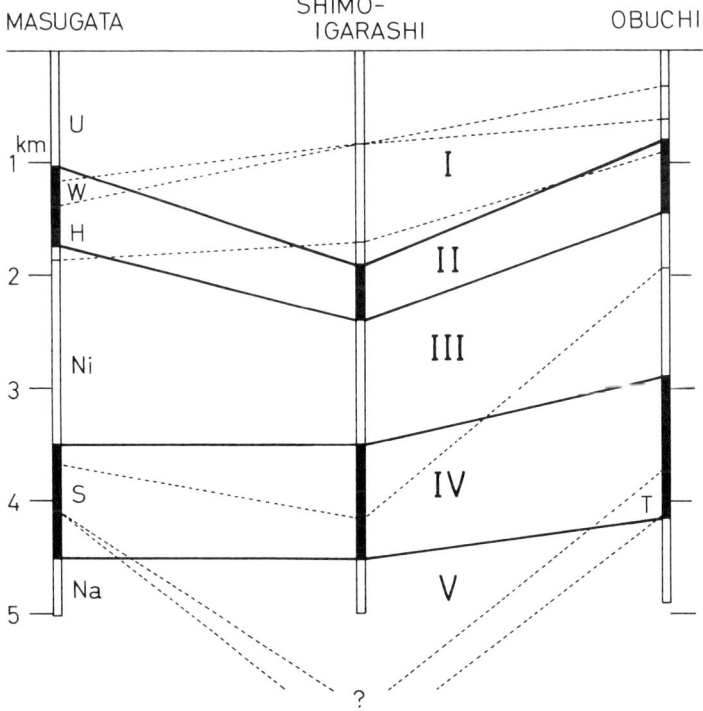

Figure 3. Relation of zeolitic zonation to stratigraphic subdivision. U: Uonuma group, W: Wanazu fm., H: Haizume fm., Ni: Nishiyama fm., S: Shiiya fm., Na: Nanatani fm. Solid lines show the boundaries of the zeolitic zonation and dotted lines represent the stratigraphic subdivision.

Table I. Ranges of Temperature, Depth, and Total Pressure of the Upper Limit of Each Diagenetic Zone

Zone	Temperature Range of the Upper Limit, °C	Depth Range, Meters	Total Pressure Range, Kb
II. Clinoptilolite	41–49	800–1900	
III. Mordenite	55–59	1450–2400	0.4–0.7
IV. Analcime	84–91	2900–3500	0.9–1.1
V. Albite	120–124	4150–4500	1.3–1.4

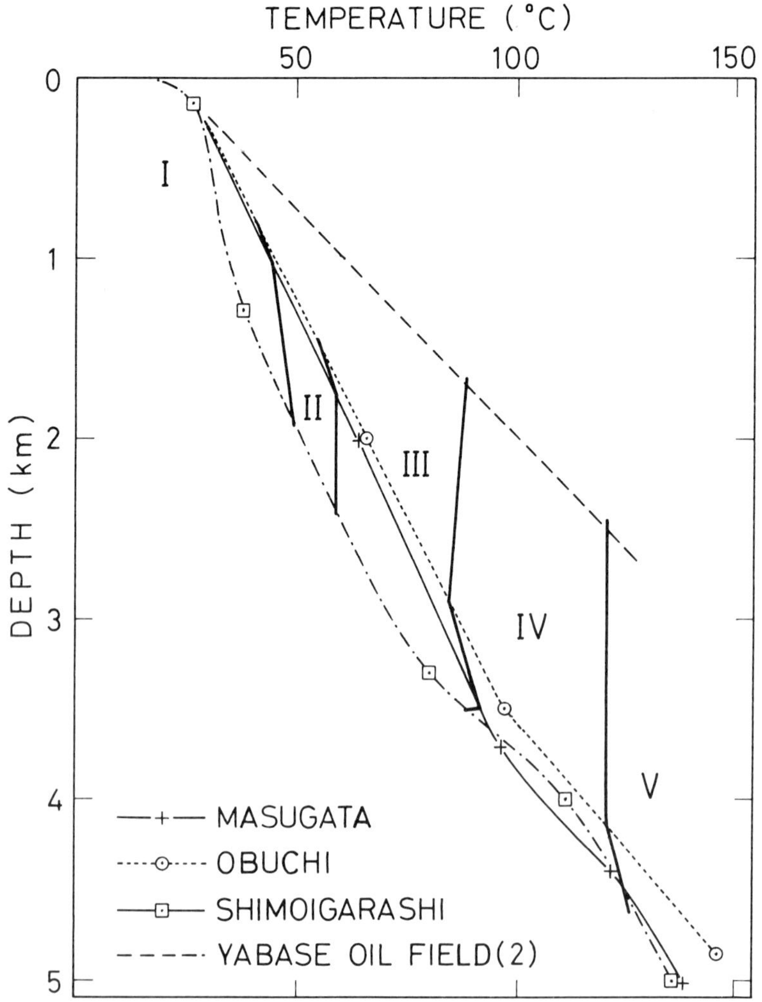

Figure 4. Relationship between geothermal gradient and zeolitic zonation. Temperatures were determined by the bottom hole temperature. Boundaries of the zonation are subparallel to each other and are influenced essentially by temperatures.

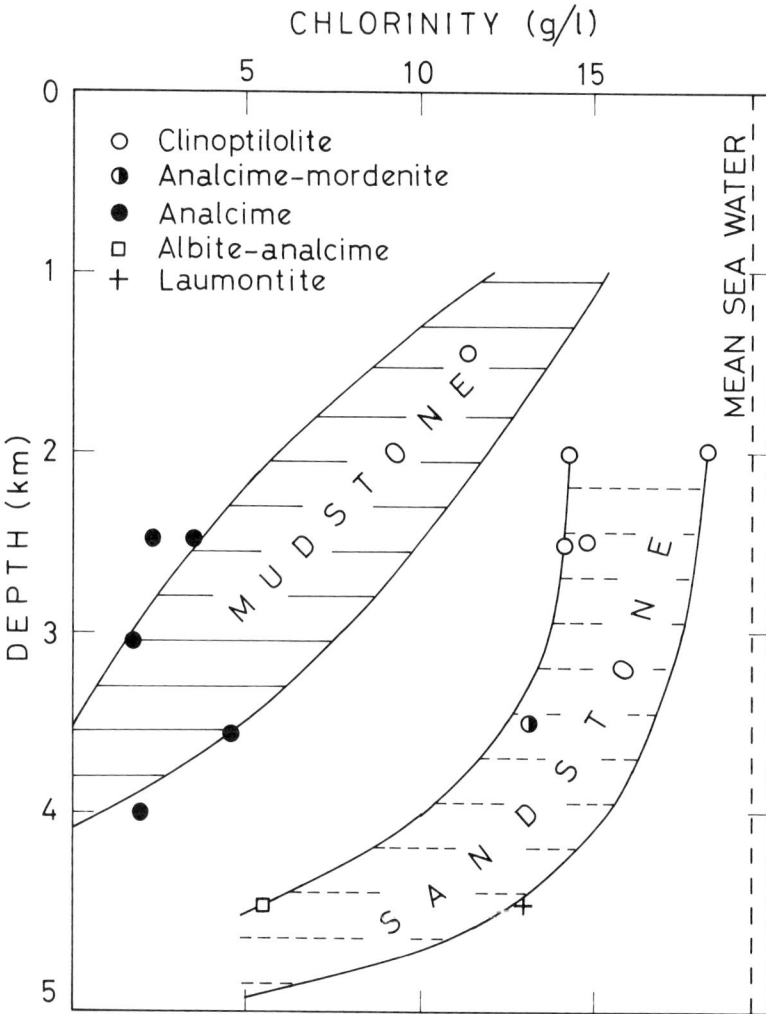

Figure 5. Relationship among depth of burial, chlorinity of interstitial waters, and zeolitic zonation. The chlorinity of the interstitial water in the normal sediments decreases with increasing depth of burial. The interstitial water in zeolite-bearing tuffs and tuffaceous sediments have chlorinity similar to that in normal sediments.

movements. The relationship between the stratigraphic sequence and the zonation is shown in Figure 3. Though it is generally oblique to the stratigraphic boundaries, the zonation progresses in all 3 cores downward from Zone I to V. These facts are interpreted to mean that authigenic minerals are growing as the tuffs are buried.

Subsurface temperatures of 3 drillings were measured by the bottom hole temperature. The mean geothermal gradient is about 20°C/km above a depth of about 3500 meters and 33°C/km below it. The temperature and depth of the upper limit of each zone is shown in Table I. Liquid water pressure is considered nearly equal to the hydrostatic pressure (8). In the Yabase oil field of Akita, having a geothermal gradient of 40°C/km, the analcime zone begins at a depth of 1700 meters and the temperature is 88°± (2); the depth and temperature of the upper limit of albite zone is extrapolated to 2500 meters and 120°C from the data (2). The temperature of the analcime reaction is not inconsistent with the analcime–albite equilibria obtained in the laboratory (1, 3), although it is much lower in nature.

Zeolitic zonation is influenced by temperature in the range of depth of burial as shown in Table I and Figure 4. The mordenite or clinoptilolite–analcime and the analcime–albite transitions in the presence of quartz and interstitial water may be used as a geothermometer in marine sedimentary piles which have not been affected by local hydrothermal alteration. The formation of clinoptilolite or mordenite from glass is affected by chemical factors such as the silica activity, pH, salinity, etc. (1, 4). The same zonation as described here has been recognized widely in felsic tuffs of the marine Tertiary and Cretaceous deposits of Japan (2, 5, 7, 10, 11, 12). The geothermal gradient in some areas that is obtained from the zeolite geothermometer is compatible with other geological evidences.

Interstitial water has an important effect on zeolitic reactions in sedimentary rocks (4, 6). Zeolitic tuffs of 3 drillings originally were saturated with sea water. The chlorinity of interstitial water in cores is generally lower than that of sea water (8, 9), and tends to decrease with increasing depth of burial, although the rate of decrease is higher in mudstone than in sandstone (Figure 5). The chlorinity in tuffs shows relationships similar to that in the normal clastic sediments. Analcime may replace clinoptilolite or mordenite in water with much lower chlorinity than sea water.

We prefer to designate the zeolitic alterations as diagenetic rather than buried metamorphism (4, 5, 6).

Acknowledgment

R. L. Hay read the manuscript.

Literature Cited

(1) Campbell, A. S., Fyfe, W. S., Am. J. Sci. **1965**, 236, 807–816.
(2) Fujioka, K., Yoshikawa, T., J. Japan. Assoc. Petrol. Technol. **1969**, 34, 145–154.

(3) Fyfe, W. S., Mackenzie, W. S., *Earth Sci. Rev.* **1969**, 5, 185–215.
(4) Hay, R. L., *Geol. Soc. Am. Spec. Paper No. 85* **1966**, 1–130.
(5) Iijima, A., ADVAN. CHEM. SER. **1971**, 101, 334.
(6) Iijima, A., Utada, M., "Commem. Papers Prof. Y. Sasa's 60th Birthday," p. 107–121, 1967.
(7) Iijima, A., Utada, M., *Sedimentology* **1966**, 7, 327–357.
(8) Japan Petroleum Development Corporation, "Report on the Core Analyses of the Obuchi, Shimoigarashi, and Nishimeoki Drillings," p. 1–46, 1968.
(9) Japan Petroleum Development Corporation, "Report on the Survey of the Masugata Drilling," p. 1–21, 1969.
(10) Nakajima, W., Koizumi, M., Nakagawa, T., *J. Geol. Soc. Japan* **1962**, 68, 173.
(11) Utada, M., *Sci. Papers Coll. Gen. Educ. Univ. Tokyo* **1965**, 15, 173–216.
(12) Yoshimura, T., *J. Geol. Soc. Japan* **1961**, 67, 578–583.

RECEIVED December 29, 1969.

28

Cation Exchange on Zeolites

HOWARD S. SHERRY

Mobil Research and Development Corp., Princeton, N. J. 08540

> *In the case of Rb^+ and Cs^+ ion exchange of NaX, 32 Na^+ ions are not exchangeable. Partial ion sieving of Tl^+ ions in zeolite Y is attributed to a slow rate of exchange of the Na^+ ions in the network of small cavities. Alkaline earth ion exchange of the Na^+ ions in the small cavities of zeolite Y is slow and thermodynamically unfavorable. The equilibrium preference of chabazite and Linde T for lower-charged ions over higher-charged ones is discussed. Kinetic models for isotopic ion exchange are analyzed and compared. Finally, an attempt to build a theory to describe ion exchange kinetics in nonideal systems is discussed, correctly predicting trends in one case but not in another.*

The ability to undergo reversible cation exchange is one of the most important properties of zeolites. It enables us to modify the electric fields inside zeolite crystals (24), which in turn modifies the sorptive and catalytic properties. The most striking modification that can be made is to change the molecular sieve properties. For example, the replacement of Na^+ ion in Linde A (NaA) by K^+ ions causes the sorption of O_2 to decrease to essentially zero (16). Propane is not sorbed by NaA and is sorbed by CaA (16). Barrer (4, 5) in his pioneering work on zeolites has shown clearly that ion exchange of Linde 13-X (NaX) profoundly affects the water and ammonia sorption isotherms and therefore the heats of sorption. Recent publications have made it abundantly clear that the nature of the cation in the zeolite affects the number of OH groups in partially dehydrated zeolites (61) and that cracking and isomerization of *n*-hexane in synthetic faujasites depends on the cations that are in the zeolite crystals (48).

It is the purpose herein to review the advances in ion exchange that have taken place from the First International Conference on Zeolites in 1967 to the present and to discuss the, as yet, unsolved problems. The review will be in 2 sections. The first section will deal with ion exchange equilibria and the second with ion exchange kinetics. Much of the discussion will be devoted to the synthetic zeolites Linde X and Y because these zeolites have been most thoroughly studied.

Ion Exchange Equilibria

Alkali Metal Ion Exchange in Zeolites X and Y. A very thorough investigation of ion exchange equilibria in the synthetic faujasites Linde X and Y was done by Barrer and Rees (6, 8, 12) and by Sherry (52). These 2 parallel studies are in essential agreement. They show that alkali metal cations with large crystal radii such as Cs^+ and Rb^+ cannot completely replace all of the sodium ions in NaX or NaY. It was concluded by Sherry (52) that the Na^+ ions that could not be replaced by Cs^+ and Rb^+ were located in the small cavities of this zeolite and that the exchangeable cations were located in the large cavities or supercages.

Early structural studies have shown by x-ray powder diffraction that in hydrated NaX, containing 80 Na atoms per unit cell, 16 are located in the hexagonal prisms in the S_I position, one in each of the 16 hexagonal prisms in a unit cell, and that the remainder are located in the 8 supercages in a unit cell (19). Of the 64 cations in the large cages, 32 are located near the rings of 6 tetrahedra that are the connecting "windows" between the sodalite and supercages, in the S_{II} sites. The remainder cannot be located by x-ray diffraction techniques and are believed to be mobile hydrated ions (19). There has been no structural study of hydrated zeolite Y reported. However, Baur's study of a hydrated single crystal of natural faujasite indicates the presence of 17 cations in the 8 sodalite cages that are in a unit cell (13). A more recent x-ray diffraction investigation of a single crystal of hydrated NaX by Olson indicates that on a unit cell basis, there are 9 Na^+ ions in the S_I position, 8 in the S_I' position, and 24 cations in the S_{II} position (44). Little evidence was found for the location of the remaining cations, and they are presumed to be mobile hydrated ions. Thus, there are still 16 sodium ions per unit cell in the network of small cages, which consists of hexagonal prisms and sodalite cages, within the error limits of this work.

The important feature for ion exchange is that in order for an entering ion to replace the Na^+ ions in the sodalite cages or hexagonal prisms it must diffuse through the ring of 6 tetrahedra that is the window between the supercage and sodalite cage. The fact that Ag^+ and K^+ ions are the largest univalent metal ions that can completely replace all of the Na^+ ions in zeolite X (52) indicates that for ion exchange this hexagonal ring has an effective diameter of 2.5 to 2.6 A. The effective diameter of the apertures of the supercage is about 9 A.

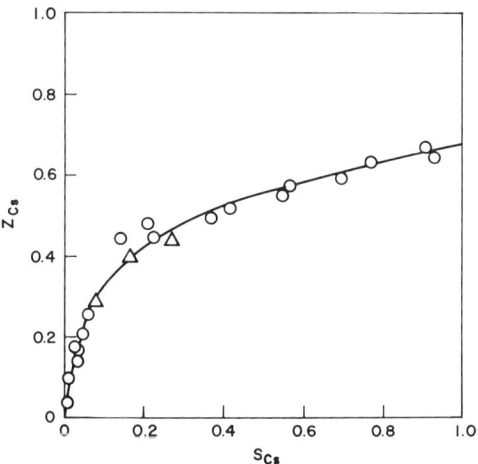

Figure 1. Cs–Na-Y isotherm at 25°C and 0.1 total normality
○ $Cs^+(S) + Na^+(Z) \rightarrow$
△ $Cs^+(Z) + Na^+(S) \rightarrow$

In NaY, it is clear from Sherry's work (12, 54) and that of Barrer, Davies, and Rees (6) that 16 Na^+ ions per unit cell cannot be exchanged by Cs^+ and Rb^+ ions. The ion exchange isotherm at 25°C in Figure 1 terminates at 68% Na replacement corresponding to the exchange of 34 of the 50 Na^+ ions that are in a unit cell (52).

Cs^+ and Rb^+ ion exchange in NaX is not so clearly understood. In NaX, with 85 Na^+ ions per unit cell, it would be expected that 16 per unit cell could not be replaced by Rb^+ or Cs^+ ions. The isotherm shown in Figure 2 clearly does not terminate at that point (52). In an attempt to arrive at 82% Cs loading, Sherry (52) prepared AgX and exchanged it with 4 M CsSCN and RbSCN. Complete Cs^+ and Rb^+ exchange still was not achieved. He concluded that this isotherm should extrapolate to the point $S = 1$, $Z = 0.82$ because only 16 out of 85 Na^+ ions per unit cell should be incapable of replacement. Barrer, Rees, and Shamsuzzoha obtained Cs^+–Na^+ and Rb^+–Na^+ isotherms (12) that are the same as Sherry's, but they did not extrapolate them to 82% Na^+ exchange. They stated that Cs^+ and Rb^+ ion exchange of NaX does not proceed beyond 65% replacement. Similar isotherms were obtained by Theng, Vansant, and Uytterhoeven (60). Recently, in an effort to resolve this disagreement, we have attempted to determine the Cs^+ and Rb^+ ion exchange capacity of NaX columns by eluting with 0.1 M CsCl and RbCl at room temperature (58). We find that only 62–65% of the Na^+ ions can be replaced by Cs^+ or Rb^+ ions, in agreement with Barrer, Rees, and Shamsuzzoha. If further elution is attempted, hydronium ion exchange for Na⁺ ion occurs. Thus, it appears that in zeolite X, 32 of the 85 Na⁺ ions in a unit cell cannot be replaced by Cs⁺ or Rb⁺ ions. It has been suggested that the replacement of Na⁺ ions in the large cages by Cs⁺ ions crowds the remaining ones into the small cages (8). Either of the 32-fold set of sites in the sodalite cages could accommodate the 32 Na⁺ that are not exchangeable.

The study of ion exchange in zeolites X and Y affords an opportunity to study alkali metal ion exchange selectivity as a function of anionic charge on the aluminosilicate framework because zeolites X and Y are isostructural, differing only in aluminum content. Focusing attention on ion exchange in the large cages of zeolites X and Y, we find that at low loading of the ingoing ion, the selectivity series Cs > Rb > K > Na > Li is observed in both zeolites (52). The thermodynamic selectivity of the negatively charged framework for alkali metal ions decreases in the order in which the ionic hydration energies increase. This selectivity series is the one that should be observed if the first ions to exchange are the mobile hydrated ones. At the 50% exchange levels, the selectivity series observed in zeolite X is Na > K > Rb > Cs > Li (12, 52) and in zeolite Y is Cs > Rb > K > Na > Li (6, 52). The selectivity series for zeolite X at 50% loading can be best accounted for by assuming that ions in or near S_{II} sites are undergoing exchange. Thus, except for Li⁺ ion, the ion selectivity decreases with increasing ionic radius because a bare ion must

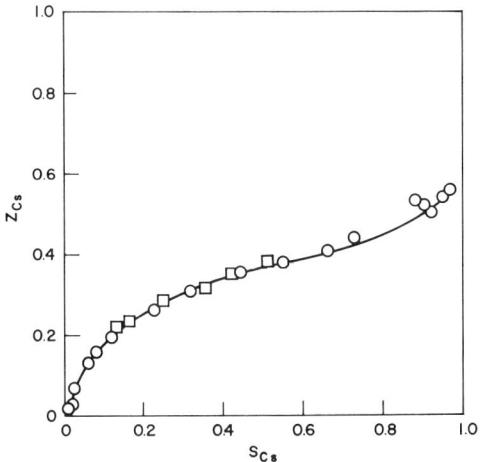

Journal of Physical Chemistry

Figure 2. Cs–Na-X isotherm at 25°C and 0.1 total normality
○ $Cs^+(S) + Na^+(Z) \rightarrow$
□ $Cs^+(Z) + Na^+(S) \rightarrow$

interact with the framework charges according to Coulomb's law. Li^+ ion is an exception because of its high hydration energy. In terms of this model, all of the ions in the large cages of zeolite Y are hydrated and not sited because the selectivity for ions at 50% loading decreases with increasing ionic hydration energy.

This picture is consistent with the numbers of water molecules and ions in the large cages of zeolites X and Y. In zeolite X there are about 3 H_2O molecules per univalent cation, and therefore there is insufficient water to fully hydrate all the ions. On the other hand, there are only 32 S_{II} sites to accommodate 69 cations in a unit cell. Thus, of necessity, some cations must coordinate to lattice oxygen atoms in S_{II} sites. The remainder must completely coordinate to oxygen atoms of water molecules and be considered hydrated ions. In zeolite Y, there are about 6 water molecules per univalent ion and only 34 ions per unit cell in the large cage. All of these ions may be sited or unsited, or they may be distributed between sited and fully hydrated ions. The selectivity series observed by Sherry (52) and Barrer, Davies, and Rees (6) at 50% loading of hydrated zeolite Y can be explained only by the absence of ion siting in or near S_{II} sites. If this conclusion is correct, only 16 out of 50 Na^+ ions should be located in each unit cell of NaY by x-ray diffraction (in the small cages). The remaining 34 ions should not be observable, or if they are observed, the cation–lattice oxygen internuclear distance

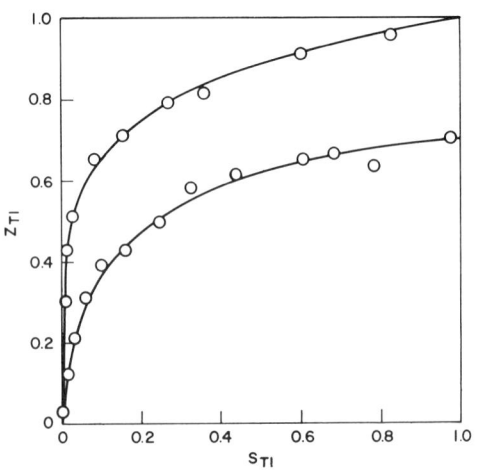

Journal of Physical Chemistry

Figure 3. Tl^+–Na^+ isotherms at 25°C and 0.1 total normality
Upper curve = Zeolite X
Lower curve = Zeolite Y

should support interposition of a water molecule between them. No x-ray diffraction study of hydrated NaY has ever been reported.

Tl^+ and Ag^+ Ion Exchange in Zeolites A, X, and Y. Thallium(I) ion exchange of zeolites A, X, and Y presents some unresolved problems. Barrer and Meier (10) and Sherry and Walton (59) have shown that when Linde Zeolite A contains more than 12 Na atoms per unit cell, the excess Na^+ can be replaced by Ag^+ ions but not by Tl^+ ions. They concluded that the excess Na^+ ions are in the sodalite cages. Sherry later studied Tl^+ ion exchange of NaX and NaY (52) and found that at 25°C and 0.1 total normality, Tl^+ ions can completely replace all of the Na^+ ions in NaX but cannot replace 16 Na^+ ions per unit cell of NaY. The ion exchange isotherms in Figure 3 very strikingly demonstrate this effect. These results were verified by Barrer, Rees, and Shamsuzzoha for zeolite X in 1966 (12) and by Barrer, Davies, and Rees for zeolite Y in 1968 (6). Thus, Tl^+ ions are able to penetrate the sodalite cages of zeolite X but not those of zeolites A and Y. Sherry (52) has observed that the lattice parameter, a_o, for the cubic faujasite type of unit cell can differ by as much as 0.05 A when comparing a zeolite X with 85 Na^+ ions per unit cell to a zeolite Y with 50 Na^+ ions per unit cell and has attributed the inability of Tl^+ ions to replace the Na^+ ions in the sodalite cages of zeolite Y to a very slow rate of exchange in the contracted Y lattice.

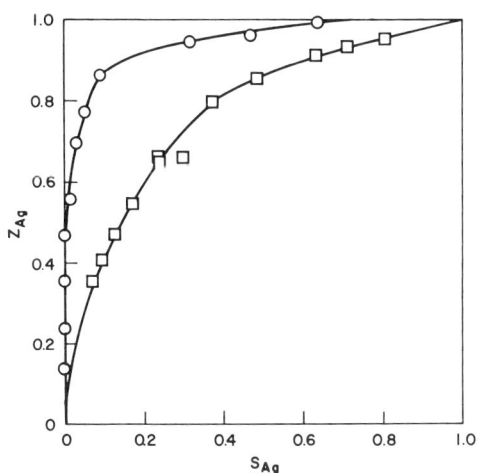

Figure 4. Ag^+–Na^+ isotherms at 25°C and 0.1 total normality
○ Zeolite X
□ Zeolite Y

On the other hand, Barrer, Davies, and Rees (8) believe that the inability of Tl⁺ ions to replace the 16 Na⁺ ions in the sodalite cages of zeolite Y results from an unfavorable equilibrium. They have stated that the higher framework charge of zeolite X must result in the S_I sites being of lower energy than in zeolite Y because of a greater polarization energy contribution with Tl⁺ ion. It is true that the free energy of Tl⁺ ions relative to Na⁺ ions on any site should be lower in zeolite X than zeolite Y owing to the stronger polarization of Tl⁺ ions in zeolite X. The same statement is true for Ag⁺ ions in these 2 zeolites. That this effect is operative is clearly demonstrated by referring to Figure 4 which shows the Ag⁺–Na⁺–X–Y isotherms at 25°C and 0.1 total normality. Ag⁺–Na⁺ ion exchange demonstrates that when the exchange is uncomplicated by partial ion-sieving, the preference for the more polarizable ion does indeed decrease with decreasing framework charge (increasing Si/Al atom ratio). This trend is evidenced by the decreasing rectangularity of the ion exchange isotherms with decreasing Al content of the zeolite.

If the lack of penetration of the sodalite cages of zeolite Y by Tl⁺ ions arises from a slow rate of diffusion through the rings of 6 tetrahedra that interconnect the supercages and sodalite cages and is not of thermodynamic origin, heating of the reaction mixture $TlNO_3$ + NaY should result in complete exchange with an equilibrium constant that favors Tl⁺ exchange because the rings of 6 tetrahedra in zeolite Y are only about 0.01 A smaller in diameter than those of zeolite X. We recently obtained data that indicate that at 100°C and 0.1 total normality, Tl⁺ ions can replace all of the ions in NaY and that Tl⁺ ions are preferred to Na⁺ ions over the complete range of Tl⁺ ion loading (58).

Rare Earth Ion Exchange in Zeolites X and Y. In 1969, Sherry (55) reported on rare earth ion exchange in zeolites X and Y. The most important point made in this work is that at 25°C La^{3+} ions cannot replace 16 Na⁺ ions per unit cell of zeolites X and Y (Figure 5)—that is, the rate is infinitely slow. For all practical purposes, equilibrium is established between the ions in the large cages of the zeolite crystals and those in the aqueous solution. Sherry reported that it takes 13 days of exchange to prepare LaX (99% replacement of the original Na⁺ ions) at 100°C. We must conclude that La^{3+} ions take up positions in the network of small cages in order to replace the Na⁺ ions located therein, else why a slow exchange step involving just 16 Na⁺ ions per unit cell? In proof, Olson (43) has studied the same LaX prepared at 100°C by Sherry using x-ray powder diffraction techniques and finds 8–10 La^{3+} ions per unit cell in the sodalite cages.

Ames (1) has obtained a Ce^{3+}–Na⁺ isotherm at 25°C and 0.5 total normality. This isotherm is very similar to Sherry's; however, he extrapolated his isotherm to the 100% exchange level, making the isotherm

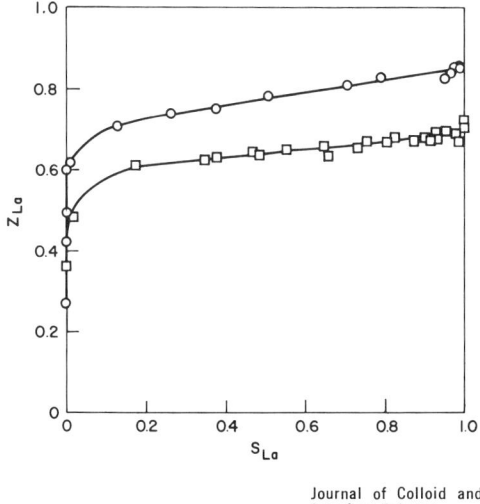

Figure 5. La–Na isotherms at 25°C and 0.1 total normality
○ Zeolite X
□ Zeolite Y

sigmoidal with a selectivity reversal. We do not believe that the inability of the La^{3+} ions to replace Na^+ ions in the network of small cages results from a small equilibrium constant because the high-temperature isotherm remains rectangular (55).

The ion sieving observed in the La–Na–X and La–Na–Y systems is quite interesting. The La^{3+} ion has a radius of 1.15 A (47) and should diffuse easily through the windows between the sodalite cages and supercages if K^+ ion with a radius of 1.33 A (47) can do so. However, the hydrated radius of La^{3+} ions is 3.96 A (42) and the network of small cages contains little water (12). Most, or all, of the water molecules must be "stripped" from the ion if it is to pass through the window between the large and small cages. The enthalpy of hydration of La^{3+} ion is about 900 to 1000 Kcal per gram mole of ions and, although diffusion into the small cages of synthetic faujasites is slow at room temperature, it should be a highly temperature-dependent process.

It was in the study of rare earth ion exchange in zeolites Y and X that Sherry (55) demonstrated that the electroselective effect should, and does, operate in zeolites as well as other exchangers (31). This effect, which is operative when ions of different charges are exchanging, states that, even in an ideal system, the zeolite preferentially selects the ion with the highest charge at low total normality and the ion with the lowest charge at high total normality. Of course, the free energy of the ion

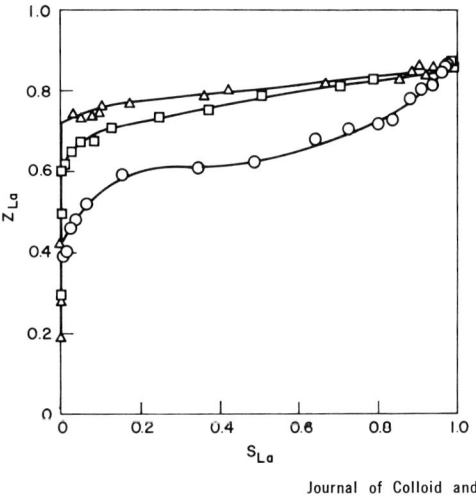

Journal of Colloid and Interface Science

Figure 6. La–Na ion exchange in zeolite X at 25°C as a function of total normality

△ 0.06 Total normality
□ 0.30 Total normality
○ 3.85 Total normality

exchange reaction is independent of total normality. A striking example is afforded by La^{3+}–Na^+ ion exchange on zeolite X and is shown in Figure 6.

Alkaline Earth Ion Exchange of Zeolites X and Y. In recent years, alkaline earth ion exchange in zeolites X and Y has been studied thoroughly. In 1966, Barrer, Rees, and Shamsuzzoha reported on alkaline earth ion exchange in zeolite X at 25°C (*12*), and in 1968, Barrer, Davies, and Rees reported on exchange in zeolite Y at 25°C (*6*). In 1968, Sherry reported on alkaline earth ion exchange in zeolites X and Y over the temperature range of 5° to 50°C (*54*).

Barrer, Rees, and Shamsuzzoha observed that the Ca^{2+}–Na^+ ion exchange reaction has a fast and a slow step. Sherry observed the same thing. His isotherm shown in Figure 7 indicates that at the end of 24 hours at 25°C, 82–85% of the original Na^+ ions are replaced by Ca^{2+} ions. At the end of 4 days, complete Ca^{2+} ion exchange has taken place. Replacement of the 69 Na^+ ions per unit cell that are in the large cages corresponds to 82% exchange because this NaX had 85 cations per unit cell. The slow step must involve the exchange of the Na^+ ions in the small cages. Barrer, Rees, and Shamsuzzoha did not observe a fast step that corresponds to a "magic number" of 82% exchange because they made their measurements in a calorimeter and arbitrarily stopped the run after

10 minutes—calling the exchange level reached at that point the fast step. Sherry concluded as in the case of La^{3+} ion exchange of NaX that the slow step involves stripping of the solvation shell from around the Ca^{2+} ions. Calcium ion with an ionic radius of 0.96 A (47) certainly should be able to pass through the rings of 6 tetrahedra that are the entrances to the network of sodalite cages and hexagonal prisms. The 4-day isotherm represents equilibrium. The equilibrium isotherm in Figure 7 is sigmoidal because of an early Ca^{2+} ion selectivity for supercage sites followed by a later Na^+ ion selectivity for small cage sites.

Ba^{2+} ion exchange presents a different picture. Sherry published an early communication (53) in which it was reported that at 25°C only 82% of the ions are replaced by Ba^{2+} ions at equilibrium. Again a "magic" exchange level was reached which corresponds to the inability of Ba^{2+} ions to replace 16 Na^+ ions per unit cell. An unpublished and incomplete powder diffraction study by Olson and Sherry (46) on BaNaX with 16 Na^+ ions per unit cell shows that all the Na^+ ions are in the small cavities. Barrer, Rees, and Shamsuzzoha reported that only about 74% Ba^{2+} ion exchange could be achieved in zeolite X at equilibrium. We can only account for the discrepancy between their and Sherry's results by citing our experience using a NaX that had 7% of the Na^+ ions replaced by H_3O^+ ions resulting from extensive washing with distilled water. All of the ion exchange isotherms were depressed by an amount corresponding to the amount of H_3O^+ exchange.

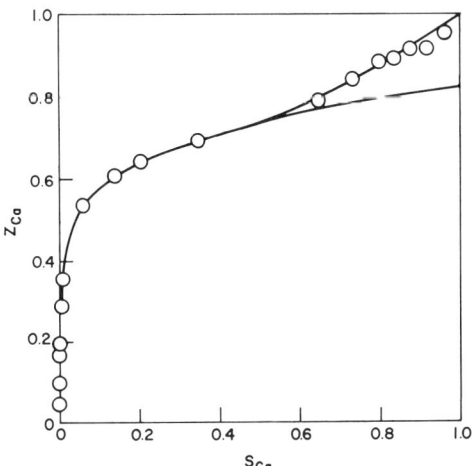

Journal of Physical Chemistry

Figure 7. Ca–Na–X isotherm at 25°C and 0.1 total normality

The inability of Ba^{2+} ions to diffuse into the sodalite cages of zeolite X can be attributed in part to the ionic radius of the bare ion being 1.35 A (47). However, K^+ ion, with an ionic radius of 1.33 A, diffuses rapidly into the small cages. Even for Ba^{2+} ion, some reduction in the size of the hydration shell must be important. Thus, both bare ion size and hydrated ion size help to determine the effective size of the rings of 6 tetrahedra that are the windows between the large and small cages. If water stripping is important in the mechanism of Ba^{2+} ion exchange, exchange at higher temperature should result in complete Na^+ ion replacement. Sherry (55) has obtained Ba–Na–X isotherms at 5°, 25°, and 50°C and finds that at 50°C Ba^{2+} ions can completely replace all of the Na^+ ions in zeolite X. The isotherm is sigmoidal, indicating, as in the case of complete Ca^{2+} ion exchange (Figure 7), that at equilibrium the sites in the small cages prefer Na^+ ions to Ba^{2+} ions. The free energy for Ba^{2+}–Na^+ ion exchange in the large cages of zeolite X was measured at 5° and 25° using the method of Gaines and Thomas (29). The free energy of exchange of the small cages at 50°C was calculated using the free energy of exchange of the large cages at 50°C, extrapolated from the 5° and 25°C values, and the free energy of exchange of the large and small cages at 50°C. The equilibrium constant for small-cage exchange was calculated to be 0.037 and for large cage exchange 72.5. Thus, divalent ion exchange of the small cages of zeolite X is very unfavorable. It will be shown later that divalent ion exchange of the small cages of zeolite Y is still more unfavorable.

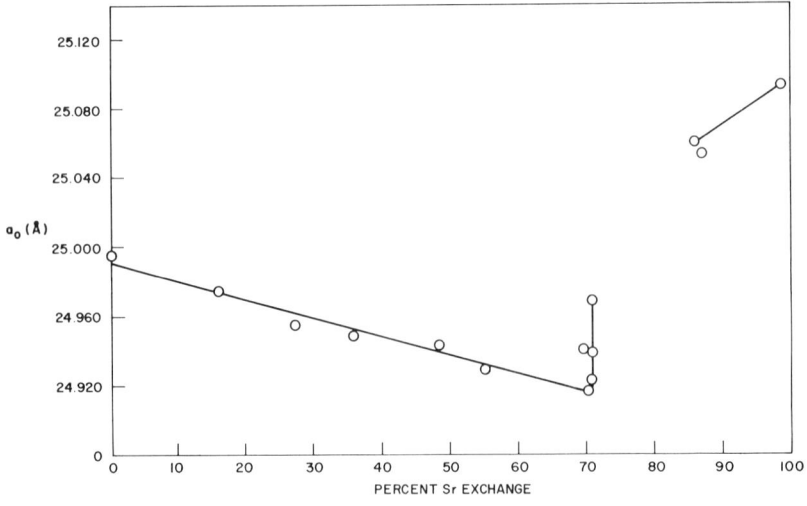

Journal of Physical Chemistry

Figure 8. Lattice parameter vs. percent Sr exchange

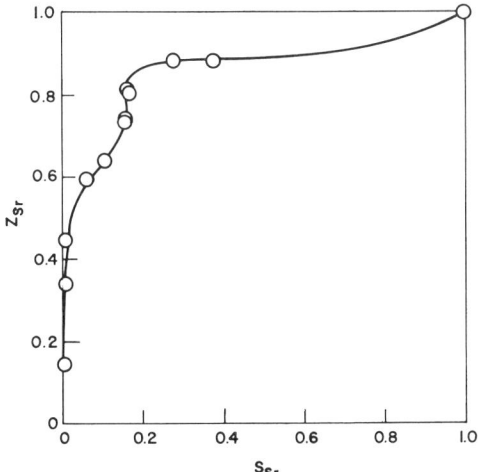

Figure 9. Sr–Na-X isotherm at 25°C and 0.1 total normality

Sr^{2+} ion exchange for Na^+ in zeolite X is an example of the occurrence of a phase reaction as a result of the limited solubility of the end members of an ion exchange series. Olson and Sherry (45) have reported an x-ray powder diffraction study that shows that as Sr^{2+} ions are exchanged into NaX, the cubic unit cell contracts. At 71% Sr loading, the unit cell suddenly expands and a new phase forms that is richer in Sr^{2+} and has a much more expanded unit cell. Evidence for the presence of 2 solid phases is the double x-ray diffraction pattern that they observed. The new phase has the faujasite structure, an Sr exchange level of 87%, and is in equilibrium with the old phase. As the miscibility gap of 71 to 87% is traversed, the Sr-poorer phase disappears and the amount of the Sr-richer phase increases until at 87% Sr loading, only the expanded phase exists. The lattice parameter data are shown in Figure 8.

This x-ray study explains the unusual ion exchange isotherm and selectivity plot (54) obtained for the Sr–Na-X system. The ion exchange isotherm shown in Figure 9 has a sudden vertical rise at about 70% Sr loading. A change in solid phase composition at constant solution phase composition appears to violate the phase rule until it is realized that, over the vertical portion of the isotherm, varying amounts of 2 solid phases each of constant composition must be in equilibrium with a solution of constant composition. The over-all Sr content of the solid phase increases because the amount of Sr-rich solid phase increases. The variation in the selectivity coefficient with Sr^{2+} ion loading shown in Figure 10 can be explained in terms of the appearance of a new solid phase. As

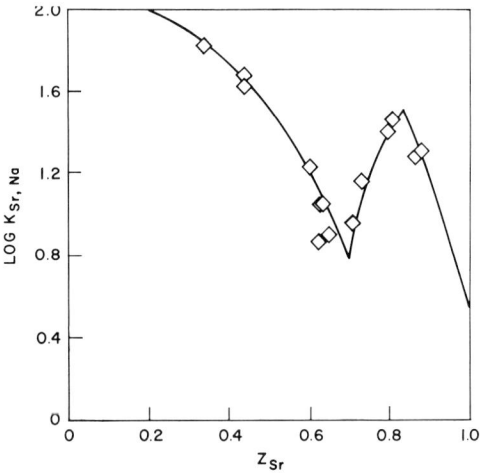

Figure 10. Selectivity vs. Sr–Na exchange in zeolite X at 25°C and 0.1 total normality

Sr^{2+} ions exchange into the NaX, the selectivity coefficient for the preferred Sr^{2+} ion decreases—not an unusual effect. At about 71% Sr loading, the selectivity coefficient suddenly begins to increase with Sr loading because the strontium-richer, and therefore strontium-selective, solid phase appears. The selectivity coefficient continues to increase with the increase in the amount of the Sr-richer phase and reaches a maximum when the miscibility gap has been traversed. The selectivity coefficient then decreases with increasing Sr loading in a normal fashion.

Olson and Sherry also reported structure analyses, using x-ray powder diffraction data, on a SrNaX at the lower side of the miscibility gap (71% Sr loading) and on an almost completely Sr^{2+} ion exchanged zeolite X containing one Na^+ ion per unit cell. They reported that a unit cell of hydrated SrNaX has 12 Na^+ ions in S_I sites, 7.3 Sr^{2+} ions in S_I' sites, and 11.5 Sr^{2+} ions in S_{II} sites. The fully exchanged hydrated SrX contains 2.1 ions in S_I sites, 11.1 in S_I' sites, and 15.0 in S_{II} sites. Thus, during the replacement of 71% of the original Na^+ ions in the zeolite, Sr^{2+} ions site in the small cages. Sr^{2+}–Na^+ ion exchange and the previously discussed Cs^+–Na^+ ion exchange in zeolite X demonstrate the danger in always assuming that the entering ion simply takes the leaving ion's place. The ingoing ion may site in new positions or may cause the remaining ions to resite.

The Sr–Na–X system is not the first example of limited miscibility of end members. Barrer and Hinds (9) reported that K^+ ion exchange of Na-analcite converts some of the crystals to K-leucite at small levels of

K⁺ loading. The 2-phase region extends to almost complete K⁺ ion exchange. Two solid phases were also obtained in the Tl–Na–, Rb–Na–, Tl–K–, and Ag–Na–analcite systems (9). It would appear that almost complete immiscibility of end members occurs when a large ion replaces a small one in a zeolite that has a fairly dense framework structure. The Sr–Na-X system is more complicated. The appearance of a new, Sr-rich, expanded unit cell may be owing to the vacating of the S_I sites. The loss of positive charge in the hexagonal prisms may cause the O atoms to move apart.

Barrer, Rees, and Shamsuzzoha (12) did not observe a region of limited miscibility of the end members NaX and SrX in their work. Perhaps insufficient points were obtained to define the isotherm, or perhaps the NaX used by them was partially hydronium ion exchanged.

Alkaline earth ion exchange of zeolite Y differs from that of zeolite X. Sherry (54) and Barrer, Davies, and Rees (6) both report that at 25°C Ca^{2+}, Sr^{2+}, and Ba^{2+} ions cannot replace 16 of the Na⁺ ions in a unit cell of NaY. Sherry has suggested that the inability of these cations to replace the last 16 Na⁺ ions in NaY is kinetic in origin, whereas it has been suggested that there is no slow step in zeolite Y and that the lack of small cage ion exchange is caused by a positive free energy of exchange of these sites resulting from the low framework charge of this zeolite compared with zeolite X (8). Sherry studied alkaline earth ion exchange at 5°, 25°, and 50°C. He reported (54) that at the highest temperature 16 Na⁺ ions per unit cell were not replaced, but isotherms at 50°C did not cover the last 20% of the solution composition range. Unpublished efforts (58) to obtain points in this region of the isotherm at 5° to 50°C by equilibrating for 1- and 2-week periods and by eluting with 0.1 M $CaCl_2$ for 1- and 2-week periods at 25° and 50°C resulted in as high as 90% Ca^{2+} ion loading. One-week exchanges gave the same results as 2-week exchanges. We now believe that Barrer, Rees, and Shamsuzzoha are correct in suggesting that the inability of divalent ions to replace Na⁺ ions in the small cages of NaY results from an unfavorable exchange equilibria. Again, as was the case with Ag⁺ ion exchange, it is found that when ions must coordinate to lattice oxygen atoms in definite crystallographic sites in the small cages, the equilibria are less favorable in zeolite Y than in zeolite X, most probably because of the lower framework charge on the more siliceous zeolite Y.

Ammonium and Alkyl Substituted Ammonium Ion Exchange in Zeolites X and Y. In 1968, Theng, Vansant, and Uytterhoeven reported (60) on their study of the exchange of ammonium ions and alkylammonium ions in zeolites X and Y. They found that, in general, not all of the Na⁺ ions could be replaced. They agreed with an earlier report (54) that 16 Na⁺ ions per unit cell of NaY could not be replaced by

NH_4^+ ions. It was found that in exchanges involving alkylammonium ions the maximum extent of exchange decreases with increasing molecular weight. This result is consistent with the volume requirements of the ingoing ions. The proportion of the Na^+ ions replaced was lower for zeolite X and also depended on the nature of the metallic ion initially present. These 2 points were taken as evidence that nonsteric factors are also important in alkylammonium ion exchange. These investigators believe that electrostatic interactions must be involved. Thus, the alkylammonium cations should interact less with the fixed negative charges of the framework as the length and the number of the alkyl groups increase. The electrostatic parameter that they correlate with maximum exchange is polarizibility. This factor may be more important in zeolite X than in zeolite Y. Zeolitic water may also play a role. The sorption of water molecules should be stronger in zeolite X than in zeolite Y because the higher cation content in the former and concomitant greater tendency to hydrate should tend to "crowd" out alkylammonium cations.

Alkylammonium Ion Exchange in Clinoptilolite. Alkylammonium ion exchange of clinoptilolite was investigated by Barrer, Papadopoulos, and Rees and reported in 1967 (*11*). They studied the exchange of Na^+ ions from clinoptilolite by obtaining ion exchange isotherms for NH_4^+, $CH_3NH_3^+$, $(CH_3)_2NH_2^+$, $(CH_3)_3NH^+$, $C_2H_5NH_3^+$, n-$C_3H_7NH_3^+$, iso-$C_3H_7NH_3^+$, and n-$C_4H_9NH_3^+$ cation exchange at 60°C. Complete replacement of all the Na^+ ions in this zeolite (empirical formula is $Na_2O \cdot Al_2O_3 \cdot 10\ SiO_2 \cdot 8.6\ H_2O$) was accomplished by all of the above cations but $(CH_3)_3NH^+$, iso-$C_5H_7NH_3^+$, and n-$C_4H_9NH_3^+$ ions, for which only partial replacement was observed. No exchange was observed to take place with still larger cations, indicating a double sieving effect was operative.

Water contents show a linear decrease with increasing loading of the organic cation. At constant exchange level, the amount of water displaced from the zeolite increases with increasing volume of the organic cation. The authors point out that the decrease in the water content of the zeolite during ion exchange should contribute a large endothermic energy term to the heat of exchange. This effect must be considered in any thermodynamic analysis of the exchange of organic cations for metal cations.

The authors have estimated the intracrystalline volume of clinoptilolite from the water content of the Na form and conclude from their estimates of the ionic volumes of the organic cations used that 100% exchange is theoretically possible for all ions. They conclude from a consideration of the heulandite structure (*41*), which is isostructural with clinoptilolite, that partial ion sieving occurs because there are 2 different networks of channels that intersect. Diffusion into and through one network is *via* rings of 10 tetrahedra having a planar projection that is elliptical with approximate free dimensions of 7.9 by 3.5 A. Diffusion through the other channels is through rings of 8 tetrahedra having a planar projection with free dimensions of about 4.4 and 3.0 A (*11*). Barrer, Papadopoulos, and Rees suspect that the inability to achieve complete Na$^+$ ion exchange with (CH$_3$)$_3$NH$^+$, iso-C$_3$H$_7$NH$_3^+$, and n-C$_4$H$_9$NH$_3^+$ ions reflects partial sieving by the network of channels having the smaller windows. Using still larger organic ions to study the sieving effect of the larger windows of the other network of channels, they have very elegantly demonstrated that the almost spherical ions, (CH$_3$)$_4$N$^+$, with van der Waals dimensions in the x, y, and z directions of 6.42, 6.10, and 6.22 A and tert-C$_4$H$_9$NH$_3^+$ with dimensions of 6.52, 6.18, and 5.6 A (*11*) are completely sieved, whereas the more elliptical ions, (CH$_3$)$_3$NH$^+$ with van der Waals dimensions of 6.42, 6.10, and 4.17 A and n-C$_4$H$_9$NH$_3$ with dimensions of 8.42, 4.88, and 4.00 A, are only partially sieved. The authors conclude that there is reasonable correspondence between the crystal structure proposed by Merkle and Slaughter (*41*) and the observed ion sieving.

Ion Exchange of Zeolite T. Sherry has studied univalent and divalent ion exchange in the synthetic Linde Zeolite T (*57*). This zeolite is essentially a synthetic offretite with small intergrowths of erionite (*14*). The offretite structure is capable of exhibiting double sieving effects because it has 2 networks of channels. The more open network consists of channels with an effective diameter of 6.27 A. The more dense network consists of stacks of cancrinite and gmelinite cages. The aperture of the gmelinite cage is a nonplanar ring of tetrahedra having an elliptical planar projection with a limiting dimension of 3.51 A (*57*). The aperture of the cancrinite cage is a very puckered ring with a planar elliptical projection having a dimension of 1.76 A. Sherry studied a batch of KT having the anhydrous unit cell contents K$_4$[(AlO$_2$)$_4$(SiO$_2$)$_{14}$]. Ion exchange with Li$^+$, Na$^+$, NH$_4^+$, Rb$^+$, Cs$^+$, Ag$^+$, Ca^{2+}, and Ba^{2+} ions stops after the replacement of 3 K$^+$ ions per unit cell. It was concluded that one K$^+$ ion is trapped in the framework and that the most likely location for this ion is in the cancrinite cage. This result is in contradistinction to the synthetic faujasites in which Na$^+$ ions are not trapped in the network of small cages. In the latter case, largely hydrated or large bare ions diffuse slowly through the apertures of the sodalite cages or are completely sieved. In the case of zeolite T, a K$^+$ ion must be occluded in a cancrinite cage during crystallization.

The thermodynamic selectivity series for zeolite T is Cs > Rb > Ag > K > NH$_4$ > Ba >> Na > Ca > Li. One of the most interesting results of this work is that this zeolite prefers many monovalent cations to divalent ones. This behavior is not often observed and is unexpected to those who follow the rule of thumb that higher charged ions are always preferred by ion exchangers. Here we see that over the whole range of loading Cs^+, Rb^+, Ag^+, K^+, and NH_4^+ ions are preferred to Ca^{2+} and Ba^{2+} ions. In the case of synthetic faujasites discussed above, the higher charged ions are preferred in the very open and hydrous network of supercages, but the lower charged ones are most preferred in the dense network of hexagonal prisms and sodalite cages. This thermodynamic preference for mono- over divalent ions is drastically higher in zeolite Y than in the more aluminous zeolite X. Zeolite T has a denser framework which contains less water in the cavities and is still lower in framework negative charge than zeolite Y. It is easy to invoke an electrostatic model (56) that takes into account the spacing between fixed negative charges (or charge density) and show that as the distance between them increases the preference for divalent ions over monovalent ones decreases until, at a sufficiently low density of negative charges on the framework, a selectivity reversal occurs.

Ion Exchange Equilibria in Chabazite. Another example of ion exchange on a more siliceous zeolite is the study of natural chabazite (7) by Barrer, Davies, and Rees using a sample of natural chabazite with a Si/Al atom ratio of 2.5. They found that this zeolite exhibited the thermodynamic selectivity series $Tl^+ > K^+ > Ag^+ > Rb^+ > NH_4^+ > Pb^{2+} > Na^+ = Ba^{2+} > Sr^{2+} > Ca^{2+} > Li^+$. This series demonstrates that chabazite also prefers many univalent ions to divalent ones. Barrer, Davies, and Rees speculate that divalent ions, from consideration of Coulomb's law, should be able to interact more strongly with anionic sites (7). They suggest that solvation affinities in the external solution play a role. This cannot be a factor because the solvation effects in the external solution phase are the same no matter what zeolite is considered and yet more aluminous zeolites prefer Ba^{2+}, Sr^{2+}, and Ca^{2+} to Cs^+, Rb^+, NH_4^+, K^+, and Na^+ ions (8, 12, 51, 55) and the reverse is true in less aluminous zeolites such as the chabazite under discussion and zeolite T (57). Interactions in the zeolite phase must be responsible for the selectivity reversal between uni- and divalent ions when moving from a more highly charged to a less highly charged anionic framework (56).

Divalent ion exchange was found to be considerably slower than univalent ion exchange in chabazite (7). The difference in rates is attributed to the necessity to strip some of the solvation shell from the more highly hydrated ions to permit diffusion through the rings of 8 tetrahedra that are the entrances to the cages in chabazite. We have here another example of slow kinetics or ion sieving by zeolites having restriction to diffusion that are only slightly larger than the size of the ions. Other examples have already been discussed (54, 55, 56).

Ion Exchange Kinetics

Particle Controlled Isotopic Ion Exchange Kinetics (Self-Diffusion). An extensive study of ion exchange kinetics in zeolites A and X is being made at the Hahn-Meitner Institute for Nuclear Research in Berlin, West Germany. It is extremely difficult to study isotope exchange in zeolites A and X because the combination of high self-diffusion coefficients and the small size of the crystals that are commercially available results in extremely short reaction times. The group at the Hahn-Meitner Institute has developed 2 techniques for studying fast ion exchange reactions. One is a pressure filtration method (38) in which an exchange solution is rapidly pressed through a thin bed of zeolite crystals. The other one is the temperature jump method (35) in which a solid is rapidly introduced into a solution at an elevated temperature and the reaction is then quenched after a suitable time with ethyl alcohol cooled by liquid air to $-110°C$.

In 1966, Hoinkis and Levi reported that Sr^{2+} self-diffusion in SrX appeared to occur by two processes—one fast and the other slow (33). In 1967, they (34) reported that Cs^+ and Rb^+ isotope exchange in zeolite A does not follow the rate law for a simple diffusion process and appears to take place with 2 steps. They plotted their isotope exchange data as $1 - U$ vs. time, where U is the fractional attainment of equilibrium at any instant in time. In order to ascertain whether or not the rate data fit the simple diffusion equation, they replotted their data as

$$\tau(U) = f(t) \qquad (1)$$

where

$$\tau = \left(\frac{16}{a^2_m}\right)\frac{Dt}{\pi}$$

a^2 is the mean edge size of the cubic zeolite A crystal, D is the diffusion coefficient, and t is the time. A plot of Equation 1 is linear, passing through the origin with a positive slope, if a simple diffusion process that obeys Fick's Laws is observed. Marked curvature of plots of Equation 1 was found for Cs^+ and Rb^+ isotope exchange in zeolite A over the temperature range 0° to 30°C. Thus isotope exchange, or self-diffusion, does not obey the simple diffusion equations.

The equation Hoinkis and Levi used to fit their data was

$$U = (1 - y)U_f + yU_s \qquad (2)$$

where y represents the fraction of the isotope exchange process that is slow, U_f is the fractional attainment of equilibrium of the fast process at any instant in time, and U_s is the fractional attainment of equilibrium of the slow process at any instant in time. U_f and U_s are expressed as functions of time through the simple rate law for diffusion, and this equation is a linear superposition of 2 simple diffusion processes that requires 2 different rate constants and a value for y to fit the isotopic exchange data. Use of Equation 2 to fit the diffusion data implies that the zeolite has 2 independent, noninterconnecting, three-dimensional networks of channels. One would expect to find 2 diffusion processes, each with a different self-diffusion coefficient, in such a structure. In zeolite A, however, there is only one three-dimensional network of channels. The network is a cubic array of large cages interconnected by rings of 8 tetrahedra with a free diameter of 4 A (49). The only determination of cation location has been made for hydrated NaA (19). Eight of the 12 cations per unit cell are located near the center of the 8 rings of 6 tetrahedra that open into the sodalite cages; the other 4 cations have not been located and are presumed to be mobile, hydrated cations. There are, therefore, 2 kinds of cations in this zeolite but only one three-dimensional network of cavities. The sodalite cages are interconnected by double rings of 4 tetrahedra through which no cations can diffuse. Although one would anticipate the possibility of observing a two-step isotopic ion exchange process because of the probable presence of 2 crystallographically distinct kinds of cations in CsA and RbA, it cannot be owing to the presence of 2 different diffusion paths. Only one diffusion path is available, and the same diffusion coefficient should be used to describe diffusion of all ions. We will discuss another model for isotopic ion exchange after reviewing Hoinkis and Levi's study of zeolite X.

In 1968, Hoinkis and Levi studied Ba^{2+} isotopic ion exchange in BaA using the temperature jump method (35). They found that the reaction rate could not be described by the simple diffusion equation over the temperature range 45° to 100°C. The reaction does obey the simple diffusion equation from 100° to 120°C. They concluded from this data that below 100°C sited and unsited Ba^{2+} ions are present and that above 100°C very few ions are sited.

Hoinkis and Levi broadened their study to include zeolite X and in 1969 reported on alkaline earth and alkali metal cation self-diffusion in zeolites A and X (36, 37). They found that Ba^{2+} and Sr^{2+} isotopic ion exchange in BaX and Cs^+ isotopic ion exchange in CsA takes place *via* 2 separate rate processes and that Ba^{2+} and Sr^{2+} isotopic ion exchange in zeolite A takes place by a process that obeys the simple diffusion equation. Again they attribute the two-step process to the 2 independent diffusion processes. In the case of zeolite X, Hoinkis and Levi believe that ions in the large and small cavities diffuse independently—those in the small cavities diffusing into, through, and out of, the crystals *via* the network of sodalite cages and hexagonal prisms and those in the large cages diffusing through the network of large cages. If this picture of ion diffusion in zeolite X were correct, the use of a linear superposition of 2 diffusion processes governed by Fick's laws would be a good mathematical description of self-diffusion. However, the networks of supercages and sodalite cages are interconnected by the rings of 6 tetrahedra that connect the supercages to sodalite cages. It is entirely possible for ions to jump from one network to the other through these rings.

Brown, Sherry, and Krambeck have proposed a new model for particle-controlled isotopic ion exchange (self-diffusion) in hydrated zeolites (20, 21). In this model, all ions in zeolite A must necessarily diffuse in one step because there is only one diffusion path—the network of large cages. The mobile, unsited ions diffuse via this path by a Fickian diffusion process with a characteristic self-diffusion coefficient, D. The bound ions can desorb from their sites and enter into the pool of mobile ions, becoming indistinguishable from them, and then diffuse along the same path with the same characteristic constant, D. Ions may leave the pool of mobile ions by adsorbing on vacant sites. The adsorption and desorption steps are assumed to obey a first-order rate law. The set of partial differential equations describing these processes in a spherical particle is

$$\frac{\partial C_1}{\partial t} = D_1 \left(\frac{1}{r^2}\right) \frac{\partial}{\partial r} \left(r^2 \frac{\partial C_1}{\partial r}\right) + k_2 (C_2 - \alpha_{12} C_1) \tag{3}$$

$$\frac{\partial C_2}{\partial t} = k_2(\alpha_{12} C_1 - C_2) \tag{4}$$

where t is the time, r is the radial coordinate, D_1 is the self-diffusion coefficient of the mobile ions, k_1 is the specific first-order rate constant for adsorption, k_2 is the specific first-order rate constant for desorption, $\alpha_{12} = k_1/k_2$, c_1 is the concentration of mobile ions, and c_2 is the concentration of bound, or sited, ions.

Brown, Sherry, and Krambeck have also proposed a model for self-diffusion in zeolite X (20, 21) in which mobile ions in the large cages diffuse by a Fickian diffusion process, ions in S_{II} sites enter the pool of mobile ions by desorbing and becoming indistinguishable from the other mobile ions, mobile ions adsorb on S_{II} sites, ions in the small cages pass through a ring of 6 tetrahedra and occupy S_{II} sites, and ions in S_{II} sites move in the other direction and enter the small cages. In addition, ions in the small cages can diffuse into, through, and out of the crystals solely through the network of small cages by a Fickian diffusion process that has a characteristic diffusion coefficient that is different from the one for large-cage diffusion. It is assumed again that siting and unsiting of ions is a first-order rate process. The equations that describe these processes are

$$\frac{\partial C_1}{\partial t} = D_1 \frac{1}{r^2} \frac{\partial}{\partial r}\left(r^2 \frac{\partial C_1}{\partial r}\right) + k_2 (C_2 - \alpha_{12} C_1) \tag{5}$$

$$\frac{\partial C_2}{\partial t} = k_2(\alpha_{12} C_1 - C_2) + k_4(C_3 - \alpha_{23} C_2) \tag{6}$$

$$\frac{\partial C_3}{\partial t} = D_3 \frac{1}{r^2} \frac{\partial}{\partial r}\left(r^2 \frac{\partial C_3}{\partial r}\right) + k_4 (\alpha_{23} C_2 - C_3) \tag{7}$$

where t refers to time, r is the radial coordinate, D_1 is the self-diffusion coefficient of ions in supercages, c_1 is the concentration of mobile ions in the supercages, c_2 is the concentration of ions in S_{II} sites, c_3 is the concentration of ions in the network of small cavities, k_1 is the specific first-order rate constant for ion adsorption on S_{II} sites, k_2 for ion desorption from S_{II} sites, k_3 is the rate constant for the jump of an ion in an S_{II} site into a sodalite cage, k_4 is the rate constant for the jump of an ion in the sodalite cage into an S_{II} site, and D_3 is the self-diffusion coefficient for the diffusion of ions in the network of sodalite cages. If we neglect transfer of ions between the 2 different networks of cages and adsorption and desorption at S_{II} sites, Equations 6, 7, and 8 reduce to those used by Hoinkis and Levi (37).

Brown, Sherry, and Krambeck (20, 21) have proposed that an ion in the network of sodalite cages is much more likely to leave a crystal by jumping through a ring of 6 tetrahedra into a large cage and then rapidly diffusing through the network of large cages than it is to diffuse through the network of sodalite cages, which involves passing through a long sequence of hexagonal prisms. In the limit of an infinitely large crystal, this assumption must be true. In the limit of a crystal containing one unit cell, ions in the network of small cages might diffuse by either mechanism. Acceptance of the proposition that ions in the small cages move into the large cages and then diffuse out of the crystal through the network of large cages removes the first term on the right side of Equation 7. Furthermore, it was pointed out (20, 21) that if the ions in S_{II} sites are only weakly sorbed, they will be in equilibrium with the mobile ions in the large cages, and they will simply serve to enlarge the pool of mobile ions. If this situation exists, Equations 5, 6, and 7 reduce to

$$\frac{\partial C_{1+2}}{\partial t} = D_1 \frac{1}{r^2} \frac{\partial}{\partial r}\left(r^2 \frac{\partial C_1}{\partial r}\right) + k_4(C_3 - \alpha_{23} C_2) \tag{8}$$

$$\frac{\partial C_3}{\partial t} = k_4(C_3 - \alpha_{23}C_2) \tag{9}$$

Equations 8 and 9, which have built into them the assumption that ions in the small cages move into large cages and not through the network of small cages and that adsorption on and desorption from S_{II} sites is rapid compared to diffusion in the large cages, are of the same form as Equations 3 and 4.

Brown and Sherry have tested their models by studying particle-controlled isotopic ion exchange of Na^+ in NaA and NaX (22). This reaction is so rapid that it would not be possible to follow using the techniques developed at the Hahn-Meitner Institute. Brown and Sherry solved this problem by using the techniques developed by Charnell (23) for growing gram-scale batches of well-formed single crystals of NaA and NaX ranging from 25 to 100μ in size. Using narrow size fractions of the batches of large crystals, they followed reactions with half-times as short as 2 seconds using the radiometric technique developed by Schwartz (51).

It was found that isotopic ion exchange or self-diffusion takes place in 2 steps in both NaX and NaA. Equations 3 and 4 fit the data for NaA well. Equations 8 and 9, which are the same form as Equations 3 and 4, fit the data for NaX well. Thus, only 2 steps are observed in NaX, one fast and one slow, and the postulate that ions in S_{II} sites are in equilibrium with mobile ions is reasonable. Since the model requires the same number of constants to fit the data as the one used by Hoinkis and Levi (37), and both models adequately describe the experimental data, more information was needed to differentiate them. In the model proposed by Brown, Sherry, and Krambeck (20, 21), the rate of the slow step is independent of crystal size, whereas in the model proposed by Hoinkis and Levi (37), the rate of the slow step is inversely proportional to the square of the particle size. Brown and Sherry (22) studied Na^+ self-diffusion in zeolite X as a function of crystal size and found no dependence of the rate of the slow step on particle size over the range of 25–75μ. Furthermore, unpublished data of Brown and Sherry show that the slow step in Ba^{2+} isotopic ion exchange of 1–10-μ crystals of BaX is as slow as in the exchange of 50-μ crystals. Thus the slow step in Ba^{2+} ion exchange must involve exchange of ions in the sodalite cages and/or hexagonal prisms with ions in the supercages and not a diffusional process.

Late in 1969, Gaus and Hoinkis (30) published a paper in which they reevaluated the model of Hoinkis and Levi for zeolite A (35). They proposed that sited ions serve as a source for mobile diffusing ions and include a term in their equation describing a first-order process. However, they do not consider that the vacant cation sites are a sink and add no term to account for loss of mobile ions to the sink. The equation they use is

$$\frac{\partial C}{\partial t} = D \frac{1}{r^2} \frac{\partial}{\partial r}\left(r^2 \frac{\partial C}{\partial r}\right) + C_0[(1-y)\,\delta(t) + ybe^{-bt}] \qquad (10)$$

where y is again the fraction of the bound cations. This equation still contains 2 diffusion terms, one for bound ions and another for the ions that are mobile at zero time.

In 1968, Dyer, Gettins, and Molyneux (28) studied Ba^{2+} isotopic ion exchange in BaA over the temperature range 19° to 65°C. They did not show plots of dimensionless time as a function of time but stated that these plots were linear and, therefore, obey the simple diffusion equation. There is some discrepancy between this work and the work of Hoinkis and Levi at the Hahn-Meitner Institute (35, 37) that is described above, in which they state that in their study of Ba^{2+} isotopic ion exchange, linear plots of dimensionless time vs. time are not obtained and fit the data by an equation for a two-step process.

In 1969, Dyer and Gettins (27) reported on isotopic ion exchange and ion exchange kinetics in the zeolites A, ZK-4, X, and Y from methanol, ethanol, and alcohol water solutions. Some of their results are shown in Table I.

Table I. Activation Energies of Isotopic Exchange Processes[a]

Zeolite	Ion, M^{2+}	Solvent	Activation Energy, Kcal/Mole
X	Ca	Water	19.9
	Sr	Water	20.0
	Ba	Water	20.1
1.86Y	Ca,Sr,Ba	Water	Rate immeasurably rapid
2.60Y	Ca,Sr,Ba	Water	Rate immeasurably rapid
X	Ca	Methanol	29.1
X	Sr	Methanol	Rate immeasurably slow
X	Ba	Methanol	24.5
1.86Y	Ca	Methanol	21.6
1.86Y	Sr	Methanol	28.1
1.86Y	Ba	Methanol	20.1
1.86Y	Ca	Ethanol	Rate immeasurably slow
1.86Y	Sr	Ethanol	Rate immeasurably slow
1.86Y	Ba	Ethanol	28.6
2.60Y	Ca	Methanol	11.9
2.60Y	Sr	Methanol	21.9
2.60Y	Ba	Methanol	16.2
2.60Y	Ca	Ethanol	13.9
2.60Y	Sr	Ethanol	25.1
2.60Y	Ba	Ethanol	21.9

[a] Taken from Reference 26.

Sr^{2+} isotopic ion exchange is immeasurably slow in zeolite X when methanol is the solvent, and Ca^{2+} and Ba^{2+} exchange rates were fast enough to measure. Two zeolite Y samples were studied—one with a Si/Al atom ratio of 1.86 (1.86Y) and one with a ratio of 2.60 (2.60Y). Exchange rates of all 3 alkaline earth ions were measured in the more aluminous Y in methanol, but in ethanol, Ca^{2+} and Sr^{2+} exchange was immeasurably slow. In the less aluminous zeolite, exchange rates were measurably fast for all 3 ions in both solvents. Thus we see that as the framework charge decreases from X to 1.86Y to 2.60Y, the rate of exchange in both alcohols increases and the activation energy for diffusion decreases. Moreover, for any given divalent ion and zeolite, the rates of isotopic ion exchange decrease in the order water > methanol > ethanol, and the corresponding activation energies for diffusion increase in the order water < methanol < ethanol. Na^+ isotopic ion exchange was immeasurably fast in all solvents except in the NaA–ethanol system at $-45°C$.

Dyer and Gettins (27) have made a more comprehensive study of self-diffusion in zeolite ZK-4. This zeolite is isostructural with zeolite A but more siliceous (39), having a Si/Al atom ratio of 1.33, and it is of interest to compare it to their study of zeolite A (25). The activation energy for diffusion and D_0 in the Arrhenius equation $D = D^0 \exp(-E_a/RT)$, along with the entropy of activation $\Delta S‡$, are shown in Table II.

Table II. Activation Energies of Isotopic Ion Exchange[a]

Zeolite	Ion, M^{2+}	E^a, Kcal	D_0, Cm^2/Sec	$\Delta S‡$, Cal/Mole/Deg
A	Ca	16.1	5.10×10^{-6}	4.0
	Sr	19.0	1.70×10^{-5}	+ 2.7
	Ba	21.6	1.48×10^{-3}	+11.6
ZK-4	Ca	20.1	3.58×10^{-7}	+ 8.6
	Sr	22.6	7.35×10^{-3}	+14.8
	Ba	15.8	7.60×10^{-7}	- 3.4

[a] Taken from Reference 27.

These data show the activation energies for Ca^{2+} and Sr^{2+} self-diffusion are higher in zeolite ZK-4 than in zeolite A, but the reverse is true for Ba^{2+} ion self-diffusion. It may be possible that diffusion of the highly hydrated ions Ca^{2+} and Sr^{2+} involves partial stripping of hydration shells, and in ZK-4, in which the ratio of water molecules to cations is greater, they are more highly hydrated and therefore more water must be stripped. Ba^{2+} ions, because of their lower hydration energy, may not have to undergo as much reduction in hydration shell in order to diffuse, and therefore the activation energy is lower in the more siliceous ZK-4 because the electric field strength in the crystals is smaller.

Particle-Controlled Ion Exchange Kinetics. Now we turn our attention to the mathematically more complex subject of ion exchange kinetics. The only significant attempt to advance in this field is the work of Brooke and Rees (17, 18, 50). They have extended the treatment of Helfferich and Plesset (32) to nonideal ion exchange systems in which the ionic activity coefficients in the exchanger phase change with ionic composition. The Helfferich-Plesset equation for the rate of exchange of ion A for B in an ideal system is given by

$$\frac{\partial C_A}{\partial t} = \frac{1}{r^2} \frac{\partial}{\partial r} \left(r^2 D_{AB} \frac{\partial C_A}{\partial r} \right) \tag{11}$$

for a particle diffusion controlled reaction where C_A is the concentration of ion A in the exchanger phase, and t and r are already defined. The interdiffusion coefficient D_{AB} is given by

$$D_{AB} = \frac{D_A D_B (Z_A{}^2 C_A + Z_B{}^2 C_B)}{Z_A{}^2 C_A D_A + Z_B{}^2 C_B D_B} \tag{12}$$

where D_A and D_B are the self-diffusion coefficient of the ions in the A and B forms of the exchanger, Z_A and Z_B are ionic charges, and C_A and C_B are the concentrations of ions A and B. Helfferich and Plesset (32) have pointed out that they assumed an ideal system in which the self-diffusion coefficient for the exchanging ions is independent of the ionic composition of the exchanger phase and ionic activity coefficients are assumed to be unity.

Brooke and Rees have extended Equation 11 to include systems in which the ionic activity coefficients are functions of the composition of the exchanger phase by expressing the inter-diffusion coefficient as

$$D_{AB} = \frac{D_A D_B [Z_A{}^2 C_A (\partial \ln a_B / \partial \ln C_B) + Z_B{}^2 C_B (\partial \ln a_A / \partial \ln C_A)]}{Z_A{}^2 C_A D_A + Z_B{}^2 C_B D_B} \tag{13}$$

where a_A and a_B are the thermodynamic activities of ions A and B, and the other quantities are defined as above. The activity gradient terms represent the effect of nonideality on the ion exchange kinetics.

Brooke and Rees have written a computer program that computes rates of exchange given by Equation 11 by a finite difference method (17). They have used this program to compute the rates of Ca^{2+}–Sr^{2+} ion exchange in chabazite and compared the computed values to the experimental ones (17). The term $\partial \ln a_B / \partial \ln C_B$ was evaluated from experimental data (2, 3) to be 0.27 for $0 \leqslant C_A \leqslant 0.4$, to discontinuously rise to 2.5 at $C_A = 0.4$ and then remain constant to $C_A = 1.0$. The theory predicts that the rate of exchange is faster when the faster ion is in the zeolite. This prediction can be clearly understood if we realize that when the exchanging ions are of equal charge, Equation 13 becomes

$$D_{AB} = \frac{D_A D_B}{(D_A - D_B)C_A + D_B} \frac{\partial \ln a_A}{\partial \ln C_A} \quad (14)$$

When

$$C_A \longrightarrow 0$$

$$D_{AB} \longrightarrow D_A \frac{\partial \ln a_A}{\partial \ln C_A} \quad (15)$$

and as

$$C_A \longrightarrow 1$$

$$D_{AB} \longrightarrow D_B \frac{\partial \ln a_A}{\partial \ln C_A}$$

In the Sr–Ca–chabazite system, D_{Ca} is approximately twice D_{Sr}. If we let $A = Ca^{2+}$ in Equations 15 and use values of $\partial \ln a_A / \partial \ln C_A$ of 0.4 when C_A approaches 0 and 2.5 when C_A approaches 1, we see that when $C_A \to 0$, $D_{AB} = 0.4 D_A$ and when $C_A \to 1$, $D_{AB} = 1.25 D_A$. Thus, when the faster ion is initially present in the zeolite, the exchange rate should be faster than when the slower ion is present. The computer-calculated rate data agree qualitatively with the experimental rate data.

Brooke and Rees have used their theory to examine a system in which the activity of the ions in the exchanger phase is governed by the Kielland equations (40):

$$\ln a_A = \ln C_A + kC_B^2$$
$$\ln a_B = \ln C_B + kC_A^2 \quad (16)$$

This system is the Na–K chabazite system, which was studied in detail by them (20). The results computed using the extended theory of Brooke and Rees disagree badly with their experimental results. The Helfferich-Plesset equation disagrees just as badly. Brooke and Rees have themselves suggested that in addition to the variation of ionic activities, the variation of ionic diffusion coefficients with concentration must be considered, and perhaps water transport.

Summary and Conclusions

During the period 1967 to 1970, our understanding of ion exchange in the synthetic zeolites X and Y has advanced considerably. It is clear that the presence of 2 independent three-dimensional networks of cavities with cations located in both networks causes double sieving effects. One network is dense and contains little water (the network of sodalite cages and hexagonal prisms) and the other network is open and highly hydrated. These 2 networks exhibit differing thermodynamic ion exchange properties. The sites in the network of large cages exhibit a thermodynamic preference for higher charged ions over lower charged ones and, among univalent ions, the thermodynamic preference depends on the aluminum content, or framework charge. The thermodynamic preference of the sites in the large cages of the more aluminous zeolite X, with the exception of Li^+, decreases with increasing ionic radius, whereas in the case of zeolite Y the thermodynamic preference decreases with increasing ionic hydration energy. In the small cages of zeolites X and Y, the sites exhibit a preference for the ion with the smallest ionic radius and lowest charge—as is true of small-pore, dense zeolites. The rejection of ions on the basis of both size and charge in the small cages is more marked in zeolite Y.

Exchange of polyvalent ions for Na^+ ions in zeolites X and Y is distinguished by 2 steps—one being many orders of magnitude slower than the other. It is clear that the slow step involves exchange of the 16 Na^+ ions per unit cell that are in the small cages of the hydrated X and Y zeolites. It is also clear from the study of Sr^{2+}–Na^+ ion exchange and Cs^+–Na^+ ion exchange in zeolite X that it is not always correct to assume that the entering ions occupy the same sites or have the same site distribution as did the leaving ions.

Until recently, particle-controlled isotopic ion exchange or self-diffusion has not been studied in zeolites A and X because the rates are too high in the 1–10-μ crystals that are available commercially. One solution to this problem has been to develop experimental techniques to measure fast reaction rates. The other solution has been to grow large batches of well-formed single crystals as large as 100μ and thereby increase the half-times of the reactions by as much as 10^4 times.

These new techniques for studying self-diffusion have enabled very detailed studies to be made. It was observed in these studies that simple diffusion laws would not predict the observed rates. Different models have been proposed by different groups of researchers. Agreement has yet to be reached as to which model is most realistic.

Isotopic ion exchange kinetics has been studied in methanol and ethanol. The solvent effects are very large and some of them can be explained in terms of ion solvation but much is yet to be done in the very large area of ion exchange from nonaqueous solvents.

An attempt has been made to develop a theoretical treatment of ion exchange kinetics in nonideal systems. This new theory considers the variation of ionic activities as the ionic concentrations change in the zeolite phase. In one case, the theory correctly predicts an effect that is contrary to what is predicted by the theory for ideal systems. In another test of the new theory, it is found that it misses the mark as badly as does the simpler theory for ideal systems. Other terms such as variation in ionic diffusion coefficients with concentration and water transport must be considered.

After summarizing 3 years of work on cation exchange in zeolites, it is clear that much progress has been made and that much more progress must still be made.

Literature Cited

(1) Ames, L. L., Jr., *J. Inorg. Nucl. Chem.* **1965**, 27, 885.
(2) Barrer, R. M., Bartholomew, R., Rees, L. V. C., *J. Phys. Chem. Solids* **1963**, 24, 51.
(3) *Ibid.,* **1963**, 24, 309.
(4) Barrer, R. M., Bratt, G. C., *J. Phys. Chem. Solids* **1959**, 12, 130.
(5) *Ibid.,* **1959**, 12, 146.
(6) Barrer, R. M., Davies, J. A., Rees, L. V. C., *J. Inorg. Nucl. Chem.* **1968**, 30, 3333.
(7) *Ibid.,* **1969**, 31, 219.
(8) *Ibid.,* **1969**, 31, 2599.
(9) Barrer, R. M., Hinds, L., *J. Chem. Soc.* **1957**, 1879.
(10) Barrer, R. M., Meier, W. M., *Trans. Faraday Soc.* **1959**, 55, 130.
(11) Barrer, R. M., Papadopoulos, R., Rees, L. V. C., *J. Inorg. Nucl. Chem.* **1967**, 29, 2047.
(12) Barrer, R. M., Rees, L. V. C., Shamsuzzoha, *J. Inorg. Nucl. Chem.* **1966**, 28, 629.
(13) Baur, W. H., *Am. Minerologist* **1964**, 49, 697.
(14) Bennett, J. M., Gard, J. A., *Nature* **1967**, 214, 1005.
(15) Breck, D. W., *J. Chem. Educ.* **1964**, 41, 678.
(16) Breck, D. W., Eversole, W. G., Milton, R. M., Reed, T. B., Thomas, T. L., *J. Am. Chem. Soc.* **1956**, 78, 5963.
(17) Brooke, N. M., Rees, L. V. C., *Trans. Faraday Soc.* **1968**, 64, 3383.
(18) *Ibid.,* **1969**, 65, 2278.
(19) Broussard, L., Shoemaker, D. P., *J. Am. Chem. Soc.* **1960**, 82, 1041.
(20) Brown, L. M., Sherry, H. S., *Proc. Intern. Conf. Ion Exchange in the Process Industries, London, July 1969,* **1970**, 349.
(21) Brown, L. M., Sherry, H. S., Krambeck, F. J., Paper No. Phys. 125, *158th National Meeting, ACS,* New York, September 1969.
(22) Brown, L. M., Sherry, H. S., Paper No. Phys. 126, *158th National Meeting, ACS,* New York, September 1969.
(23) Charnell, J. F., Mobil Research and Development Corp., unpublished data.
(24) Dempsey, E., *in* "Molecular Sieves," p. 293, Society of the Chemical Industry, London, 1968.
(25) Dyer, A., Fawcett, J. M., *J. Inorg. Nucl. Chem.* **1966**, 28, 615.
(26) Dyer, A., Gettins, R. B., *Proc. Intern. Conf. Ion Exchange in the Process Industries, London, July 1969,* **1970**, 357.
(27) Dyer, A., Gettins, R. B., *J. Inorg. Nucl. Chem.* **1970**, 32, 319.
(28) Dyer, A., Gettins, R. B., Molyneux, A., *J. Inorg. Nucl. Chem.* **1968**, 30, 2823.

(29) Gaines, G. L., Thomas, H. C., *J. Chem. Phys.* **1953**, 21, 714.
(30) Gaus, H., Hoinkis, E., *Z. Naturforsch.* **1969**, 24a, 1511.
(31) Helfferich, F., "Ion Exchange," McGraw-Hill, New York, 1962.
(32) Helfferich, F., Plesset, M. S., *J. Chem. Phys.* **1958**, 28, 418.
(33) Hoinkis, E., Levi, H. W., *Z. Naturwissenschaften* **1966**, 53, 1.
(34) Hoinkis, E., Levi, H. W., *Z. Naturforsch.* **1967**, 22a, 226.
(35) *Ibid.*, **1968**, 23a, 813.
(36) *Ibid.*, **1969**, 24a, 1672.
(37) Hoinkis, E., Levi, H. W., *Proc. Intern. Conf. Ion Exchange in the Process Industries, London, July 1969,* **1970**, 339.
(38) Hoinkis, E., Levi, H. W., Lutze, W., Miekeley, N., Tamberg, T., *Z. Naturforsch.* **1967**, 22a, 220.
(39) Kerr, G. T., *Inorg. Chem.* **1966**, 5, 1537.
(40) Kielland, J., *J. Soc. Chem. Ind.* **1935**, 34, 232.
(41) Merkle, A. B., Slaughter, M., *Conf. Geol. Soc. Am.* (1965).
(42) Nightingale, E. F., Jr., *J. Phys. Chem.* **1959**, 63, 1381.
(43) Olson, D. H., *J. Coll. Interface Sci.* **1968**, 28, 305.
(44) Olson, D. H., *J. Phys. Chem.* **1970**, 74, 2758.
(45) Olson, D. H., Sherry, H. S., *J. Phys. Chem.* **1968**, 72, 4095.
(46) Olson, D. H., Sherry, H. S., Mobil Research and Development Corp., unpublished data.
(47) Pauling, L., *J. Am. Chem. Soc.* **1927**, 49, 765.
(48) Pickert, P. E., Rabo, J. A., Dempsey, E., Shoemaker, V., *Proc. Intern. Conf. Catalysis, 3rd, Amsterdam, 1964,* VI, p. 714, Wiley, New York, 1965.
(49) Reed, T. B., Breck, D. W., *J. Am. Chem. Soc.* **1956**, 78, 5972.
(50) Rees, L. V. C., Brooke, N. M., *Proc. Intern. Conf. Ion Exchange in the Process Industries, London, July 1969,* **1970**, 352.
(51) Schwartz, A., Marinsky, J. A., Sprigler, K. S., *J. Phys. Chem.* **1964**, 68, 918.
(52) Sherry, H. S., *J. Phys. Chem.* **1966**, 70, 1158.
(53) *Ibid.*, **1967**, 71, 780.
(54) *Ibid.*, **1968**, 72, 4086.
(55) Sherry, H. S., *J. Coll. Interface Sci.* **1968**, 28, 288.
(56) Sherry, H. S., "Ion Exchange," Vol. 2, Chap. III, J. A. Marinsky, Ed., Marcel Dekker, New York, 1969.
(57) Sherry, H. S., *Proc. Intern. Conf. Ion Exchange in the Process Industries, London, July 1969,* **1970**, 329.
(58) Sherry, H. S., unpublished data.
(59) Sherry, H. S., Walton, H. F., *J. Phys. Chem.* **1967**, 71, 1457.
(60) Theng, B. K., Vansant, E., Uytterhoeven, J. B., *Trans. Faraday Soc.* **1968**, 64, 3370.
(61) Ward, J. W., *J. Catalysis* **1968**, 10, 34.

RECEIVED July 14, 1970.

Discussion

M. S. Goldstein (American Cyanamid, Stamford, Conn. 06902): You have reported that the La-Na exchange isotherms for types X and Y sieves are reversible at 25°C. Are they still reversible in your higher temperature experiments at 82°?

H. S. Sherry: No. We observe irreversibility at the higher temperatures.

G. V. Tsitsishvili (Academy of Sciences of the Georgian SSR, Tbilisi, USSR): Interesting results are presented in the paper about sodium ion replacement by thallium ions. You observed that total replacement of Na^+ by Tl^+ in NaY proceeds at 100°C. Did you study the temperature dependence of this ion exchange process? If the 6-membered rings in zeolite Y are only about 0.01Å smaller in diameter than those of zeolite X, can one expect the strong temperature dependence of the sodium–thallium exchange process?

H. S. Sherry: No. A strong temperature dependence would not be expected. However, I have not studied this temperature dependence.

R. M. Barrer (Imperial College, London): Your Equations 5 to 7 are set up for a constant exchange diffusion coefficient. However, exchange diffusion coefficients are functions of crystal composition, and to make adequate use of such relations in interpreting experimental data, this concentration dependence would have to be taken into account. Have you considered this?

H. S. Sherry: We have thus far only considered isotopic ion exchange reactions. Our equations are written to describe this process and not ion exchange. Thus, our diffusion coefficients are for self-diffusion and are constant.

29

Infrared Spectroscopic Studies of Zeolites

JOHN W. WARD

Union Oil Co. of California, Union Research Center, Brea, Calif. 92621

> *Studies of the surface of and molecules adsorbed on zeolites are reviewed. Pure sodium zeolites contain no structural hydroxyl groups but such groups can be introduced by hydrolysis and by exchange with ammonium or multivalent cations. The nature and population of hydroxyl groups depend on the cation and the treatment temperature. Hydroxyl groups associated with silicon atoms and cations have been observed. Dehydroxylation exposes tricoordinated aluminum atoms. Hydroxyl groups can be reformed in some cases but not in others by readsorption of water. Some of the hydroxyl groups are acidic. Results from adsorption of organic and inorganic molecules are discussed. Hydroxyl groups, cations, and aluminum atoms can function as adsorption sites. The properties and locations of the exchanged cations influence the adsorbate–adsorbent interactions.*

In most applications of zeolites, surface properties and reactivity are of major importance. Infrared spectroscopy can give useful information on the constitution and surface properties of zeolites and how these are modified by various treatments. Changes in the spectra of the zeolite and of molecules adsorbed on the surface can yield direct information about the surface, how adsorbed molecules interact, and where molecules adsorb.

Although many molecular sieve zeolites are known, most of the spectroscopic studies have involved the commercially important large-pore X and Y zeolites. Isolated studies of other zeolites, such as A, L, and mordenite have been reported. Discussions of the application of infrared spectroscopy to zeolites appeared recently (7, 24, 39, 74, 75). In this review, recent studies of the nature of surface structural groups and the interactions of various molecules will be discussed.

Structural Hydroxyl Groups

Cation-Containing Zeolites. Zeolite surface structural groups have been studied extensively because of their importance in adsorption and catalysis. From a consideration of the theoretical structure, only absorption bands due to —Si—O— and —Al—O vibrations would be expected. The earliest work has been reviewed previously (7, 74, 75). In general, except for the initial study of Bertsch and Habgood (9), at least 3 types of structural hydroxyl groups were detected.

Contrary to the early studies, Eberly (17) reported no hydroxyl groups on Na and CaY zeolites after dehydration at 427°C. Ward (63, 70) and Hall et al. (13, 59) also reported no hydroxyl groups on Group IAY zeolites but detected structural hydroxyl groups with frequencies similar to those reported by Angell et al. (5) on Group IIAY zeolites. In their first report, Hall et al. (13) listed hydroxyl frequencies for Mg, Ba, Zn, and CeY zeolites and assigned them to the types of hydroxyl groups observed in hydrogen Y. They found that the band frequencies near 3640 and 3540 cm^{-1} were a function of the electron affinity of the cations. Later (59), they assigned a band near 3690 cm^{-1} to water bound to residual sodium ions in the structure and a band near 3605 cm^{-1} to $CaOH^+$ groups. A systematic study of Group IA and IIAY zeolites was made by Ward (70). No hydroxyl groups were detected on Group IAY, BaY, and NaX zeolites, apart from Li, after evacuation at 360°C. For the other Group IIA zeolites, hydroxyl bands were detected in agreement with other workers (5, 13, 17, 26, 59). The frequencies of the latter 3 bands increased with increasing electrostatic field, electrostatic and ionization potential calculated for the cation form of the zeolite. Typical spectra for Na and CaY zeolites are shown in Figure 1. Frequencies of hydroxyl groups on rare earth zeolites were similar to those reported by Hall et al. (13) for cerium and Rabo et al. for cerium and lanthanum Y zeolites (47). Recently, Hattori and Shiba (26) studied Ca, Zn, and LaX. For Ca and ZnX, only a band at 3750 cm^{-1} was detected after evacuation at 450°C. Bands at 3650 and 3570 cm^{-1} were detected at lower temperatures. An additional band was seen at 3530 cm^{-1} in the case of LaX.

Allowing for minor differences in technique, the studies of cation-exchanged zeolites can be rationalized to give a coherent description of the surface groups. Although detected a few times on the Group IA zeolites, it appears that noncation-deficient monovalent zeolites contain no structural hydroxyl groups. Hydroxyl groups on some Group IA zeolites probably result from cation deficiency caused by partial hydrolysis and slight amounts of siliceous impurities. The only hydroxyl groups expected from structural considerations would be those terminating the

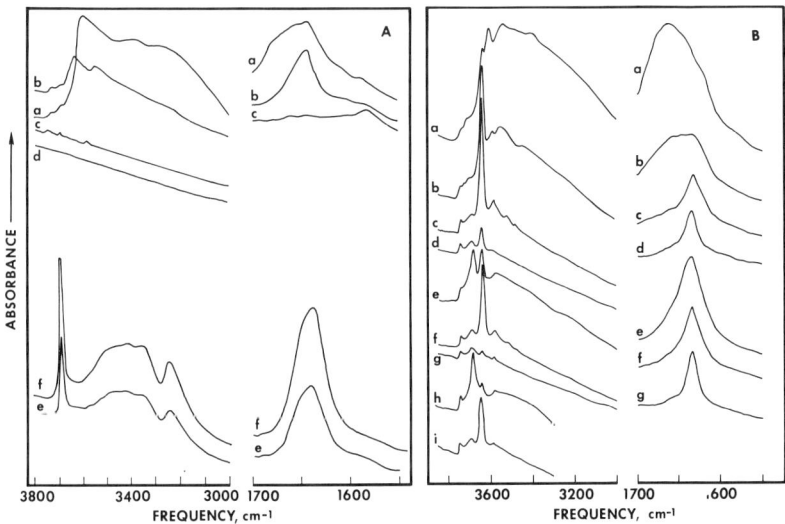

Figure 1. (A) Spectra of the hydroxyl stretching and water bending vibration of NaY zeolite

    (a) Evac. at 100°
    (b) 225°
    (c) 380°
    (d) 480°
    (e) 6 Micromoles of water readsorbed
    (f) 12 Micromoles of water readsorbed

(B) Spectra of CaY zeolite

    (a) Evac. at 130°
    (b) 245°
    (c) 325°
    (d) 470°
    (e) 10 Micromoles of water added
    (f) Evac. at 200°
    (g) 470°
    (h) 10 Micromoles of water added
    (i) Evac. at 200°

giant lattice with a frequency near 3740 cm^{-1}. These conclusions are confirmed by the data of Habgood (23) for cation-deficient samples. Barium appears to behave like Group IA zeolites, although some structural hydroxyl groups must be present since it exhibits Bronsted acidity whereas Group IA zeolites do not. All other zeolites studied, including silver, appear to have structural hydroxyl groups with frequencies near 3690, 3650, and 3540 cm^{-1}. These include the alkaline earth, transition metal, and rare earth cation zeolites. The absorption bands near 3650 and 3540 cm^{-1} are at frequencies similar to those observed in hydrogen Y zeolite (below) and are probably caused by the same type of hydroxyl groups. Furthermore, pyridine and ammonia adsorption studies have shown that these hydroxyl groups interact with the bases analogous to the hydroxyl groups of hydrogen zeolites. The concentration of these groups is much less than in hydrogen Y zeolite but is much greater than that observed for cation-deficient Group IA zeolites. It therefore seems that the presence of these groups is not a result of cation deficiencies. A simple divalent cation apparently cannot satisfy the charge distribution requirements of the zeolite lattice in the absence of much water. During dehydration, the multivalent cation becomes localized and its associated electrostatic field may induce dissociation of coordinated water molecules to produce MOH^+ and H^+ species. The proton would react then with a lattice oxygen at a second exchange site to produce the type of hydroxyl present in the hydrogen zeolites with absorption frequencies near 3650 and 3540 cm^{-1} (13, 17, 29, 63).

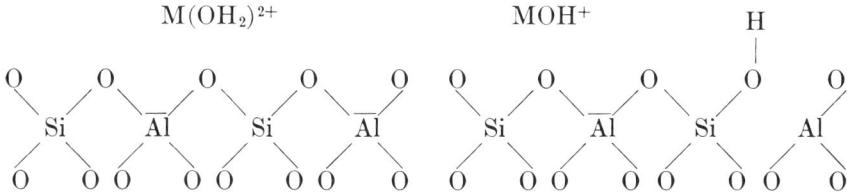

Other bands found in the spectra of the divalent cation zeolites are not found in the spectra of the alkali cation and hydrogen zeolites, and must be specific to the presence of divalent cations. A band usually is observed near 3690 cm^{-1}. This band was assigned originally to AlOH groups (5), although assignment to physically adsorbed water is probably more correct (59, 70). A second band occurring between 3600 and 3570 cm^{-1} varied in frequency with the cation, increasing with decreasing cation size. This type of hydroxyl group has been attributed to MOH^+ groups (59, 70). It is very sensitive to the level of hydration, and it disappears on mild dehydration, possibly with the formation of MO or M^+—O—M^+ groups (44, 59). The only trivalent ions reported so far are the rare earth, cerium, and lanthanum ions. After removal of physically adsorbed water, hydroxyl frequencies are observed near 3740, 3650, and 3530 cm^{-1} (13, 47, 48, 70). The band near 3530 cm^{-1} has been attributed to MOH^{2+} or $M(OH)_2^+$ groups formed according to the scheme (60, 70).

$$M(OH_2)^{3+} \rightleftarrows MOH(OH_2)^{2+} + H^+ \rightleftarrows M(OH)_2^+ + 2H^+$$

Alternatively, Rabo et al. (48) suggested formation of M—OH—M groups which on dehydration form M—O—M groups. The groups are relatively abundant at 500°C but are eliminated at 700°C.

In contrast to X and Y zeolites, only 1 investigation of L zeolite has been reported. For the monovalent group IA zeolites, no structural hydroxyl groups are detected apart from the usual band near 3740 cm^{-1}

(69). Zhdanov et al. detected broad OH bands at 3450 and 3290 cm^{-1} on NaA zeolite after evacuation at 400°C for 4 hours. No band was detected near 3690 cm^{-1} (77).

Hydrogen Zeolites. Because of their catalytic properties, a great deal of attention has been given to the ammonium zeolites and their decomposition products. As the ammonium zeolites are heated at progressively higher temperatures, the ammonium ions are decomposed and hydroxyl groups formed (7, 74, 75). This is shown by the disappearance of absorption bands near 3450 to 3000 cm^{-1} and near 1450 cm^{-1} because of NH$_4^+$ and the formation of hydroxyl group bands near 3740, 3650, and 3550 cm^{-1}. The hydroxyl band intensities decrease on further heating. These phenomena can be represented by the following scheme.

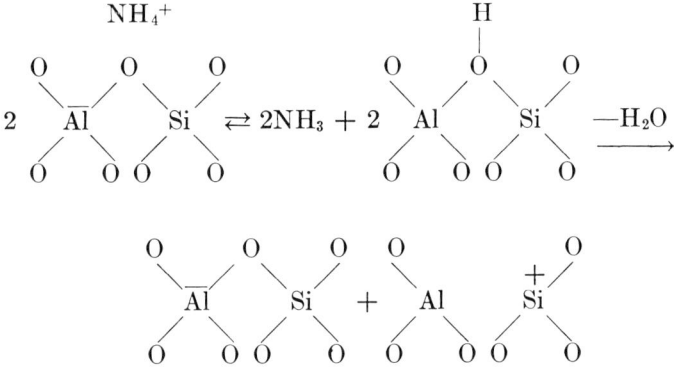

Typical spectra of the decomposition of ammonium Y are shown in Figure 2.

Although all workers agreed that the 2 strongest bands near 3650 and 3540 cm^{-1} represented silanol groups formed by interaction of protons formed during deammination with the lattice oxygen, there has been much discussion as to the location of these groups. Since the 3650 cm^{-1} band interacts with many adsorbents, most investigators have located it in accessible parts of the structure. On the other hand, the 3550 cm^{-1} band does not interact with many molecules. Angell and Schaffer (5) suggested that the 3550 cm^{-1} band was caused by interaction between 2 neighboring hydroxyl groups, possibly through hydrogen bonding. Hall and coworkers (38, 56) suggested that the 2 bands represented groups in different locations, such as in the sodalite cages and in the supercages. White et al. reached a similar conclusion from the study of physically adsorbed molecules (73). Eberly (16) concluded initially that the 3550 cm^{-1} band represented hydroxyl groups inaccessible to large molecules, and these groups were located in the hexagonal prisms or near the S$_I$ positions in the structure. He also concluded that the 3650 cm^{-1} hydroxyl

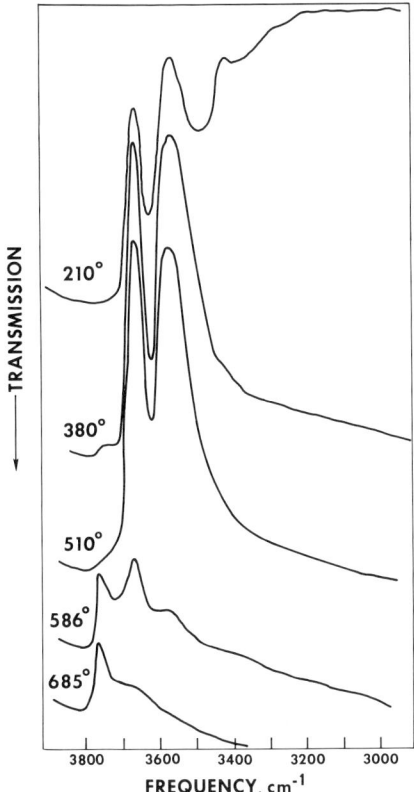

Figure 2. Spectra of hydroxyl group on ammonium (hydrogen) Y zeolite as a function of calcination temperature

groups represented hydroxyl groups near the S_{II} positions. Because both types of hydroxyl groups interact with piperidine, Hughes and White (31) concluded that the 2 types of hydroxyl group were accessible to large molecules and probably were located on the O_1 or O_4 oxygen atoms. Eberly (17), Uytterhoeven et al. (57), and Ward (61) reached similar conclusions. In contrast, Kermarec et al. (33) concluded that the 3650 cm^{-1} band was associated with S_I or S_{II} exchange positions and a band near 3580 cm^{-1} was associated with the S_{II} positions. Recently, Olson and Dempsey (46) reported single-crystal x-ray diffraction data for hydrogen Y zeolite and concluded that the original deductions of Eberly (16) were correct. They located the hydrogen as being attached to the O_1 and O_3 oxygen atoms. By studying the growth of the hydroxyl bands as a function of ion exchange and utilizing the ion exchange data of Sherry (51), Ward (70) provided additional spectroscopic evidence to support the conclusions of Eberly (16) and Olson et al. (46). This study showed that the 3650 cm⁻¹ band is formed on initial exchange, presumably by removal of the most accessible sodium ions, and that the 3550 cm⁻¹ band is formed only in the latter parts of the exchange. By forming the CsHY zeolite, it was shown that the 3650 cm⁻¹ band almost could be eliminated, whereas the 3550 cm⁻¹ was unchanged in comparison with hydrogen Y zeolite. Cesium ions are too large to enter the small-pore system of the zeolites through the 2.2–2.5A diameter windows. Hence, exchange of ammonium Y by cesium is occurring at the more accessible sites and not influencing the sites responsible for the formation of the 3550 cm⁻¹ hydroxyl groups. If these locations are correct, the 3550 cm⁻¹ hydroxyl groups should be inaccessible to most molecules. Although a number of molecules which interact with the 3650 cm⁻¹ band do not interact with the 3550 cm⁻¹ band, polar basic molecules such as piperidine, pyridine (31), ammonia (56), and cumene (66) do interact. This apparent contradiction can be rationalized in terms of proton mobility, as suggested for the piperidine interaction (46).

The question as to whether hydrogen on the surface of molecular sieves is mobile or localized has led to a number of investigations (12, 58, 62, 69). The studies have involved the deammination of NH_4^+Y zeolite at 450° to 500°C, followed by the study of the intensities and frequencies of the hydroxyl groups as a function of sample temperature. All of the detailed studies showed a decrease in integrated intensity of the 2 hydroxyl bands as the temperature was increased. The magnitude of the changes varied amongst the various studies. The 3650 cm⁻¹ band was more temperature-dependent than the 3550 cm⁻¹ band. The band frequencies also decreased with increasing temperature. These observations have been interpreted in terms of delocalization of the protons (58, 62, 69) and can be depicted by the following scheme.

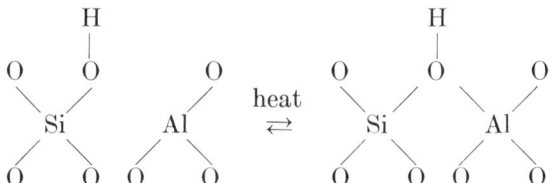

This type of equilibrium is supported by the known details of the dehydration of hydrogen Y and the reconstitution of ammonium Y zeolite. The decrease in frequencies with increasing temperature is probably an indication of the increasing interaction between neighboring atoms in the structure. The situation is not completely clear since if the proton is visualized as moving from oxygen to oxygen atom, the lifetime of a given OH band must be very short ($< 10^{-14}$ sec). This time scale led Cant and Hall (12) to consider that their observations did not represent delocalization. However, using wide-line magnetic resonance measurements, Dollish and Hall (15) demonstrated proton mobility. The time scale in such magnetic measurements is much longer than in infrared measurements. Whether or not infrared measurements of hydroxyl groups indicate proton delocalization, the reconciliation of hydroxyl group locations and interactions with adsorbates requires proton mobility (46, 68). Furthermore, the number of protons represented by the decrease in hydroxyl group intensities is of the same order of magnitude as the number of active sites estimated for these materials in acid-catalyzed reactions (62, 68). The deammination and dehydration of NH_4^+Y stabilized with Mg ions has been reported (65). The results were essentially the same as for ammonium Y except that the 3750 cm^{-1} band did not increase in intensity during the dehydration. This is expected from the increased thermal stability of the magnesium form.

In order to examine the surface during a reaction, hydrogen Y zeolite has been studied while the cracking of cumene was taking place on the surface (66). Studies were made between 250° and 450°C while a dilute stream of cumene in helium passed through the infrared cell and over the catalyst. The 3650 cm^{-1} band interacted at a lower temperature with the cumene than the 3550 cm^{-1} band. At 250°C, only the 3650 cm^{-1} band interacted. At the higher temperatures, the 3550 cm^{-1} hydroxyl groups are sufficiently activated or mobile to interact with the cumene and function as active centers.

Ammonium Y zeolite can be decomposed into 2 different products depending upon the calcination conditions used (34, 42). Conditions which allow rapid escape of ammonia and water favor the formation of the expected hydrogen Y zeolite, whereas calcination in an NH_3 or H_2O vapor atmosphere results in a product of enhanced stability—the so-called "ultrastable" zeolite. The 2 thermal decomposition products can be differentiated by means of their thermograms by differential thermal analysis, and by differences in lattice constants. It is therefore of interest to compare the spectra of hydroxyl groups on these materials. In Figure 3, a comparison is made between a hydrogen Y and 2 ultrastable Y zeolites prepared by the procedure of Hansford (25) using steaming temperatures of 500° and 650°C, respectively. In Figure 3, spectra are shown for samples prepared by the procedure of McDaniel and Maher (42), with the second calcination being at 600°, 700°, and 815°C, respectively. Comparison of the spectra of the hydrogen and stabilized Y zeolites indicate that there are fewer hydroxyl groups on the latter, particularly of the 3650 cm^{-1} type. The resolution of the spectral bands is poor for the stabilized zeolite, suggesting that there are hydroxyl groups in various

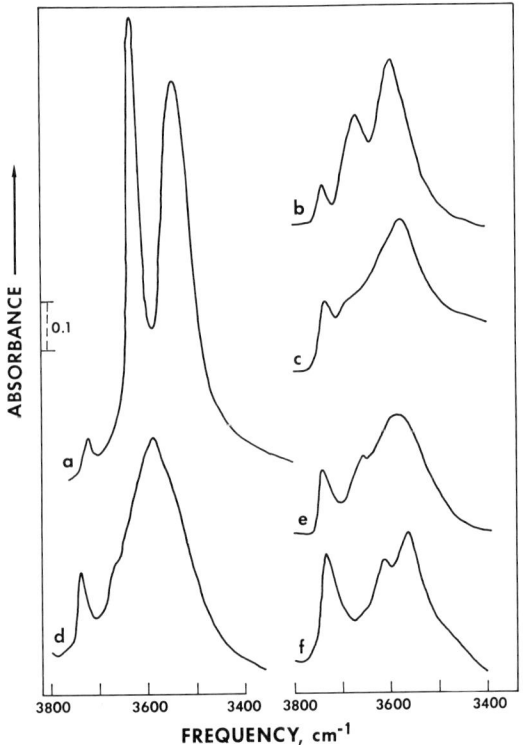

Figure 3. Spectra of hydrogen Y zeolite after evacuation at 500°C

(a) Simple evacuation
(b) Steamed at 500°C (32)
(c) Steamed at 650°C (32)
(d) Calcinated by procedure of Ref. 49; final calcination, 600°C
(e) Same as (d), but calcination at 700°C
(f) Same as (d), but calcination at 815°C

environments. Since Angell and Schaffer (5) and Uytterhoeven *et al.* (57) have noticed changes in the hydroxyl stretching bands using various calcination procedures, it is possible that some of their conditions produced ultrastable Y zeolite.

Spectra of hydrogen L zeolite are considerably different from those of X and Y, as shown by Figure 4 (69). On progressive heating, fairly sharp bands are observed at 3740 and 3630 cm^{-1} together with a very broad band centered at about 3200 cm^{-1}. After evacuation at 540°C, only the 3740 cm^{-1} band remained. No discrete hydroxyl bands were formed on readdition of water. The results suggest that extensive hydrogen-bonding of the surface hydroxyl groups must be involved. Since the broad absorption band appears at low degrees of exchange ($\approx 40\%$), it seems that hydroxyl groups in the large-pore structure must be involved. Similar but more intense spectra are observed at higher degrees of exchange ($> 80\%$).

Hence, the zeolite surface groups which are probably important as adsorption and catalytic sites are the structural hydroxyl groups, the tricoordinated aluminum sites, and the exchangeable cations.

Interaction of Zeolites with Water

Cation Zeolites. Bertsch and Habgood (9) studied the adsorption of small amounts of water on Group IA X zeolites. They observed a sharp band between 3720 and 3648 cm^{-1} depending on the cation and broad bands near 3400 and 3200 cm^{-1}. Because the frequencies depended

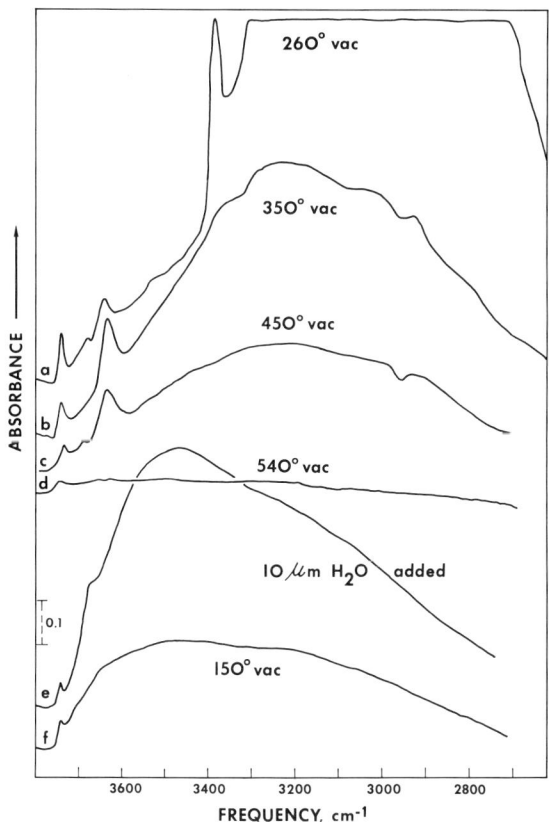

Figure 4. Spectra of OH stretching vibrations of ammonium (hydrogen) L zeolite

on the cation and the band intensities depended on the degree of hydration, these bands were considered to represent water directly bonded to the cation *via* the oxygen and to the oxygen ions of the surface by a hydrogen atom.

Zhdanov (78) showed that the hydroxyl frequencies decreased with increasing cation radius, indicating the polarizing influence of the cations. With divalent cation Y zeolites, a broad band appeared near 3540 cm^{-1} with little effect on the 3640 cm^{-1} band (5). Eberly (17) reported the formation of bands at 3650 and 3550 cm^{-1} when water was added to a well-dehydrated CaY zeolite. Ward (70) studied the readsorption of water on Group IA, IIA, and rare earth Y zeolites. For the group IA Y zeolites, the readsorption of water was similar to that reported for NaX—that is, a sharp cation-dependent band and several broad bands were observed. The frequency of the sharp band near 3720–3640 cm^{-1} varied in a systematic manner with the electrostatic field and potential of the zeolite and with the ionization potential of the cation. The water bending frequency near 1640 cm^{-1} decreased as the electrostatic and ionization potentials increased. The intensity of all the bands observed increased uniformly, indicating no preferential adsorption sites. The water is adsorbed in the same manner as that proposed by Bertsch and Habgood (9) and was completely reversible. The 3 broad bands were attributed to hydrogen of the water molecule bound to crystallographically different oxygen atoms or to adsorption of water on cations located in crystallographically different positions. Spectra observed during initial dehydration were different from those observed during rehydration and subsequent dehydration, suggesting that changes in position or coordination of the cations or extensive hydrogen bonding to the surface occurs. Because the band near 3690 cm^{-1} was dependent on the degree of hydration and only occurred in the sodium zeolite, it probably represented water bonded to sodium ions and not AlOH groups. Uytterhoeven *et al.* (59) reached a similar conclusion.

Spectra for water readsorbed on Group IIA Y zeolites were more complicated, except for barium which resembled the Group IA cation on rehydration behavior. When water is readsorbed on Group IIA zeolites below 100°C, bands near 3690 and 1640 cm^{-1} are observed. Upon heating to 200°C, these bands decrease in intensity and new bands are formed at 3640 and 3600–3570 cm^{-1}, depending on the cation. These observations can be interpreted in terms of the scheme —SiO— + M(OH$_2$)$^{2+}$ ⇌ MOH$^+$ + SiOH suggested above. The intensity of the band caused by MOH groups is very sensitive to the degree of hydration and temperature, and upon mild dehydration the band disappears, suggesting the formation of MO or M$^+$ —O—M$^+$ groups (44, 59, 70). In contrast, the 3640 cm^{-1} is relatively unchanged. When water was added back to barium zeolite, these hydroxyl bands were not observed, suggesting that the electrostatic field associated with the barium ion is insufficient to bring about the dissociation of adsorbed water. Similar results were obtained with transition ion zeolites (69).

Readsorption of water on rare earth Y zeolite did not give rise to a band at 3690 cm^{-1} but to bands at 3610 and near 3560–3550 cm^{-1}. Subsequent dehydration simply removed the water reversibly. If the surface hydroxyl groups are eliminated by evacuation at 700°C, readdition of water at room temperature does not result in the formation of new hydroxyl groups but subsequent heating at 200°C restores the 3640 and 3520 cm^{-1} bands (48). Zhdanov et al. (78) found that the frequency and ease of removal of adsorbed water depended on the silica-to-alumina ratio of the zeolite.

Several groups have studied the readsorption of water on hydrogen zeolites. Most of these studies have concerned the influence of water on the surface acidity of the zeolite.

Surface Acidity of Zeolites

Cation Zeolites. Since zeolites are important in acid-catalyzed reactions, much effort has been spent determining if zeolites are acidic and what the nature of the acidity is. Although many reactions catalyzed by zeolites are those catalyzed by amorphous silica–alumina, early theories of their action (49) suggested that the electrostatic field of the zeolites was responsible for their catalytic action. There is evidence to support this concept since the activity of zeolites in several reactions increases with the calculated electrostatic field (49, 50, 63). However, further studies have suggested that if the electrostatic fields are involved, their role may be to promote the formation of acid sites.

Most acidity studies have been made using basic molecules such as ammonia, pyridine, and piperidine as probes. These molecules have the property that their interaction with Bronsted acid sites, Lewis acid sites, and cations and their hydrogen-bonding interactions give rise to different species detectable by infrared spectroscopy. Thus, adsorption on Bronsted acid sites gives rise to ammonium, pyridinium, and piperidinium ions with characteristic absorption frequencies of 1475, 1545, and 1610 cm^{-1}, respectively. Adsorption on Lewis acid sites—tricoordinated aluminum ions—occurs *via* a coordinate bond $\mathrm{N{-}Al}$ and gives rise to bands near 1630, 1450, and 1450 cm^{-1}. The exchangeable cations can also act as a Lewis acid and give rise to similar bands. Fortunately, the frequencies for and strengths of adsorption on aluminum centers differ from those attributed to adsorption on cations. Similarly, hydrogen bonding can be separated by means of frequencies and ease of removal of adsorbed species.

Several studies of the surface acidity of cation zeolites have been made (13, 17, 26, 63, 72). In general, the group IA zeolites are nonacidic (13, 17, 63). Although pyridine is adsorbed on these zeolites, it is removed easily by evacuation at about 200°C. An absorption band near 1440–1450 cm^{-1} could be attributable to Lewis acidity. However, the frequency varies with cation and is always less than that observed for adsorption on Lewis acid sites of alumina, silica–alumina, and dehydroxylated molecular sieves (13, 63). Although some workers have attributed this absorption band to adsorption on Lewis acid sites, the band is probably caused by adsorption on cations, which, since they are electron acceptors, can function as Lewis acids in the general sense (13, 63, 72). The frequency of the band is a function of the ionic radius of the cation, electrostatic potential, and electrostatic field of the zeolite (13, 63). In mixed-cation zeolites, 2 bands often are observed, the relative intensities depending on the ratio of the cations. For sodium Y zeolite, addition of water resulted in an increase in the intensity of the band near 1438 cm^{-1} reflecting the cation interaction. It is possible that this change is caused by movement of cations into more accessible positions on hydration. No new Bronsted acid sites were generated. Watanabe and Habgood (72) detected Bronsted acidity by pyridine chemisorption on partially dehydrated NaY and on washed NaX samples. The Bronsted acidity might reflect cation deficiency but there were indications that possible metastable structures or cation mobility may be important factors. Bronsted acidity on washed group IA zeolites has been reported, presumably resulting from hydrolysis. The absence of Bronsted and Lewis acidity on pure Group IA zeolites is expected from the theoretical structure of the zeolite and the absence of hydroxyl groups. Studies of the alkaline earth zeolites have shown that there are Bronsted and/or Lewis acid sites on the surface depending on calcination conditions (13, 17, 26, 63). Eberly (17) used the spectra of pyridine adsorbed on Mg, Ca, and Cd Y zeolites to detect Bronsted and Lewis acidity and showed that addition of trace amounts of water increased the concentration of Bronsted acidity and decreased the concentration of Lewis acidity. At the same time, an OH band was observed near 3580 cm^{-1}. This band could represent CaOH$^+$ groups. The concentration of Bronsted acid sites varied with cation and seemed to be a function of the ionic charge-to-radius ratio. It was not clear whether the band at 1440–1450 cm^{-1} represented adsorption on Lewis acid sites or interaction with cations. The latter probably is occurring, at least in part, since the band is much stronger than on hydrogen zeolites calcined at the same temperature. Ward (63) studied the alkaline earth Y zeolites. After calcination at 500°C, no

Lewis acidity was observed. However, a band appeared near 1440–50 cm^{-1} whose frequency depended on the cation. It was attributed to pyridine coordinately bonded to the cations. The frequency and difficulty of removal increased with decreasing ionic radius and increasing electrostatic field and potential. This frequency change is similar to that reported for carbon monoxide (6) and carbon dioxide (7). The concentration of Bronsted acid sites also increased with increasing electrostatic field. On calcination at higher temperatures—e.g., 650°C—the concentration of Bronsted acid sites is less but Lewis acid sites are detected. Rehydration eliminates the Lewis acid sites and increases the concentration of Bronsted acid sites. New hydroxyl bands are also formed at 3595 and 3588 cm^{-1} for Mg and Ca, respectively. These bands are believed to reflect MOH$^+$ groups and do not interact with pyridine, indicating that they are non—or weakly—acidic. Christner et al. (13) observed Bronsted acidity and pyridine adsorbed on the cations in Mg, Ba, and ZnY zeolites. They also found that Bronsted acidity could be generated on Li, Mg, and Ba by addition of small amounts of carbon dioxide. This observation correlates with the work of Frilette and Munn (19), who found that the catalytic activity of NaY increased on addition of small amounts of carbon dioxide.

More recently, Hattori and Shiba (26) and Ward (68) studied the acidity of X zeolites. For Mg, Mn, and ZnX, Hattori and Shiba (26) report a small amount of Bronsted acidity and Lewis acidity which is too weak to be converted into Bronsted acidity by water. On Ca and SrX, they reported strong Lewis acidity which could be converted into Bronsted acidity. In contrast, Ward (68) observed Bronsted acidity on all Group II A zeolites, the concentration of sites increasing with decreasing cation radius or increasing field. The concentration of Bronsted acid sites was increased by hydration. No Lewis acid sites were detected, although pyridine interaction with the cations was observed. Ignat'eva et al. (32) reported the presence of Bronsted acid sites on CaY but not on NaY. Bronsted acidity but no Lewis acidity was observed on the transition metal ions, Mn, Co, Zn, Ag, Cd, but not on Cu (68). The concentration was increased in all cases by hydration. There appeared to be no relationship between the concentration of acid sites and the physical properties of the zeolites. Studies of the same series of transition metal ion Y zeolites yielded similar results (69).

Of the trivalent ions, cerium, lanthanum, and mixed rare earth X and Y zeolites have been studied (13, 47, 48, 67). The 3640 cm^{-1} band, but not the 3520 cm^{-1} band, reacts with pyridine and piperidine. In all cases, Bronsted acidity was detected. La, Ce and rare earth X zeolites also showed Lewis acidity after heating to 450°C, which was decreased by addition of water. Christner et al. (13) detected both Bronsted and

Lewis acidity on CeY zeolite after heating to 460°C, whereas Ward (67) detected only Bronsted acidity on rare earth Y zeolite after heating to 480°C but both forms of acidity after heating to 680°C.

Hydrogen Zeolites. Early studies of Uytterhoeven et al. (56) showed that addition of ammonia to hydrogen zeolite, deamminated by heating to about 400°C, eliminated the hydroxyl bands and reformed NH_4^+ ions. No absorption bands due to NH_3 adsorbed on Lewis acid sites were observed. The hydroxyl groups on the hydrogen X and Y zeolites are Bronsted acids and are accessible to molecules of the size and basic strength of ammonia. Heating of the zeolites to higher temperatures—e.g., 600°C—followed by NH_3 adsorption resulted in the detection of NH_3 adsorbed on Bronsted and Lewis acid sites. Readdition of a small amount of water reconverted some of the Lewis acid sites into Bronsted acid sites. On the other hand, after evacuation at 400°C, Geodakyan et al. (22) observed NH_3 adsorbed on both Lewis and Bronsted acid sites. Hughes and White (31) extended these results with an extensive study of the adsorption of pyridine and piperidine on ammonium Y zeolite calcined at temperatures between 300° and 700°C. Both the 3650 and 3550 cm^{-1} bands interacted with piperidine but only the 3650 cm^{-1} hydroxyl group protonated the weaker base, pyridine. These results demonstrate that both hydroxyl groups are acidic to the strong base, piperidine, but not to pyridine and show that the 3650 cm^{-1} band represents the stronger acidic groups. With increasing calcination temperature, the concentration of Bronsted acid sites decreased and the concentration of Lewis acid sites increased to a maximum near 600°C. Above this temperature, both site concentrations decreased, probably resulting from loss of structure. Readdition of water to the 650°C calcined zeolite reformed Bronsted acid sites at the expense of Lewis acid sites, but did not intensify the 3650 cm^{-1} band. Hughes and White concluded that the reformed Bronsted acid sites may be different from the original sites. Up to 600°C, the sum (Bronsted acid sites) + 2 (Lewis acid sites) remained constant, as would be expected from structural considerations. Zhdanov et al. (76) investigated NH_4X and Y after heating to 400°C. They detected both Bronsted and Lewis acid sites.

Ward (61) carried out a study similar to that of Hughes and White but simultaneously followed the concentration of hydroxyl groups and acid sites. The concentration of Bronsted acid sites, as detected by pyridine chemisorption, closely followed the concentration of the 3650 and 3550 cm^{-1} hydroxyl groups and is shown in Figure 5. The rapid decay in hydroxyl group concentration was accompanied by a decline in the Bronsted acidity and the buildup of Lewis acidity. Although the 3550 cm^{-1} hydroxyl groups did not form pyridinium ions, they did hydrogen bond at room temperature. Liengme and Hall (38) reported similar

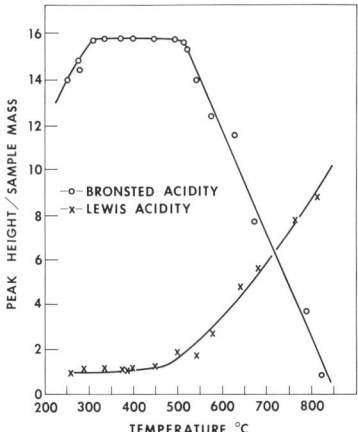

Figure 5. Acidity of hydrogen Y zeolite as a function of calcination temperature

results for HY calcined at 480° and 600°C. Eberly (17) carried out experiments in which the NH_4Y was calcined at 427°C and then the pyridine adsorption and spectral measurements were made at elevated temperature (150°–260°C). Under these conditions at 0.1 torr pressure of pyridine, the 3650 cm^{-1} band was eliminated and most of the acidity was of the Bronsted type. On evacuation at 260°C, some desorption of chemisorbed pyridine occurs, as indicated by the decrease in intensity of the 1545 cm^{-1} band and a partial restoration of the 3650 cm^{-1} hydroxyl band. This suggests that the 3650 cm^{-1} band may represent hydroxyl groups of different acid strengths. The difference in acidity of the 2 major hydroxyl groups is shown further by the adsorption of olefins (17, 38) and hexane (38), which interact with the 3650 cm^{-1} band but not the 3550 cm^{-1} band. In order to clarify the behavior at high calcination temperatures, Ward studied a magnesium-stabilized Y zeolite which showed no loss of structure at temperatures up to 800°C (64). The basic results were essentially the same. The major exceptions were that the sum (Bronsted acid sites) + 2 (Lewis acid sites) remained constant up to 800°C and 3740 cm^{-1} band did not increase in intensity above 600°C, confirming that structural decay did not occur. Ammonium X zeolite has both types of acid sites after calcination at 270°, 300° (26), and 480°C (68). However, structural collapse occurred. After calcination at 450°C, NH_4 L zeolite possesses both Bronsted and Lewis acid sites (69). Treatment of both X and L zeolites with water converted Lewis acid sites into Bronsted sites.

Ward et al. (65, 71) reported on the NaHY and MgHY systems. The 2 systems behave as a mixture of the 2 components. Sodium hydrogen Y was studied as a function of the sodium content from 100 to 0% exchanged (71). For the calcination conditions used (480°C), the series of samples were all in the Bronsted acid form. The concentration of acid sites detected by pyridine chemisorption increased linearly as the sodium was removed until only about 16 sodium ions were left. This corresponds to the number of easily removed sodium ions which can be exchanged with ammonium ions and subsequently form hydroxyl groups of the 3650 cm^{-1} type. Further exchange yields hydroxyl groups in the small-pore structure which are not accessible and/or not sufficiently acidic to react with pyridine.

For all cases studied, the Y zeolites are considerably more acidic than the X zeolites. This may be unexpected since there are more ion exchangeable cations in X zeolites and consequently more potential acid sites. It is possible that the weaker electrostatic fields in X zeolites do not result in the production of the maximum number of hydroxyl groups. In fact, for Ca and Mg, the Bronsted acidity increased linearly with decreasing aluminum content of the zeolite (69). Quinoline chemisorption also has been used. However, the results indicate that only Lewis acid sites are detected (10).

Pyridine chemisorption on hydrogen mordenite has been studied (11). As the calcination temperature was increased from 200° to 600°C, the population of Bronsted acid sites decreased and of Lewis acid sites increased. Above 300°C, 2 absorption bands (1462 and 1455 cm^{-1}) were observed, suggesting the presence of 2 types of Lewis acid sites. Cannings suggested the following structures:

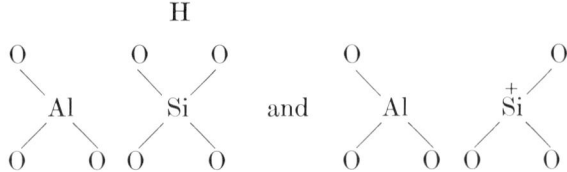

The same structures would be expected in X and Y zeolites but only 1 absorption band is observed. The Lewis acidity could be converted into Bronsted acidity by water. The total acidity decreased markedly on calcination at 600°C, suggesting a collapse or major change of structure.

In summary, Group IA zeolites are nonacidic. Other zeolites examined so far are Bronsted and/or Lewis acids, depending on the calcination temperature. The Bronsted acid sites on the cation and hydrogen zeolites are the same and are believed to be the 3650 and 3550 cm^{-1} hydroxyl groups. The 2 bands represent acid sites of different strengths.

The Lewis acid sites are also the same, possibly tricoordinated aluminum atoms. The electrostatic field appears to promote the formation of Bronsted acid sites on the cation zeolites by the reaction $M(OH_2)^{2+} \rightleftharpoons MOH^+ + H^+$ as discussed above.

Interaction of Zeolites with Inorganic Molecules

The study of carbon monoxide adsorption by Angell and Schaffer (6) has been discussed elsewhere (7, 74, 75). An interesting observation not previously emphasized is the appearance of cation specific absorption bands in the spectra of the multivalent cation zeolites because of the CO interacting with the exchanged cation *via* the carbon atom. They showed that the band frequency was a function of the electrostatic field strength which polarized the CO molecules.

Because the CO molecule is unable to enter the hexagonal prisms of the structure and since cation-specific CO absorption bands occur, the spectra of adsorbed CO can be used to give information about cation locations in zeolites. Thus, for NaY zeolite 35% ion exchanged with Ca, no CO-cation band was observed, indicating that all the Ca ions were located preferentially in the inaccessible hexagonal prism sites. In contrast, zeolites 20–22% exchanged with Co and Ni gave rise to cation-specific bands, indicating nonselective exchange. X-ray diffraction studies confirmed these findings (45).

Recently, Fenelon and Rubalcava (18) studied the interaction of CO with Na and Ca A and X zeolites at pressures of about 10 torr. Using isotopic CO and by analyzing the absorption band contours of the adsorbed species, gaseous and liquid CO, they concluded that the CO molecules freely rotate in the Na zeolites until they collide with the cage walls. For Ca zeolites, the absorption bands of adsorbed CO indicate strongly hindered rotation. This is plausible since CO adsorbs preferentially on the multivalent ions and is held more strongly than on univalent ions.

The adsorption of carbon dioxide on X and Y zeolites has been reviewed previously (7, 74, 75). Part of the CO_2 is held by an ion–dipole interaction while part of it reacted to give a carbonate type species. Angell showed that the frequency of the former species was a linear function of the electrostatic field.

Recently, Angell and Howell (2) found that heating divalent cation zeolites to 500°C in air or CO_2, in contrast to vacuum, resulted in absorption bands in the 1600–1300 cm^{-1} region. Evacuation at 500°C had no effect. Heating at 700°C for long periods did not remove the bands, indicating high stability. Ni, Na, and AgY did not produce the absorption bands. The band frequencies suggest that unidentate carbonate species

($-O-CO_2$) are formed together with carbonate ions in some cases. The interaction is believed to result from the chemisorption of CO_2 in a bent configuration on surface oxygen atoms. Since the stable species are not formed on the monovalent cation zeolites, the divalent cations must promote the chemisorption. Furthermore, since the frequencies observed are a function of the exchanged cation, the adsorption must occur in the vicinity of the cations. Adsorption of water and ammonia removed the 1575 and 1380 cm^{-1} bands and formed bands at 1620 and 1450 cm^{-1}, respectively. The carbonate bands were restored by evacuation at 300°C. Benzene had no effect. The observations are believed to reflect hydrogen bonding.

Addition of hydrogen chloride to fully dehydrated Mg zeolite results in the formation of hydroxyl bands with frequencies at 3643 and 3533 cm^{-1} (5). These bands are the same as those present on decationized Y zeolite, showing the reaction of HCl with the lattice to give hydroxyl groups. A similar reaction occurred with DCl, and bands appeared at 2684 and 2605 cm^{-1}. No band was formed near 1605 cm^{-1}, indicating that water was not produced. Similar reactions occurred with hydrogen cyanide.

Interaction of Zeolites with Sulfur-Containing Molecules

In order to characterize the types of acid sites, Lygin *et al.* (41) studied the adsorption of thiophene on Ca and decationized Y zeolites heated in air at 550°C and then in vacuum at 400°C. Thiophene adsorbs on the zeolite with partial decomposition, and strongly-held chemisorbed species are formed. Methyl, methylene, and CH-groups are detected up to 400°C. Bands indicative of C=S and thiophene rings are observed up to 300°C. The similarity of the spectra of thiophene chemisorbed on alumina and HY zeolites suggest similar adsorption centers on the 2 surfaces, namely Lewis acid sites.

As part of a study of the mechanism of and intermediates in the Claus Reaction, Deo, Dalla Lana, and Habgood (14) studied the adsorption of H_2S and SO_2 on Na and HY zeolites. H_2S at 98 torr was physically adsorbed on NaY zeolite. A small amount of water was formed, possibly by oxidation of the H_2S by chemisorbed O_2. On heating to 400°C, more water was formed. The H_2S could be removed by room-temperature evacuation. Hydrogen bonding to the hydroxyl groups of HY occurred but only deuterium exchange of the 2689 cm^{-1} OD (3650 OH) band occurred. The 3650 cm^{-1} band represents the most acidic and accessible OH group and hence the groups involved in the strongest hydrogen bonding interaction. No evidence of chemisorption or oxidation of H_2S was detected on HY zeolite. Adsorption of sulfur dioxide on NaY produced a single band at 1330 cm^{-1} owing to physically adsorbed SO_2. No change occurred on heating to 400°C. When H_2S was added, a rapid reaction occurred, producing water. Adsorption of SO_2 on HY resulted in hydrogen bonding to the 3650 cm^{-1} groups. No chemisorbed species were detected. Hence, the role of the zeolite must be to bring the molecules together.

Interaction of Zeolites with Organic Molecules

Geodakyan *et al.* (*21*) studied the adsorption of acetaldehyde and acetone on Group IA and IIA X zeolites. The C=O frequency decreased with increasing cation radius from Li to Cs and from Ca to Sr, indicating that the adsorption involves interaction of the C=O group with the cations. A second band, observed in the case of acetone and removed by evacuation at 200°C, is believed to indicate interaction with the zeolite lattice. Both molecules hydrogen bond to the 3650 cm^{-1} band of hydrogen Y zeolite. Adsorbed acetone has bands at 1715 (hydrogen bonded C=O) and a pair of bands at 1640 and 1575 cm^{-1} possibly caused by adsorption on dehydroxylated sites. After evacuation of the acetone, bands remain at 1455 and 1410 cm^{-1} owing to the formation of a carbonate structure. Apart from bands resulting from physically adsorbed acetaldehyde, an absorption band showing C=O vibrations reflects interaction of the CO group with aprotic centers on the zeolite surface. Removal of the physically adsorbed acetaldehyde by evacuation leaves a surface species with bands near 1645, 1600, and 1460 cm^{-1}. It appears that a surface reaction is occurring in which carboxylate groups are formed by interaction with the lattice oxygen. However, the band frequencies are removed somewhat from those observed when acetic acid is adsorbed on NaX zeolite. Phenol, methanol, and aniline also adsorb by means of interaction of the lone pair electrons of the OH and NH_2 groups with the exchangeable cations or the structural hydroxyl groups (*35, 37, 74, 75*). In contrast, nitrobenzene interacts *via* the aromatic ring. The results can be interpreted in terms of electron availability (*36*).

Hydrogen cyanide, acrylonitrile, and acetonitrile adsorption on NaX and HY have been observed (*20*). Stronger interactions occur on the zeolite than on silica since the frequency shifts are greater. The nitriles interact with the cations in NaX and with the hydroxyl groups and dehydroxylated sites on HY zeolite. Acetonitrile, acetonitrile-d_3, and benzonitrile have been adsorbed on various cation and HY zeolites (*4*). For acetonitrile, the C≡N bond frequency is higher than that of the liquid phase and varies with cation, indicating that the adsorption of molecules is associated with the cation. A linear correlation is found between the cation electrostatic field and the CN bond frequency.

Acetonitrile interacts, *via* the CN group, with the 3650 but not the 3550 cm^{-1} hydroxyl groups of HY zeolite. Benzonitrile behaved similarly. Acetonitrile interacts with the 3650 but not the 3520 cm^{-1} hydroxyl groups of LaY, indicating that the lanthanum ions are inaccessible.

Abramov, Kiselev, and Lygin (*1*) first reported the adsorption of benzene on Na and CaX after vacuum calcination at 450°C. They observed that the C—C stretch vibration near 1486 cm^{-1} was much more intense than the C—H stretch vibrations. Several other bands changed in position and intensity relative to the liquid phase. The spectra were interpreted in terms of benzene adsorbed on the walls of the large cavities and the changes in intensity and position were attributed to the changes in the electron distribution by the zeolite electrostatic field. Ultraviolet spectra confirmed that the adsorption involved the π electron system of adsorbed benzene (*36*). $C_6H_6H^+$ species were formed by irradiation of adsorbed benzene on HY zeolite. Angell and Schaffer (*3*) found that benzene hydrogen-bonded to 3650 cm^{-1} OH groups of Ni and MgY. Similar spectra were found when benzene and benzene-d_6 were adsorbed on Mg, Ni, Zn, Ag, La, Ce, and CoY zeolites (*3*). The CH stretching frequencies were the same as those for liquid benzene. Examination showed that frequencies due to vibrations in the plane of the ring were unchanged while those due to out-of-plane vibrations were shifted to higher frequencies. The spectra were interpreted in terms of interaction between the surface and the π orbitals of benzene, assuming the molecule adsorbed parallel to the surface. Toluene behaved similarly. Benzene, cyclohexane, cyclohexene, and cumene have been studied recently on hydrogen Y zeolite (*53*). Benzene could be desorbed at 200°C and cyclohexane at room temperature. No H—D exchange was observed when either molecule was adsorbed on DY. Absorption bands caused by =CH and C=C groups were absent from the spectra of adsorbed cyclohexene which resembled the spectrum of liquid cyclohexane. H—D exchange occurred on all 3 OD groups. Adsorption of cumene resulted in strong interaction with the 3650 cm^{-1} OH groups. At 150°C or higher, H—D exchange of the isopropyl group occurs. The isopropyl group underwent larger perturbations than the aromatic ring on adsorption. The adsorption of olefins has been studied by several workers (*7, 74, 75*). Spectroscopic studies indicate that adsorption occurs by interaction of the π-electron system of the double bond with the cations or structural hydroxyl groups. These interpretations are supported by magnetic resonance methods (*43*).

Recently, in order to gain an insight into the catalytic reaction, Tempere *et al.* (*52*) studied the interaction of butene-1 with A, X, and Y zeolites. A relationship exists between the activity for butene-1 isomerization and an absorption band near 3570–3610 cm^{-1}. These hydroxyl groups were produced by exchange of the monovalent zeolite with divalent ions or by washing the zeolite with water. The properties of these hydroxyl groups seem to indicate that they are the same as those reported by others with frequencies of 3630–3650 cm^{-1}.

Adsorption of acetylene on Co, Ni, Mn, Ca, Na, and H A and X zeolites also indicate a charge transfer mechanism of adsorption (54).

Possible Relationships Between Catalytic Activity and Surface Acidity

Correlations of catalytic activity and zeolite properties have been discussed in terms of electrostatic fields, Bronsted acidity, and Lewis acidity (49, 55, 60). In many cases, the combination of spectroscopic and reactor data has been considered. Since bases such as ammonia, pyridine, and quinoline (10, 55) poison the catalysts, it seems that acid sites are involved. For many reactions over NH_4Y catalysts, the activity decreases with increasing Lewis acid concentrations as the calcination temperature is increased, suggesting that Lewis acidity probably is not the primary active site (8, 26, 27, 30, 60, 61). It also has been shown spectroscopically that cation zeolites are not Lewis acids and that the activity is promoted by proton donors. On the other hand, spectral studies show that active catalysts contain structural hydroxyl groups and are Bronsted acids. In general, it appears that as the hydroxyl group concentration increases, the Bronsted acidity and catalytic activity for both cation and hydrogen zeolites increase. For cation zeolites, several reactions correlate with Bronsted acidity concentrations (26, 48, 50, 52, 60, 63, 67, 68, 71). However, there are a number of exceptions to this simple picture, particularly with transition elements (68). Furthermore, the maximum activity of hydrogen Y zeolite does not occur at the maximum Bronsted acidity or hydroxyl content but after some dehydroxylation has occurred (28, 30, 55). It is possible that only certain acid site strengths are important and that these strengths are influenced by the cations or dehydroxylated sites (29, 30, 40, 50, 67, 68, 71). Alternatively, a dual site mechanism may be involved, utilizing both Bronsted and Lewis acid sites (67, 71).

Although the electrostatic field may be involved, it appears that its role is to dissociate water so as to form acidic protons and, at higher temperatures, dehydroxylated sites (59, 61).

Literature Cited

(1) Abramov, V. N., Kiselev, A. V., Lygin, V. I., *Russ. J. Phys. Chem.* **1963**, 37, 613.
(2) Angell, C. L., Howell, M. V., *Can. J. Chem.* **1969**, 47, 3831.
(3) Angell, C. L., Howell, M. V., *J. Colloid Interface Sci.* **1968**, 28, 279.

(4) Angell, C. L., Howell, M. V., *J. Phys. Chem.* **1969**, 73, 2551.
(5) Angell, C. L., Schaffer, P. C., *J. Phys. Chem.* **1965**, 69, 3463.
(6) *Ibid.*, **1966**, 70, 1413.
(7) Basila, M. R., *Appl. Spectry. Rev.* **1968**, 1, 289.
(8) Benesi, H. A., *J. Catalysis* **1967**, 8, 368.
(9) Bertsch, L., Habgood, H. W., *J. Phys. Chem.* **1963**, 67, 1621.
(10) Boreskova, E. G., Lygin, V. I., Topchieva, K. V., *Kinetics Catalysis* **1964**, 5, 991.
(11) Cannings, F. R., *J. Phys. Chem.* **1968**, 72, 4691.
(12) Cant, N. W., Hall, W. K., *Trans. Faraday Soc.* **1968**, 64, 1093.
(13) Christner, L. G., Liengme, B. V., Hall, W. K., *Trans. Faraday Soc.* **1968**, 64, 1679.
(14) Deo, A. V., Dalla Lana, I. G., Habgood, H. W., *Canadian Chemical Engineering Conference, 19th, Edmonton, Alberta*, October 1969.
(15) Dollish, F. R., Hall, W. K., unpublished results.
(16) Eberly, P. E., Jr., *J. Phys. Chem.* **1967**, 71, 1717.
(17) *Ibid.*, **1968**, 72, 1042.
(18) Fenelon, P. J., Rubalcava, H. E., *J. Chem. Phys.* **1969**, 51, 961.
(19) Frilette, V. J., Munns, G. W., *J. Catalysis* **1965**, 4, 504.
(20) Geodokyan, K. T., Kiselev, A. V., Lygin, V. I., *Russ. J. Phys. Chem.* **1966**, 40, 857.
(21) *Ibid.*, **1967**, 41, 476.
(22) *Ibid.*, **1969**, 43, 106.
(23) Habgood, H. W., *J. Phys. Chem.* **1965**, 69, 1764.
(24) Hair, M. L., "Infrared Spectra in Surface Chemistry," Ch. 5, Dekker, New York, 1967.
(25) Hansford, R. C., U. S. Patent **3,354,077** (November 21, 1967).
(26) Hattori, H., Shiba, T., *J. Catalysis* **1968**, 12, 111.
(27) Hickson, D. A., Csicsery, S. M., *J. Catalysis* **1968**, 10, 27.
(28) Hildebrandt, R. A., Skala, H., *J. Catalysis* **1968**, 12, 61.
(29) Hirschler, A. E., *J. Catalysis* **1963**, 2, 428.
(30) Hopkins, P. D., *J. Catalysis* **1968**, 12, 325.
(31) Hughes, T. R., White, H. M., *J. Phys. Chem.* **1967**, 71, 2192.
(32) Ignat'eva, L. A., Moskovskayan, I. F., Oppengeim, V. D., Spozhakima, A. A., Topchieva, K. V., *Kinetics Catalysis* **1968**, 9, 111.
(33) Kermarec, J., Tempere, J. F., Imelik, B., *Bull. Soc. Chim. France* **1969**, 3792.
(34) Kerr, G. T., *J. Catalysis* **1969**, 15, 200.
(35) Kiselev, A. V., Kubelkova, L., Lygin, V. I., *Russ. J. Phys. Chem.* **1964**, 38, 1480.
(36) Kiselev, A. V., Kupcha, L. A., Lygin, V. I., *Kinetics Catalysis* **1966**, 7, 621.
(37) Kiselev, A. V., Lygin, V. I., Starodubcseva, R. V., *Kolloidn. Zh.* **1969**, 31, 68.
(38) Liengme, B. V., Hall, W. K., *Trans. Faraday Soc.* **1966**, 62, 3229.
(39) Little, L. H., "Infrared Spectra of Adsorbed Species," Ch. 14, Academic, New York, 1966.
(40) Lunsford, J. H., *J. Phys. Chem.* **1968**, 72, 4163.
(41) Lygin, V. I., Romanovskii, B. V., Topchieva, K. V., Tkhoang, K. S., *Russ. J. Phys. Chem.* **1968**, 42, 156.
(42) McDaniel, C. V., Maher, P. K., "Molecular Sieves," p. 186, Society of the Chemical Industry, London, 1968.
(43) Muha, G. M., Yates, D. J. C., *J. Chem. Phys.* **1968**, 49, 5073.
(44) Olson, D. H., *J. Phys. Chem.* **1968**, 72, 1400.
(45) *Ibid.*, **1968**, 72, 4366.
(46) Olson, D. H., Dempsey, E., *J. Catalysis* **1969**, 13, 221.
(47) Rabo, J. A., Angell, C. L., Kasai, P. H., Schomaker, V., *Discussions Faraday Soc.* **1966**, 41, 328.

(48) Rabo, J. A., Angell, C. L., Schomaker, V., *Intern. Congr. Catalysis, 4th, Moscow,* **1968**, Preprint 54.
(49) Rabo, J. A., Pickert, P. E., Stamires, D. N., Boyle, J. E., *Actes Congr. Intern. Catalyse, 2nd, Paris, 1960,* **1961**, 2055.
(50) Richardson, J. T., *J. Catalysis* **1967**, 9, 182.
(51) Sherry, H. S., *J. Phys. Chem.* **1966**, 70, 1158.
(52) Tempere, J. F., Kermarec, J., Imelik, B., *Compt. Rend.* **1969**, 269, 77.
(53) Topchieva, K. V., Kubasov, A. A., Ratov, A. N., *Dokl. Akad. Nauk SSSR* **1969**, 184, 383.
(54) Tsitsishvili, G. V., Bagratishvili, G. D., Oniashvili, N. I., *Zh. Fiz. Khim.* **1969**, 43, 950.
(55) Turkevich, J., Nozaki, F., Stamires, D. N., *Proc. Intern. Congr. Catalysis, Amsterdam, 1964,* **1965**, 1, 586.
(56) Uytterhoeven, J. B., Christner, L. G., Hall, W. K., *J. Phys. Chem.* **1965**, 69, 2117.
(57) Uytterhoeven, J. B., Jacobs, P., Makay, K., Schoonheydt, R., *J. Phys. Chem.* **1968**, 72, 1768.
(58) Uytterhoeven, J. B., Schoonheydt, R., Fripiat, J. J., *Intern. Symp. Reaction Mechanisms of Inorganic Solids, Aberdeen, 1966.*
(59) Uytterhoeven, J. B., Schoonheydt, R., Liengme, B. V., Hall, W. K., *J. Catalysis* **1969**, 13, 425.
(60) Venuto, P. B., Landis, P. S., *Advan. Catalysis* **1968**, 18, 259.
(61) Ward, J. W., *J. Catalysis* **1967**, 9, 225.
(62) *Ibid.,* **1967**, 9, 396.
(63) *Ibid.,* **1968**, 10, 34.
(64) *Ibid.,* **1968**, 11, 238.
(65) *Ibid.,* **1968**, 11, 251.
(66) *Ibid.,* **1968**, 11, 259.
(67) *Ibid.,* **1969**, 13, 321.
(68) *Ibid.,* **1969**, 14, 365.
(69) *Ibid.,* in press.
(70) Ward, J. W., *J. Phys. Chem.* **1968**, 72, 4211.
(71) Ward, J. W., Hansford, R. C., *J. Catalysis* **1969**, 13, 364.
(72) Watanabe, Y., Habgood, H. W., *J. Phys. Chem.* **1968**, 72, 3066.
(73) White, J. L., Jelli, A. W., Andre, J. M., Fripiat, J. J., *Trans. Faraday Soc.* **1967**, 63, 461.
(74) Yates, D. J. C., *Catalysis Rev.* **1968**, 2, 113.
(75) Yates, D. J. C., "Molecular Sieves," p. 334, Society of the Chemical Industry, London, 1968.
(76) Zhdanov, S. P., Kiselev, A. V., Lygin, V. I., *Russ. J. Phys. Chem.* **1966**, 40, 560.
(77) Zhdanov, S. P., Kiselev, A. V., Lygin, V. I., Ovepyan, M. E., Titova, T. I., *Russ. J. Phys. Chem.* **1965**, 39, 1309.
(78) Zhdanov, S. P., Kiselev, A. V., Lygin, V. I., Titova, T. I., *Russ. J. Phys. Chem.* **1964**, 38, 1299.

RECEIVED February 4, 1970.

Discussion

G. V. Tsitsishvili (Academy of Sciences of the Georgian SSR, Tbilisi, USSR): In principle, hydrolithic ion exchange when a sodium ion is replaced by an ion $[Me^{n+}(OH)_{n-1}]^+$ is possible, but then one should expect an increased content of Me^{n+} in the composition of a unit cell of a zeolite at the transition from NaZ to MeNaZ. Indeed, for calcium faujasite (*see* data of J. K. Bennet and J. V. Smyth, *Mater. Res. Bull.* **1968,** 3, N8, 633), the composition is given by the formula $Ca_{27}Al_{57}Si_{123}$...; that is, for each two atoms of aluminum (an atom of aluminum is a carrier of one negative charge) there is one Ca^{2+}.

J. W. Ward: One is replacing, in most cases, two positive charges by two positive charges which satisfy two negative charges. With divalent ions, whether one introduces Ca^{2+} or $Ca(OH)^+ + H^+$, the same amount of calcium is introduced so that electrical neutrality is maintained. Hence, two exchangeable sodium ions (or other monovalent ions) are replaced by one calcium ion. In general, n Na$^+$ ions are replaced by one M^{n+} ion.

J. B. Uytterhoeven (University Leuven, 3030 Heverlee, Belgium): Is the OH band at 3600 cm^{-1} in ultrastable zeolite acidic, and have you an assignment for that band?

J. W. Ward: In another paper (*J. Catalysis* **1970,** 18, 348), we assign tentatively the absorption band to $Al(OH)_3$, $Al(OH)_2^+$, and/or $AlOH^{3+}$. The data do not seem to justify more detailed interpretation. The absorption bands in the spectra between approximately 3660 and 3620 cm^{-1} interact with adsorbed pyridine but the bands around 3600 cm^{-1} do not appear to interact extensively.

30

Kinetics of Ion Exchange from Partially Exchanged Starting Materials

N. M. BROOKE[1] and L. V. C. REES

Physical Chemistry Laboratories, Imperial College of Science and Technology, South Kensington, London, S.W.7, England

> *The rates of ion exchange from a mixed (Ca^{2+}/Sr^{2+})-chabazite starting material have been computed by a finite difference method, assuming the exchanger to be (a) ideal and (b) nonideal. In the latter case, the differential interdiffusion coefficient, D_{AB}, contains an activity correction term. The exchange kinetics are calculated for starting materials containing various ratios of Ca to Sr and the results are discussed in terms of the 2 theoretical models. The kinetics of exchange of partially exchanged starting materials can be used to throw light on the nonideality of the exchanger and to indicate heterogeneity among the cation exchange sites.*

The diffusion equation for a binary, particle diffusion controlled, ion-exchange system was solved by a finite difference method in 2 previous papers (4, 5) using an interdiffusion coefficient, D_{AB}, which includes activity gradient terms and is given by

$$D_{AB} = \frac{D_A^* D_B^* \left(C_A Z^2_A \frac{\partial \ln a_B}{\partial \ln C_B} + C_B Z^2_B \frac{\partial \ln a_A}{\partial \ln C_A} \right)}{D_A^* C_A Z^2_A + D_B^* C_B Z^2_B} \quad (1)$$

where D_A^*, D_B^* are the self-diffusion coefficients of the ions A and B in their respective pure forms of the exchanger; C_A, C_B are the concentrations of the ionic species in the exchanger phase; Z_A, Z_B are their valencies and a_A, a_B their activities, which may be functions of C_A and C_B.

---

[1] Present address: U.K.A.E.A., A.W.R.E., Aldermaston, Berkshire, England.

The activity terms $\partial \ln a/\partial \ln C$ allow for the nonideality of the exchanger material (3), but cross-coefficients L_{ij} where $i \neq j$ have been assumed to be zero and water fluxes have been neglected. The activity terms markedly alter the computed kinetics of exchange when 2 very different types of activity functions were investigated (4, 5). In these treatments, the kinetics have been computed taking the appropriate homoionic forms of the exchangers as starting materials. This paper considers the effect of starting with binary ionic forms of the exchanger. In order to investigate these kinetics, the Ca^{2+}/Sr^{2+}-chabazite system was chosen as a suitable test system. The kinetics were computed using interdiffusion coefficients appropriate to ideal (7) and nonideal (4) exchangers. Several ratios of Ca^{2+} to Sr^{2+} in the starting material were taken which covered the range pure Ca^{2+} to pure Sr^{2+}.

Theory and Computations

The diffusion equation to be solved for C_A in systems having spherical symmetry is

$$\frac{\partial C_A}{\partial \tau} = \frac{1}{\rho^2} \cdot \frac{\partial (\rho^2 \partial S)}{\partial \rho} \qquad (2)$$

where C_A is the concentration of the ion species initially resident in, but to be removed from, the exchanger phase (expressed in mole fractions)

$$\tau = \frac{\left(\int_0^1 D_{AB} dC_A\right) t}{r_0^2} \quad \text{and} \quad \rho = \frac{r}{r_0} \qquad (3a,b)$$

where t is time, r_0 is the radius of the exchanger particle, r is the distance from the center of the sphere, and (6)

$$S = \frac{\int_0^{C_A} D_{AB} dC_A}{\int_0^1 D_{AB} dC_A} \qquad (4)$$

The appropriate starting and boundary conditions for a system with infinite solution phase volume are

$$C_A = x, \; 0 \leqslant \rho \leqslant 1, \; \tau = 0 \qquad (5a)$$

$$C_A = 0, \; \rho = 1, \; \tau > 0 \qquad (5b)$$

and, due to symmetry,

$$\left(\frac{\partial S}{\partial \rho}\right)_{\rho=0} = 0 \qquad (5)$$

x represents the initial mole fraction composition of the exchanger.

Since the thermodynamic treatment used to obtain the activity terms (3) gives

$$\frac{\partial \ln a_A}{\partial \ln C_A} = \frac{\partial \ln a_B}{\partial \ln C_B} = g \tag{6}$$

and in the case of this test system

$$Z_A = Z_B \tag{7}$$

the interdiffusion coefficient, D_{AB}, simplifies to

$$D_{AB} = \frac{D_A^* D_B^*}{(D_A^* - D_B^*)C_A + D_B^*} \cdot g \tag{8}$$

For an ideal exchanger

$$g = 1, \; 0 \leq C_A \leq 1 \tag{9}$$

and for the nonideal test system, data taken from earlier work (1, 2) gives

$$\begin{aligned} g &= 0.27, \; 0 \leq C_{Ca} \leq 0.4 \\ g &= 2.50, \; 0.4 < C_{Ca} \leq 1.0 \end{aligned} \tag{10a,b}$$

In addition, $D_{Ca}^* = 1.15 \times 10^{-18}$ m² sec⁻¹; $D_{Sr}^* = 7.4 \times 10^{-19}$ m² sec⁻¹ when $T = 75.8°C$ and $r_o = 1.74$ μm.

To avoid any possible embarrassment with the cusp thus formed in the function $S(C_A)$, the factor g was taken to change linearly from the first value to the second over a range of ± 0.1 about the critical value of $C_{Ca} = 0.4$, as before (4). The solution of the partial differential Equation 2 by a finite-difference method and the integration of the function $D_{AB}(C)$ to obtain the parameter S of Equation 4 have been described (4). The quantity of the ionic species A remaining in the exchanger phase at time τ is given by

$$q_A(\tau) = 3 \int_0^1 C_A(\rho, \tau) \cdot \rho^2 \cdot d\rho \tag{11}$$

and the kinetics of the exchange process are represented as a function of τ (or t) by the fractional attainment of equilibrium

$$F(\tau) = \frac{q_A(\tau = 0) - q_A(\tau)}{q_A(\tau = 0) - q_A(\tau = \infty)} \tag{12}$$

where in this test system

$$q_A(\tau = 0) = x \text{ and } q_A(\tau = \infty) = 0 \tag{13a,b}$$

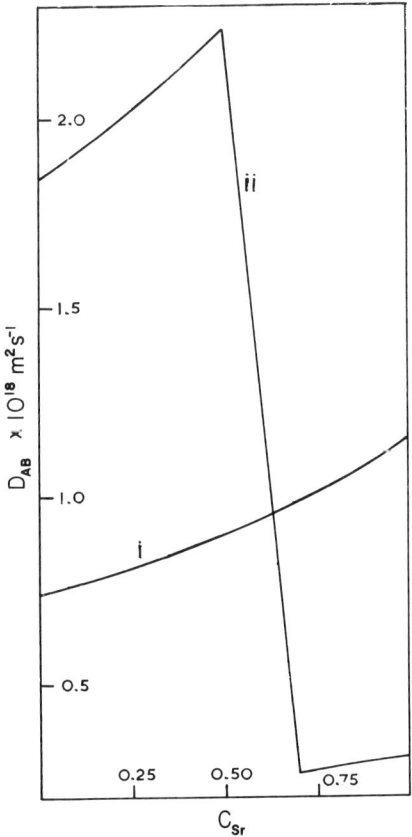

Figure 1. Variation of D_{AB} with composition for (i) an ideal exchanger and (ii) a nonideal exchanger

Results

The kinetics of the 2 exchange reactions

$$(Ca/Sr)\text{-Chabazite} + Sr^{2+} \rightarrow Sr\text{-Chabazite} + Ca^{2+}$$

and

$$(Ca/Sr)\text{-Chabazite} + Ca^{2+} \rightarrow Ca\text{-Chabazite} + Sr^{2+}$$

were computed for both ideal and nonideal exchangers.

The interdiffusion coefficients, D_{AB}, for both these exchangers is shown as a function of the composition in Figure 1. The rates of exchange for an ideal exchanger when C_{Ca} and C_{Sr} at $\tau = 0$ is 1.00, 0.75, 0.50, 0.25, 0.10, and 0.00, respectively, are shown in Figure 2 for each of the above

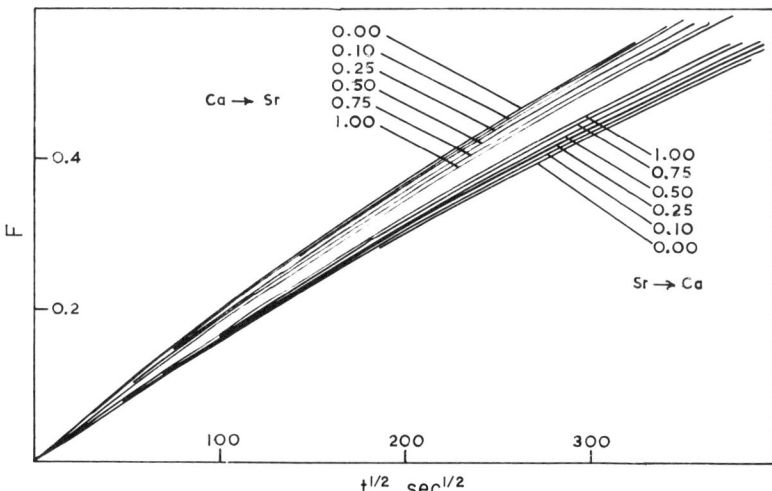

Figure 2. Fractional attainment of equilibrium, F, plotted against $\sqrt{t}$ for Ca and Sr exchange in (Ca/Sr)-chabazite considered here as an ideal exchanger; initial mole fraction concentration of the ion to be exchanged is indicated

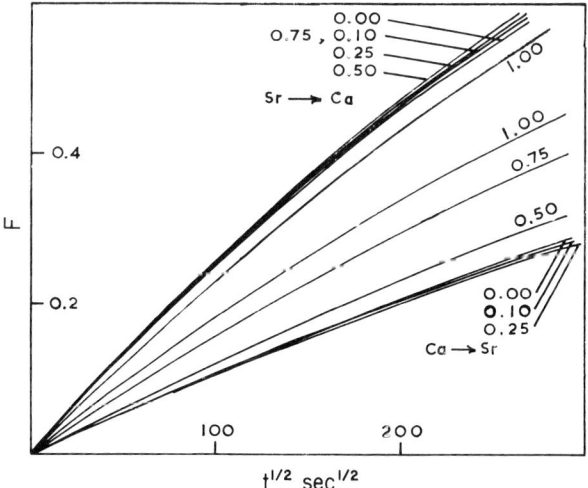

Figure 3. Effect of the introduction of the activity factors quoted in Equations 10 on the rate curves shown in Figure 2

2 exchange reactions. Figure 3 shows the exchange kinetics of the corresponding series of nonideal exchangers. The accuracy of the computations has been discussed previously (4).

Discussion

During the course of a binary ion exchange process, layers enriched with the incoming ionic species move inwards from the periphery while the central regions of the exchanger tend to retain their initial composition. The interdiffusion coefficient governing the kinetics, therefore, approaches the value appropriate to the pure incoming ionic form at the peripheries and retains the value appropriate to the starting form at the center of the particles. In an ideal system, the complete exchange A(Zeol) → B(Zeol) where (Zeol) represents the anionic exchanger framework is more rapid than the reverse exchange if $D_A^* > D_B^*$. Since the $\partial \ln a / \partial \ln C$ terms are unity in this case, it is easy to see from Equation 1 that

$$D_{AB} \rightarrow D_A^* \text{ as } C_A \rightarrow 0 \text{ and } D_{AB} \rightarrow D_B^* \text{ as } C_A \rightarrow 1$$

At the rate-controlling peripheral layers (7), therefore, A ions are replaced most rapidly because D_{AB} in these regions is approaching its maximum value of D_A^* since $C_A \rightarrow 0$. If $D_B^* > D_A^*$, the exchange rates are reversed since D_{AB} now increases with increasing C_A and the peripheral layers are the least readily depleted of the resident ionic species.

These results are illustrated in Figure 2 for the ideal (Ca/Sr)-chabazite test system where $D_{Ca}^* > D_{Sr}^*$. The complete exchange $\overline{Ca} \rightarrow \overline{Sr}$ (*i.e.*, C_{Ca} at $\tau = 0$ is 1) is faster than the complete reverse exchange (*i.e.*, C_{Sr} at $\tau = 0$ is 1). As the Ca^{2+} content of the starting form of the exchanger is reduced, the rate of the exchange $\overline{Ca/Sr} \rightarrow \overline{Sr}$ increases; conversely, the rate of the exchange $\overline{Ca/Sr} \rightarrow \overline{Ca}$ decreases as the Sr^{2+} content of the starting form is reduced. These changes in the rates result from the restricted range of D_{AB} applicable to the process. Figure 1 shows that as the initial Ca^{2+} content decreases, only larger and larger D_{AB} are involved in the exchange reaction. Thus, the replacement of Ca^{2+} ions by Sr^{2+} ions becomes increasingly faster at all points within the exchanger as its initial concentration is reduced, although the peripheral layers are still the ones which are most rapidly depleted of Ca^{2+}. The converse arguments apply to the exchange process $\overline{Ca/Sr} \rightarrow \overline{Ca}$. The maximum and minimum limits to these rates of exchange occur when exchangers containing only trace quantities of Ca^{2+} or Sr^{2+}, respectively, are used as starting materials. The trace quantities of Ca^{2+} and Sr^{2+} are removed at rates which are controlled by a D_{AB} which is equal to D_{Ca}^* for traces of Ca^{2+} and D_{Sr}^* for the traces of Sr^{2+}. Therefore, for an ideal exchanger, the rates of exchange for all possible starting compositions must remain within the limits set by the self-diffusion rates.

The interesting behavior of these exchange kinetics in the case of a nonideal exchanger (Figure 3) is a direct consequence of the more complex form of the interdiffusion coefficient as a function of the exchanger composition for this specific system. D_{AB} is not now a monotonic function like that for the ideal exchanger. The rate of the exchange $\overline{Ca/Sr} \rightarrow \overline{Sr}$ first decreases and subsequently increases as the Ca^{2+} content of the starting material is reduced; the $\overline{Ca/Sr} \rightarrow \overline{Ca}$ exchange rate first increases, then decreases with reducing Sr^{2+} content in the starting material. The over-all range in the rates of the $\overline{Ca/Sr} \rightarrow \overline{Sr}$ exchange is much greater than occurred with the ideal exchanger.

The $\overline{Ca/Sr} \rightarrow \overline{Sr}$ exchange for starting materials with $C_{Ca} < 0.3$ shows a steady increase in rate as the initial Ca^{2+} concentration decreases below this value. Similarly, the $\overline{Ca/Sr} \rightarrow \overline{Ca}$ exchange shows a steady decrease in rate with decreasing initial Sr^{2+} content below $C_{Sr} = 0.5$. In both these examples, Figure 1 shows that in these ranges of composition, the values of D_{AB} applicable to the exchange process form a monotonic function of the exchanger composition. In this respect, therefore, the exchange behaves similarly to that for an ideal exchanger. For all other starting compositions, the D_{AB} function applicable includes as its most significant feature the changeover region from one value of activity correction term to the other.

Among the effects of the inclusion of the activity terms in the interdiffusion coefficient is the prediction that the rate of the complete exchange $\overline{Ca} \rightarrow \overline{Sr}$ is slower than that of the complete exchange $\overline{Sr} \rightarrow \overline{Ca}$, despite the fact that $D_{Ca}^{*} > D_{Sr}^{*}$. This feature, which is in agreement with the experimentally determined rates, has been discussed previously (4) but is shown again in Figure 3. In the exchange $\overline{Ca/Sr} \rightarrow \overline{Sr}$, the rate of exchange decreases rapidly at first with decreasing Ca^{2+} content of the starting material. Figure 1 shows that as the Ca^{2+} content decreases, larger and larger fractions of the exchange process must occur in regions of low resident Ca^{2+} concentration where D_{AB} is low; i.e., the low D_{AB} values in the range $0 \leqslant C_{Ca} \leqslant 0.3$ become more and more important and rate controlling. When the starting material contains $C_{Ca} \leqslant 0.3$, D_{AB} increases slightly as the Ca^{2+} content decreases, and Figure 3 shows that in this range only a very slight increase in the rate of exchange occurs. In the reverse exchange, $\overline{Ca/Sr} \rightarrow \overline{Ca}$, the initial increase in the rate of exchange is much smaller as the composition of the starting material increases in Ca^{2+}. Figure 1 shows that D_{AB} is governed mainly by the high values which exist at $C_{Ca} > 0.5$. The low D_{AB} values which operate in the range $0 \leqslant C_{Ca} \leqslant 0.3$ have only a slight effect, and the removal of these D_{AB} values when the starting material contains $C_{Ca} > 0.3$ has only a small effect.

The activity correction terms obtained from the ion exchange equilibrium isotherm (2) suggest the presence of at least 2 different types of cation sites in chabazite. When the activity terms are included in the interdiffusion coefficient, the computed kinetics of the complete exchanges are in more sensible agreement with the experimental results than when the ideal model, which assumes uniform sites, is applied. Indeed, the predicted forward and backward rates of exchange are reversed (4). The extension of the computations to the partially exchanged starting material reveals further striking differences between ideal and nonideal models, and it is likely that experimental investigations of such exchanges could be used to reveal whether inhomogeneities in cation sites are likely to be present in a material or not. The behavior of these kinetics for a nonideal exchanger at varying degrees of initial exchanger composition also could be used to indicate the correctness of the postulated activity correction terms. Although difficult to carry out experimentally, the rate of removal of trace quantities of an ion, A, from exchangers containing only trace amount of A in the starting material could be used to establish nonideality in the exchanger, since D_{AB} now would not tend to be equal to $D_A{}^*$, and to evaluate the activity correction function in this concentration range since $D_{AB} \to D_A{}^* \frac{\partial \ln a_A}{\partial \ln C_A}$ (see Equation 1). Figure 3 shows an example of this latter type of behavior. For the chabazite system, $D_{AB} \to 2.5\ D_{Sr}{}^*$ as $C_{Ca} \to 1.0$ and $D_{AB} \to 0.27\ D_{Ca}{}^*$ as $C_{Ca} \to 0.0$.

In the treatment of nonideal exchanger where at least 2 different types of sites probably exist, it is assumed that during exchange the 2 cations involved in the exchange process assume the same distribution among the available sites as would occur under equilibrium conditions.

Conclusions

The initial purity of an ideal exchanger does not have a great effect on the kinetics of the ion exchange process if the self-diffusion coefficients of the 2 ions in the exchanger are similar in magnitude. The initial composition of a nonideal exchanger, however, can have a striking effect on the kinetics. No experimental results are available as yet for the direct testing of these predictions of the computations, but such results would be very interesting.

A series of results for exchange kinetics at varying initial exchanger compositions would be more revealing of the nonidealities of the exchanger than the results for the complete forward and backward exchanges of a binary system alone. Although it is not yet possible to invert the computations and use the partial exchange kinetics data as a means to set up activity correction term functions, if such an inversion could be accomplished, a most useful diagnostic method would become available with which to investigate the process of ion exchange.

Literature Cited

(1) Barrer, R. M., Bartholomew, R., Rees, L. V. C., *J. Phys. Chem. Solids* **1963**, 24, 51.
(2) *Ibid.*, **1963**, 24, 309.
(3) Barrer, R. M., Rees, L. V. C., *J. Phys. Chem. Solids* **1964**, 25, 1035.
(4) Brooke, N. M., Rees, L. V. C., *Trans. Faraday Soc.* **1968**, 64, 3383.
(5) *Ibid.*, **1969**, 65, 2728.
(6) Crank, J., "Mathematics of Diffusion," Oxford Univ. Press, London, 1956.
(7) Helfferich, F., Plesset, M. S., *J. Chem. Phys.* **1958**, 28, 418.

RECEIVED February 4, 1970.

Discussion

D. J. C. Yates (Esso Research Co., Linden, N. J. 07036): It is interesting that you find such a large difference in calcium and strontium exchange in chabazite. Would you care to comment on the reason for this?

L. V. C. Rees: In chabazite there are probably at least two sets of sites—one with preference for Ca^{2+} and the other for Sr^{2+}. These calculations show a method of gaining information about the cation site selectivities. One could guess that Ca or Sr prefers a site in the hexagonal prism, while the other ion prefers another site, say in the large cavity. By studying partially exchanged starting material kinetics, some estimation can be made of the ratio of the number of these two sites.

H. S. Sherry (Mobil Research & Development Corp., Princeton, N. J. 08540): From the chabazite ion exchange kinetics, you deduce that two kinds of sites must be involved and that one of these sites is in the hexagonal prism. I believe that J. V. S. Smith and coworkers of the University of Chicago showed that the site in the hexagonal prism is occupied only after dehydration of hydrated CaA. This site is empty in the hydrated zeolite and thus is not likely to be involved in Sr-Na ion exchange.

L. V. C. Rees: I seem to remember that Smith showed that Ca^{2+} ions are contained in the hexagonal prism of chabazite when hydrated and if so, my explanation could apply. However, if this is not correct, these calculations show the effect of at least two different sites in chabazite—one which prefers Ca^{2+} ions and the other selective toward Sr^{2+}.

31

Properties of Linde A in Aqueous, Nonaqueous, and Mixed Media

R. B. BARRETT,[1] J. A. MARINSKY, and P. PAVELICH

The State University of New York, Buffalo, N. Y. 14214

> *The solvent and ion selectivity behavior of Na^+ and K^+ forms of Linde A have been measured in nonaqueous and mixed media. An osmotic pressure model accurately describes ion exchange in concentrated aqueous electrolyte solutions and is extended to solvent selectivity in mixed media. Water is preferentially adsorbed over alcohols, and ethylene glycol is preferred over ethanol in accord with the derived equation*
>
> $$\ln K_1^2 = \frac{\pi}{RT}(\bar{V}_1 - \bar{V}_2)$$
>
> *The strongly hydrophyllic nature of the exchanger and the resulting constancy of the internal environment permit assessment of electrolyte activity coefficient ratios in the external mixed solvent through the equation*
>
> $$\ln K_M^N = \frac{\pi}{RT}(\bar{V}_M - \bar{V}_N) + \ln \frac{\bar{\gamma}M}{\bar{\gamma}N} - 2\ln \frac{\gamma_{\pm}MX}{\gamma_{\pm}NX}$$
>
> *The first 2 terms on the right retain their known aqueous values until 70 wt% external alcohol, when serious alcohol invasion first occurs.*

Platek and Marinsky (*13*) first suggested that a zeolite may be considered a highly cross-linked ion exchanger and that a relationship of the type introduced by Gregor (*7, 8*) and Glueckauf (*6*) for organic resins also may apply for the zeolite; namely

$$\ln a_j = \ln \bar{a}_j + \frac{\pi}{RT} V_j \qquad (1)$$

[1] Present address: Rosary Hill College, Buffalo, N. Y. 14226.

where a_j and V_j represent the activity and partial molar volume of component j, π is the difference in osmotic pressure between the interior of the zeolite and the external solution, and the bar placed above the symbol is used to differentiate the resin phase from the aqueous phase. Equation 2 is the thermodynamic representation of the uni-univalent exchange reaction in 1:1 electrolyte solutions with the A-zeolite in the M^+ form and the exchange carried out in solutions of NX and MX (3)

$$\ln\left[\frac{m_N \bar{m}_M}{\bar{m}_N m_M}\right] = \ln K_M^N = \frac{\pi}{RT}(\bar{V}_M - \bar{V}_N) + \ln \frac{\bar{\gamma}_M}{\bar{\gamma}_N} - 2\ln \frac{\gamma_\pm MX}{\gamma_\pm NX} \quad (2)$$

where K is the experimentally determined selectivity coefficient, m is the molal concentration of the species, $\bar{\gamma}$ is the activity coefficient of the ion in the zeolite phase, and $\gamma_\pm$ is the mean molal activity coefficient of the electrolyte in the external phase.

Bukata and Marinsky (4) considered the zeolite's structural rigidity, high resistance to electrolyte intrusion (a consequence of the high negative charge provided by the rings of oxygen atoms in its unit cubic cell), and the constancy of solvent uptake until very low external solvent activity values sufficient to maintain the $\ln \frac{\bar{\gamma}_M}{\bar{\gamma}_N}$ term of Equation 2 invariant at any external electrolyte concentration so long as the internal ion composition remained fixed. They also suggested that the osmotic pressure, π, could be evaluated by use of Equation 1 if a_j of this equation is constant at every experimental situation. Since the third term of Equation 2 is available as well, by utilization of the Harned-Cooke equation (17), the value of K_M^N as a function of external electrolyte concentration was expected to be calculable after evaluation of the $\frac{\pi}{RT}(\bar{V}_M - \bar{V}_N) + \ln \frac{\bar{\gamma}_M}{\bar{\gamma}_N}$ term from a single measurement of K_N^M at any external electrolyte concentration.

To demonstrate the validity of this model, a series of experiments was performed (4). The exchanging ion N was kept at radioactive tracer level concentrations in solutions of MX (0.05m and greater). Since the ion-fraction of M was essentially unity in both the zeolite and external solution phases and since the ion-concentration of the zeolite was constant in a fixed geometry by this experimental arrangement, the value of $\ln \frac{\bar{\gamma}_M}{\bar{\gamma}_N}$ was presumed to remain constant. The value of π was obtained as a function of electrolyte composition from Equation 1 by considering the change in the activity, a_j, of the solvent component of the solution phase with experimental conditions. The value of $\bar{a}_j$ was obtained from adsorption isotherm data as described in Ref. 4. Briefly, that solvent activity value below which solvent adsorption showed a marked decrease was presumed to identify $\bar{a}_j$. It was assumed that the partial molar volume of solvent (18 ml) and ions (10) were constant. The activity coefficient of MX, since NX was present in trace quantities, was identical with the pure MX solutions and was from the literature (17). The activity coefficient for trace NX in the presence of MX was calculated by use of the Harned-Cooke equation (17) in the form

$$\log \gamma_{0(NX)} = \log \gamma_{NX(0)} + \alpha m + \beta m^2 \qquad (3)$$

Table I. Selectivity Data

System: NaA–$NaCl$–$CsCl$

External NaCl Molality	$K_{Na}^{Cs}{}_{(exp)}$	$K_{Na}^{Cs}{}_{(pred)}$
0.053	2.77	2.81
0.106	2.78	2.83
0.537	2.56	2.55
1.085	2.22	2.31
2.255	1.85	a
3.383	1.61	1.52
4.510	1.37	1.32
6.068	1.18	1.09

System: KA–KCl–$CsCL$

External KCl Molality	$K_K^{Cs}{}_{(exp)}$	$K_K^{Cs}{}_{(pred)}$
0.049	2.92	2.80
0.098	2.77	2.80
0.499	2.67	2.72
1.000	2.58	2.66
2.090	2.48	a
3.220	2.45	2.39
4.414	2.45	2.36

System: KA–KCl–$NaCl$

External KCl Molality	$K_K^{Na}{}_{(exp)}$	$K_K^{Na}{}_{(pred)}$
0.109	3.42	3.41
0.439	3.52	3.48
0.891	3.49	3.51
1.831	3.59	a
2.829	3.75	3.69
3.885	3.80	3.77

[a] Computation base.

where $\gamma_{0(NX)}$ is the activity coefficient of a trace of NX in the presence of MX at molality m, $\gamma_{NX(0)}$ is the activity coefficient of pure NX at molality m, and α and β are experimentally determined parameters.

Representative results of these studies are presented in Table I, where experimental $K_M{}^N$ values are compared with the $K_M{}^N$ values predicted from Equation 2 on the basis of a single selectivity measurement. The $\gamma_\pm$ values and interaction parameters for use in Equation 3 were obtained from the data of Robinson (14, 15, 16, 17).

The excellent agreement over the large concentration range examined of $K_M{}^N$ (exp) and $K_M{}^N$ (pred) in these and the other uni-univalent systems studied strongly support the validity of the osmotic model that was employed.

Solvent-Selectivity of the Synthetic A-Zeolite in Mixed Media

The osmotic model has been shown by Barrett, Marinsky, and Pavelich (2) to be applicable as well for the interpretation of the solvent-selectivity properties of the synthetic A-zeolite in mixed media. They studied the competitive sorption of several alcohols and water by the A-zeolite from two-component mixtures. The results of these solvent distribution studies are reported in Table II as stoichiometric distribution coefficients, $^SK_2{}^1$, defined by Equation 4

$$^SK_2{}^1 = \frac{\bar{x}_1}{\bar{x}_2} \cdot \frac{x_2}{x_1} \tag{4}$$

where x is the mole fraction of solvent, and subscripts 1 and 2 identify the competing solvent components. As before, the bar over the symbol designates the internal phase.

It was not possible to obtain a direct measurement of the solvent composition of the internal phase at equilibrium in these studies. A material balance needed to be effected by accurate analysis of the initial phases and the final equilibrium solution to yield this information. This experimental restriction blunted the accuracy and range of the experimental program as described below.

Experiments were confined to alcohol-rich solutions because of the high affinity of the water component for the zeolite phase. By limiting experiments to this composition range, the internal solvent ratio was maintained near unity to assure reasonably reliable results. With one internal component significantly in excess of the other, this ratio is susceptible to sizeable distortion by small errors in analysis of the external phase.

In spite of these precautions, the determination of $^SK_2{}^1$ remained subject to large uncertainty. In a given experiment, the initial composi-

Table II.

No.	x_w	$\bar{x}_w$	$^sK_m^w$
System: NaA–Water–Methanol			
A-1	0.0203	0.34	4 <25.0 < 67
2	0.0270	0.55	24 <44.0 < 76
3	0.0242	0.56	51.3
4	0.0253	0.49	37.0
5	0.0262	0.53	41.9
6	0.0262	0.52	40.3
7	0.0262	0.54	43.6
8	0.0301	0.675	35 <68.5 <137
9	0.0520	0.635	32.0
10	0.0700	0.57	18.0
11	0.0700	0.78	30 <47.0 <100
12	0.0900	0.675	21.5
13	0.0900	0.834	33.0
14	0.1340	0.877	46.0
System: KA–Water–Methanol			
B-1	0.0130	0.62	123
2	0.0130	0.68	163
3	0.0143	0.62	114
4	0.0154	0.66	125
5	0.0177	0.63	96
6	0.0552	0.845	96
7	0.0770	0.83	58
8	0.0820	0.86	67
9	0.121	0.964	152
10	0.124	0.92	93

tion of both phases was precisely known by their controlled preparation. Assay of the solution phase at equilibrium was based on density and Karl Fischer determination at the lowest water concentrations, with good agreement. At the higher water concentrations, the determination of $^sK_2^1$ was based solely on density measurements. An error limit of ±0.1% in the density measurements, which is believed to be realistic, leads to the representative error limits in $^sK_1^2$ that are listed. Since the Karl-Fischer method at low water concentrations is more accuate than the ±0.1% uncertainty ascribed to the density measurements, the solvent selectivity coefficients presented are believed to fall within the error limits because of the good agreement between the 2 methods of measurement when both were employed.

Applicability of the osmotic model for interpretation of these solvent selectivity data was demonstrated (2) as follows: The thermodynamic expression for the exchange reaction of 2 solvent components between zeolite and solution phases (Equation 1) is according to the osmotic model

Solvent Selectivity

No.	x_w	$\bar{x}_w$	$^sK_e^w$
System: NaA, Ka–Water–Ethanol			
NaA 1	0.0184	0.74	156
2	0.0209	0.76	155
Ka 1	0.0130	0.89	806
2	0.0104	0.87	631
System: NaA–Water–Ethylene Glycol			
C-1	0.0425	0.667	44.9
2	0.0451	0.604	32.3
3	0.0601	0.667	27.2

No.	x_{eg}	$\bar{x}_{eg}$	$^sK_e^{eg}$
System: Ethylene Glycol–Ethanol–NaA			
D-1	0.045	0.637	36.5
2	0.048	0.594	28.5
3	0.017	0.451	50.6
4	0.018	0.427	38.8
5	0.019	0.405	34.9
6	0.019	0.360	29.4
7	0.043	0.520	23.7
8	0.036	0.554	33.3
9	0.032	0.634	51.2
10	0.032	0.578	40.8

$$\ln \frac{\bar{a}_2}{\bar{a}_1} \frac{a_1}{a_2} = \ln {}^TK_1^2 = \frac{\pi}{RT} (\bar{V}_1 - \bar{V}_2) \qquad (5)$$

In order to use Equation 5 successfully, the $\bar{a}$ parameter must be known for each component. In the case of water, it was possible to assign a value of 0.0127 for the Na-form zeolite from recent adsorption isotherm data (1) by assuming the validity of the Bukata and Marinsky (4) analysis of such data. The value for $\bar{a}_m$ (methanol) was less certain (5), and similar data were unavailable for ethanol and ethylene glycol.

Because of the unavailability of pertinent adsorption isotherm data for the various solvents studied, the interpretation of solvent selectivity by the model proposed could not be explicit. Resort to the model itself was consequently made to estimate the chemical composition term ($RT \ln \bar{a}$) contributing to the chemical potential (μ) of each solvent component for which adsorption isotherm data were lacking in the zeolite phase of equilibrium mixtures. This approach to the evaluation of $\bar{a}$, because it was not explicit, had to be justified in an unambiguous manner. This was attempted in the following fashion.

Table III. System: Sodium

No.	Mole % H_2O	Mole % H_2O	$^sK_m^w$	$\bar{\gamma}_w$	$\bar{\gamma}_m$	$\bar{\gamma}_w$
A-1	2.03	34	25	1.47	1.00	2.80
2	2.70	55	44	1.28	1.03	2.70
8	3.0	67.5	68.5	1.13	1.16	2.60
11	7.0	78	47	1.10	1.30	2.41

By using an experimental selectivity point obtained with methanol–water–NaA and the adsorption isotherm deduced $\bar{a}_w$ value, the internal pressure of the experimental mixture was evaluated first with Equation 6a.

$$\ln \frac{x_w\, \gamma_w}{\bar{a}_w^\circ\, \bar{\gamma}_w\, \bar{x}_w} = \frac{\pi \bar{V}_w}{RT} \qquad (6a)$$

Since the internal methanol component must be subject also to the same pressure, Equation 6b could be employed to evaluate $\bar{a}_m^\circ$ for methanol.

$$\ln \frac{x_m\, \gamma_m}{\bar{a}_m^\circ\, \bar{\gamma}_m\, \bar{x}_m} = \frac{\pi \bar{V}_m}{RT} \qquad (6b)$$

Equations 6a and 6b are based on Equation 1. In the mixed solvent system, the a_j parameter of Equation 1 has to be modified to account for the fact that there are 2 solvent components occupying the zeolite matrix. It has been considered, *a priori,* that $\bar{a}_{j(Eq.1)} = \bar{a}_j^\circ \bar{x}_j \bar{\gamma}_{j(Eq.\,6a,b)}$ where $\bar{a}_j^\circ$, the activity parameter obtained in the pure solvent medium, must be modified to account for its fractional constitution of the zeolite phase (x_j) and for solvent–solvent interaction (γ_j). In the limiting case of $\bar{x}_j = 1$, $\bar{\gamma}_j = 1$, and Equations 6a and 6b reduce to Equation 1.

To facilitate the computation of a_m°, the assumption was made that the activity coefficients of the 2 components as a function of composition were the same in both phases. Data from experiment A–2 in Table II were used for the computation. Solution of Equation 6a yielded a value of 2850 atm for π. The value of $\bar{a}_m^\circ$ that resulted from Equation 6b was 0.020, in reasonable agreement with the $\bar{a}_m^\circ$ value of 0.03 deduced from the only adsorption isotherm data (5) available for methanol. As a consequence of this agreement, it was thought justifiable to suggest that the adsorption isotherm data were reliable and supported the use of this approach to the evaluation of $\bar{a}$. By applying these parameters to the data listed in Table II, the selectivity that was predicted by multiplying $^sK_m^w$ by the ratio $\dfrac{\bar{a}_w^\circ \bar{\gamma}_w \bar{\gamma}_m}{\bar{a}_m^\circ \bar{\gamma}_m \bar{\gamma}_w}$ to obtain $^TK_m^w$ is compared in Table III with the value of $^TK_m^w$, the thermodynamic selectivity constant predicted by

A–H$_2$O–MeOH

$^sK_m^w$ $\left(\dfrac{\bar{a}_w^°\bar{\gamma}_w\gamma_m}{\bar{a}_m^°\bar{\gamma}_m\gamma_w}\right)$	π	antilog $\dfrac{\pi(V_m - V_w)}{2.3\,RT}$
1.4 < 8.4 <23	2960	15.5
7 <12.8 <22	2850	14
8.3 <16.2 <33	2830	13.6
6.9 <10.7 <23	3700	31.2

Table IV. Evaluation of $\bar{a}^°$ Parameter

γ_w	γ_e	$\bar{\gamma}_w$	$\bar{\gamma}_e$	$^sK_e^w$	π	$\bar{a}_e$
System: EtOH–H$_2$O–NaA						
2.00	1.00	1.09	1.48	156	1670	0.0586
2.00	1.00	1.09	1.48	155	1872	0.0334
System: Et(OH)–H$_2$O–NaA						
2.32	1.82	1.72	2.40	32.3	2770	0.00215
2.32	1.80	1.61	1.54	44.9	2700	0.00586
2.32	2.00	1.61	1.54	27.2	3170	0.00163

the antilog $\dfrac{\pi(\bar{V}_m - \bar{V}_w)}{2.3RT}$. Apparent molar volumes (9, 11) were substituted for partial molal volumes without detriment to this treatment. Although there are strong interactions in the methanol–water system, the difference ($V_{MeOH} - V_{H_2O}$) remains fairly constant at 22.6 cc/mole over the entire mole fraction composition range (11). Since the sorbates were envisioned to be under considerable pressure, molar compressabilities were considered and found roughly equal at 2500 atm (9). Ethanol–water mixtures were amenable to the same treatment (9).

At the higher alcohol concentrations, the agreement between the 2 computations of ^TK was good. However, in the less alcohol-rich systems, considerable discrepancy between the 2 computations suggested either failure of the model or failure of the assumption that activity coefficients in the internal phase are environment-independent. We consider the second alternative more likely.

Additional support for the proposed model was provided from the correlation obtained between the 2 computations of ^TK for the ethylene glycol, ethanol, NaA system when solvent selectivity data for the 2 systems, ethanol, water, NaA and ethylene glycol, water, NaA were employed in the model to evaluate $\bar{a}_e$ and $\bar{a}_{eg}$. The $\bar{a}^°$ values for ethylene glycol and ethanol, respectively, that were obtained (Table IV) were averaged for application to the prediction of thermodynamic selectivity constants for the anhydrous system, ethylene glycol, ethanol, and NaA. In the absence of activity coefficient data for ethylene glycol–ethanol mixtures, ideal behavior, observed in methanol–ethanol mixtures, was assumed. The results of these computations are given in Table V, together with the ln $^{T}K_{s2}{}^{s1}$ values predicted with Equation 5.

The internal consistency between the 2 computed values of $^{T}K_{e}{}^{eg}$ is believed sufficient to provide additional meaningful support for the validity of the osmotic treatment in view of the error possible in evaluation of $^{S}K_{e}{}^{eg}$. The implication that the general shape of experimentally difficult adsorption isotherms can be approximated by the analysis of solvent selectivity data is of great importance.

Ion-Exchange Selectivity of the Synthetic A-Zeolite in Mixed Media

An important observation that derives from the solvent selectivity data is that water is preferred highly over alcohols by the zeolite phase. As a consequence, the zeolite solvent is essentially aquatic even when in contact with alcohol-rich solutions. Since this is the case, the osmotic model should be applicable as well to the examination of ion-exchange phenomena in mixed media which contain a high proportion of the nonaqueous material.

The terms $\frac{\pi}{RT}(V_M - V_N) + \ln \frac{\bar{\gamma}_M}{\bar{\gamma}_N}$ of Equation 2 are available from a single ion-selectivity measurement in an aqueous medium, as described earlier. By altering the solution composition with the controlled addition of alcohol, only π is expected to be affected in these 2 terms as long as alcohol incursion is minimal. The variation of π can be accurately assessed by application of Equation 1. The variation of the ion-exchange distribution measurement with solvent composition should be identifiable with the change in $-2 \ln \frac{\gamma_{\pm} NX}{\gamma_{\pm} MX}$, the third term of Equation 2.

Table V. Prediction of Solvent Selectivity Constants

Expt. No.	$log\ {}^{S}K_{e}{}^{eg}$	$log\ {}^{S}K_{e}{}^{eg} \frac{\bar{a}_{eg}{}^{\circ}}{\bar{a}_{e}{}^{\circ}} \frac{\bar{\gamma}_{eg} \gamma_{e}}{\bar{\gamma}_{e} \gamma_{eg}}$	$\pi\ (atm)$	$log\ {}^{T}K_{e}{}^{eg} = \pi \cdot (V_1 - V_2)/2.3RT$
		System: Ethylene Glycol–EtOH–NaA		
1	1.464	0.310	1490	0.531
2	1.70	0.546	1555	0.554
3	1.607	0.453	1660	0.591
4	1.557	0.403	1740	0.619

Table VI. Ion Selectivity in Mixed Media

System: KA–$0.5m$ KI, $Na^{22}I$–H_2O, $MeOH$

Wt % MeOH	0	17	50	80	85	90	98.8	99.2	99.6	100
$K_K^{Na^{22}}$		3.52	3.72	4.47	5.25	5.40	5.60	10.3	27	33	43

Table VII. Computation of the Mean Molal Activity Coefficient Ratio of Simple Electrolytes in Mixed Media

Wt. % MeOH	$\dfrac{\pi \Delta V}{RT} + \ln \dfrac{\bar{\gamma} K}{\bar{\gamma} Na}$	$\ln K_{exp}$	$-2 \ln \dfrac{\gamma_\pm KI}{\gamma_\pm NaI}$
0	1.091	1.235	0.142
10	1.077	1.280	0.203
20	1.058	1.344	0.286
30	1.037	1.385	0.348
40	1.009	1.440	0.431
50	0.970	1.500	0.530
60	0.917	1.549	0.632
70	0.842	1.610	0.768

Ion selectivity data were obtained as function of solvent composition with the system KA–0.5m KI, $Na^{22}I$–H_2O, MeOH. These data are presented in Table VI and the K_K^{Na} values interpolated from these data have been used in Equation 2 to evaluate the third term of Equation 2 at the selected solvent composition values. The results of this computation are presented in Table VII.

The increasing deviation between the mean molal activity coefficient of KI and NaI with methanol content is to be expected. As the dielectric of the medium decreases with larger methanol content, differences in the ion pair formation capability of Na^+ and K^+ are believed to be enhanced with the more highly polarizable iodide ion.

On the basis of the above result, it is suggested that the osmotic pressure approach may be as useful for the estimate of thermodynamic properties of simple electrolyte mixtures in mixed solvents as in aqueous media. Additional research with alkali chlorides in alcohol–water mixtures is in progress to substantiate this conclusion. The mean molal coefficient data that are published for the alkali chlorides (12) are expected to facilitate a meaningful prognosis of the osmotic pressure model.

At present, this model cannot be employed usefully to interpret ion selectivity in the essentially pure nonaqueous solvent. Various parameters essential to the model are not yet available in the literature. However, since ion selectivity is so sensitive to the solvent content of the exchanger, it appears that the most important selectivity term in Equation 2 must be $\ln \frac{\bar{\gamma} M^+}{\bar{\gamma} N^+}$. A number of speculative approaches are presently under consideration to facilitate a quantitative interpretation of the ion-selectivity of the A zeolite in anhydrous media.

Acknowledgment

Financial support through Contract No. AT(30-1)-2269 with the United States Atomic Energy Commission is gratefully acknowledged.

Literature Cited

(1) Barrer, R. M., Denny, A. F., *J. Chem. Soc.* **1964**, 4677.
(2) Barrett, R. B., Marinsky, J. A., Pavelich, P., "Studies in Solution and Nuclear Chemistry," Research Progress Report, AEC Contract No. At(30-1)2269, July, 1969.
(3) Boyd, G. E., Soldano, B. A., *Z. Elektrochem.* **1953**, 57, 162.
(4) Bukata, S., Marinsky, J. A., *J. Phys. Chem.* **1964**, 68, 994.
(5) Flanigen, E. M., Linde Division, Union Carbide Corp., private communication, 1969.
(6) Glueckauf, E., *Proc. Roy. Soc. (London)* **1952**, A 214, 207.
(7) Gregor, H. P., *J. Am. Chem. Soc.* **1948**, 70, 1923.
(8) *Ibid.*, **1951**, 73, 642.
(9) Lewis, G. N., Randall, M., "Thermodynamics," 2nd ed., McGraw-Hill, New York, 1961.
(10) Mukerjee, P., *J. Phys. Chem.* **1961**, 65, 740.
(11) Neidig, H. A. *et al.*, *J. Chem. Educ.*, **1965**, 42, 309.
(12) Parsons, R., "Handbook of Electrochemical Constants," Academic, London, 1959.
(13) Platek, W. A., Marinsky, J. A., *J. Phys. Chem.* **1961**, 65, 2118.
(14) Robinson, R. A., *J. Am. Chem. Soc.* **1952**, 74, 6035.
(15) Robinson, R. A., *J. Phys. Chem.* **1961**, 65, 662.
(16) Robinson, R. A., *Trans. Faraday Soc.* **1953**, 49, 1147.
(17) Robinson, R. A., Stokes, R. H., "Electrolyte Solutions," 2nd ed., Butterworth, London, 1959.

RECEIVED February 4, 1970.

Discussion

A. Dyer (University of Salford, Salford, Lancs., England): Have you estimated what the concentration of water per ion is at the point of change of slope in the graph of K vs. % methanol?

R. B. Barrett: It appears that the variation of the trace ion selectivity coefficient is a linear function of internal alcohol concentration.

H. S. Sherry (Mobil Research & Development Corp., Princeton, N. J. 08540): Successful use of the osmotic approach depends on the fact that you have chosen a very special system—zeolite A at 25°C and trace concentration of the exchanging ion. Thus, two terms in your equations are fixed—the water activity in the zeolite phase and the term containing the zeolite phase ionic activity coefficients. Have you ever worked in a system in which salt imbibement takes place like zeolite X and at macroscopic levels of ion exchange?

R. B. Barrett: Solvent exchange (H_2O–MeOH) studies on Na-13-X have been investigated but since the external phase is electrolyte-free, the limitations you mention are not serious. Our results are in accord with the osmotic pressure model.

32

Thermodynamics of the Exchange of Alkylammonium Ions in Synthetic Faujasites

E. F. VANSANT and J. B. UYTTERHOEVEN

University of Leuven, de Croylaan 42, 3030 Heverlee, Belgium

> *The thermodynamics of exchange of alkylammonium ions in synthetic faujasites is studied. A correlation exists between the maximum limit to exchange and the selectivity. Both properties are controlled by the same factors. An adapted form of the Gaines and Thomas formula is proposed, allowing the calculation of the equilibrium constant when the exchange reaction is incomplete. Values of ΔG, ΔH, and ΔS were calculated. Interaction with the lattice is the most important factor in the exchange with low aluminum faujasites (Y). In the X-like samples, exchange seems to be ruled by the change in hydration state of the large cavities. A possible redistribution of the exchangeable cations during the exchange reaction is assumed.*

Ion sieve effects are observed in ion exchange reactions on synthetic faujasites when the dimensions of the ingoing ion are such that they cannot diffuse inside the small cavities. These phenomena were reported by Sherry (8, 9) for exchange with inorganic ions and by Barrer and coworkers (1) for the exchange of organic ions. Barrer emphasized the influence of space requirements, which he considered to be the main factor determining the maximum limit to exchange.

Theng et al. (12) studied the exchange of several alkylammonium ions on a typical X and a typical Y zeolite. These ions could only exchange in the large cavities. They concluded that the exchange is lower than the limit imposed by possible space requirements. The limit to exchange was different for the X and Y zeolites although the available volume is almost the same.

The maximum limit to exchange was a linear function of the molecular weight of the ingoing ions. The decrease of that function with increasing molecular weight was more pronounced in X than in Y. The exchange limit was influenced also by the nature of the ions initially present. From the correlation between the limit to exchange and the ΔG of exchange, Theng *et al.* (*12*) concluded that the maximum limit to exchange was determined by the same factors as the exchange selectivity. They assume that the distribution of the ions over different exchange sites in the small and large cavities is not fixed but controlled by the nature of the exchanging ions. These hypotheses were confirmed by Vansant *et al.* (*13*) in a study of the exchange of aromatic and heterocyclic ammonium derivatives. A very pronounced selectivity in favor of these organic ions was observed and, in agreement with the hypothesis of Theng *et al.* (*12*), the amount of ingoing ions was also very high. Packing calculations proved a maximum filling of the zeolite cages with the aromatic and heterocyclic ammonium ions.

In this work, a study is made of the exchange of the *n*-alkylammonium ions, going from NH_4^+ to butylammonium, on 4 synthetic faujasites differing by their Al content. An extensive thermodynamic treatment of the exchange with the propylammonium ion is included.

Experimental

Materials. Four different samples with Al/Si ratios ranging from an extreme Y to an extreme X were supplied by the Linde Co., New York. As they constitute a complete substitutional series, the indications X and Y will not be used further but will be replaced by the symbol *F* followed by the name of the saturating ion and a figure indicating the number of ions per unit cell. In order to ensure complete saturation, the samples as received were treated with a 1N NaCl solution and washed free from excess electrolyte. The chemical composition was determined and expressed in terms of unit cell formula, as follows:

Sample	Composition
F Na 49	$Na_{49} (AlO_2)_{49} (SiO_2)_{143}$
F Na 55	$Na_{55} (AlO_2)_{55} (SiO_2)_{137}$
F Na 71	$Na_{71} (AlO_2)_{71} (SiO_2)_{121}$
F Na 85	$Na_{85} (AlO_2)_{85} (SiO_2)_{107}$

The symbols Me, Et, Pr, and Bu will be used to symbolize the ions methylammonium, ethylammonium, propylammonium, and butylammonium, respectively.

Exchange Procedure. Zeolite samples of 0.1 gram were equilibrated overnight in 40 ml of a solution containing appropriate proportions of Na^+ and alkylammonium ions. A fixed temperature (4°, 17°, 25°, 40°, or

55°C) was maintained during the equilibration. The amount of alkylammonium ions adsorbed was calculated from the variation in the concentration of the equilibrium solution. The validity of this method was tested in a few cases by making a complete stoichiometric balance. More details of the analytical procedures have been given elsewhere (*11*).

The results of the exchange can be expressed as exchange isotherms plotting the equivalent fraction of the exchange capacity of the zeolite (Z) saturated with alkylammonium as a function of S, the equivalent fraction of alkylammonium ions in the solution. The isotherms for the propylammonium exchange are given in Figure 1. The maximum limits to exchange were determined on separate samples. These were equilibrated with $0.5N$ solutions of the appropriate alkylammonium chlorides. The equilibrated solution was renewed twice; then the samples were washed free from excess alkylammonium, and the adsorbed amount of organic ions was analyzed. The maximum limits to exchange obtained

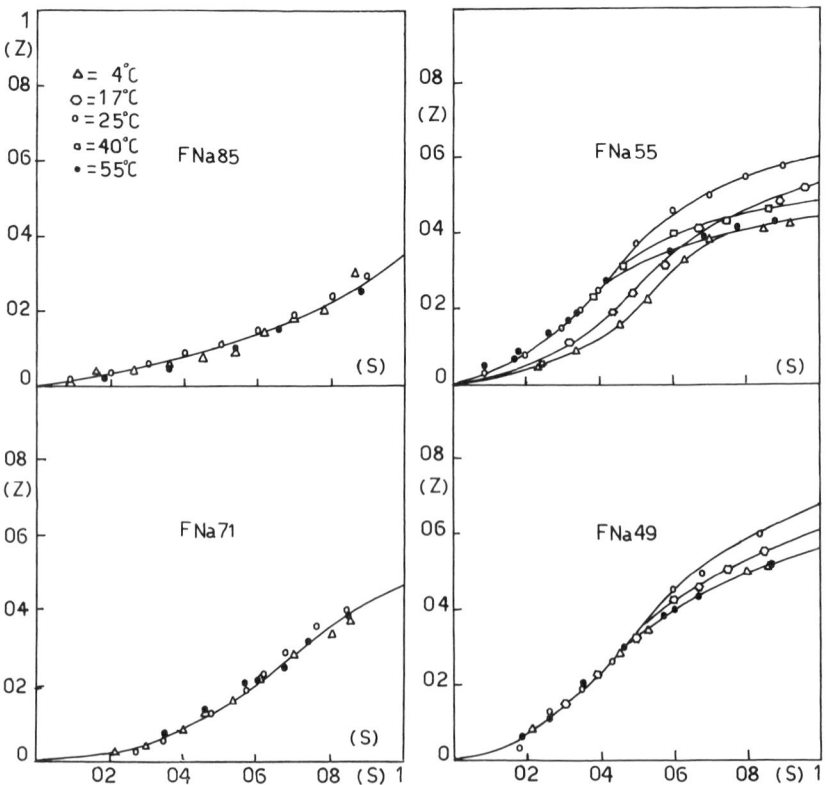

Figure 1. Exchange isotherms of the propylammonium ions in the different samples at different temperatures

in this way corresponded to the limits that could be derived from the isotherms using the extrapolation procedure proposed by Barrer and Meier (3). This procedure is based on the linear relationship between Z_M and $\frac{S_{Na}}{S_M} Z_M$.

Estimation of Thermodynamic Functions for Partial Exchange. An exchange of ions between a solid and a liquid phase can be represented by: $Na^+_{(z)} + M^+_{(s)} \rightleftharpoons M^+_{(z)} + Na^+_{(s)}$. The subscripts z and s refer to the zeolite and the liquid phase. According to Gaines and Thomas (6), the thermodynamic equilibrium constant is expressed as follows:

$$\ln K = \int_0^1 \ln K_c \, d Z_M,$$

where K_c is the "corrected" selectivity coefficient defined as

$$K_c = \frac{Z_M S_{Na}}{Z_{Na} S_M} \cdot \frac{\gamma_{Na}}{\gamma_M}$$

Z and S are the equivalent fractions of the ions on the zeolite and in solution, and γ is the activity coefficient of the ions in solution.

With a total ion concentration of 0.05N, the ratio of the activity coefficients does not deviate significantly from unity (7). This expression of the equilibrium constant can be used only if there is a complete exchange. A number of workers applied this equation on incomplete exchanges in zeolites (8–12) after a normalization putting the experimental maximum exchange equal to unity ($Z_M^{max} = 1$).

Such a normalization assumes that the residual Na^+ ions are inaccessible for exchange and have no influence on the selectivity coefficient. That this assumption is incorrect results from the fact that the maximum limit to exchange is influenced by the nature of the outgoing and ingoing ions, and also by the Al/Si ratios of the zeolites. In order to obtain comparable values of the equilibrium constant, we modified the Gaines and Thomas equation. The corrected and normalized selectivity coefficient is defined as

$$K_c^N = \frac{Z_M f \, S_{Na}}{(1 - Z_M f) S_M}$$

where $f = 1/Z_M^{max}$. Introduction of this K_c^N in the Gaines and Thomas equation requires a correction term taking into account the influence of the residual Na ions. The Gaines and Thomas equation now can be written as

$$\ln K = \frac{1}{f} \int_0^1 \ln K_c^N \frac{1 - Z_M f}{f - Z_M f} \, d Z_M f$$

K represents the real equilibrium constant for an exchange between $Z = 0$ and $Z = Z^{max}$. The standard free energy of exchange is obtained as $\Delta G = -RT \ln K$. The enthalpy of exchange is derived from the Van't Hoff equation, and the entropy change is calculated from the ΔG and ΔH values:

$$\frac{\partial \Delta G/T}{\partial T} = -\frac{\Delta H}{T^2} \text{ and } \Delta G = \Delta H - T\Delta S$$

Results

The exchange isotherms, established with the ion Pr on the 4 zeolite samples, are given in Figure 1. The exchange was independent of the temperature on the samples with high aluminum content: FNa85 and FNa71. On the other samples, especially FNa55, the extent of exchange increases from 4° to 25°C, indicating an endothermic process, and decreases again at higher temperatures, which is typical for an exothermic process. The values of ΔG, ΔH, and ΔS derived from these isotherms are collected in Table I.

The over-all order of affinity of the zeolites for the propylammonium ion, derived from ΔG values in Table I, is:

$$F \text{ Na } 85 < F \text{ Na } 71 < F \text{ Na } 55 < F \text{ Na } 49.$$

The affinity order is reflected also in the maximum limits to exchange, which are collected in Figures 2 and 3 for all the ions, including propyl,

Table I. Thermodynamic Values of the Exchange Na$^+$–Propylammonium

Sample	Temp., °C	ΔG, Cal Mole^{-1}	ΔH, Cal Mole^{-1}	ΔS, Cal °K^{-1} Mole^{-1}
F Na 85	4	+505	0	−1.82
	25	+544	0	−1.82
	55	+598	0	−1.82
F Na 71	4	+465	0	−1.67
	25	+500	0	−1.67
	55	+550	0	−1.67
F Na 55	4	+393	+393	0
	17	+393	+393	0
	25	+303	−1313	−5.42
	40	+315	−1313	−5.20
	55	+393	−1313	−5.20
F Na 49	4	+382	+1835	+5.11
	17	+349	+1127	+2.6
	25	+333	+676	+1
	55	+381	−1127	−4.7

at room temperature. In Figure 2 is plotted the variation of the exchange capacity as a function of the molecular weight of the alkylammonium ions on the different samples. The decrease is steeper for the samples with high aluminum content. When these data are replotted (Figure 3)

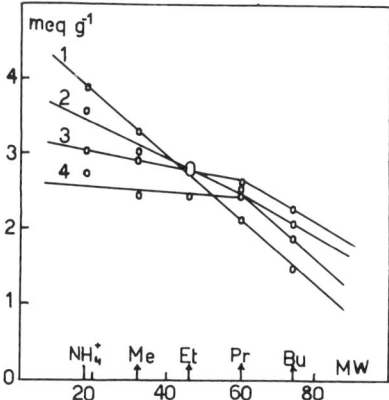

Figure 2. *Maximum limit to exchange (meq gram^{-1}) as a function of the molecular weight of the ingoing ions*

1 = FNa 85, 2 = FNa 71, 3 = FNa 55, 4 = FNa 49

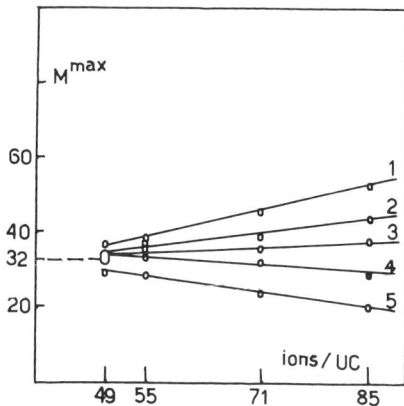

Figure 3. *Maximum limit to exchange (ions per unit cell) for the alkylammonium ions as a function of the total amount of exchangeable ions per unit cell*

1 = NH$_4^+$, 2 = Me, 3 = Et, 4 = Pr, 5 = Bu

in terms of maximum limit to exchange as a function of the total exchange capacity of the samples, the data converge to a limit of 32 ions per unit cell, independent of the nature of the ions.

Discussion

Since the alkylammonium ions cannot penetrate in the cubooctahedra, the exchange reaction can take place only in the large cavities. The present results confirm the hypothesis of Theng et al. (12) that the available space is not the limiting factor. Indeed, the free space in the large cavities is almost the same in the 4 samples, and important variations of the maximum limit to exchange (M^{max}) are observed, depending on the zeolite sample and the ingoing ion (Figures 2 and 3). The variability of the shape of the isotherms from sample to sample (Figure 1), but also from one alkylammonium ion to another, indicates that M^{max} is related to the selectivity.

The continuity of the lines in Figure 2 further indicates that M^{max} cannot be interpreted in terms of a fixed cation distribution. It is possible that the introduction of organic ions into the large cavities provokes a migration of Na^+ ions and a redistribution of the cations between the large and the small cavities. Moreover, it is not evident that the exchange reaction necessarily replaces all the Na^+ ions in the large cavities. If this were the case, one would expect a minimum exchange of 32 ions per unit cell, corresponding to the number of available six-rings in the large cavity. This limit is indicated in Figure 3. M^{max} for the different ions converges to that limit in the sample $FNa49$. All the ions, except Bu, exchange more than 32 ions per unit cell. The exchange limit for Bu is below 32 for the samples with high exchange capacity.

The behavior of the propylammonium ion was considered characteristic for the series of ions, and a more extensive thermodynamic investigation was made with this ion. The values in Table I show that ΔG is positive in any case, and that the affinity for the Pr ion decreases with increasing lattice charge. This variation in ΔG is not caused by the amount of organic ions, because the number of Pr per unit cell is almost constant in the different samples (Figure 3).

The total changes in entropy and enthalpy (Table I) can be considered to be composed of 2 terms, one accounting for the changes in the energy of interaction with the lattice, the other for the changes in the hydration of the ions. This was discussed earlier by Sherry et al. (10) and by Barrer et al. (2). This gives us the following equations where the index 1 indicates the interaction terms with the lattice, and the index h the hydration terms.

$$\Delta S = \Delta S^l + \Delta S^h = (S_{Pr}^l - S_{Na}^l) + (S_{Na}^h - S_{Pr}^h)$$

$$\Delta H = \Delta H^1 + \Delta H^h = (H_{Pr}^1 - H_{Na}^1) + (H_{Na}^h - H_{Pr}^h)$$

The lack of information on the thermodynamic properties of the alkylammonium makes it impossible to calculate absolute values for the ΔH^1, ΔH^h, ΔS^1, ΔS^h terms. Taking into account the hydrophobic nature of the Pr ion, it seems reasonable to assume that H_{Pr}^h is smaller in absolute values than H_{Na}^h. The hydration of ions being an exothermic effect, the ΔH^h is expected to yield a negative value. The entropy of hydration is also negative and increases in absolute value when the ion is more hydrated. Thus, ΔS^h also will be a negative quantity.

H^1 is equal to the potential energy of the ions with respect to the lattice. This electrostatic interaction is exothermic. According to Coulomb's law and the relative dimensions of the ions Na⁺ and Pr, H_{Na}^1 will be more important than H_{Pr}^1. ΔH^1 is expected, therefore, to be positive, the absolute value depending on the electrostatic charge associated with the exchange site. On pure electrostatic grounds, the ΔS^1 also is expected to be positive, although it can be influenced by the configuration of the organic ion on its exchange site. On the basis of these qualitative considerations, the data in Table I can be understood as follows.

It was found by x-ray diffraction experiments (4) that in hydrated zeolite XNa, an important fraction of the exchangeable Na⁺ ions cannot be located. These ions probably are not in close interaction with the lattice and most likely are located in the large cavities. Exchange of such weakly bound ions is expected to give small values of the ΔH^1 and ΔH^h terms. These 2 terms having opposite signs, a very small total enthalpy effect can be expected. The ΔH observed for the FNa85 and FNa71 is actually zero, which is consistent with the proposed explanation. On the basis of the same considerations, a small entropy effect also would be expected. However, the induction of large organic ions in the zeolite cages can reduce the hydration state of the remaining Na⁺ ions and thus cause a negative entropy effect. This is observed.

The situation is different for the samples FNa55 and FNa49. From the comparison of the ΔG values, it is apparent that the affinity of the Pr ions for these 2 samples is greater than for those with higher total exchange capacities. The probable reason for this is that in the FNa55 and FNa49 samples a higher fraction of the lattice charge is neutralized by ions on well defined sites. This is in line with Sherry's observation (8, 9) that the "ionic character" of the exchange is more pronounced in Y than in X zeolites. A tighter binding with the lattice will increase the importance of the enthalpy effects. The positive ΔH values obtained for the FNa55 and FNa49 samples indicate an increased binding energy of the ions to the lattice; replacement of Na⁺ by Pr requires energy, as explained earlier in this note. At higher temperature, the enthalpy effects become negative. This implies that the interaction with the lattice becomes less important, and that the negative effect of the hydration term predominates. These variations cannot be explained easily by only the effect of increased thermal motion. This would cause a gradual change in the thermodynamic values rather than a discontinuous change like that observed in FNa55. We believe that the increase of the temperature delocalizes a fraction of the cations and may cause a redistribution of the residual Na ions initially present inside the small cavities. Such a redistribution can affect the charge associated with the exchange sites and explain important changes in the exchange properties. In the FNa49, where the number of ions is small, the localization of the lattice charges will be the most pronounced, which is effectively reflected in the higher ΔH values. The inversion of ΔH from endothermic to exothermic is shifted also to higher temperatures for the FNa49 sample as compared with the FNa55. The evolution of the ΔS values confirm the variation in contribution of the interaction phenomenon with the lattice and the hydration term derived from the ΔH values.

Our interpretations are based on speculations about the distribution of Na^+ ions in hydrated Y. No x-ray data are available on this subject. If the situation is comparable to this in dehydrated Y samples, we can conclude from Eulenberger's data (5) that approximately 30 ions per unit cell are inside the large cavities on six-membered rings of oxygen; the rest is in the hexagonal prism and inside the cubooctahedra. It seems reasonable to assume that this ion distribution in hydrated samples is temperature-dependent, but more detailed speculations are impossible.

The thermodynamic data presented here are confirmed by more extensive experiments with the other alkylammonium ions. It is impossible to give an extensive review of these complementary data in the limited length of this article. The propylammonium system was the easiest to interpret because the number of ingoing ions was almost the same for the different samples.

From a practical viewpoint, it may be important to know that the NH_4^+ exchange is an exothermic process at room temperature or higher. In the preparation of so-called decationated samples, the NH_4^+ exchange is often made at higher temperatures. However, the extent of the NH_4^+ exchange is maximum at room temperature.

Acknowledgment

E. F. Vansant is indebted to the National Science Foundation (NFWO–Belgium) for a research grant. We acknowledge the gift of zeolite samples by the Linde Co.

Literature Cited

(1) Barrer, R. M., Buser, W., Grutter, W. F., *Helv. Chim. Acta* **1956**, 39, 518.
(2) Barrer, R. M., Falconer, J. D., *Proc. Roy. Soc.*, A **1956**, 236, 227.
(3) Barrer, R. M., Meier, W. M., *Trans. Faraday Soc.* **1959**, 55, 130.
(4) Broussard, L., Shoemaker, D. P., *J. Am. Chem. Soc.* **1960**, 82, 1041.
(5) Eulenberger, G. R., Shoemaker, D. P., Keil, J. G., *J. Phys. Chem.* **1967**, 71, 1812.
(6) Gaines, G. L., Thomas, H. C., *J. Chem. Phys.* **1953**, 21, 714.
(7) Helffrich, F., "Ion Exchange," pp. 151–200, McGraw-Hill, New York, 1962.
(8) Sherry, H. S., *J. Phys. Chem.* **1966**, 70, 1158.
(9) *Ibid.*, **1968**, 72, 4086.
(10) Sherry, H. S., Walton, H. F., *J. Phys. Chem.* **1967**, 71, 1457.
(11) Theng, B. K. G., Greenland, D. J., Quirk, J. P., *Clay Minerals Bull.* **1967**, 7, 1.
(12) Theng, B. K. G., Vansant, E., Uytterhoeven, J. B., *Trans. Faraday Soc.* **1968**, 64, 3370.
(13) Vansant, E., Theng, B. K. G., Maes, A., Uytterhoeven, J. B., *Bull. Groupe Franc. Argiles* **1969**, XXI, 46.

Received February 16, 1970.

Discussion

H. S. Sherry (Mobil Research & Development Corp., Princeton, N. J. 08540): Have you considered the volume of the Na^+ ions and water molecules in the zeolite? In your F-49 zeolite there should be only 33 Na^+ ions per unit cell in the large cages. You do get essentially complete replacement of these Na^+ ions with alkylammonium ions. Furthermore, the number of Na^+ ions replaced in the X type zeolites is higher, and complete replacement of the ions in the large cages should not be expected.

J. B. Uytterhoeven: The propylammonium ion was chosen for the thermodynamic study because it replaced a constant number of ions ($\sim$32) per unit cell for the different samples. This could be taken as an indication that space requirement is the predominant factor, but it is not. This is very clear when other ions are considered, especially the butylammonium ion, which exchanges much less in Y than in X zeolite. Some estimates were made taking into account the water of hydration of the Na^+ molecules, but the hydration state of the Na^+ ions in the zeolites with organic ions can only be approached by speculation.

R. B. Barrett (Rosary Hill College, Buffalo, New York): Do your enthalpy changes perhaps reflect changes in solvation energetics? Have you followed the extent of hydration as the ions become increasingly hydrophobic?

J. B. Uytterhoeven: We are trying to clarify the variation in the hydration state during the exchange with organic ions. However, it is difficult to obtain meaningful data about the hydration properties of the organic ions.

33

Stability of Heteroionic Forms of the Synthetic Zeolite A

A. DYER, W. Z. CELLER,[1] and M. SHUTE[2]

University of Salford, Salford M5 4WT, England

The barium form of zeolite A is of moderate stability, unlike other alkali metal and alkaline earth forms which can be stable to about 1000°C. This work uses the techniques of differential thermal analysis (DTA) and thermogravimetric analysis (TGA) in conjunction with isotopic labelling and x-ray powder photography to investigate the thermal stability of heteroionic forms of zeolite A. Reasons for the instability of BaA and Na/BaA zeolites are suggested and comparisons made with A zeolites containing Na and Sr cations.

First studies (1, 2) of ion-exchanged forms of zeolite A reported that the exchange of Na for certain ions caused the breakdown of the zeolite lattice. However, Sherry and Walton (9) reported the existence of a hydrated BaA and concluded that the earlier reports of the nonexistence of this ion-exchanged form were based upon x-ray examination of calcined samples. Dyer, Gettins, and Molyneux (7) confirmed the existence of BaA and were able to measure Ba cation self-diffusion parameters in A. They also concluded that removal of water even at temperatures below 100°C caused lattice collapse. Recently, Radovanov, Gacinovic, and Gal (8) have reported the preparation of Co(II)A in hydrated form, again contrary to the original studies.

This work forms part of a program to investigate the reasons for the observed instability in certain ion-exchanged forms by examining the properties of zeolites into which proportions of the ion apparently causing thermal instability have been introduced.

[1] Present address: Institute of General Chemistry, Warsaw, Poland.
[2] Present address: Allied Chemical Corp., Petersburg, Va.

Experimental

Zeolite Materials. Linde Molecular Sieve 4A in powder form was sedimented. Particles of approximately uniform size were collected and used throughout the experimental procedures.

Isotope Experiments. ^{22}NaA was prepared and 1-gram samples placed in sealed tubes at 40°C for 2 days in the presence of appropriate amounts of 0.5M barium chloride solution. The extent of exchange which occurred was measured by determining the amount of ^{22}Na in solution and by flame photometry. Samples of ^{22}Na/BaA were equilibrated for 1 week in a desiccator over saturated NaCl. Aliquots (0.4-gram) were heated for 2 hours at 300°C, allowed to cool, and left in contact with 2.0 ml of 1M NaCl for 24 hours. The amount of ^{22}Na which self-exchanged into solution then was measured. Further determinations showed very little increase in the extent to which ^{22}Na can be exchanged from the calcined samples after 1 week. A determination on an uncalcined sample V showed that all the ^{22}Na initially in the zeolite was recovered by treatment with 0.1M NaCl at 22°C after 20 min had elapsed. All isotopic determinations were by liquid scintillation counting (5).

Thermal Analyses. Samples of Na/BaA were weighed into sample tubes and subjected to DTA up to 500°C. Great care was taken to ensure even packing of sample and reference material. Two series of experiments were carried out with alumina and NaA as reference material, respectively. Representative samples were also examined by TGA. Thermal analyses were performed using a DuPont 900 thermal analysis unit with a 950 TGA attachment.

X-Ray Investigations. X-ray powder diffraction photographs (Ievins-Straumanis mounting) were taken of samples as initially prepared and after heating. Some Guinier photographs were also obtained. Visual comparisons of line intensities were made.

Zeolites Containing Strontium. Where appropriate, experiments similar to those described were carried out on samples in which Na had been partially exchanged by Sr.

Results and Discussion

Isotope Experiments. The constitutions of the samples used are shown in Tables I and II. These tables also show the extent to which the initial ^{22}Na content could be reexchanged by unlabelled sodium ions after calcination (% Na recovered). Despite repeated attempts, no samples in the range 2.7–3.7 Ba p.u.c. were prepared.

Table I shows that, for the Ba-containing samples, the amount recovered suddenly decreases when more than 2 Ba p.u.c. are present and increases when the average concentration of Ba rises above 4 p.u.c. For samples containing Sr, only in those with high Sr contents was the recovery of ^{22}Na apparently hindered.

Thermal Analyses. The TGA results are in Tables I and II. Inconsistencies were observed in Na/Ba samples, whereas those for Na/Sr zeolite showed a small regular increase of water content with increasing

Table I. Constitution of Na/BaA Zeolite Samples

% Na Exchanged by Ba

Sample Number	Isotopic Determination	Flame Photometry	Na Ions P.u.c.[a]	Ba Ions P.u.c.[a]	Moles H$_2$O P.u.c.[a]	% Na "Recovered"
I	8.3	—	11.0	0.5	22.7	95
II	31.1	—	8.3	1.8	22.3	96
III	36.3	—	7.7	2.1	25.4	94
IV	43.3	48.7	6.8	2.6	25.7	89
V	46.2	—	6.5	2.7	20.6	34
VI	62.0	62.9	4.6	3.7	23.4	17
VII	63.5	68.7	4.4	3.8	26.1	68
VIII	65.6	—	4.1	3.9	22.7	25
IX	73.0	—	3.2	4.4	23.9	65
X	75.2	75.1	3.0	4.5	25.3	44
XI	83.5	—	2.0	5.0	29.0	70

[a] P.u.c. = per unit cell.

Table II. Constitution of Na/SrA Zeolite Samples

Sample Number	% Na Exchanged by Sr Isotopic Determination	Na Ions P.u.c.[a]	Sr Ions P.u.c.[a]	Moles H$_2$O P.u.c.[a]	% Na "Recovered"
I	11.8	10.6	0.7	27.3	74.2
II	35.0	7.8	2.1	27.4	81.6
III	44.1	6.7	2.6	27.6	71.2
IV	56.3	5.2	3.4	29.6	66.6
V	62.5	4.5	3.7	29.7	50.0

[a] P.u.c. = per unit cell.

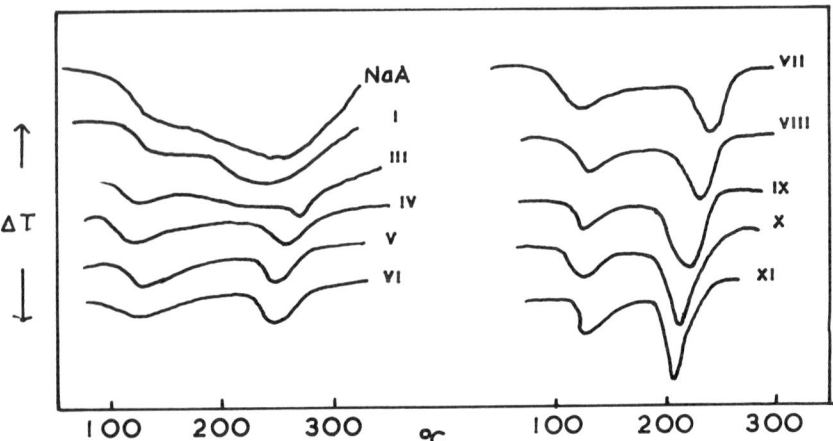

Figure 1. Sketch of DTA curves obtained for Na/BaA samples

divalent cation content. The inconsistent water contents of samples containing Ba may indicate that some lattice collapse occurred when the initial exchanges of Na for Ba were made (7).

The DTA curve obtained for Na/BaA samples are shown in Figure 1. NaA shows 2 broad endothermal water loss peaks at about 125° and 230°C, respectively. The introduction of Ba into the structure increasingly resolves the curve into 2 more discernible peaks, one of which remains at about 125°C and a second whose position varies with Ba content. This second peak appears when there are >2 Ba p.u.c. Attempts to "balance out" the first peak by using NaA as a reference were unsuccessful, thus implying that water loss from both Na and Ba contributes to this peak. The areas under the second peak were measured from alumina reference experiments and plotted against Ba content (Figure 2). These are a function of change in heat content occurring on lattice collapse and show a maxima at approximately 4 Ba p.u.c. The DTA curves for Na/SrA are complicated and seem to indicate a variety of ion–water environments. Similar curves have been observed in other heteroionic zeolite samples (4).

X-Ray Results. The powder patterns of uncalcined samples containing Ba show a slight decrease in intensity with increasing Ba content. This may be an indication of loss of crystallinity or it could be an anomaly caused by the presence of Ba which has a high absorption coefficient for the Cu Kα radiation used to obtain the powder patterns (6). X-ray patterns of calcined samples (Figure 2) showed that when Ba is present at >2 Ba p.u.c. extensive loss of structure occurred, but only sample XI showed a complete absence of structure. Only sample V of the Sr series showed any loss of structure on calcination as determined by the absence of X-ray powder diffraction pattern, but again a slight general decrease

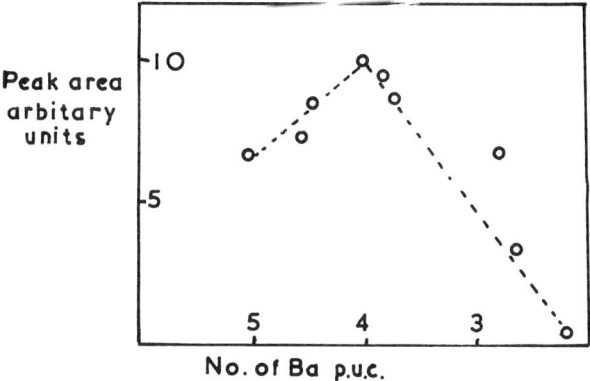

Figure 2. Peak area vs. number of Ba ions p.u.c. for second DTA feature (Ba-containing samples)

in intensity with divalent ion introduction was observed in patterns from uncalcined samples.

Discussion and Conclusions

For the Na/BaA samples, the x-ray results indicate that the thermal stability of the A lattice is reduced considerably when there are >2 Ba p.u.c. That lattice collapse occurs on calcination of samples containing this Ba concentration is demonstrated also by a sudden decrease in the ability to recover ^{22}Na by self-exchange and the appearance of an endotherm on the DTA curve. However, the x-ray results show that up to a concentration of at least 4.5 Ba p.u.c. some structure is retained after calcination. The isotope experiments also demonstrate that above concentrations of about 3.7 Ba p.u.c. there is a relative increase in the amount of ^{22}Na recoverable from calcined samples. These 2 points, taken in conjunction with the observed maximum in Figure 2 and the inability to prepare samples in the concentration range 2.7–3.7 Ba p.u.c., may indicate that the presence of 4 Ba ions p.u.c. is one of relative stability. Presumably, the first 2 divalent ions entering the crystal replace the 4 monovalent ions which are not located by x-ray structural analysis (3) and

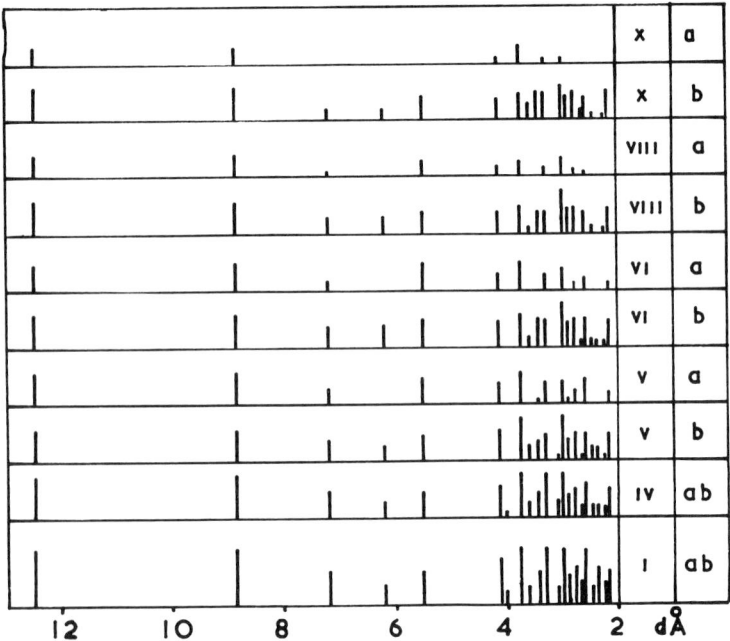

Figure 3. Lattice spacings (dA) for samples of Na/BaA (b) before and (a) after calcination

are assumed, therefore, to be present as ions "dissolved" in the occluded water of the large cavities. The ingress of further divalent ions requires the replacement of Na ions located in the hexagonal windows of the constituent sodalite units. This seems to promote lattice thermal instability. The introduction of more than 4 Ba into the unit cell seems to stabilize the structure, perhaps by a rearrangement whereby hydrated Ba ions occupy the octagonal windows, but lattice collapse occurs when water is removed. This premise is consistent with the observations of Sherry and Walton (9), who conclude from their ion-exchange investigations of the Na–BaA system that the Ba ions are equally and weakly bound to the anionic lattice. The recovery of ^{22}Na after heat treatment is relatively high at high Ba content p.u.c. This may result from the formation of Na-rich parts of the crystal, as the temperature rises, because of the higher mobility of the Na compared with that of the large heavy Ba. Thus, a type of topotactic change occurs, and the lines observable in x-ray patterns of calcined high-Ba-content samples may represent preservation of the NaA structure (Figure 3).

The foregoing comments do not apply to the presence of Sr in the A structure. The greater stability of Sr-containing samples is probably caused by the ability of the Sr to occupy a position close to the hexagonal windows, which is denied the larger Ba. On dehydration, the Sr ion "fits" into the window and preserves the A lattice. Thus, in the absence of water, the presence of a cation in the hexagonal windows is concordant with the stabilization of the A structure; this is certainly the cation position recognized in the very stable Ca form of this zeolite (3).

Acknowledgment

One of us (W. Z. C.) thanks the United Nations for a fellowship, during the duration of which this work was carried out.

Literature Cited

(1) Barrer, R. M., Meier, W. M., *Trans. Faraday Soc.* **1958**, 54, 1074.
(2) Breck, D. W., Eversole, W. G., Milton, R. M., Reed, T. B., Thomas, T. L., *J. Am. Chem. Soc.* **1956**, 78, 5963.
(3) Broussard, L., Shoemaker, D. P., *J. Am. Chem. Soc.* **1960**, 82, 1041.
(4) Celler, W. Z., Dyer, A., unpublished results.
(5) Dyer, A., Fawcett, J. M., Potts, D. U., *Intern. J. Appl. Radiation Isotopes* **1964**, 15, 377.
(6) Dyer, A., Gettins, R. B., Brown, J. G., *J. Inorg. Nucl. Chem.*, in press.
(7) Dyer, A., Gettins, R. B., Molyneux, A., *J. Inorg. Nucl. Chem.* **1968**, 30, 2823.
(8) Radovanov, P. D., Gacinovic, O. M., Gal, I. J., *J. Inorg. Nucl. Chem.* **1969**, 31, 2981.
(9) Sherry, H. S., Walton, H. F., *J. Phys. Chem.* **1967**, 71, 1457.

RECEIVED January 14, 1970.

Discussion

D. J. C. Yates (Esso Research Co., Linden, N. J. 07036): Do you think that useful information in these cases of breakdown can be obtained *via* surface area measurements? Also, under what conditions was the calcination done? For example, if fast heating was employed, the breakdown might be due to the water evolved rather than to temperature as such.

A. Dyer: I agree that surface area measurements would be of use, but we did not pursue this. Calcination was carried out under D.T.A. conditions—*i.e.*, static air—and we did not investigate the effects under either flowing air or nitrogen. Our previous publication on BaA (*see* Ref. 7) presented definite evidence that water loss under vacuum, at room temperature and at 70°C, promoted loss of structure as evidenced by loss of intensity of x-ray powder diffraction patterns.

J. D. Sherman (Union Carbide, Tarrytown, New York): How was the H_2O content determined? Could you comment on the reasons for the variability of the H_2O content in Table I? How were the samples prepared for x-ray examination, especially with regard to their state of hydration?

A. Dyer: The water content was determined by T.G.A. We assess the variable H_2O content shown in Table I as possibly being caused by breakdown of crystal structure occurring at the initial exchange step. All samples were allowed to equilibrate over a saturated solution of sodium chloride for one week before x-ray examination.

34

Evolution of the Structure and Texture of a Type 4A Molecular Sieve in the Course of Thermal Treatments between 400° and 800°C

J. L. THOMAS, M. MANGE,[1] and C. EYRAUD

Laboratoire de Chimie Appliquee et de Genie Chimique de la Faculte des Sciences de Lyon, France

> *The experimental techniques used are water adsorption, x-ray diffraction, and densimetry. Starting from 550°C, the zeolite undergoes a transformation. After thermal treatment at 670°C, half of the zeolite phase disappears. The water adsorption capacity of the product is practically nonexistent. It resumes a noticeable value following a grinding stage, indicating that the solid–solid transformation develops from the surface of the particles. Since the zeolite density is much lower than that of the solid product formed, the transformation is effected with the appearance of a closed macroporosity which is later eliminated by sintering. At each stage of the transformation, it is possible to determine the microporosity (residual zeolite) and the macroporosity (fissures within the particles).*

Wolf (6, 7) set forth that the transformation of Type 4A zeolite by thermal treatment at 600°C begins at the periphery of the crystal. This opinion is related to the decrease of water adsorption capacity at 600°C observed by Milton (2) and the abrupt disappearance of activity in the sample at 670°C pointed out by Piguzova (5), while an x-ray crystallographic analysis indicates the presence of a large quantity of untransformed zeolite. In order to show clearly the mechanism of thermal transformation of a Type 4A zeolite, we decided to determine the quantity of residual zeolite and the possible macroporosity at each step of the

---

[1] Present address: Laboratoire d'Etude des Matériaux, Institut National des Sciences Appliquées de Lyon, France.

transformation. Since neither x-ray crystallographic analysis nor differential thermal analysis furnish quantitative analyses, we made use of more specific techniques: grinding of the particles, giving rise to the residual zeolite phase, water adsorption, and densimetry.

Experimental Techniques

Thermal Treatment. A sample-holder of quartz containing 30 grams of Type 4A zeolite is introduced into a stove in which the temperature is regulated at ±1°C. The temperature of the zeolite is checked by a thermocouple placed inside the sample and protected by a quartz sheath. The heating and cooling rates are 300°C per hour. The sample is maintained at the chosen temperature for 2 hours in air at atmospheric pressure.

Grinding. Three grams of sample are dry-ground for 3 hours in an agate ball-mill.

Water Adsorption. The activation of samples under vacuum at 450°C and the study of water vapor adsorption at 25°C and under 15 torr were effected by means of an Ugine-Eyraud electronic thermobalance with a continuous recording device.

Densimetry. The picnometric liquid, cyclohexane, is introduced under vacuum and does not enter the cavities of Type 4A zeolite. One gram of sample placed in the picnometer is desorbed at 450°C, in air, for 2 hours. The picnometer equipped with the capillary and its stopper

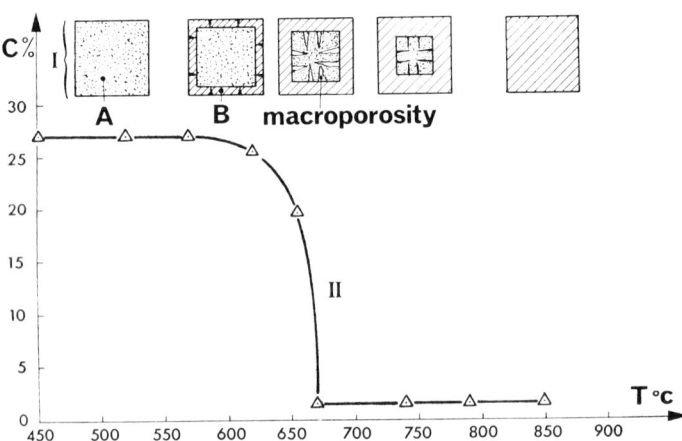

Figure 1. Topochemical evolution of the Type 4A zeolite crystal (I) and plot of values of water adsorption capacity $\left(C\ \% = \dfrac{\Delta p\ mg\ H_2O\ adsorbed}{p\ mg\ dry\ Type\ 4A\ zeolite} \times 100 \right)$ (II) as a function of the thermal treatment temperature

is cooled to room temperature in a container holding dry nitrogen, then weighed on a precision balance. No weight gain is noted in the course of the measurement. After the capillary has been unstopped, the picnometer is placed quickly in the cyclohexane-introduction device under vacuum. After thermostatization at 21 ± 0.01°C and adjustment of the cyclohexane to a constant level, the picnometer is weighed. The experiment shows that the amount of water which can be adsorbed during the placement of the unstopped picnometer in the cyclohexane-introduction device is negligible. Another less rigorous method consists in introducing into the picnometer a sample equilibrated with the air humidity, determining the water which the sample contains. In this case, cyclohexane must not be introduced under vacuum. Both methods give concordant data. The density value is the average of 6 measurements, 3 effected by each of the 2 methods. They are separated by less than 1% of the average value.

X-ray Diffraction. The x-ray powder diffraction patterns of the samples, cooled to room temperature, were obtained with the help of a Phillips PW 1009/30 generator, using cobalt Kα1 radiation.

Results

Figure 1 shows the adsorption capacity of a Type 4A zeolite at the end of thermal treatment. The water adsorption capacity lessens starting at 550°C and falls abruptly from 620° to 670°C. The x-ray diffraction patterns of the samples treated from 450° to 670°C are characteristic of pure Type 4A zeolite. When a sample is treated at 740°C, lines characteristic of low form carnegieite appear. After treatment at 790°C, the pattern indicates the presence of a small amount of zeolite, a large proportion of low form carnegieite, and a very small quantity of nepheline. After treatment at 850°C, nothing remains except low form carnegieite and nepheline in small proportions.

The thermal transformation of zeolite produced during heating and cooling of the sample results in a solid, B, according to the reaction

$$\text{Zeolite Type 4A} \rightarrow \text{solid B}$$

The density measurements of the unground samples and the water quantity measurements contained in the ground samples lead to the results presented by Figure 2.

The volume of each cubical particle of a dimension close to 1 micron is made up of 3 parts in which the proportions change as a function of the temperature: the solid B defined earlier, the residual Type 4A zeolite, and a closed macroporosity. It is assumed that the residual zeolite is proportional to the water retention capacity of the ground sample. The weight of the water desorbed under vacuum at 450°C by a sample is measured, after having been saturated at 25°C under a water vapor pressure of 15 torr.

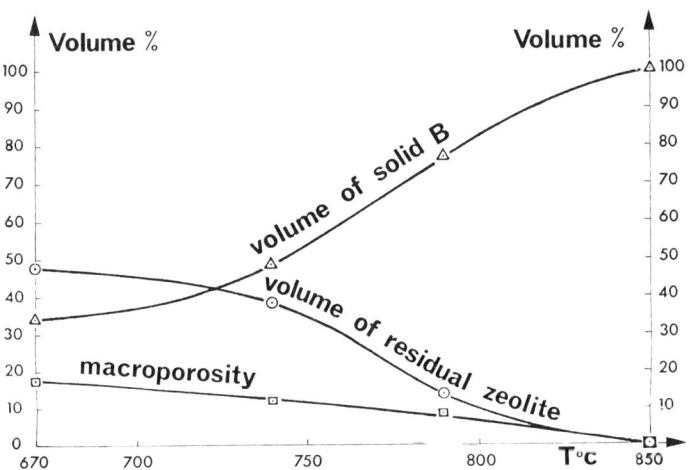

Figure 2. Values of volumes in % of residual Type 4A zeolite, of solid B, and of macroporosity as a function of thermal treatment temperature

If x is the percentage in weight of water contained in the ground sample in relation to the weight of an anhydrous sample and y is the water adsorption capacity of the initial zeolite in relation to the weight of an anhydrous zeolite,

$$P_1 = \frac{x \cdot 100}{y}$$

is the percentage in weight of the residual zeolite in the ground sample. The percentage in weight, P_2, of solid B is obtained by subtraction.

$$P_2 = 100 - P_1$$

The volumes V_1 of residual zeolite and V_2 of solid B are obtained by dividing P_1 and P_2 by their respective densities, ρ_1 and ρ_2. We took for ρ_1 the value 1.67 grams/cm³, determined by Novikova (4) and attributed to ρ_2 the value obtained for the product treated at 850°C, that is, 2.55 grams/cm³.

The macropososity, V_3, is obtained by the difference between the apparent volume, V_4, of the unground sample and the sum of the volumes V_1 and V_2

$$V_3 = V_4 - (V_1 + V_2)$$

Figure 2 shows the respective percentages of residual zeolite, of solid B, and of macroporosity.

Table I. Macroporosity, Volume %

Temp., °C	670	740	790	850
Method 1	17.7	13.7	8.4	0
Method 2	12.0	8.0	2.3	0

Discussion

The 3 essential points of our study are the following: The solid B formed by thermal treatment masks the adsorption properties of the residual zeolitic phase. The adsorption capacity reappears with the grinding of the sample. The transformation of the zeolite occurs with the appearance of a closed macroporosity.

For the purposes of our study, the exact nature of the solid B is only a secondary consideration. In fact, when passing from a structural microporosity phase like zeolite to a dense phase, the volume contraction depends to a relatively small extent on the nature of the final phase. However, we used x-ray diffraction to reexamine the solid product resulting from the thermal treatment. The presence of low form carnegieite is not observed until after a thermal treatment at 740°C. At 790°C, a small amount of nepheline appears (1). The D.T.A. was only done for a thermal treatment at 850°C. The cooling stage exhibits an exothermal peak of weak intensity, attributed to the formation of low form carnegieite. Since there is no reason to think that thermal treatment at lower temperature leads to the same dense phases, we decided to calculate the macroporous volume using the experimental density of the product resulting from a treatment at 850°C. In determining the density of the products before and after grinding, the macroporous volume can be measured. This new method of determining the macroporous volume yields lower results than those of Figure 2, as indicated in Table I.

This difference can be attributed to two factors. The density for pure zeolite is that indicated by Novikova, 1.67 grams/cm^3 (4). It is higher than our products, 1.57 grams/cm^3 after activation treatment at 450°C. The grinding permits reaching all the residual zeolite but only part of the macroporosity.

The residual zeolite does not manifest the same thermal properties as the initial zeolite. Its x-ray diffraction pattern is nevertheless the same. It loses its adsorption properties when an attempt is made to desorb it at 450°C for 2 hours. We limited ourselves to determining its water loss after saturation at 25°C under 15 torr. Since grinding of the untreated zeolite does not change its adsorption properties and its structure as determined by x-ray diffraction, we shall attribute the lack of thermal stability of the residual zeolite to a change in composition owing to the formation of solid B which *a priori* has no reason to possess the exact composition of the initial zeolite. Zeolites having the same structure as identified by x-ray diffraction can have very different chemical composition.

To explain the total suppression of adsorption capacity from the moment of disappearance of half of the zeolitic phase and the appearance of a closed macropososity, it seems necessary to invoke the hypotheses of Wolf (6, 7): the solid–solid transformation begins at the periphery of the particle and progresses toward the center. It is sufficient to cause the shell of product B to break in order to bring about the residual zeolitic phase and thus find the properties of the microporous solid. After thermal treatment at 670°C, calculation indicates the presence of 37 volume % of solid B. If a cubic particle with an edge equal to 1 micron is considered, the thickness attained by the layer of solid B impervious to water is close to 0.1 micron.

Conclusion

The thermal transformation of Type 4A zeolite begins at 550°C, as indicated by the decrease of water vapor adsorption capacity. This capacity is nonexistent after a thermal treatment at 670°C. We have demonstrated that, after grinding the samples which have been heated previously at this temperature, half of the zeolitic phase, characterized by its water retention capacity, remained. The residual zeolite is thermally unstable. It has the same x-ray diffraction pattern as the initial zeolite, but should not have the same chemical composition. We have shown that the solid–solid transformation is accompanied by a closed macroporosity which disappears gradually with sintering. There is every reason to believe that the solid–solid transformation begins at the periphery of the particles and progresses towards the center.

Literature Cited

(1) Berger, A. S., Yakoblev, L. K., *Zh. Prikl. Khim.* **1965**, 38 (6), 1240.
(2) Milton, R. M., Fr. Patent, **1,117,776** BO1 d, (1956).
(3) Ni, L. P., Khalyapina, O. B., Perekhrest, G. L., *Zh. Prikl. Khim.* **1966**, 39 (12), 2639.
(4) Novikova, O. S., Rassonskaya, I. S., Ryabova, N. D., *Uzbeksk. Khim. Zh.* **1964**, 8 (6), 61.
(5) Piguzova, L. I., *Khim. i. Technol. Topliva i Masel* **1961**, 4.
(6) Wolf, F., Fuertig, H., *Tonind. Zgt. Keram. Rundschau* **1966**, 90 (7), 297, 303, 310.
(7) Wolf, F., Fuertig, H., Nemitz, G., *Chem. Techn. (Berlin)* **1967**, 19 (2), 83–87.

RECEIVED January 30, 1970.

Discussion

J. D. Sherman (Union Carbide, Tarrytown, New York): Please describe the sequence of operations involving grinding and sorption measurements which you carried out after calcination, to show that the sorption capacity is increased by grinding.

J. L. Thomas: After calcination and cooling, the sample is dry-ground for three hours. Then it is saturated at 25°C under 15 torr inside an electronic thermobalance, where the desorption is carried out at 450°C for two hours. The water quantity desorbed corresponds to a water retention capacity of 17% in weight for the sample heated at 670°C, while the water adsorption capacity of the same unground sample heated at the same temperature is 2%. Thus, by grinding the samples heated at 670°–850°C, it is possible to recover a noticeable value of the sorption capacity.

R. B. Barrett (Rosary Hill College, Buffalo, New York): After heating beyond 650°C, will your samples readsorb water?

J. L. Thomas: Yes. The decrease in water adsorption capacity begins at 550°C but at 650°C, the water adsorption capacity is still about 20% in weight.

35

Recovery and Purification of Cesium-137 from Purex Waste Using Synthetic Zeolites

LANE A. BRAY and HAROLD T. FULLAM

Battelle Northwest, Richland, Wash. 99352

> *The recovery and purification of cesium-137 from Purex acid waste using a synthetic zeolite has been studied. Zeolite capacity and selectivity for cesium were determined. Stability of the synthetic zeolite to high radiation fields and chemical attack was adequately demonstrated. Kilocurie quantities of cesium-137 of 98+% chemical purity were prepared using zeolite ion exchange.*

In the reprocessing of nuclear fuels, disposition of the radioactive waste is a serious problem. At Hanford, the Purex process is used to reprocess spent fuel. The radioactive fission products leave the process as an aqueous nitric acid stream. Current waste management planning at Hanford calls for the separation of the strontium-90 and cesium-137 from the acid waste, with subsequent packaging and long-term storage of each element as individual compounds in small high-integrity containers. Cesium chloride and strontium fluoride were selected as the optimum compounds for storage. The final storage sites have not been selected yet, but salt mines are a possible choice. Interim storage (30–50 years) will be in concrete canyons on the Hanford Reservation.

The Atlantic Richfield Hanford Co. (ARHCO) operates the Purex Plant at Hanford for the USAEC, and has responsibility for design, construction, and operation of the Waste Packaging Plant. The Pacific Northwest Laboratory (PNL, operated for the USAEC by Battelle Memorial Institute) has responsibility for developing the technology required for the packaging plant.

For cesium, the waste packaging process calls for the separation of the cesium from the acid waste, purification to remove metallic contaminants, conversion to anhydrous chloride, and subsequent encapsulation in double-walled metal cans. The cesium currently is separated from the acid waste by phosphotungstic acid precipitation and partially purified by ion exchange using an inorganic exchanger AW-500 (a synthetic zeolite produced by the Linde Division of the Union Carbide Corp.). The partially purified cesium product from the AW-500 column contains large amounts of sodium and potassium, and additional purification is required before the cesium is converted to the chloride for packaging. Inorganic ion exchange was selected as the best method for obtaining the needed purification.

The cesium product stream from the AW-500 column after concentration for ammonia removal is a carbonate solution having the approximate composition:

^{137}Cs — >1000 curies/liter   Rb$^+$ — 0.004M
Cs$^+$— 0.3M NH$_4^+$— 0.0024M
Na$^+$— 2.0M
K$^+$ — 1.0M pH— 10.7

Radioactive heat generation ~ 6 watts per liter

Packaging requirements dictate that the purity of the cesium chloride be at least 95 wt %. Earlier work at PNL by Mercer and others (1, 2) using a nonradioactive cesium solution indicated that ion exchange using the synthetic zeolite Zeolon (3) (produced by the Norton Co.) offered the best chance of obtaining the required cesium purification. Accordingly, a series of experiments was carried out in a PNL High Level Radiochemical Facility (hot cell) to purify radioactive cesium solution obtained from the AW-500 column. The solution was diluted three-fold prior to loading on the Zeolon column.

Two Zeolon columns were used in the study: one was a 1.9-cm. diameter column containing 40 ml of exchanger (L/D = 7), and the second was a 5-cm diameter column containing 1000 ml of exchanger (L/D = 10). The small column was used to evaluate the performance of the Zeolon exchanger, while the second was used to prepare large quantities of high-purity cesium solution for further processing.

The operation of each column was essentially identical. First the column was loaded with the cesium feed at the rate of 2 column volumes per hour (2 Cv/hr). The column then was scrubbed with 8 Cv of a 0.15M (NH$_4$)$_2$CO$_3$–0.1M NH$_4$OH solution at 2 Cv/hr. Next the column was eluted with a 3M (NH$_4$)$_2$CO$_3$–2M NH$_4$OH solution at 2 Cv/hr. Finally, the column was washed with 4 Cv of water before the next loading cycle. Downflow was used for all column operations, and the column temperature was maintained at approximately 25°C. Each column volume of effluent was analyzed for cesium by in-cell gamma energy analysis. The entire eluent effluent was combined after each run and sampled for sodium, potassium, and rubidium analysis by flame photometry.

Variations in column operating conditions were limited because of the high cost involved in hot cell operations. The operating conditions used for the hot cell tests were selected on the basis of tests carried out using a nonradioactive cesium solution.

Twenty-one exchange cycles were carried out using the 40-ml column. Cesium breakthrough curves were obtained for Runs 2, 7, and 20 (see Figure 1). In each case, a 5% Cs breakthrough occurred at the 10th column volume of feed. This corresponds to a cesium loading of 1.25 milliequivalents per gram of dry Zeolon at 5% breakthrough and shows that the Zeolon is quite stable to radiation and chemical attack, and little loss of cesium capacity should result from extended use.

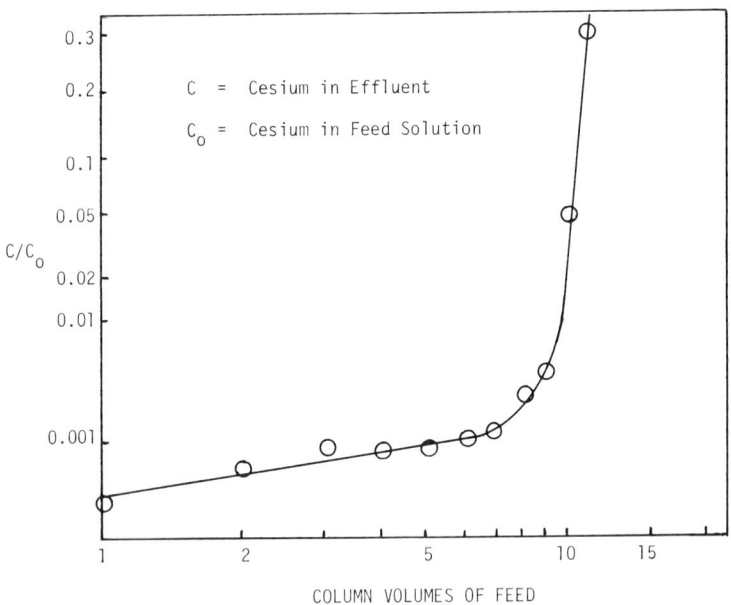

Figure 1. *Cesium breakthrough curve for Zeolon exchanger, Cycle 20*

Decontamination factors (DF's) for sodium, potassium, and rubidium were obtained at various loading levels using the 40-ml column. The results obtained (Table I) show that for sodium and potassium the DF's decreased with increased column loading. For rubidium, the DF is essentially independent of column loading.

The data presented indicate that low cesium loadings should be used in order to obtain high sodium and potassium DF's. However, from operating considerations, the loading level should be as high as possible

Table I. Effect of Cesium Loading on Na, K, and Rb Decontamination

Element	Decontamination Factors		
	6 Cv Feed	7 Cv Feed	9 Cv Feed
Na	216	85	31
K	39	25	9
Rb	2.5	3.5	2.5

Table II. Comparison of DF's Obtained in 40- and 1000-Ml Columns at 6 Cv Feed Loading

Element	40-Ml Column	1000-Ml Column
Na	216	600
K	39	87
Rb	2.5	9

Table III. DF's for Na, K, and Rb Using a Nonradioactive Feed Solution

Element	Decontamination Factors		
	2 Cv Feed	6 Cv Feed	8 Cv Feed
Na	1.3	160	1000
K	4	88	–
Rb	1	4	7

to reduce the number of exchange cycles required. Six column volumes of feed per cycle was selected as the minimum loading level which would be acceptable in a plant operation.

Three runs were made in the 1000-ml column using a loading of 6 Cv of feed per run. The decontamination factors obtained in the large column were significantly better than those obtained in the 40-ml column (Table II), possibly because of decreased wall effects and the greater length-to-diameter ratio. The cesium solution from the large column was sufficiently low in sodium and potassium that an anhydrous cesium chloride of 98+% was obtained on further processing. Several kilocuries of the chloride were prepared.

It is of interest to compare results obtained with the radioactive cesium feed with results obtained using simulated nonradioactive feed solutions. Two nonradioactive solutions were tested: one was a nitrate feed of pH 5.9; the second a carbonate feed of pH 11.2. The cation composition of each simulated feed was the same as that of radioactive cesium feed. A 40-ml Zeolon column was used and the operating conditions were similar to those used in the radioactive tests.

In the nonradioactive studies, the decontamination factors for Na^+, K^+, and Rb^+ increased with increasing cesium loading (Table III) for both the nitrate and carbonate feed solutions. In addition, the cesium capacity of the exchanger decreased substantially (up to 30%) with extended use and also varied substantially from lot to lot (Figure 2). However, lots of exchanger obtained within the past year showed more consistent loading behavior.

Reasons for the differences in behavior of the radioactive and nonradioactive feed solutions are currently under investigation.

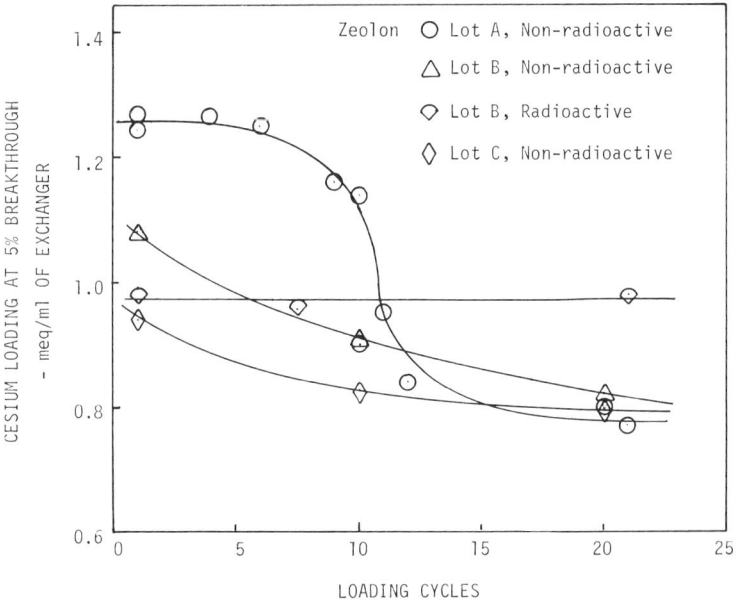

Figure 2. Variations in cesium capacity with exchanger lot and loading cycle

One problem was encountered in the elution phase of the exchange cycle. In about 10% of the cycles, complete elution of the cesium from the exchanger could not be effected. This occurred in both the radioactive and nonradioactive tests. Analyses of exchanger samples from the nonradioactive tests after 20–30 exchange cycles indicated that the residual cesium amounted to 0.1–0.15 milliequivalents per gram of exchanger. This amounted to about 10% of the initial loading capacity of the exchanger under process operating conditions. Various eluting agents were tested to remove the residual cesium. Only hot (70°C) concentrated nitric acid was effective in stripping the cesium from the exchanger.

Literature Cited

(1) Mercer, B. W., Ames, L. L., "The Adsorption of Cs, Sr, and Ce on Zeolites from Multication Systems," USAEC Report **HW-78461** (1963).
(2) Nelson, J. L., Mercer, B. W., "Ion Exchange Separation of Cs from Alkaline Supernatant Solutions," USAEC Report **HW-76449** (1963).
(3) Norton Co., "Zeolon Synthetic Zeolites—Technical Data Sheet," Norton Co., Worcester, Mass.

RECEIVED February 4, 1970.

36

Dielectric Study of Synthetic Zeolites X and Y

R. A. SCHOONHEYDT and J. B. UYTTERHOEVEN

Laboratorium voor Oppervlakteschéikunde, University Leuven,
42 de Croylaan, 3030 Heverlee, Belgium

The electrical conductivity and the capacitance of dehydrated zeolites X and Y, saturated with different monovalent cations (Li^+, Na^+, K^+, Ag^+, Rb^+, and Cs^+), were measured as a function of temperature and frequency. At higher temperatures ($>350°K$) an ionic conduction mechanism was observed. At low temperature ($<350°K$) a dielectric absorption is superposed on the ionic conduction. The activation energies of both ionic conductivity and dielectric absorption are comparable in magnitude. The observed phenomena are explained by the migration of the exchangeable cations over the hexagonal rings and square faces in the large cavities. The experimental activation energies are compared with theoretical values estimated on the basis of an ionic interaction between the exchangeable cations and the zeolite lattice.

The mobility of the exchangeable cations has been demonstrated in various types of zeolites, hydrated or dehydrated. Barrer and Rees (4) studied the self diffusion of monovalent cations in analcite. The activation energy of self diffusion increases with increasing radius. A linear relationship was found between the activation energy and the polarizability of the exchangeable cations.

Beattie (6, 7) investigated the electrical conductivity of dehydrated analcites and chabazite. Freeman and Stamires (19) confirmed the conclusions of Barrer and Rees (4) and Beattie (6, 7) by electrical conductivity measurements at 200 Hz on various ion-exchanged forms of dehydrated synthetic zeolites of type A, X, and Y. They found a purely ionic conduction with a strong dependence of the activation energies on the nature of the zeolite and the kind of cation. The decrease of the activation energy in X and Y zeolites with increasing monovalent cationic radius contrasts with the results found for analcite, and illustrates the openness of the zeolitic framework in X and Y zeolites. In the case of A zeolites, a steric effect is obvious: the activation energy decreases from Li^+ to Na^+ and increases from Na^+ to K^+. From the difference in activation energy between X and Y, Freeman and Stamires (19) concluded that at least 2 types of cation sites exist in that lattice. In the X samples, the lower activation energies are explained by the population of the less energetic sites.

However, a difference in cation sites between X and Y has not been confirmed by x-ray diffraction (8–12, 18). The crystallographic data available so far merely reveal a different occupancy of the sites in X and Y instead of a difference in the nature of the sites. Recently, Schoonheydt and Uytterhoeven (33) ascribed the electrical conductivity of dehydrated X and Y zeolites to the migration of the exchangeable cations located in the large cavities. The difference between the 2 types of zeolite was explained by the possible higher occupancy of the sites inside the cubo-octahedron in X zeolites. A dipole absorption process was observed at low temperatures. Following an hypothesis advanced by Barrer and Saxon-Napier (5), this dipole absorption was ascribed also to the migration of the cations. Morris (28, 29, 30) reported a dielectric study of zeolite 5A in the presence of various adsorbed molecules. The dielectric absorption in the range of 5 Hz to 148 kHz was assigned to a jumping process involving "interstitial" cations.

Stamires (35) investigated the effect of various adsorbed phases on the electrical conductivity of Linde synthetic crystalline zeolites type X and Y. He concluded that the potential energy barrier for cationic conduction is decreased by interaction with the adsorbed phases, but he did not investigate the dielectric response of the material.

In the present work, a more detailed mechanism of the dipole relaxation and ion migration processes in zeolite X and Y will be proposed. Theoretical calculations based on the proposed model will be made. The semi-quantitative agreement between the experimental and calculated values favors the proposed model.

Experimental

Samples. X and Y samples were obtained from the Linde Co. in the Na^+ form. The preparation of the samples containing other cations follows a conventional ion exchange method described elsewhere (37). The full details for the preparation of the decationated samples is reported also in an earlier paper (34). The procedure exists in an exchange with NH_4^+ ions followed by vacuum heating. The samples lose NH_3 and protons enter the lattice. However, according to the work of Kerr (26), our pretreatment conditions are such that so-called "ultrastable zeolites" are obtained which suffered almost complete dehydroxylation. The results of the chemical analysis, expressed in unit cell composition of the dry material, are reported in Table I.

Sample Preparation and Electrical Equipment. The conductivity measurements were performed on compacted cylinders with a diameter of 10 mm and a thickness between 1 and 2 mm. In order to create reproducible conditions, the compaction of the pellets was investigated as a function of the applied pressure. Above 3000 kg cm^{-2}, the conductivity and capacitance of the pellets were independent of the pressure. Therefore, a standard pressure of 3000 kg cm^{-1} has been adopted in the preparation of all the pellets. The pellets were placed between 2 Pt electrodes in a vacuum cell, allowing pretreatment at temperatures up to 500°C under constant high vacuum. The conductivity and capacitance were measured at various temperatures between -20 and $+500$°C, while continuous evacuation (10^{-5} torr) prevented rehydration of the samples. Measurements have been made over the frequency range 200 to 20,000 Hz using an AC Wayne Kerr Bridge B221 in connection with a Wayne Kerr Waveform Analyzor A231, and a Wayne Kerr Audio Signal Genera-

Table I. Analytical Data of the Zeolite Samples[a]

Sample	Unit Cell Composition		
LiX	$Li_{67.38}$	$Na_{17.6}$	$(AlO_2)_{85}(SiO_2)_{107}$
NaX		Na_{85}	$(AlO_2)_{85}(SiO_2)_{107}$
AgX	Ag_{84}		$(AlO_2)_{84}(SiO_2)_{108}$
KX	$K_{82.43}$	$Na_{2.57}$	$(AlO_2)_{85}(SiO_2)_{107}$
$K_{77}X$	$K_{64.62}$	$Na_{19.38}$	$(AlO_2)_{84}(SiO_2)_{108}$
$K_{47}X$	$K_{40.51}$	$Na_{43.49}$	$(AlO_2)_{84}(SiO_2)_{108}$
$Rb_{66}X$	$Rb_{56.4}$	$Na_{28.6}$	$(AlO_2)_{85}(SiO_2)_{107}$
$Cs_{54}X$	$Cs_{45.9}$	$Na_{39.1}$	$(AlO_2)_{85}(SiO_2)_{107}$
$Cs_{40}X$	$Cs_{31.15}$	$Na_{48.85}$	$(AlO_2)_{80}(SiO_2)_{112}$
$Li_{50}Y$	$Li_{27.28}$	$Na_{27.72}$	$(AlO_2)_{55}(SiO_2)_{137}$
NaY		Na_{55}	$(AlO_2)_{55}(SiO_2)_{137}$
AgY	Ag_{52}		$(AlO_2)_{52}(SiO_2)_{110}$
KY	$K_{45.97}$	$Na_{9.03}$	$(AlO_2)_{55}(SiO_2)_{137}$
$K_{83}Y$	$K_{45.88}$	$Na_{9.12}$	$(AlO_2)_{55}(SiO_2)_{137}$
$K_{45}Y$	$K_{24.96}$	$Na_{29.04}$	$(AlO_2)_{54}(SiO_2)_{138}$
$Rb_{73}Y$	Rb_{40}	Na_{15}	$(AlO_2)_{55}(SiO_2)_{137}$
$Cs_{66}Y$	$Cs_{36.6}$	$Na_{18.4}$	$(AlO_2)_{55}(SiO_2)_{137}$
$Cs_{54}Y$	$Cs_{29.07}$	$Na_{24.93}$	$(AlO_2)_{54}(SiO_2)_{138}$
$H_{50}Y$	$H_{24.83}$	$Na_{30.17}$	$(AlO_2)_{55}(SiO_2)_{137}$
$H_{90}Y$	$H_{49.58}$	$Na_{5.42}$	$(AlO_2)_{55}(SiO_2)_{137}$

[a] The symbols used in this table and in all figures indicate the type of zeolite (X or Y) and the nature of the saturating ions. Some samples were only partially exchanged. The exchange levels are included in the label of each sample by a number which indicates the per cent exchange. The unexchanged fraction of the ions is constituted by NaY ions. For samples where this indication is missing, the exchange level was 100% or nearly so.

tor S121. The source delivered a 30-volt signal, but the effective voltage on the sample depended on the frequency and never exceeded 2V rms. More details on the construction of the cell and the measurements were given elsewhere (29, 31).

Electric Measurements. The measurements with the Wayne Kerr equipment provide the conductivity and the capacitance of the pellets. From these we computed the specific conductivity (σ), the real and imaginary part (ϵ' and ϵ'') of the dielectric constant, and the loss factor $\left(\text{tg } \delta = \frac{\epsilon''}{\epsilon'}\right)$ at different frequencies (200–20,000 Hz) and temperatures ($-20°$ to $+400°C$).

The electrical conductivity measurements on powdered compacts suffer from 2 major difficulties: the boundaries between the microcrystals introduce a supplementary energy barrier to current transport known as the interfacial polarization or Maxwell-Wagner effect, and the current is limited by an electrode polarization caused by the imperfect contact between the electrode and pellet surfaces and by the rate of discharge of the cations at the electrodes.

The electrode polarization could be reduced greatly by improving the electrode contacts with vacuum sputtered gold or silver powder films pressed on the pellets. Also, blocking the electrodes with thin mica sheets prevented the discharge of the cations and enabled us to recognize in our results the effects of the electrode polarization.

In the interpretation of the loss factor tg δ, it is not easy to make a distinction between a dipole relaxation and the interfacial polarization. With metallic electrodes, both effects are superposed on the ionic part of the dielectric loss and not necessarily distinguishable from it. With blocking electrodes, the relative intensity of the dipole relaxation and the Maxwell-Wagner effect depends on the ratio of the thickness of the blocking layers and the zeolite pellets (15).

However, the Maxwell-Wagner effect is inherent to powder compacts and if it were present in the range of temperature and frequency covered by our investigations, it should be observed with all the samples regardless of the nature of the exchangeable ions in the zeolite. Our results indicate that dielectric absorption peaks are found in the samples containing 2 different exchangeable cations. Moreover, the intensity of the absorption increases with increasing difference in the radius of the 2 cations. Because of this indirect experimental evidence, we think that the dielectric loss observed at low temperature can be ascribed to a dipole relaxation process, and that Maxwell-Wagner effects are not present. Equivalent circuit calculations were attempted to strengthen this conclusion but were not entirely successful (2, 27).

Results

The conductivity of the samples increases with the temperature. The relationship can be expressed by an Arrhenius equation, an example of which is shown in Figure 1. At high temperatures, the relation between log σ and $1/T$ produces straight lines. At lower temperatures, strong deviations from straight line behavior occur, especially for samples with 2 kinds of exchangeable cations. For samples with only 1 saturating cation, the Y species show a full straight line behavior. For the corresponding X samples, a knee will occur in the Arrhenius plot at low temperature. The transition temperature between the 2 regions increases slightly with increasing frequency. Generally, it occurs in the neighborhood of 400°K.

The specific conductivity depends slightly on the nature of the electrode contacts, as shown in Figure 1, and an interpretation of the

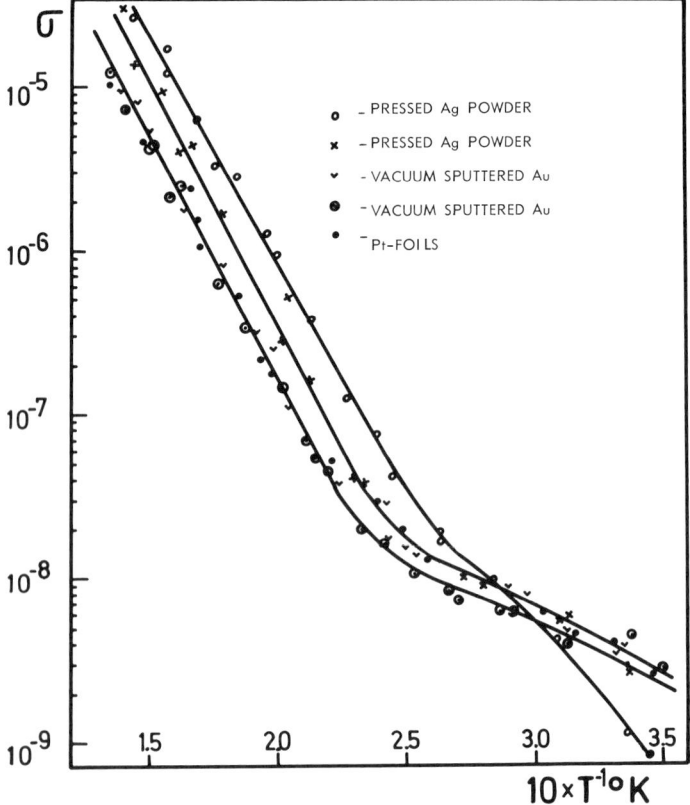

Figure 1. Specific conductivity (ohm^{-1} cm^{-1}) of RbX at 1000 Hz as a function of $1/T$ with different kinds of electrodes

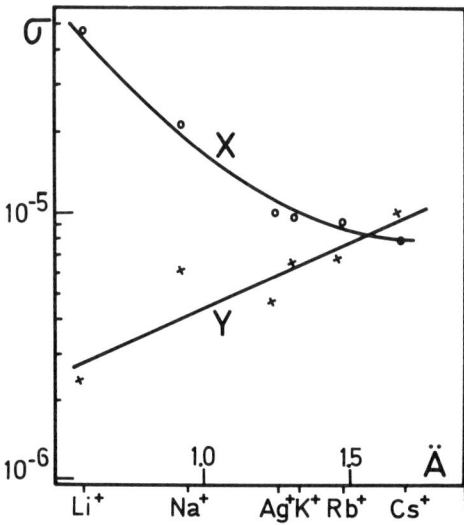

Figure 2. Specific conductivity (ohm^{-1} cm^{-1}) at 700°K and 19,000 Hz as a function of ionic radius

absolute values requires great care. However, the variations of σ as a function of the frequency, the temperature, the nature of the zeolite, and the exchangeable cations is nearly independent of the electrode material and characteristic for the samples studied. Thus, comparisons can be made between zeolite pellets fitted with the same type of electrodes. Conclusions about the charge transport mechanism can be drawn from the relationship $\sigma = \omega\epsilon''$. Here σ is the specific conductivity, ω the angular frequency, and ϵ'' the imaginary part of the complex dielectric constant. In the high-temperature range, plots of log ϵ'' as a function of log ν produce straight lines with slope -1, which is characteristic for a DC conduction mechanism. Important deviations from linear behavior occur in the low-temperature region, indicating that an AC conduction mechanism predominates. The 2 phenomena will be analyzed separately and indicated further as HT (high temperature) and LT (low temperature) mechanisms.

The HT Mechanism. Figures 2, 3, and 4 resume the main characteristics of the HT conduction mechanism. The conduction can be ascribed to the mobility of the exchangeable cations. This is in agreement with conclusions by other authors (3, 6, 7, 19). Thus, in Figure 2, the conductivity at 19 kHz and 700°K is represented as a function of the radius of the exchangeable ions for the X and Y zeolites. The specific conductivity of X is considerably higher than that of Y for samples saturated with small ions. The differences decrease with increasing radius of the exchangeable cations. The conductivity as a function of the Na^+ content is plotted in Figure 3 for the series of NaX, NaY, and the decationated samples. The break in these curves suggests that not all the Na^+ ions contribute in the same way to the conductivity.

From the slope of the Arrhenius plots (Figure 1), the activation energy E of the HT mechanism can be calculated. The results are presented in Figure 4 as a function of the radius of the cations. These data are obtained at 19 kHz. For Y samples, a linear decrease of E with increasing radius of the cations is observed. Points obtained with different electrode material differ by about 2 kcal, and the E values obtained at 1 kHz are higher than those obtained at 19 kHz by about 1 kcal. The accuracy of the determination of the activation energy is thus of the order of 2 kcal. For X samples, the same trends are observed but the scattering of the points is higher. The average values are lower for X than for Y, but the difference is approximately 2 kcal and cannot be considered significant. In a similar comparison, Freeman and Stamires (19) found a difference of 4 kcal.

The activation energies of the decationated samples are strongly dependent on the number of residual Na^+ ions and on the frequency. For the $H_{50}Y$ samples, 25.06 and 15.76 kcal mole^{-1} are obtained, respectively, at 19 and 1 kHz. The corresponding values for the $H_{90}Y$ samples are 20.83 and 17.62 kcal mole^{-1}.

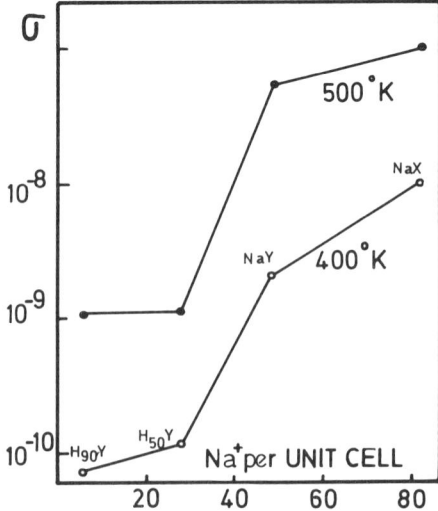

Figure 3. Specific conductivity (ohm^{-1} cm^{-1}) as a function of residual Na^+ content at 1000 Hz

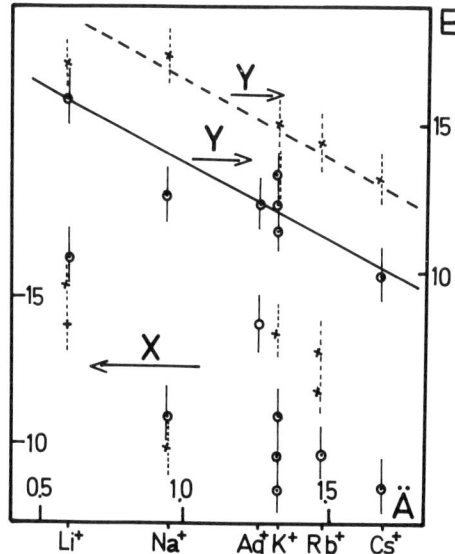

Figure 4. Activation energy (kcal mole⁻¹) of the DC conduction vs. the ionic radius, obtained with pressed silver electrodes (×) or with platinum electrodes (○ and ⊙)

The Arrhenius equation can be interpreted in terms of the theory of the absolute reaction rates (20). Taking 4.65 Å as the jump distance, and estimating the number of charge carriers from the total ion exchange capacity of the zeolite, we can calculate the entropy change of conduction. The values range between −5 and − 20 eu. Freeman and Stamires reported ΔS values between −2 and −11 eu. Our values have a tendency to decrease with increasing cationic radius. However, all the uncertainties of the measurements accumulate in this factor. Indeed, the influence of the electrode contact caused variations of equal importance as the size of the cations. For this reason, we prefer not to insist on the interpretation of ΔS values.

The LT Mechanism. The dielectric behavior of the zeolites in the low-temperature region is more complicated than at high temperatures. A frequency-dependent process is superposed on the ionic conductivity, and the relative intensity of the 2 phenomena varies from sample to sample. The understanding of the LT mechanism requires the interpretation of the dissipation factor tg δ, which is obtained experimentally as the ratio between the imaginary and the real part of the complex dielectric constant: ϵ''/ϵ'. For a purely ionic conduction mechanism, the plot of tg δ as a function of the temperature at a fixed frequency would show a

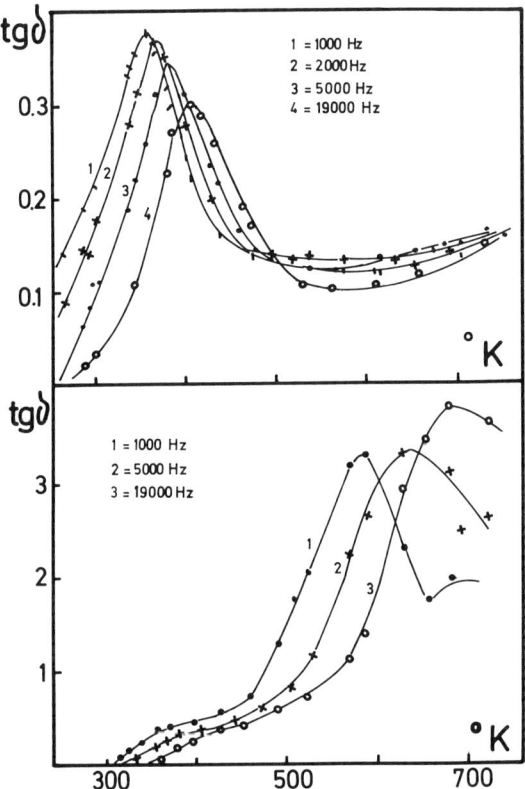

Figure 5. tgδ For $K_{83}Y$ as a function of the temperature at different frequencies
Upper Part: blocking electrodes
Lower part: Pt electrodes

continuous exponential rise. The existence of maxima in this graph indicates the contribution of other effects. Examples are given in Figure 5. This figure also illustrates the suppression of an important electrode effect by the technique of blocking electrodes. The maxima observed as shoulders between 300° and 400°K (lower part) are resolved well with measurements in experiments with blocking electrodes (upper part). Not all samples show LT absorption maxima. The phenomenon is more pronounced when 2 different cations are present and these 2 cations differ considerably in size. Thus, the intensity is lowest for $Li_{79}X$ (tg $δ_{max} < 0.1$) and highest for $Cs_{66}Y$ (tg $δ_{max} \simeq 0.4$). The intensities of the dispersion peaks with and without blocking electrodes are comparable. Also, the intensity of the maxima are nearly independent of the frequency. These points, together with the arguments given in the experimental section, are considered sufficient support for an assignment of the dielectric dispersion to a dipole relaxation phenomenon. The frequencies at which the maxima occur are related to the temperature: $v_m = v_o \exp(-E'/RT)$ (16) where v_m is the frequency at which tg δ reaches a maximum for a given temperature T. E' is the activation energy for the relaxation process. The E' values are summarized in Table II.

Table II. Activation Energies of the Dipole Absorption Process (±1 Kcal)

Sample	E', Kcal Mole^{-1}	Sample	E', Kcal Mole^{-1}
Li$_{79}$X	15.80	Li$_{50}$Y	–
NaX	–	NaY	–
KX	–	KY	–
K$_{77}$X	16.56	K$_{45}$Y	17.00
K$_{47}$X	16.70	K$_{83}$Y	11.53
AgX	–	AgY	–
Rb$_{66}$X	15.53	Rb$_{73}$Y	16.40
Cs$_{40}$X	11.57	Cs$_{54}$Y	14.02
Cs$_{54}$X	11.57	Cs$_{66}$Y	14.62
		H$_{50}$Y	24.20
		H$_{90}$Y	20.13

These energies are very close to the values calculated for the HT mechanism. Barrer and Saxon-Napier (5) found similar results for analcites. These activation energies are average values for a very heterogeneous relaxation phenomenon. This is deduced from the so-called "Cole-Cole" plots relating ϵ'' to ϵ'. An example of such a plot is given in Figure 6. As the plot is constructed from only 4 points, resulting from an extrapolation on the tg δ and ϵ''-curves, we cannot expect a very accurate estimate of the angle α. Nevertheless, its value is high, which indicates an important heterogeneity in the dipole relaxation phenomenon.

Discussion

In agreement with previous workers, we explain the dielectric properties of the synthetic faujasites by the migration ability of the cations

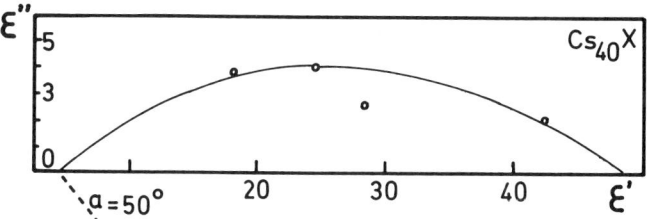

Figure 6. Example of a "Cole-Cole" plot

(5, 6, 7, 19). In addition, we believe that the conduction mechanism can be ascribed almost exclusively to the ions in the large cavities. In the following paragraphs, we attempt to substantiate this hypothesis by focusing on a few experimental points, and by comparison of the obtained activation energies with the results of some theoretical estimations.

Experimental Evidence. The activation energy of conduction represents the highest energy barrier that the migrating cations have to overcome. The continuous decrease of E with increasing radius shown in Figure 4 reflects the diminishing coulombic interaction between the cations and the oxygen lattice, as already pointed out by Freeman and Stamires. The E values of Figure 4 can be explained as corresponding to the energy barriers between sites in the large cavities. The influence of the activation energy on the specific conductivity, according to the Arrhenius function, would cause at a given temperature an increase of the conductivity with increasing cationic radius for both X and Y samples. Figure 2 shows that such an increase is real for Y, but for X a decrease is observed instead of an increase. To explain this, we assume that only the ions located in the large cavities contribute to the conductivity. In that case, 2 factors must be considered. The negative charge of the oxygen framework along the walls of the large cavities will be influenced by the number of ions in the small cavities. The number of ions in the large cavities constitutes the number of charge carriers. A complete interpretation thus would require detailed information on the distribution of the ions over the different sites for X and Y samples saturated with the different cations. Unfortunately, the available data on the distribution of monovalent cations are scarce.

For the KY, NaY, and AgY samples, this information can be taken from the work by Eulenberger *et al.* (*18*). The number of ions in the large cavities is between 28 and 30 ions per unit cell. At room temperature, the bulky ions Rb^+ and Cs^+ are not in the small cavities; 40 Rb^+ and 36.5 Cs^+ ions per unit cell are located initially in the large cavities. However, during the pretreatment at high temperature, a fraction of these large ions may penetrate into the small cavities. After heating a RbY sample to 300°C, an important fraction of the Rb^+ ions became unavailable for ion exchange. In any case, the number of ions in the large cavities is not very different from one Y sample to the other (min. 28, max. 40) and the variation of the activation energy explains the change in conductivity almost completely as a function of the size of the cations.

We have no data on the distribution of the cations in X zeolites. From the work by Eulenberger *et al.* (*12*) a tendency appears towards a higher population of the sites inside the small cavities for the larger ions (K^+ and Ag^+). The Rb^+ and Cs^+ ions, although excluded from the small cavities at low temperatures, could obey the same trend at high temperatures. An increase of the population of the sites inside the small cavities would reduce the number of charge carriers as compared with the samples saturated with the smaller ions, Li^+ and Na^+. These assumptions explain qualitatively the evolution of the conductivity of the X samples given in Figure 2, and the different evolution for the Y samples. A real quantitative interpretation would require more detailed information on the distribution of the cations, especially in the X zeolites.

The evolution of the conductivity as a function of the Na content, as shown in Figure 3, is another argument in favor of our hypothesis. The structure of the decationated samples $H_{50}Y$ and $H_{90}Y$ is not known accurately. Nevertheless, if all the Na^+ ions were contributing in the same way to the conductivity, we would expect a linear relationship between the conductivity and the Na^+ content. In the preparation of the decationated samples, the NH_4^+ exchange is restricted to the large cavities (37). Therefore, the residual Na^+ ions probably are located inside the small cavities. This is consistent with the break in the conductivity curves of Figure 3. Also, the activation energy of conduction in decationated samples is 5 kcal higher than in the NaY sample. This supplementary energy barrier probably represents the energy required to pass through a six-membered ring of oxygen ions towards the large cavities.

The LT Conduction Mechanism. Barrer and Saxon-Napier (5) explained the dipole absorption effect in analcites by a migration of the exchangeable cations. They considered the combination of a vacated negative site and a neighboring positive ion as a dipole. The migration of the ions has the effect of a reorientation of that dipole. Morris (28, 29, 30) observed also a dipole relaxation in zeolite A, and ascribed it to the exchangeable ions without explicit description of the specific mechanism. We agree with these ideas, and attempted a theoretical estimation of the expected activation energy based on the picture described by Barrer. To do this, we used a formula proposed by Breckenridge (13, 14):

$$tg\ \delta_{(max)} = \frac{n\ (Zed)\ (\varepsilon'_\infty + 2)}{18\varepsilon_o\ kT\ \varepsilon'_\infty}$$

In our experiments, the observed intensity of the absorption phenomenon was between 0.1 and 0.3. We considered the distance between the centers of a six-membered ring of oxygens and the adjacent four-ring as the average length d of the dipole. Introducing our experimental values for the dielectric constant, we could deduce from the formula of Breckenridge the number of ions per unit volume contributing to the phenomenon. This was of the order of magnitude of 10^{19}. Comparison with the total amount of exchangeable ions and application of Boltzmann statistics resulted in an activation energy of $\simeq 5$ kcal. This is only between a third and one half of the experimental values given in Table II.

A better approximation cannot be expected because of our oversimplified assumptions. Nevertheless, the correct order of magnitude is obtained, and this is in favor of the assignment of the dielectric loss to a dipole relaxation associated with the migrating cations. The fact that the dipole absorption is particularly pronounced for the samples containing Rb^+ and Cs^+ ions, which are mainly located in the large cavities, is in favor of the hypothesis that the cations in the large cavities offer the main contribution to the dielectric phenomena. Again the relaxation in the decationated samples has activation energies considerably higher than in the samples containing ions in the large cavities.

Migration in the large cavities implies that an ion leaves a six-membered ring of oxygen ions and passes over a sequence of 3 square faces, considered as possible interstitial sites, before arriving at another six-ring. Energetically, this migration path corresponds to a sequence of 4 potential wells. A treatment inspired by Hoffmann (*21, 22, 23*) and adapted to the zeolite system shows that such a migration produces 3 relaxation times. Such a multiplicity is in agreement with the information of the Cole-Cole plot (Figure 6) although the relaxation mechanism is still more complex. The local distribution of the ions can increase the heterogeneity of the potential well system and broaden the spectrum of relaxation times.

The HT Conduction Mechanism. As stated before, the six-rings in the large cavities are separated from each other by a series of 3 square faces. The square face in the middle of this series belongs to a cubooctahedron; the 2 others are part of hexagonal prisms. Geometrically, the square face of the cubooctahedron seems to be the most probable interstitial site. The samples with more than 32 ions per unit cell are thought to have the excess of ions located on this site. The limiting step in the ion migration is the passage from a six-ring towards an interstitial site. The activation energy represents the energy required to extract the ions from a normal lattice site and to induce the migration. The mechanism is identical in X and Y samples and the slight difference in activation energy between X and Y reflects the difference in electrostatic interaction between the cations and the oxygen lattice. This interpretation is at variance with the ideas of Freeman and Stamires (*19*), who attributed the difference between X and Y to the existence of low-energy sites in X which are not present in Y.

On the basis of our model, the activation energy for ion migration can be calculated (*31*) from the expression $E = W - W_p(i)$. W is the Madelung energy at a six-membered ring of oxygen ions. W_p is the polarization energy occurring when a cation leaves a six-membered ring and migrates to an interstitial site. If the zeolitic system is considered as a homogeneous dielectric with dielectric constant ϵ, the formula of Jost

(25) can be applied to calculate this polarization energy: $W_p = \left(\frac{\epsilon - 1}{2\epsilon}\right)\frac{e^2}{R}$. R is the radius of the exchange site and e the charge of the migrating cation. ϵ Is taken from our experimental values and is 8.71 and 15.2 for Y and X zeolites, respectively. Introducing in (i) the Madelung energies reported by Dempsey (17), we estimated activation energies, which are unrealistically high. Therefore, we recalculated W using the method of Yun-Yan-Huang (38). In this method, the negative charge is considered to be spread uniformly over all the oxygen ions. The local modulation of the electrostatic potential is ruled by the distribution of the cations. We used for the Y samples the distribution of the cations reported by Eulenberger (18). For the X and Y samples containing Rb^+ and Cs^+ ions, we assumed that these large ions were inside the large cavities. For the X samples saturated with Li^+, Na^+, and K^+ ions, we assumed that 48 ions were inside the small cavities. These figures are not presented as corresponding to the real situation, but are used only as a basis of computation. In this way, the calculation of W at a six-ring is restricted to the summation of the Coulombic energy over the oxygen ions and the exchangeable cations immediately surrounding the large cavities:

$$W = e \sum_i \frac{e_i}{r_i}$$

where e = charge of the cation on a six-ring
e_i = charge of any other ion i
r_i = distance of ion i to the exchange site on the six-membered ring.

The activation energies estimated in this way are listed in Table III. They agree qualitatively with the experimental values and account for the dependence on the size of the cation, at least for the Y samples.

Finally, an estimation of activation energies was attempted following an empirical method proposed by Anderson and Stuart (1). It considers the activation energy as the difference in electrostatic interaction between the lattice and the exchangeable cation on its normal lattice site and halfway between the six-ring of oxygen ions and the interstitial square face. The Anderson and Stuart (1) method takes into account the Coulombic interaction (Ec), the repulsion energy, the polarization energy, and the Van der Waals interaction. We neglected the Van der Waals energy in our calculations. The first 2 terms are included in the formula:

$$E = \left(1 - \frac{1}{m}\right)\frac{A}{\epsilon} Z Z_0 e^2 \left(\frac{1}{r + r_0} - \frac{1}{d/2}\right)$$

To this the polarization energy, E_p, must be added, which can be calculated from the ratio

$$\frac{E_p}{E_c} = \frac{1}{2}\frac{\alpha}{(r+r_o)^3}$$

In these formulas, Z and r are the valency and the radius of the exchangeable ions, Z_o and r_o the valency and the radius of the oxygen ions; $d/2$ is half the jump distance between a six-ring and an interstitial site; e is 4.774×10^{-10} ese; α is the polarizability of the cations; m results from the repulsion energy and was given the extreme value 10. A, the Madelung constant, is not known for the zeolites but was estimated using a method proposed by Templeton (36). 1.50 and 1.49 were the values adopted for X and Y, respectively. The value of ϵ was 15.2 for X and 8.71 for the Y zeolites.

The results of these calculations are included in Table III. Agreement with the experimental values is fairly good for the Y samples. The values for X are somewhat low. This can be attributed to the value of $\epsilon = 15.2$ which was determined by extrapolating our experimental values to infinite frequency.

Table III. Estimated Activation Energies

			Anderson and Stuart (1)		$E = W - W_p$	
	$(r + r_o)$	$d/2$	E, Kcal Mole^{-1}		E, Kcal Mole^{-1}	
	(A)	(A)	X	Y	X	Y
Li$^+$	2.00	3.42	12.34	21.40	20.8	34.6
Na$^+$	2.35	3.65	9.13	15.83	27.4	30.7
Ag$^+$	2.66	3.81	7.50	13.00	14.0	19.4
K$^+$	2.73	3.85	6.64	11.50	17.3	21.4
Rb$^+$	2.88	3.92	5.86	10.16	–	16.4
Cs$^+$	3.09	4.02	4.90	8.48	–	7.85

Conclusions

The dielectric properties of the zeolites X and Y can be explained by the migration of the ions in the large cavities. The activation energies of conduction represent the energy necessary to extract an ion from the hexagonal windows of the cubooctahedra. Different model calculations based on this hypothesis account for the experimental activation energies. A better knowledge of the distribution of the exchangeable cations, especially for the X zeolites, is an urgent need.

Acknowledgment

The authors thank J. J. Fripiat for the use of the Wayne Kerr equipment. One of the authors (R. A. S.) is indebted to the National Science Foundation (Belgium) for a grant as aspirant. The gift of zeolite samples by the Linde Co. is acknowledged.

Literature Cited

(1) Anderson, O. L., Stuart, D. A., *J. Am. Ceram. Soc.* **1954**, 37, 573.
(2) Arulanandan, K., Mitchell, J. K., *Clays Clay Minerals* **1968**, 16, 337.
(3) Barrer, R. M., Rees, L. V. C., *Nature* **1960**, 187.
(4) Barrer, R. M., Rees, L. V. C., *Trans. Faraday Soc.* **1960**, 56, 709.
(5) Barrer, R. M., Saxon-Napier, E. A., *Trans. Faraday Soc.* **1962**, 58, 156.
(6) Beattie, I. R., *Trans. Faraday Soc.* **1954**, 50, 581.
(7) *Ibid.*, **1955**, 51, 712.
(8) Bennett, J. M., Smith, J. V., *Mater. Res. Bull.* **1968**, 3, 633.
(9) *Ibid.*, **1968**, 3, 865.
(10) *Ibid.*, **1968**, 3, 933.
(11) *Ibid.*, **1969**, 4, 7.
(12) *Ibid.*, **1969**, 4, 77.
(13) Breckenridge, R. G., *J. Chem. Phys.* **1948**, 16, 459.
(14) *Ibid.*, **1950**, 18, 913.
(15) Dansas, P., Sixou, P., Arnault, R., *XIVe Coll. AMPERE, Ljubljana*, **1967**.
(16) Debey, P., "Polar Molecules," Dover, New York, 1929.
(17) Dempsey, E., *J. Phys. Chem.* **1969**, 73, 3660.
(18) Eulenberger, G. R., Shoemaker, D. P., Keil, S. G., *J. Phys. Chem.* **1967**, 71, 1812.
(19) Freeman, D. C., Jr., Stamires, D. N., *J. Chem. Phys.* **1961**, 35, 799.
(20) Glasstone, S., Laidler, K. J., Eyring, H., "The Theory of Rate Processes," McGraw-Hill, New York, 1941.
(21) Hoffmann, J. D., *J. Chem. Phys.* **1952**, 20, 541.
(22) *Ibid.*, **1955**, 23, 1331.
(23) Hoffmann, J. D., Pfeiffer, H. G., *J. Chem. Phys.* **1954**, 22, 132.
(24) Huggins, C. M., Sharbaugh, A. H., *J. Chem. Phys.* **1963**, 38, 393.
(25) Jost, W., *J. Chem. Phys.* **1933**, 1, 466.
(26) Kerr, G. T., "Chemistry of Crystalline Aluminosilicates. VII. Thermal Decomposition Products of Ammonium Zeolite Y," in press.
(27) Miliotis, D., Yoon, D. N., *J. Phys. Chem. Solids* **1969**, 30, 1241.
(28) Morris, B., *J. Phys. Chem. Solids* **1969**, 30, 73.
(29) *Ibid.*, **1969**, 30, 89.
(30) *Ibid.*, **1969**, 30, 103.
(31) Mott, N. F., Gurney, R. W., "Electronic Processes in Ionic Crystals," Dover, New York, 1940.
(32) Schoonheydt, R. A., Ph.D. Thesis, 1970.
(33) Schoonheydt, R., Uytterhoeven, J. B., *Clay Minerals Bull.* **1969**, 8, 71.
(34) Schoonheydt, R. A., Uytterhoeven, J. B., *J. Catalysis*, submitted.
(35) Stamires, D. N., *J. Chem. Phys.* **1966**, 36, 3174.
(36) Templeton, A., *J. Chem. Phys.* **1953**, 21, 2097.
(37) Theng, B. K. G., Vansant, E., Uytterhoeven, J. B., *Trans. Faraday Soc.* **1968**, 64, 3370.
(38) Yun-Yan-Huang, Benson, J. E., Boudart, M., *Ind. Eng. Chem. Fundamentals* **1969**, 8, 346.

RECEIVED February 16, 1970.

Discussion

D. A. Hickson (Chevron Research Co., Richmond, Calif. 94802): Please contrast and compare the dielectric behavior of decationated and ultrastable faujasite. How does the dielectric behavior of these materials change on filling the cavities with polar molecules such as water?

R. Schoonheydt: Our dielectric measurements are performed on tightly compacted pellets. Therefore, we believe that the heating process, producing H-zeolites and removing NH_3, gives a sample which is more likely to be an ultrastable than a decationated faujasite. From our results, a distinction is not possible.

The dielectric behavior on water adsorption is heterogeneous. First, a regular small increase in conductivity occurs up to about three water molecules per cation in the large cavities. Adsorption of more water results in a sudden increase in conductivity, probably because of the high mobility of the exchangeable cations in the hydrated state.

N. G. Parsonage (Imperial College, London): Can you give a simple picture of the process whereby the activation energy decreases with increasing size of the cations?

R. Schoonheydt: The interaction between the exchangeable cations and the zeolitic lattice is mainly electrostatic. Therefore, the binding energy follows the simple law $F = \dfrac{e_c R_r}{r_{rc}}$ where e_c and e_r are the charges of the cation and lattice, respectively, and r_{rc} is the distance between the cation and the lattice. An increase in cationic radius gives an increase in r_{rc} and thus a decrease of F. In other words, the activation energy for migration decreases with increasing size of the cation.

37

NMR Relaxation of Water in Zeolite 13-X

H. A. RESING and J. K. THOMPSON

Naval Research Laboratory, Washington, D. C. 20390

> *The nuclear magnetic resonance relaxation times for the protons of water adsorbed to saturation in a high-purity specimen of zeolite 13-X have been measured between 200° and 500°K. The data can be accounted for by the model of an intracrystalline fluid which is about 30 times as "viscous" as bulk water at room temperature, shows a broad distribution of molecular mobilities, and is about as dense as liquid water. The median correlation time (time between molecular flights) is*
>
> $$\tau^* = 2.8 \times 10^{-12} \exp[417/(T - 189)]$$

"**I**ntracrystalline fluid" is the term chosen by Barrer to describe the state of water and other molecules condensed in the pores of zeolites (*1*). This report presents a determination of the fluidity of water adsorbed to saturation in zeolite 13-X. As a measure of the fluidity, the time between molecular jumps or correlation time, τ, is employed; it has been derived from measurements of nuclear magnetic resonance (NMR) relaxation times *via* the theory of Bloembergen, Purcell, and Pound (BPP) (*2*). For a bulk liquid, this correlation time is related to the viscosity, η, by an equivalent of the Debye equation (*2*):

$$\tau = \pi \eta a^3/(6 kT) \qquad (1)$$

where a is the molecular diameter. Formally, a viscosity for the intracrystalline fluid could be defined thus, but would be without operational significance; further discussion of fluidity, therefore, will be carried out in terms of τ.

The zeolite specimen used in the present studies was prepared from zone-refined aluminum and silicon; its ferric ion impurity content of < 6 ppm is low enough to ensure that only "intrinsic" NMR relaxation is observed (*11*) and that the effects of paramagnetic impurities, which troubled previous studies (*4, 6, 8, 10*), are completely suppressed. The broad-line NMR absorption studies of Kvlividze (5), which we call upon for support in this paper, were done on samples of unspecified purity. Arguments given previously (10), however, indicate that at the low temperatures of her experiments, the presence of paramagnetic impurities would not affect her interpretations to any significant extent. The specimen used in this study was allowed to come to equilibrium with water vapor at a relative pressure $P/P_0 = 1$. The total weight was 0.2886 gram, of which 0.0766 gram is H_2O.

The pulsed NMR apparatus (12 MHz) and methods of procedure have been described previously (10, 13). The spin–spin relaxation time, T_2, equivalent to the inverse absorption line width, is a time constant for the exponential decay toward internal equilibrium in the nuclear spin system. The spin–lattice relaxation time, T_1, is the time constant for the exponential decay toward equilibrium between the nuclear spins and all other degrees of freedom of the system. The data are presented in Figure 1. Note that at higher temperatures T_2 is much lower than T_1.

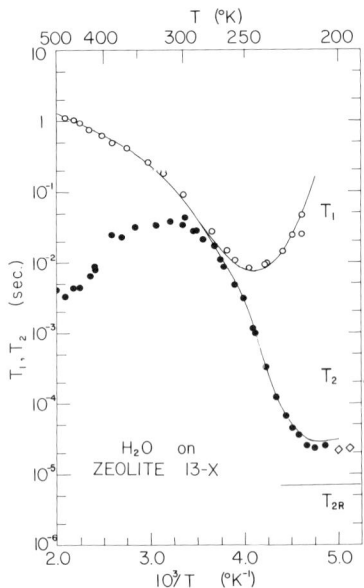

Figure 1. T_1 and T_2 vs. reciprocal temperature at $\omega = 12$ MHz for water in zeolite 13-X

● = T_2
○ = T_1
◇ = T_2 of Kvlividze
—— = Best fit of theoretical expressions to data

This effect is general in solid–water adsorption systems (8) and is caused at least in some cases by an exchange process between the protons of surface hydroxyl groups and the protons of water molecules (15). At about 400°K, the lifetime of a water molecule proton with respect to exchange into this other phase is about $= T_2$, i.e., 10 msec. We intend to discuss this more thoroughly in a future publication, and in this paper we restrict our discussion of T_2 to temperatures such that $10^3/T > 3.6$. In the low-temperature region, around $10^3/T = 5$ (Figure 1) 2 values of T_2 are indicated for a given temperature; here the "T_2" process was describable as a superposition of 2 decays, the relative intensity varying with temperature. Equivalently, the absorption line of Kvlividze (5) (given as T_2 in Figure 1) was composed of a broad and narrow component in this temperature interval. This is the "apparent phase transition effect" caused by a broad distribution of correlation times (8), as will be explained.

The fluidity of associated liquids in general is represented well by the free volume model (3), according to which

$$\tau^* = \tau_0 \exp [A/(T - T_0)] \qquad (2)$$

where τ_0 and A depend on molecular parameters, and where T_0 is the temperature at which free volume disappears. Such a model is consistent with the data of this study. An added assumption which must be made, however, is that at any temperature, τ is spread over a broad range of values; the distribution function assumed is the often used log-normal distribution (9)

$$P(\tau)d\tau = (B/\sqrt{\pi}) \exp (- B^2 Z^2) dZ \qquad (3)$$

$$Z = \ln (\tau/\tau^*) \qquad (4)$$

$$B = \alpha (T - T_0) \qquad (5)$$

where B is a spread parameter dependent on temperature and τ^* is the median of the distribution. This broad distribution is necessary to explain the apparent "two-phase" relaxation described above (9).

BPP (2) showed that the modulation of the internuclear magnetic field by the translation and rotation of molecules is an effective mechanism for relaxation. The other important parameter in their theory besides the correlation time is the second moment or mean squared nuclear dipolar field seen by a nucleus (σ_0^2) (14):

$$\sigma_0^2 = \frac{3}{5} \gamma^4 h^2 I (I + 1) \sum_i r_{ij}^{-6} \qquad (6)$$

where the sum is taken over all i neighboring nuclei. Because the nuclear separation r_{ij} is raised to the sixth power, the second moment is a sensitive measure of the density of molecular packing. We assume that it is not necessary to distinguish between rotational jumps (which modulate the intramolecular field) and translational jumps (which modulate the extramolecular field). In an associated liquid, both processes should have roughly the same rate; thus, both fields are included in Equation 6. The relaxation time data show several features in accord with the BPP theory: there is a minimum in T_1, and at low temperatures T_2 decreases as the temperature is lowered. However, the ratio of T_1 to T_2 and the value of T_1 at the minimum are inconsistent with their theory. Further, their theory does not predict the two-phase T_2 behavior. All of these departures from the BPP theory can be taken into account by allowing a broad distribution of correlation times (9). At high temperatures, exchange between sites of different τ is sufficiently rapid that an "average" T_2 is observed (16). At low temperatures, the averaging is no longer effective; there are 2 values of T_2—the mobile T_2 and the rigid lattice T_{2r}—and as the temperature is lowered further, the fraction of nuclei in the mobile state (f_m) decreases with a complementary increase in the fraction $(1 - f_m)$ of nuclei in the rigid state. The effect of the distribution on T_1 is only to broaden the minimum and to raise the expected value of T_1 at the minimum; no break-down in averaging occurs, in contrast to T_2. The equations used are (9)

$$T_2^{-1} = \frac{1}{3}\sigma_0^2 \left\{ 3 f_m^{-1} \int_{-\infty}^{\tau_c} \tau P(\tau)d\tau + \int_{-\infty}^{\infty} \left[\frac{5\tau}{1 + \omega^2\tau^2} + \frac{2\tau}{1 + 4\omega^2\tau^2} \right] P(\tau)d\tau \right\} \quad (7)$$

$$f_m = \int_{-\infty}^{\tau_c} P(\tau)d\tau \quad (8)$$

$$T_1^{-1} = \frac{2}{3}\sigma_0^2 \int_{-\infty}^{\infty} \left[\frac{\tau}{1 + \omega^2\tau^2} + \frac{4\tau}{1 + 4\omega^2\tau^2} \right] P(\tau)d\tau \quad (9)$$

$$T_{2r} = (\sigma_0^2)^{-1/2} \quad (10)$$

$$\tau_c = (\sigma_0^2)^{-1/2} \quad (11)$$

where τ_c is the cut-off above which averaging in T_2 no longer occurs. The total number of parameters in the model is 5: 3 to fit the free volume model for τ, 1 to characterize the distribution width, and the second moment. The introduction of the distribution of correlation times introduces only 1 new parameter, the width parameter, but it explains several departures from the BPP theory. Equations 7 and 9 were fitted simul-

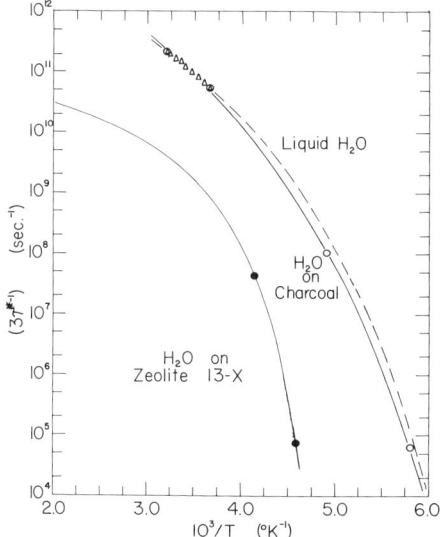

Figure 2. Median molecular jump rate $(3\tau^*)^{-1}$ vs. reciprocal temperature for liquid water, water in charcoal pores, and water in zeolite 13-X

Circles at about 10^7 represent the respective T_1 minima. Those at about 10^5 represent the points at which only half the water remains in the respective mobile phases.

taneously to the data of Figure 2 by a least squares routine which also calculated numerically the necessary integrals (*10*). The "best-fit" theoretical curves are given in Figure 2 and agree quite well with the experimental data. The second moment extracted from the data is 2.0×10^{10} rad^{2}sec^{-2}, which is roughly equal to the 2.29×10^{10} rad^{2}sec^{-2} found by Kvlividze (*5*) from the low-temperature broad-line NMR studies; these results concur in showing that the number of nearest neighbor water molecules is, on the average, less than in ice. The values of the median jump time, τ^*, are presented in Figure 2 in the form $(3\tau^*)^{-1}$—*i.e.*, as jumping rates—and these are compared with the same quantities for water in charcoal pores and for bulk water (hypothetical) at various temperatures. The hypothetical curve for bulk water was obtained by fitting room temperature dielectric data and an estimated glass transition temperature to Equation 2 (*7, 12*). At room temperature, the jumping rate (proportional to fluidity) is about 100 times higher for bulk water and for water in charcoal pores (*dia* 25 A) than for the zeolitic "intra-crystalline-fluid" water. Nevertheless, the zeolitic water protons are at least 1000 times more mobile than the protons in ice at 0°C. This and the fact that the free volume model applies to this system show that the water adsorbed in zeolite 13-X is truly an intracrystalline fluid.

Literature Cited

(1) Barrer, R. M., *Ber. Bunsen Ges. Phys. Chem.* **1965**, 69, 786.
(2) Bloembergen, N., Purcell, E. M., Pound, R. V., *Phys. Rev.* **1948**, 73, 679.
(3) Cohen, M. H., Turnbull, D., *J. Chem. Phys.* **1963**, 39, 2783.
(4) Cohen-Addad, J. P., Farges, J. P., *J. Phys. (Paris)* **1966**, 27, 739.
(5) Kvlividze, V. I., Kiselev, V. F., Serpinski, V. V., *Dokl. Akad. Nauk SSSR* **1965**, 165, 1111.
(6) Pfeifer, H., Przyborowski, F., Schirmer, W., Stach, H., *Z. Phys. Chem. (Leipzig)* **1967**, 236, 345.
(7) Pryde, J. A., Jones, G. O., *Nature* **1952**, 170, 685.
(8) Resing, H. A., *Advan. Mol. Relaxation Processes* **1968**, 1, 109.
(9) Resing, H. A., *J. Chem. Phys.* **1965**, 43, 669.
(10) Resing, H. A., Thompson, J. K., *Ibid.*, **1967**, 46, 2876.
(11) Resing, H. A., Thompson, J. K., *Proc. Colloque Ampere XV, North Holland*, Amsterdam, 1969, p. 237.
(12) Saxton, J. A., "Meteorological Factors in Radio Wave Propagation," p. 278, The Physical Society, London, 1946.
(13) Thompson, J. K., Krebs, J. J., Resing, H. A., *J. Chem. Phys.* **1965**, 43, 3853.
(14) Van Vleck, J. H., *Phys. Rev.* **1948**, 74, 1168.
(15) Woessner, D. E., *J. Chem. Phys.* **1963**, 39, 2783.
(16) Zimmerman, J. R., Brittin, W. E., *J. Phys. Chem.* **1957**, 61, 1328.

RECEIVED February 13, 1970.

Discussion

N. G. Parsonage (Imperial College, London): Is there any useful information which can be obtained by deliberately using zeolites which have been ion-exchanged so as to contain almost stoichiometric amounts of paramagnetic ions?

H. A. Resing: Effects of paramagnetic ions in zeolites on relaxation times have been only cursorily examined (Ducros, Cohen, -Addad— theses, University of Grenoble) with not too much success. Large effects certainly occur, but much interpretive work appears to be required.

R. A. Munson (Bureau of Mines, College Park, Md. 20740): You have compared your mobilities in zeolites with those in pure water and ice. Would it not be more realistic to compare with electrolyte solutions of analogous concentration? What results would you expect for this case?

H. A. Resing: I believe that electrolytes of similar concentration would not lead to such a 20-fold lowering of mobility, but only to a 2- or 3-fold effect at most.

D. M. Ruthven (University of New Brunswick, Fredericton, N.B., Canada): I was interested in the conclusion that the density of intracrystalline water is similar to that of the bulk liquid. This agrees rather well with the conclusions of Dubinin *et al.* which are based on equilibrium sorption studies in NaX zeolite and also in other systems.

H. A. Resing: Such information based on adsorption and crystallography is certainly more definitive than that adduced in our paper. This "external" information is in the nature of a boundary condition on our interpretation of the NMR results; *i.e.*, it is a check on the correctness of our interpretations.

P. B. Venuto (Mobil Oil, Paulsboro, N. J. 08034): I would like to reinforce your comment concerning the decreasing mobility of water at very low intrazeolite concentrations by citing the extremely high isosteric heats of adsorption, @ 30–45 kcal/mole, reported at low θ some time ago for sodium faujasite by Barrer and coworkers.

38

Stereospecific Adsorption of Nitrous Oxide, Cyclopropane, Water, and Ammonia on the Co(II)A Synthetic Zeolite

K. KLIER[1]

Institute of Physical Chemistry, Czechoslovak Academy of Sciences, Prague

> *Electronic reflectance spectra of the Co(II)-exchanged dehydrated Type A zeolite, Co(II)A, show that the cobalt ions enter the S-II positions, where they form complexes with nitrous oxide, cyclopropane, water, and ammonia. These molecules represent ligands of increasing bond strength and, with the exception of ammonia, form reversible complexes with Co(II)A. In the case of nitrous oxide and cyclopropane, these complexes have a C_{3v} symmetry; water and ammonia complexes are tetrahedral. On a long exposure to water and ammonia, the Co(II) ions become highly coordinated to these ligands.*

The properties of transition metal ions in molecular sieve zeolites have been investigated by various methods, including reflectance spectroscopy (4). In an earlier work (6, 7, 8), Ni(II)A sieves underwent changes of color and spectra during the adsorption of molecules having permanent dipoles or unsaturated bonds. These changes were interpreted as reflecting the differences in electronic structure of the Ni(II) ions in different ligand environments. The spectrum of dehydrated Ni(II)A sieve was that of Ni(II) ions in D_{3h} (D_{6h}) field of oxygen atoms, which represents the S-II type position in the large cavity of the Type A sieve. On adsorption of molecules such as water, cyclopropane, and olefins, a bathochromic shift of the main visible spectral band (22,760 cm^{-1}) occurred which was caused by the formation of complexes of C_{3v} sym-

---

[1] Present address: Center for Surface and Coatings Research, Lehigh University, Bethlehem, Pa.

metry containing 1 nickel ion, 1 adsorbed molecule, and 3 oxygen atoms of the zeolite.

In the present work and in (5), evidence is reported that similar complexes are formed with divalent cobalt. All complexes described, with the exception of ammonia, are reversible.

Experimental

The reflectance spectra were recorded in the geometry $R_{0,D}$ (4) in the frequency range 5000–40,000 cm^{-1} in a previously described apparatus (4, 6). The absorbance is represented by the logarithm of the Schuster-Kubelka-Munk (SKM) function, $F(R_x) = (1 - R_x)^2/2R_x$, where R_x is the reflectance measured against a white standard. Since the comparison of the reflectance spectra with the transmission spectra of complexes in molecular sieves revealed that the scattering coefficient is a constant independent of the wavelength (4), the logarithm of the SKM function is, but for an additional constant, a correct representation of absorbance.

The samples were ion-exchanged Type A molecular sieves in which the final concentration of the Co(II) ions approximately corresponded to 1 cobaltous ion in 1 large cavity. All adsorbing gases were admitted at pressures between 400 and 760 torr at 20° to ensure completeness of the adsorbed complex formation.

Results and Discussion

The spectrum of the hydrated Co(II)A sieve after ion exchange was identical with that of the hexaquo-complexes of divalent cobalt, which is a situation completely analogous to that in the Ni(II)A sieves (4).

The dehydrated Co(II)A sieve (350°C in vacuum for several hours) was pale blue, and its spectrum is demonstrated in Figure 1, curve 1. Heating *in vacuo* up to 500°C did not change the spectrum between 5000 and 30,000 cm^{-1}, but an increased continuous absorption occurred in the ultraviolet region between 30,000 and 40,000 cm^{-1} which did not recede on rehydration. This phenomenon is explained as being caused by an irreversible reaction of the Co(II) ions with the aluminosilicate skeleton, which eventually results in the destruction of the zeolitic structure. Therefore, the temperature range of this beginning destruction reaction, indicated by the increased ultraviolet absorption, was avoided by dehydration at lower temperatures in vacuum and for prolonged periods of time. The residual content of water was similar to the nickel sieves (6), indicating that dehydration was essentially complete in 2 hours at 350°C in vacuo. Under these conditions, the Co(II) ions were rehydratable into the origi-

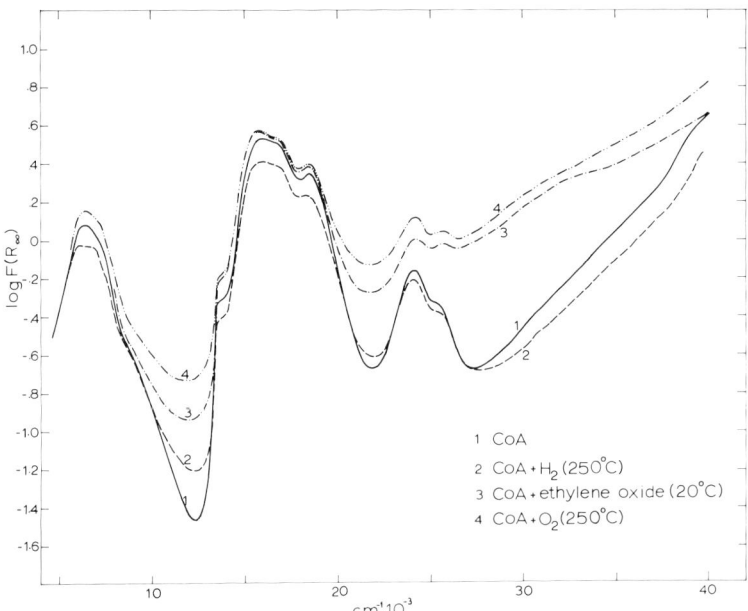

Figure 1. Spectra of dehydrated Co^{2+}-exchanged Type A molecular sieve zeolite pure (1) and after exposure to hydrogen at 250°C (2), to ethylene oxide at 20°C (3), and to oxygen at 250°C (4)

nal pink form, having a spectrum identical with the original spectrum of the ion-exchanged sieve, and they were back-exchangeable with sodium ions.

A quantitative interpretation of spectrum 1 (Figure 1) depends upon the determination of the electron repulsion integrals in the Co(II) (d^7) system in proper ligand field. The computations of the repulsion matrix elements are in progress. At the present stage, a qualitative interpretation of spectrum 1 can be offered on the basis of previous analyses and results concerning the d^1 (d^9) and the d^2 (d^8) electron systems. First, judging from the effects of adsorbing molecules shown below, the Co(II) ions in dehydrated Co(II)A are in a low-coordinated complex prior to adsorption. From these, the tetrahedral complexes are ruled out on the grounds that their spectra are different (Figure 2) from spectrum 1 of Figure 1. Therefore, the Co(II) ions are probably placed, like the nickel ions, in the S-II trigonal sites. Since it was shown earlier that the trigonal, hexagonal, and circular planar ligand symmetries generate the same effective ligand fields (6, 7), the ambiguity in the negative charge distribution on the oxygen atoms of the aluminosilicate skeleton does not affect the term splitting, the ligand fields in the S-II site being represented by one having the D_{3h} symmetry of negative charges on the ligands. Secondly, a complete configuration interaction treatment (within the d-configurations) revealed that the D_{3h} fields act as strong ligand fields upon the d^2 and d^8 systems (6, 8). Assuming that this is also true for the Co(II) (d^7) ions, one can interpret the Co(II)A spectra as electron jumps among the real one-electron d-functions in the following way. Since the one-electron (d^1) system is split in the D_{3h} fields (the high-symmetry axis being in the z-direction) into the low-lying level d_{z^2} (a_1'), the degenerate levels d_{zx} and d_{yz} (e''), and the highest degenerate in-plane orbitals, d_{xy} and $d_{x^2-y^2}$ (e') (7), the ground high-spin state of the Co(II) ion in the D_{3h} field will be realized by the following occupation of the a_1', e'', and e' orbitals.

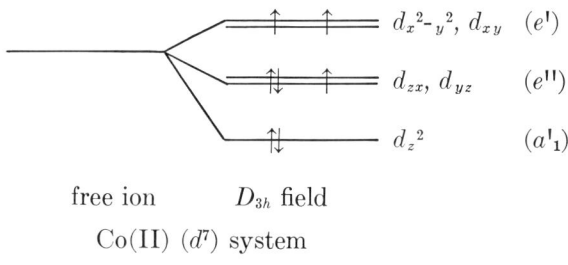

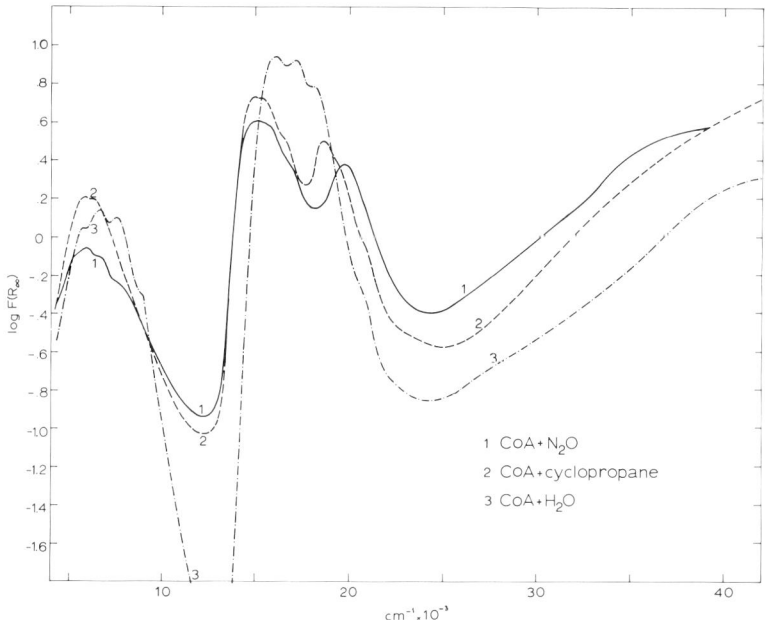

Figure 2. Spectra of dehydrated Co^{2+}-exchanged Type A molecular sieve zeolite with adsorbed nitrous oxide (1), cyclopropane (2), and water (3) at 20°C

The symbols in parentheses for the irreducible representations of the D_{3h} symmetry group are those used in (7), the lower case indicating that they relate to the single-electron system. The spin-allowed one-electron transitions can take place between the a_1' and e'', a_1' and e', and the e'' and e' orbitals. From the two-electron transitions, the only spin-allowed one is that in which 1 electron jumps from the e'' into the e' orbitals and 1 electron from the a_1' into the e' orbitals. This transition requires a considerable amount of energy and probably is not observed in our spectrum, except perhaps in the region of increasing continuous absorption between 30,000 and 40,000 cm^{-1}.

Based on the above considerations and on the effects of adsorbed molecules, the band at 24,000 cm^{-1} is assigned to the highest energy one-electron transition $a_1' \to e'$—i.e., the transition from the d_{z^2} orbital with maximum electron density on the axis perpendicular to the plane of oxygen ligands, to the in-plane partially occupied orbitals d_{xy} and $d_{x^2-y^2}$. The intense bands at 15,000–20,000 cm^{-1} and at 7000 cm^{-1} are assigned to the remaining one-electron transitions, $e'' \to e'$ and $a_1' \to e''$, without specifying, at the present stage, their absolute energies.

The Effects of Hydrogen, Oxygen, and Ethylene Oxide on Spectrum 1. Hydrogen and oxygen produced no effects on the CoA spectrum (Figure 1, curve 1) when admitted at room temperature. Heating of Co(II)A in hydrogen atmosphere at 250°C overnight resulted in slight changes of the spectrum (curve 2) and heating in oxygen increased the background (curve 4), leaving the band maxima frequencies unchanged. These results mean that the divalent cobalt ions are resistant toward both reduction and oxidation and represent stable chemical species in the Type A sieves, unlike in the oxides and hydroxides, where they are easily oxidized. The effect of ethylene oxide adsorption (Figure 1, curve 3) is also to alter only the background and not the positions of the bands. It is concluded, therefore, that hydrogen, oxygen, and ethylene oxide molecules do not adsorb selectively on the Co(II) ions.

The Spectra of Co(II)A with Adsorbed Nitrous Oxide, Cyclopropane, Water, and Ammonia. The main effects of adsorbing N_2O, C_3H_6, H_2O, and NH_3 are to eliminate or shift to lower frequencies the band at 24,000 cm^{-1} and to produce some splitting in the band at 7000 cm^{-1} (Figures 2 and 3). These changes are so profound that they are accompanied by a visible change of color from pale blue to violet immediately following adsorption. After desorption of any adsorbate except ammonia, the original spectrum of Co(II)A is completely regenerated, the molecules of nitrous oxide desorbing without decomposition and cyclopropane desorbing without isomerization to propylene. Desorption of water and ammonia requires elevated temperatures (120°C for water) and pro-

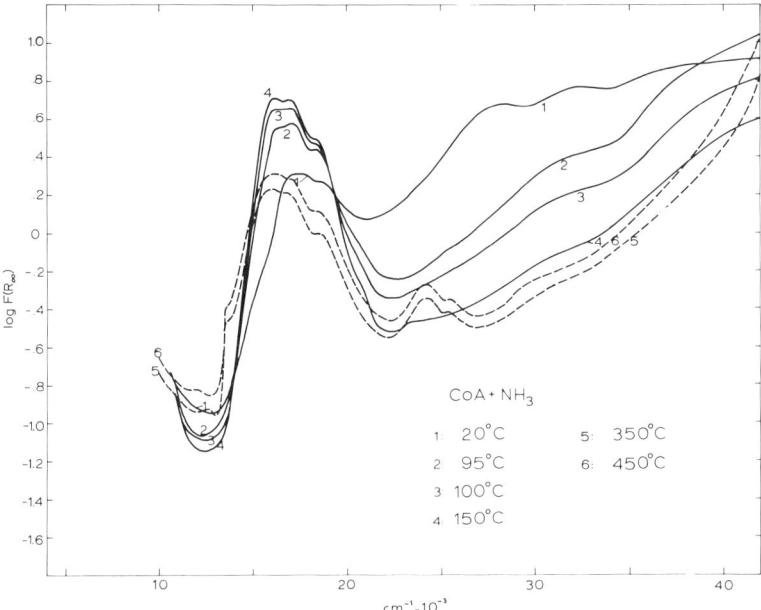

Figure 3. Spectra of dehydrated Co^{2+}-exchanged Type A molecular sieve zeolite with adsorbed ammonia at temperatures between 20 and 450°C (1-6)

longed pumping. The spectral changes accompanying ammonia desorption are represented in Figure 3.

The qualitative interpretation of the spectra of Co(II)A with adsorbed molecules is the following. The original band at 24,000 cm^{-1} reflecting the $a_1' \rightarrow e'$ transition shifts to lower frequencies due to the repulsive interactions between the lone pair orbitals (or the $\sigma-\pi$ system in the case of C_3H_6) and the out-of-plane d_{z^2} orbital of the Co(II) ion, whose energy is raised by this interaction. The magnitude of the spectral shift is determined by the strength of the repulsive field generated by the ligand. From the observed shifts, the weakest ligand is nitrous oxide (the 24,000 cm^{-1} band is shifted to 20,000 cm^{-1}), then cyclopropane (the shift is to 18,000 cm^{-1}), and eventually water and ammonia, which produce tetrahedral complexes. Both molecules are strong ligands, as is evident from their desorption properties. The spectrum of the water complex is almost identical with that of the $Co(OH)_4^{2-}$ ion, which has a tetrahedral structure (2). The solution spectra of $Co(OH)_4^{2-}$ are compared with the reflectance spectra of Co(II)A–H_2O, Co(II)X, and Co(II)Y sieves in Figure 4, showing that all these compounds contain tetrahedrally coordinated cobalt. The result for the Co(II)X and

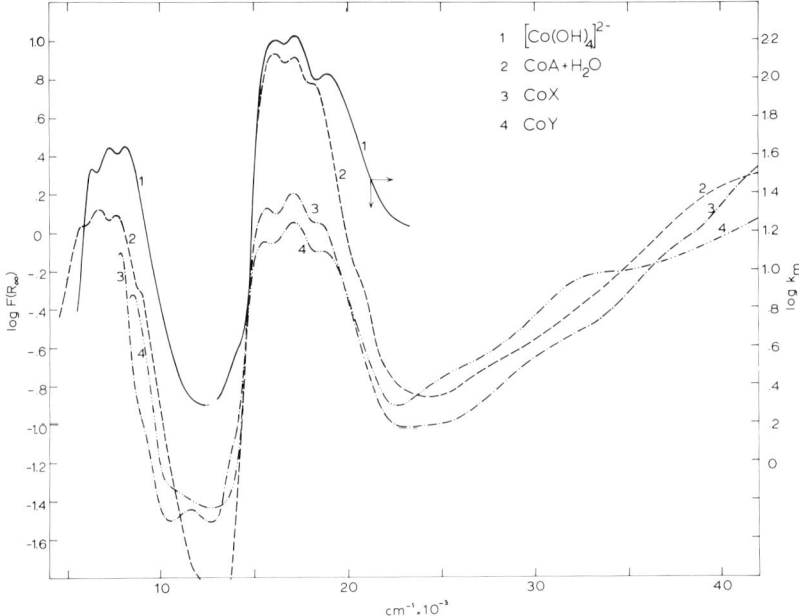

Figure 4. Spectra of the complex ion $Co(OH)_4^{2-}$ from (2) (k_m, the molar extinction coefficient, curve 1), of molecular sieve Co(II)A with adsorbed water (2), and of dehydrated Co^{2+}-exchanged Type X and Y zeolites (3,4)

Co(II)Y sieves agrees well with the observations of Barry and Lay (1) and Stone (9); the question of tetrahedral positions in the X and Y sieves was discussed in (1) and (3), seems to be well resolved, and will not, therefore, be discussed in the present work. Spectral assignments of transitions in tetrahedral cobaltous complexes have been made in the work of Cotton, Goodgame, and Goodgame (2), the high-intensity bands at 7000 cm^{-1} corresponding to the $^4A_2(F) \rightarrow {}^4T_2(F)$ and $^4T_1(F)$ transitions, and the band centered at 18,000 cm^{-1} corresponds to the $^4A_2(F) \rightarrow {}^4T_1(P)$ transition.

By analogy with the Ni(II)A sieves, in which the complex of Ni(II)A with 1 water molecule per nickel ion was identified by combining the spectroscopical evidence with chemical analysis, it is likely that the Co(II)A–H_2O spectrum corresponds to an analogous monoaquo-complex. In order to confirm this hypothesis, the content of the water in Co(II)A–H_2O must be determined. A different possible tetrahedral complex might be one in which 4 water molecules are bound to 1 cobalt ion. Since complexes containing similar ligands have spectra which reflect their symmetry rather than the chemical identity of the ligands (2), our measured spectra do not allow one to distinguish between the two possibilities without analysis for water.

In the case of ammonia, the situation is analogous to the Co(II)A–H_2O complex except that after heating there remains an increased background absorption (Figure 3) indicating that, contrary to all preceding cases, adsorption of ammonia is not completely reversible. At room temperature, the 15,000–20,000 cm^{-1} band is shifted to higher frequencies (Figure 3, curve 1) possibly reflecting the beginning of the formation of highly coordinated ammo-complexes such as the hexammo-cobaltous ion.

Conclusions

To summarize the present results, Co(II) ions can be introduced into the Type A molecular sieves into the S-II type positions, where they are stable and resist both oxidation and reduction. These ions bind nitrous oxide, cyclopropane, water, and ammonia as additional ligands, their spectra being simultaneously changed in a defined fashion. The spectrochemical series of these ligands is, in the order of increasing ligand strength,

$$N_2O < C_3H_6 < H_2O < NH_3$$

Tetrahedral complexes are formed with water and ammonia. Schematically, the complex symmetries qualitatively explaining the observed spectral changes are represented by the following diagram, although some uncertainty remains as to the number of ligands in the aquo- and ammo-complexes.

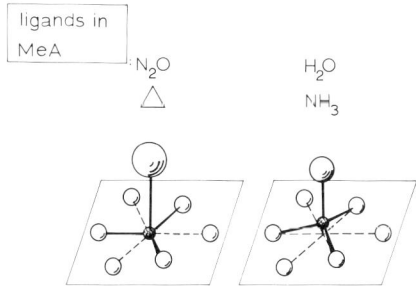

Literature Cited

(1) Barry, T. I., Lay, L. A., *J. Phys. Chem. Solids* **1968**, 29, 1395.
(2) Cotton, F. A., Goodgame, D. M. L., Goodgame, M., *J. Am. Chem. Soc.* **1961**, 83, 4690.
(3) Delgass, W. N., Garten, R. L., Boudart, M., *J. Phys. Chem.* **1969**, 73, 2970.
(4) Klier, K., *Catalysis Rev.* **1967**, 1(2), 207.
(5) Klier, K., *J. Am. Chem. Soc.* **1969**, 91:19, 5392.
(6) Klier, K., Ralek, M., *J. Phys. Chem. Solids* **1968**, 29, 951.
(7) Polak, R., Cerny, V., *J. Phys. Chem. Solids* **1968**, 29, 945.

(8) Polak, R., Klier, K., *J. Phys. Chem. Solids* **1969**, 30, 2231.
(9) Stone, F. S., private communication.

RECEIVED January 23, 1970.

Discussion

P. B. Venuto (Mobil Oil, Paulsboro, N. J.): Do you think that the lack of selectivity in hydrocarbon oxidation over the Co-zeolite to which you alluded in your paper may derive from (conventional) free radical chains initiated by cobalt cations located at or near the crystal external surface?

K. Klier: I think that free radical formation can be initiated on cobalt ions on the internal surface of zeolites (Type A) similarly to the suggested O atom formation in N_2O decomposition by these ions. Small amounts of reversible Co^{3+} intermediate (not detectable in the spectra) may be involved according to the reaction

$$Co^{2+} + N_2O \rightarrow Co^{2+}-N_2O \rightarrow Co^{3+}-O^- + N_2 \rightarrow Co^{2+} + O\uparrow$$

If the external surface contains sites similar to those in the cages, radical reaction could occur more effectively because of better accessibility of the gas phase. However, I doubt that an irregular external surface could hold Co^{2+} in surface positions because of the tendency of Co^{2+} to find tetrahedral or octahedral positions. The present spectra, of course, show only complexes formed on internal surfaces because there is a linear correspondence between the absorption coefficient and concentration and because all ions observed on Type A are surface ions.

Concerning selectivity in oxidation reactions, I think we must seek a system which does work through valency switching of the transition metal ion and where the active oxygen species is localized on the internal surface. Only in these systems can the red-ox potential of the catalyst be regulated by the ion identity (atomic number), Madelung energy be influenced through the Si/Al ratio and perhaps the field generated by other ions in the cavity. According to my preliminary results, an interesting system of this kind is the Type A–Fe^{II}/Fe^{III} system.

J. P. Thomas (Southwest Minnesota State College, Marshall, Minn. 52658): George McIntyre and myself have also examined the electronic spectra of Type A zeolite containing Co^{2+}, Ni^{2+}, and Cu^{2+} ions, not only in a single low concentration but also for a range of concentrations (George McIntyre, PhD thesis, Illinois Institute of Technology, 1970). Many similarities exist between your results and ours. There are, however, some pronounced differences. We have both observed small shifts in the band frequencies for the fully hydrated zeolite samples from their energies observed for aqueous solutions. Our spectra correspond closely to those observed by Drickamer and coworkers at the University of Illinois for Co^{2+}, Ni^{2+}, and Cu^{2+} hydrated salts under an applied external pressure of 20,000–40,000 atm. Perhaps these numbers delineate the range to be expected for the osmotic pressure and hydrostatic tension existing within the zeolite cavities. The effects of such internal pressure upon the many diverse properties of the zeolites should, perhaps, be more generally considered and explored. The pronounced differences in our spectra and those of Klier are difficult to understand at present. Perhaps they are attributable to unrecognized slowness in the attainment of equilibrium within the zeolite after ion exchange and after gas adsorption. In general, what was the time interval between the preparation of your samples (fully hydrated, etc.) and the recording of the spectra? Have you observed any variations of spectra with time?

K. Klier: No time dependence of spectra was observed, the complexes formed being stable for weeks. Concerning the small shifts in frequencies of fully hydrated sieves, I refer to our work in *J. Phys. Chem. Solids* **1968**, 29, 951, where similar shifts were observed with hydrated NiA and NiX sieves. From these results, it would seem that the hexahydrate Ni^{2+} ion in Type A sieve is in a positive pressure field whereas that in Type X is in a negative pressure field, expanding the complex and lowering effectively the crystal field splitting parameters. I should be interested to see your $[Co^{2+}(H_2O)_6]$ spectra, which I have not followed closely, in order to learn how pronounced the differences are.

39

Thermal Decomposition Patterns in Methylammonium Cation-Exchanged Y-Type Faujasites

E. L. WU, G. H. KÜHL, T. E. WHYTE, JR., and P. B. VENUTO

Mobil Research and Development Corp., Paulsboro, N. J. 08066

The mono-, di-, tri-, and tetramethylammonium exchanged forms of a Y-type faujasite have been prepared. The cationic size determined the extent of exchange; hydronium ion incorporation was indicated. Thermal decomposition patterns of the 4 forms are generally comparable. Detailed transformations were analyzed for the tetramethylammonium (TTMA) form by in situ *infrared observations and identification of decomposition products. These cations decomposed by a combination of reactions involving low-temperature decomposition in the presence of the intrazeolitic pool of hydroxylic sorbate, "methyl disporoportionation" between methyl-containing species and possible surface methoxy intermediates, and high-temperature ylide or carbene mechanisms. The protonic sites generated and associated zeolitic changes were compared with those of NH_4 faujasite.*

The thermochemistry of NH_4 cation-exchanged zeolites has been the subject of numerous studies (4, 22, 24, 27, 28). The transformations involved are relatively simple: deammoniation to give the hydrogen form and dehydroxylation by elimination of water at higher temperatures. Other structural changes may occur, depending on processing conditions (10). Ion exchange of alkyl-substituted ammonium cations into zeolites (2, 26) has been reported; thermal transformations in such systems are expected to be considerably more complex and interesting.

The importance of the anion in determining the decomposition pathway of quaternary ammonium salts was recognized by Ingold in 1927 (7). At 135°–140°C, tetramethylammonium (TTMA) hydroxide decomposes to give trimethylamine, dimethylether, or methanol, depending on experimental conditions (8, 15). With the weaker base, chloride, decomposition occurs at a higher temperature (360°C) to form methyl chloride (16). Salts with other organic anions decompose between 100° and 250°C, also *via* simple displacement of trimethylamine (16, 32). More complex pathways involving ylide or carbene mechanisms have been postulated in other TTMA–cation systems (6, 9, 15, 31, 33, 34, 35).

We now report results of a study of the thermal decomposition patterns in the 4 methyl-substituted ammonium (MA) exchanged Y-type faujasites, detailed transformation of the TTMA form, and the resulting generation of protonic acidity.

Experimental

Materials. The methylammonium Y-zeolites (Table I) were prepared by ion-exchanging NaY (5 grams) with 4 100-ml portions of 1 N alkylammonium salt solution at 80°C. Monomethylammonium (MMA) chloride, dimethylammonium (DMA) chloride, and tetramethylammonium (TTMA) bromide (Matheson, Coleman, and Bell) and the trimethylammonium (TMA) chloride (Eastman Organic Chemicals) were used as received.

Table I. Unit Cell Composition of Mono-, Di-, Tri-, and Tetramethylammonium Y and Parent NaY Faujasites

Sample	Na	N	H[a]	AlO_2	SiO_2
Na Y	52.2	—	—	52	140
MMA Y	9.3	36.5	6.2	52	140
DMA Y	11.8	34.0	6.2	52	140
TMA Y	20.8	25.5	5.7	52	140
TTMA Y	28.4	17.7	6.0	52	140

[a] Calculated by difference, assuming Na + N + H = AlO_2.

The ammonia (Matheson anhydrous) was purified by fractional condensation and distillation in vacuum. Research grade helium (Matheson) was dried over 5A molecular sieve prior to use.

Apparatus and Procedures. Thermal analyses were performed using DuPont 900 differential thermal (DTA) and 950 thermogravimetric (TGA) analyzers. Basic decomposition products in the TGA effluent were titrated with sulfamic acid at regular temperature intervals (11).

TTMA-Y was examined *in situ* using a high-temperature infrared cell (36) in which simultaneous zeolite treatment and spectral observation could be effected. The cell was connected to a conventional vacuum and gas-handling system (ultimate vacuum, 10^{-6} torr). Spectra were recorded on a Perkin-Elmer 421 grating spectrophotometer, modified by the addition of a presample chopper to eliminate the effect of spurious radiation originating from the hot sample and cell. Spectral resolution was 4 cm^{-1} at 3600 cm^{-1} and 2 cm^{-1} at 1600 cm^{-1}.

Samples (2-gram) of TTMA-Y were calcined in a tubular reactor under vacuum and the decomposition products (Table II) over 2 major temperature ranges were collected. After gas fractionation where necessary, components in the various fractions were analyzed by infrared and mass spectroscopy (CEC Model 21-104, ionization voltage 10 ev).

Table II. Analysis of Organic Decomposition Products from Thermal Decomposition of TTMA Y in Vacuum[a]

Product, Estimated Mole %[b]	Temp. Range, °C		Total
	150°–275°[c]	275°–450°[d]	
$(CH_3)_3N$	50	5	9
CH_4	11	24	23
NH_3	—	23	21
C_2H_4	Tr	14	13
$(CH_3)_2O$	10	—	1
CO	9	—	1
CH_3OH	6	—	1
H_2	4	15	14
C_3H_6	Tr	9	8
C_4H_8	4	5	5

[a] 10^{-6} Torr, held at 275° for 3 hr and 450° for 2½ hr. Noncondensables were allowed to expand into an evacuated bulb and condensables collected in liquid nitrogen traps.
[b] Estimated, based on relative pressure of the fractions and mass spectral analysis.
[c] Trace (Tr)—<2 mole %, including $(CH_3)_2NC\equiv N$ and $(CH_3)_2NCH_2C\equiv N$.
[d] Trace of C_5H_{10}, C_2H_6, C_3H_8, and H_2O.

Results

Methylammonium Ion Exchange. The observed extent of exchange (70%) of MMA cations was to be expected since 16 of the Na cations (31%) initially present are located in the small-pore system (23), into which these MA cations cannot pass. Bulkiness of the cations further limits the number of Na cations that can be replaced (2) in the large-pore system, decreasing with increase in size of the MA ion. The degree of exchange of various MA cations in this study agrees with earlier results (2, 26).

Since the MA cations cannot occupy exchange sites within the sodalite cages, removal of more than 70% of the Na ions must result from hydronium ions exchanged into these cages. Since the degree of hydronium ion exchange (Table I) was about the same for all 4 MA cations at 80°C, it seems likely that all the hydronium ions, even in cases of lower degree of MA exchange, migrated into the small cages. The presence of hydronium ions in the small-pore system of the faujasite structure has been indicated by a recent x-ray and ion-exchange study (18); high affinity of the sodalite cages for hydronium ions is suggested.

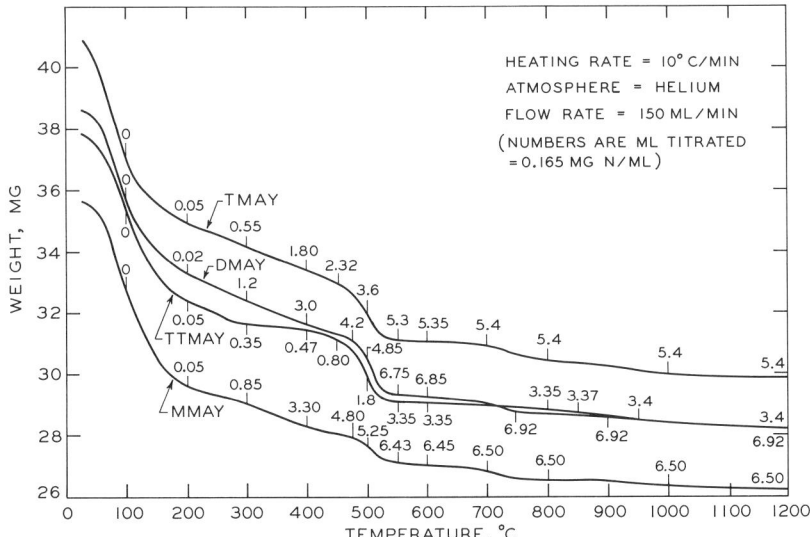

Figure 1. Thermogravimetric analysis of methylammonium Y faujasite

Thermogravimetric Analysis. All the MA forms (Figure 1) exhibited a rapid weight loss at about 100°C, caused by removal of physically sorbed water. Over the temperature range of 250° to 450° (for TMA- and TTMA-Y) or 475°C, a slow, steady weight loss occurred with basic products in the effluent. From the observed weight loss and amount of base found by titration, it was possible to infer the type of base liberated. It appeared that for each methyl-substituted ammonium cation, the next lower substituted amine was eliminated (ammonia from MMA cation, methylamine from DMA, etc.). The amount of basic product found in this slow step, in percentage of the total base titrated up to 1200°C, decreasing with increasing methyl substitution, was 74, 61, 43, and 24% for MMA-, DMA-, TMA-, and TTMA- cations, respectively. Double peaks were observed in the derivative thermograms (in all but TMA-Y), suggesting the presence of more than one mode of decomposition.

When sample temperature was raised further, a rapid weight loss occurred, accompanied by increased emission of basic gases. By 550°C, more than 95% of the basic amines had been titrated.

There was a slight but distinct weight loss near 720°C. The step was not as sharp in TTMA-Y. The observed weight loss (600°–800°C) was more than 70% of that expected for dehydroxylation, as calculated from the total amount of MA cations initially present. However, since the weight of the sample prior to dehydroxylation indicated that the evolution of decomposition products was not complete and some of the residue might desorb over this temperature range, a quantitative determination of the extent of dehydroxylation was not possible.

Differential Thermal Analysis. Details revealed in the differential thermograms of the 4 MA-Y zeolites (Figure 2) are consistent with the TGA results. They all exhibited an endotherm (122°–190°C) caused by loss of adsorbed water. Over the region where slow decomposition occurred, there were weak endotherms or inflections in the curve, and the rapid decomposition was confirmed by a sharp endotherm (516°–577°C). Dehydroxylation was clearly revealed by the endotherm at 785°C in MMA-Y, but was not as well resolved in the others. The exotherm (977°–1026°C) was owing to "mullite" transformation (1). However, all the samples were found to be amorphous (by x-ray) after 1 hour at 900°C.

Decomposition of TTMA Cations. Decomposition of TTMA-Y under vacuum was monitored in detail by infrared analysis (Figure 3). All bands arising from the TTMA cation decreased in intensity above 200°C, reflecting decomposition of the quaternary ion. With further evacuation, absorption bands became resolved in the region expected for stretching vibration of lattice-associated OH groups (ν_{OH}). In addition to a band at 3735 cm^{-1} associated with surface silanol groups, 2 bands at 3637 (ν_{HF}) and 3550 cm^{-1} (ν_{LF}) appeared. Relative intensity (I_{HF}/I_{LF}—based on peak absorbance) was one-half (Figure 3, 275°C, 3/4 hour).

The intensity of ν_{HF} and ν_{LF} gained as those of the TTMA bands decreased with further calcination. These data showed that OH groups

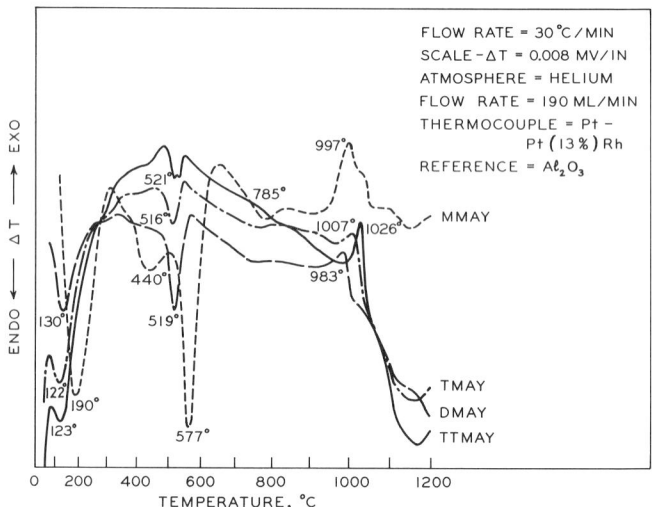

Figure 2. Differential thermal analysis of methylammonium Y faujasite

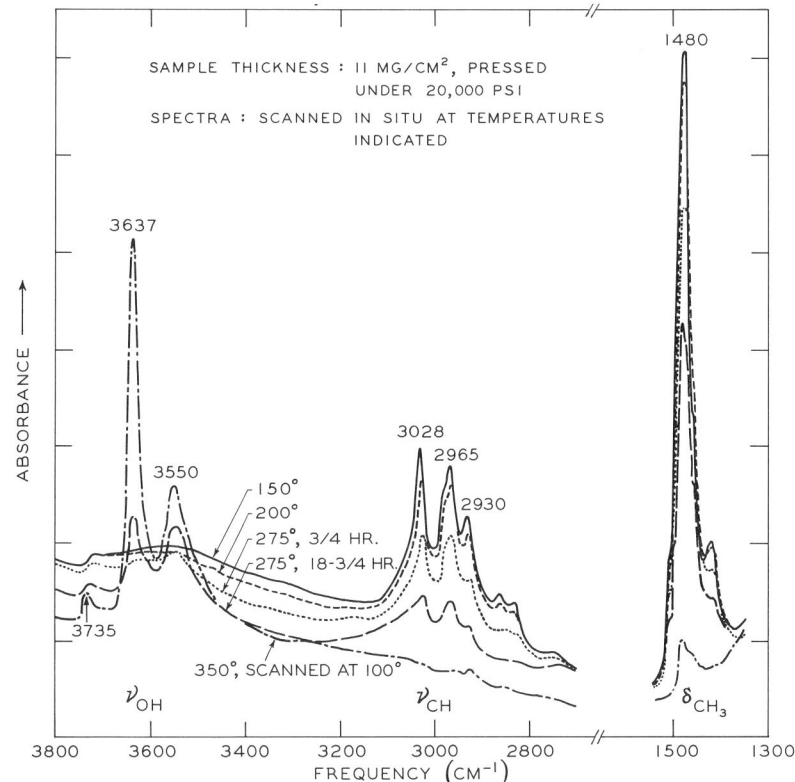

Figure 3. Infrared spectrum of TTMA-Y after evacuation at various temperatures

were generated as a direct consequence of the TTMA cation decomposition. The intensity ratio of ν_{HF} to ν_{LF} also increased with increasing total OH content. The spectrum after 350°C resembles that of HY (27).

Protonic Acidity. The protonic nature of the hydrogen on these OH groups was confirmed by their interaction with ammonia to give ammonium ions, as evidenced by the characteristic infrared bands—NH stretching (ν_{NH}) at about 3200 and 3000 cm^{-1} and deformation (δ_{NH_4}) at 1430 cm^{-1} (Figure 4). Some of these acidic OH groups exhibited a rather weak interaction with ammonia; they could be regenerated by evacuation at 100°C (Figure 4). These weaker groups amounted to about half of the total acidity (based on the relative intensity ratio of δ_{NH_4} and sum of ν_{HF} and ν_{LF} in spectrum prior to desorption and after 19 hours). Regeneration of the hydrogen form was accomplished by evacution at 300°C to desorb the remaining ammonia.

Decrease in the intensity of these acidic OH bands occurred upon calcination at higher temperature, and 90% dehydroxylation was achieved by evacuation at 650°C.

Lewis Acidity. Lewis acidity was generated on the present sample by dehydroxylation, as shown by interaction with ammonia to give characteristic infrared bands of ammonia coordinated to electron-accepting sites (at 3375, 3275, and 1615 cm^{-1}). Interaction of ammonia adsorbed on these Lewis sites was comparatively weak and could be removed by evacuation at 300°C.

Decomposition Products. A complex mixture of gases was obtained from vacuum calcination of TTMA-Y zeolite. Relative amounts of components in the mixture varied with the decomposition temperature (Table II). Because of variation in experimental conditions, the amount of products found in the 2 steps cannot be directly correlated to weight loss over the same temperature ranges in Figure 1. Trimethylamine (accounting for half of the products), methane, and oxygenated products were found in the low-temperature step. Decomposition at higher temperature gave mainly ammonia, hydrogen, methane, and low-molecular-weight olefins. Decomposition in the absence of air left a deep purple organic residue.

Discussion

Generation of Acidity from TTMA-Y Zeolite. During vacuum calcination of TTMA-Y, ν_{LF} appeared first, even prior to any significant de-

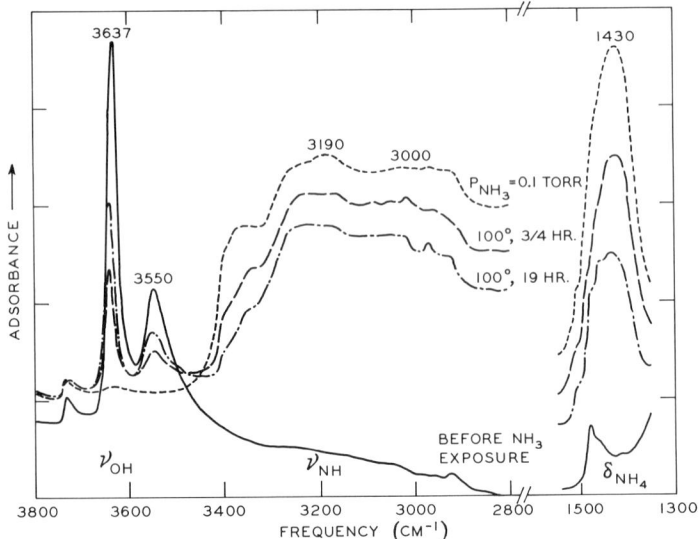

Figure 4. Interaction with NH_3 and subsequent desorption, both at 100°C

composition of the TTMA cations. This Y zeolite contained 6 hydronium cations per unit cell, presumably located in the sodalite cages, introduced during the TTMA ion exchange. Therefore, the LF band can be attributed mainly to OH groups formed by an interaction of protons—from dehydration of these hydronium ions—with specific lattice oxygens within the small-cage system. This is consistent with the low frequency expected for OH groups located on O_3-oxygens of the small-cage system (19, 30) in the NH_4Y–HY system.

The HF band of the NH_4Y–HY system has been correlated with separate OH groups located on O_1-oxygens of the supercage system (19, 30). Since the TTMA cations were located in the supercages, any protons derived from these probably would interact with O_1-oxygens to give a higher frequency band. Consistent with this, intensity of ν_{HF} increased as those of TTMA cations decreased.

Ammonia adsorption confirmed that both hydronium and TTMA cations were protonic precursors in TTMA-Y. Generation of protons from the latter involved 2 steps, as shown by TGA and infrared analysis. TTMA cations could be decomposed completely by either prolonged evacuation at 350°C or heating to 550°C at 10°C/min. Although only the TTMA system was studied in detail, the 4 MA-Y zeolites have sufficiently similar DTA and TGA patterns to suggest that information on the source of protons and their generation is generally applicable to all 4 systems.

The present acid Y-zeolite dehydroxylated at 650°C (*in vacuo*), more than 100°C higher than the NH_4–HY system (4), probably because of the difference in site density between the 2 Y samples. A similar difference in thermal stability (toward dehydroxylation) was reported (28) between 50 and 90% ammonium-exchanged NaY.

TTMA Cation Decomposition Pathway. Product analyses from the decomposition of intracrystalline TTMA cations over 2 temperature ranges indicate that different reaction pathways were involved.

Low-Temperature Decomposition. Over the range of 150°–275°C, a certain amount of intracrystalline water, whether physically adsorbed or hydrating the hydrogen ions, was present. The amount was minimized by pre-evacuating the sample at 150°C. A displacement reaction of water nucleophile on the TTMA cations forming methanol, trimethylamine—and zeolite hydroxyl groups—can occur.

$$(CH_3)_3 \overset{\oplus}{N} - CH_3 \quad \overset{H}{\underset{H}{O}} \quad \longrightarrow \quad (CH_3)_3 N + CH_3 OH + H^{\oplus} \quad \overset{\ominus}{O}\text{-ZEOL} \quad (1)$$
$$\underset{O\ ZEOL}{\ominus}$$

A similar attack by methanol on TTMA cation generates dimethylether and trimethylamine. Such substitution reactions in the TTMA hydroxide system have been proposed by Tanaka *et al.* (25).

Formation of methane, carbon monoxide, and hydrogen can be explained by a sequence (Equation 2–3a) involving collapse of TTMA cation to trimethylamine and a surface methoxy species (I), analogous to that observed on Aerosil and alumina (3, 20).

$$(CH_3)_3\overset{\oplus}{N} - CH_3 \longrightarrow (CH_3)_3N + CH_3-O-ZEOL \quad (2)$$
$$\underset{O-ZEOL}{\ominus} \qquad\qquad\qquad\qquad I$$

Reaction of I with another C_1-fragment follows in what best can be termed a "methyl disproportionation reaction." Equation 3 depicts such an intermolecular hydride-transfer event between I and methanol.

$$\underset{\underset{I}{ZEOL}}{H-\overset{H}{\underset{O}{C}}\leftarrow \overset{H}{\underset{H}{C}}{\overset{H}{\diagdown}}_H} \longrightarrow CH_4 + \underset{II}{(HCHO)} + HO-ZEOL \quad (3)$$
$$\qquad\qquad\qquad\qquad\qquad \downarrow$$
$$\qquad\qquad\qquad\qquad H_2 + C \equiv O \quad (3a)$$

Decomposition of formaldehyde (II) or a closely related species would produce hydrogen and carbon monoxide (Equation 3a). Instability of formaldehyde in the intrazeolitic environment is well known (29). Similar hydride transfers from methanol have been reported (5, 14, 17).

HIGH-TEMPERATURE DECOMPOSITION. At higher temperatures (275° to 450°C), decomposition may be affected by the protonic sites already generated in the vicinity of the remaining TTMA cations, and the intracrystalline environment is relatively anhydrous. Here, ylide formation is possible (Equation 4), with at least enough stability to allow for characteristic ylide reactions—assuming that the zeolitic anionic oxygens act as (weak) bases or as a stabilizing "pseudosolvent" (21). Additional protonic acidity is generated as shown.

$$(CH_3)_3\overset{\oplus}{N} - CH_2 \longrightarrow (CH_3)_3\overset{\oplus}{N} - \overset{\ominus}{CH_2} \quad (4)$$
$$\qquad\qquad |\!\nearrow H$$
$$\underset{O-ZEOL}{\ominus} \qquad\qquad \underset{III}{\overset{\oplus\;\ominus}{H\;\;O-ZEOL}}$$

A carbanionic mechanism similar to that proposed by Wittig and Krauss (33) involving a Steven's rearrangement can reasonably follow (Equation 5), i.e., an intramolecular attack on ylide (III) methyl forming dimethylethylamine. Protonation of dimethylethylamine followed by Hoffmann elimination (Equation 5a) would yield ethylene.

$$(CH_3)_2\overset{\oplus}{\underset{|}{N}}-\overset{\ominus}{CH_2} \longrightarrow (CH_3)_2N\;CH_2CH_3 + \overset{\oplus\;\ominus}{H\;O-ZEOL} \quad (5)$$
$$\quad H_3C\swarrow$$
$$\overset{\oplus\quad\ominus}{H\quad O-ZEOL} \qquad \downarrow$$
$$\underset{III}{} \qquad (CH_3)_2\overset{\oplus}{N}H\;CH_2CH_3\;\overset{\ominus}{O}-ZEOL$$
$$\qquad\qquad\qquad\qquad \downarrow$$
$$\qquad\qquad (CH_3)_2NH + CH_2=CH_2 + \overset{\oplus\;\ominus}{H\;O-ZEOL} \quad (5a)$$

Repetition of such processes or acid-catalyzed polymerization of lower olefins could afford a whole spectrum of hydrocarbon products, $(CH_2)_n$. The presence of the protonic sites, generated in the low-temperature step, effectively intercepts diffusion of the amines from the zeolitic pores until most of them are degraded to ammonia. Only traces of trimethylamine escaped and were found in the product.

It is conceivable that the TTMA cation could decompose to form trimethylamine, acidic lattice hydroxyls, and a bivalent "carbenoid" or methylene species. Polymerization of methylene could lead then to the observed olefinic products. However, such carbene polymerizations have been accorded low probability (12, 13, 33).

The "methyl disproportionation reaction" appeared to occur at the higher temperatures. Hydrogen-rich species—methane and hydrogen—were observed in the gaseous products. In the absence of carbon monoxide or other multiple-bonded gaseous molecules, it may be assumed that an amine or nitrogenous species was the hydride donor, and that hydrogen-deficient moieties (nitriles, polyunsaturates, etc.) remained as adsorbed "coke" (0.2 wt % N, 2.2 wt % C) on the zeolite surface.

Decomposition of the 3 lower-substituted MA cations (based on the TGA data) may not be visualized as easily as occurring in 2 stages. Nevertheless, the various mechanisms (which coherently rationalized the observed products obtained from the TTMA cations) probably functioned in these other systems as well, although the extent to which they contributed may have differed.

In conclusion, examination of the MA cation–zeolite system reveals a significantly greater temperature requirement and considerably more complex pathways for decomposition of intrazeolitic MA cations than those for pyrolysis of most simple quaternary salts (15, 16, 32, 35). In addition, the MA cations function as acidity precursors in zeolites. This suggests an important role for the zeolitic lattice in determining the course of intracrystalline quaternary fragmentations, including both pore–channel geometry (steric and mass transport aspects) and chemical factors (anionic oxygen base strength, site density, and type).

Acknowledgment

We thank C. Craig and J. H. Wuertz for excellent technical assistance, J. G. Mollet for mass spectral analyses, and acknowledge helpful discussions with W. O. Haag of Mobil Research & Development Corp. and D. S. Noyce of the University of California.

Literature Cited

(1) Ambs, W. J., Flank, W. H., *J. Catalysis* **1969**, 14, 118.
(2) Barrer, R. M., Buser, W., Grütter, W. F., *Helv. Chim. Acta* **1956**, 39, 518.
(3) Borello, E., Zecchina, A., Morterra, C., *J. Phys. Chem.* **1967**, 71, 2938.
(4) Cattanach, J., Wu, E. L., Venuto, P. B., *J. Catalysis* **1968**, 11, 342.
(5) Detar, D. F., Kosuge, T., *J. Am. Chem. Soc.* **1958**, 80, 6072.
(6) Franzen, V., Wittig, G., *Angew. Chem.* **1960**, 72, 417.
(7) Hanhart, W., Ingold, C. K., *J. Chem. Soc.* **1927**, 997.
(8) Hofmann, A. W., *Ber.* **1881**, 14, 494.
(9) Johnson, A. W., "Ylide Chemistry," p. 251, Academic Press, New York, 1966.
(10) Kerr, G. T., *J. Catalysis* **1969**, 15, 200.
(11) Kerr, G. T., *J. Phys. Chem.* **1969**, 73, 2780.
(12) Kirmse, W., "Carbene Chemistry," p. 8, Academic Press, New York, 1964.
(13) Kirmse, W., *Angew. Chem. Intern. Ed.* **1965**, 4, 1.
(14) Landis, P. S., Haag, W. O., *J. Org. Chem.* **1963**, 28, 585.
(15) Musker, W. K., *J. Am. Chem. Soc.* **1964**, 86, 960.
(16) Musker, W. K., *J. Chem. Educ.* **1968**, 45, 200.
(17) Olberg, R. C., Pines, H., Ipatieff, V. N., *J. Am. Chem. Soc.* **1944**, 66, 1096.
(18) Olson, D. H., private communication.
(19) Olson, D. H., Dempsey, E., *J. Catalysis* **1969**, 13, 221.
(20) Papera, J. M., Figoli, N. S., *J. Catalysis* **1969**, 14, 303.
(21) Pines, H., Manassen, J., *Advan. Catalysis* **1966**, 16, 49.
(22) Rabo, J. A., Pickert, P. E., Stamires, D. N., Boyle, J. E., *Actes Congr. Intern. Catalyse, II, Paris, 1960,* **1961**, 2, 2055.
(23) Sherry, H. S., *J. Phys. Chem.* **1966**, 70, 1158.
(24) Stamires, D. N., Turkevich, J., *J. Am. Chem. Soc.* **1964**, 86, 749.
(25) Tanaka, J., Duuning, J. E., Carter, J. C., *J. Org. Chem.* **1966**, 31, 3431.
(26) Theng, B. K. G., Vansant, E., Uytterhoeven, J. B., *Trans. Faraday Soc.* **1968**, 64, 3370.
(27) Uytterhoeven, J. B., Christner, L. G., Hall, W. K., *J. Phys. Chem.* **1965**, 69, 2117.
(28) Uytterhoeven, J. B., Jacobs, P., Makay, K., Schoonheydt, R., *J. Phys. Chem.* **1968**, 72, 1768.
(29) Venuto, P. B., Landis, P. S., *Advan. Catalysis* **1968**, 18, 259.
(30) Ward, J. W., *J. Phys. Chem.* **1969**, 73, 2086.
(31) Weygand, F., Daniel, H., Schroll, A., *Chem. Ber.* **1964**, 97, 1217.
(32) Wittig, G., Heintzeler, M., Wetterling, M.-H., *Ann.* **1947**, 557, 201.
(33) Wittig, G., Krauss, D., *Ann.* **1964**, 679, 34.
(34) Wittig, G., Polster, R., *Ann.* **1956**, 599, 1.
(35) Wittig, G., Wetterling, M.-H., *Ann.* **1947**, 557, 193.
(36) Wu, E. L., unpublished work.

RECEIVED February 4, 1970.

Discussion

W. H. Flank (Houdry Laboratories, Marcus Hook, Pa. 19061): All of your samples, after a 1-hour treatment at 900°C, were found by x-ray examination to be amorphous. The Na_2O content reported for your samples ranged from about 3–13% by weight. The mullite transformation you mentioned, which really occurs in two parts, is a kinetically-controlled process. This product is also rather difficult to detect by the usual x-ray methods. Furthermore, the DTA mullitization peak size is strongly dependent on the sodium content, as shown by the data published by Ambs and Flank. Considering these factors, your data are not at all surprising or inconsistent for the materials and techniques used. One further point is that the exotherm identified in Figure 2 as the mullite transformation is more likely the collapse of the zeolitic structure which, with the relatively high sodium content in these samples, is likely followed by a much smaller mullitization peak at a somewhat higher temperature. The data in Figure 2 are somewhat suggestive of this.

E. L. Wu: Your alternative assignment to the high-temperature exothermic peaks in Figure 2 may be reasonable.

J. W. Ward: (Union Oil Co. of California, Brea, Calif. 92621): You state that your samples were amorphous, as determined by x-ray diffraction after heat treatment at 900°C. Is it possible that the use of other techniques, such as electron diffraction, might show some retention of crystallinity?

E. L. Wu: We have not carried out such experiments and do not want to venture a guess as to the possibility of such techniques to determine residual crystallinity.

W. M. Meier (Eidgenössische Technische Hochschule, Zurich): In connection with the postulated formation of a surface methoxy species on heating TTMA-Y, it might be of interest to note that we have observed pronounced methyl–oxygen interaction in the structures of TTMA-sodalite and TTMA-gismondine (C. Baerlocher and W. M. Meier, *Helv. Chim. Acta* **1969**, 52, 1853, and in press).

R. L. Wadlinger (Niagara University, N. Y. 14109): Have you ever re-exchanged your samples after the cation decompositions?

E. L. Wu: No. We have not examined the ion exchange properties after decomposing the organic cations.

40

Properties of Aluminum-Deficient Large-Port Mordenites

W. L. KRANICH, Y. H. MA, L. B. SAND, A. H. WEISS, and I. ZWIEBEL

Department of Chemical Engineering, Worcester Polytechnic Institute, Worcester, Mass. 01609

> *A series of aluminum-deficient mordenites, ranging in Si/Al ratio from 6 to greater than 600, has been prepared by thermal and acid treatment of H-mordenite. The series has been characterized for cumene adsorption, catalytic activity for cumene cracking and hydrocracking, and catalytic activity for butene isomerization. Removal of aluminum from the lattice results in diminished cumene adsorptive capacity at a given pressure for cumene pressures above 0.15 torr, but increased capacity at low pressures. Both cracking and hydrocracking activity decrease with decreasing aluminum content at high Si/Al ratios. Initial leaching of H-mordenite gives increased isomerization activity, but continued decationization results in diminished activity.*

Several authors have reported on changes in characteristics of mordenites as aluminum is removed from their structure by leaching with mineral acid. Sand (6), Belen'kaya *et al.* (3), and Frilette and Rubin (5) described the acid leaching of mordenites producing Si/Al ratios up to about 15. Beecher, Voorhies, and Eberly (2) reported on the catalytic activity of palladium supported on a mordenite which had a Si/Al "several-fold greater than 5." Weller and Brauer (8) recently described catalytic experiments on mordenite leached to a ratio of about 9.

Although Barrer and Makki (1) succeeded in reducing the aluminum content of clinoptilolite essentially to zero, the characterization of a fully decationized mordenite has not been reported.

In the present study, by use of a combination of thermal and acid treatments, a series of large-port mordenites has been prepared down to a material with essentially no residual aluminum but which retains the mordenite crystal structure as determined by x-ray powder diffraction and single-crystal electron diffraction. Large-port mordenites are defined as those which sorb molecules larger than 5A into the 12-membered ring channels, as contrasted with those which sorb only smaller molecules such as methane (6). The series has been systematically investigated for its adsorptive and catalytic properties at both high and low temperatures. Preliminary results are reported in this paper.

As aluminum is removed, its place in the lattice is taken by hydrogen in hydroxyl groups as shown by infrared absorption spectra. In the original H-mordenite, the structure in the vicinity of the aluminum may be represented by

$$\begin{array}{c} -\text{Si}- \\ | \\ \text{O} \\ | \quad\quad H_3^+O \\ -\text{Si}-\text{O}-\text{Al}^--\text{O}-\text{Si}- \\ | \\ \text{O} \\ | \\ -\text{Si}- \end{array}$$

As the aluminum is removed from the structure, the unbalanced charges which result in the exchangeable cations are successively removed. The fully decationized mordenite (aluminum-free) may be represented as

$$\begin{array}{c} -\text{Si}- \\ | \\ \text{O} \\ \text{H} \\ -\text{Si}-\text{OH} \quad\quad \text{HO}-\text{Si}- \\ \text{H} \\ \text{O} \\ | \\ -\text{Si}- \end{array}$$

This is a similar structural alteration to that proposed by Barrer and Makki (1) for clinoptilolite.

Although no major disruption of the crystal lattice occurs, as evidenced by stability of the x-ray pattern, there are significant changes in ionic character and charge distribution. These may be expected to lead to changes in the adsorptive and catalytic behavior of the mordenite.

Table I. Data on Hydrogen-Exchanged Large-Port

Sample No.	Wt. % Al_2O_3	Partial Unit Cell Composition
1 (starting material)	11.24 —not heat treated	$Na^{ex}_{0.48}H^{ex}_{6.19}Al_{6.67}Si_{40}$
2	6.90	$H^{ex}_{4.00}Al_{4.00}Si_{40}$
3	1.99	$H^{ex}_{1.14}Al_{1.14}Si_{40}$
4	1.42	$H^{ex}_{0.80}Al_{0.80}Si_{40}$
5	0.10	$H^{ex}_{0.06}Al_{0.06}Si_{40}$

Materials Preparation

The basic material for preparation of the series was a hydrogen-exchanged mordenite (Lot No. HB-33-36) obtained from the Norton Co. Its chemical analysis is as follows: 72.28% SiO_2, 9.8% Al_2O_3, 0.26% TiO_2, 0.08% Fe_2O_3, 0.37% Na_2O, 12.09% L.O.I. No other phases were detected by x-ray diffraction analysis; the presence of 5–10% of amorphous material was estimated by microscopic examination. The treatments on this starting material had negligible effect on the TiO_2 content and reduced the Fe_2O_3 and Na_2O content to less than 0.01%.

The Al_2O_3 content, reported on an anhydrous basis, of the aluminum-deficient mordenites was obtained by x-ray fluorescence analysis using fused lithium tetraborate discs with a working curve established on mixtures of 2 chemically analyzed H-mordenites of end-member composition (11.24 and 0.10% Al_2O_3). Glass discs were prepared by calcining the mordenite powder samples at 1100°C for 1 hour, fusing a mixture of 2 grams of mordenite and 10 grams of lithium tetraborate at 1300°C, and pouring the liquid into a graphite mold kept at red heat in a kiln. The bottom of the glass disc was polished flat before analysis. No significant changes in lattice parameters or degradation of the crystal structure was noted by x-ray diffraction analysis. Increases in diffraction peak intensities of 110, 200, and 020 resulted from the treatments. Summations of peak intensities of the major diffractions from 110, 200, 020, 111, 510, 022, 600, and 620 were obtained and compared with those of the starting material ($\Sigma = 100$).

The data on sample preparation and analysis are summarized in Table I. A 700°C thermal treatment was used for all samples except Sample No. 3 which was treated at 500°C. Details on the preparation of aluminum-deficient mordenites are given by Chang and Sand (4). The particle size range of the hydrogen-exchanged mordenite used as the starting material is 5–10μm with an average mean diameter of 7μm. The particles consist of 40% of single crystals and 60% of crystal aggregates.

Mordenite and Aluminum-Deficient Derivatives

Si/Al	Σ XRD Peaks	Acid Treat.
6	100	$6N$ H_2SO_4 ambient
10	98	$1N$ HCl boil 2 hrs.
35	118	$6N$ HCl boil 6 hrs.
50	141	$6N$ HNO_3 boil 12 hrs.
>600	102	$6N$ HNO_3 boil 6 hrs.

The severe acid treatments separated the aggregates into single crystals in a size range of 0.5–10μm with an average mean diameter of 3μm. Twenty per cent of the crystals are euhedral acicular and the remainder are equant and anhedral.

In order to investigate changes in the characteristics of the mordenite as aluminum was extracted, studies were made of the adsorption of cumene, the cracking and hydrocracking of cumene, and the isomerization of 1-butene.

Experimental Methods

Cumene adsorption experiments were carried out in a microbalance apparatus equipped with a quartz spring whose deflections were measured with a cathetometer. The unit also included a vacuum system capable of reducing the pressures to below 0.01 micron, and a Baratron continuous pressure-sensing device with a sensitivity of 1 micron. The amount adsorbed was measured by spring deflection, and checked by material balance on the gas phase based on an approximate chamber volume of 600 cc. The pressure measurements were noted at regular intervals until "equilibrium" was reached (less than 1 micron change in 5 minutes). These pressure vs. time data were used for diffusivity calculations.

Cumene Cracking and Hydrocracking. Cracking studies were conducted in a tubular stainless steel microreactor inserted in a bronze block. The entire assembly was constructed within the oven of a Perkin Elmer Model 880 gas chromatograph.

With this construction, the reactor could be fed in 5-μl pulses directly from the injection block into a stream of hydrogen or helium flowing at a rate of 282 standard cc per minute. A fraction of the product gases was conducted directly to the chromatograph column.

The catalyst, diluted 1000:1 with glass sand, was pretreated at 538°C for 15 hours with hydrogen or helium prior to use.

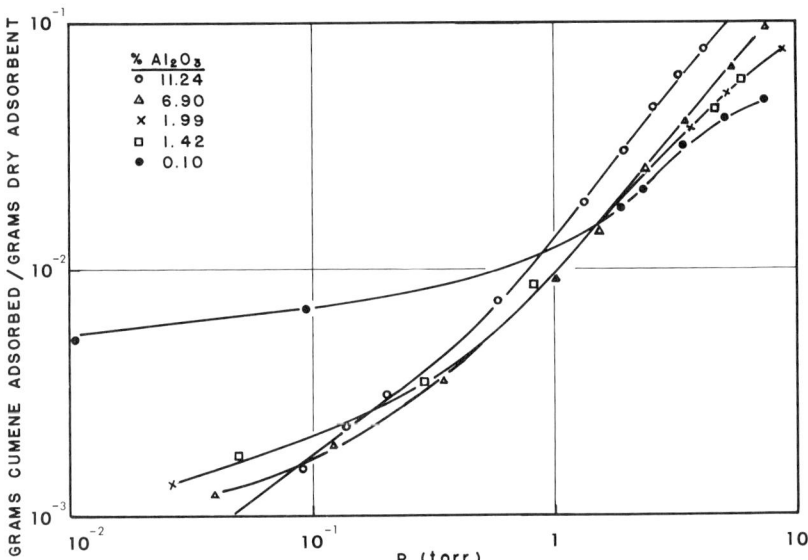

Figure 1. Cumene adsorption isotherms

Typical loading for cumene conversion of 0.1 to 1.0% was 4×10^{-4} gram of catalyst at a temperature of about 371°C and atmospheric pressure. Without catalyst, the benzene yield from thermal cracking at 421°C was only about 0.002% of the cumene fed. The very small charge of catalyst allowed study of intrinsic activity decline at low conversion levels while operating at temperatures of practical interest which were high enough to avoid desorption problems. Unfortunately, this same small catalyst charge resulted in an uncertainty in mass as great as ±40%, with a corresponding uncertainty in comparison of different individual catalyst charges.

Successive pulses of cumene were fed to the reactor until the initial high activity level had been reduced to approximately steady state.

Butene Isomerization. A steady-flow tubular reactor was used for studies of isomerization of 1-butene to *cis-* and *trans-*2-butene.

The catalyst powder was pressed into a thin disk, crushed, screened, and dried at 200°C. After being diluted about 5 to 1 with glass beads and packed into a cylindrical flow reactor, the catalyst was activated in place by heating in a stream of dry helium for 2 hours at 350°C.

The catalyst mass of about 0.75 gram was then brought to the reaction temperature and exposed to a mixture of 1-butene (132 cc/min) and helium (260 cc/min). Temperature was maintained constant by immersion of the reactor in a fluidized sand bath.

The rapid decrease in activity characteristic of fresh, activated catalysts was followed by trapping and analyzing successive gas samples at measured intervals in a gas chromatograph. The butene–helium mixture was allowed to flow over the catalyst for several hours, until successive samples showed essentially no change in composition. Steady-state activities were measured at several temperatures.

Results

Cumene Adsorption. Cumene adsorption equilibrium data are presented in the form of isotherms at room temperature in Figure 1.

At the high-pressure end of the graph (above $P/P_o = 0.15$) the adsorption capacity at a given pressure decreases markedly as aluminum is leached from the mordenite. In mid-range, however, the curves cross, so that at low pressures (below $P/P_o = 0.01$) the adsorption capacity increases as aluminum is removed. This effect is particularly marked in the fully decationized mordenite. This material adsorbs as much cumene at a pressure of 0.005 torr as the other mordenites adsorb at pressures about 50 times as great.

Effective diffusion coefficients were calculated from adsorption rate data on 3 materials based on a spherical model. The crystallites were assumed to be spheres with a diameter of 3.0 microns.

For comparison, the diffusion coefficients were calculated from each adsorption rate curve at the point where the cumene uptake was half of its equilibrium value. The results are shown in Table II.

Table II. Effective Cumene Diffusion Coefficients at 50% Uptake

% Al_2O_3	$D \times 10^{12}\ Cm^2/Sec$
11.24	10.0
6.90	5.2
1.42	3.4

The results indicate that the original mordenite permits significantly easier diffusion of cumene than does leached material. Further work is underway to clarify this rather unexpected relationship.

Cumene Cracking. Figures 2 and 3 show cumene conversion per gram of catalyst for runs in He and in H_2, respectively, at 360°C as a function of the cumulative quantity of cumene injected over each catalyst studied. Activity loss could be partially recovered to as much as 50% of original by heating in H_2 at 538°C, indicating that both the formation of dimers and higher, as well as irreversible coking, contributed to activity loss. The activities of the fully-leached samples appear to be almost the same in H_2 and in He; the apparently enhanced reactivity in

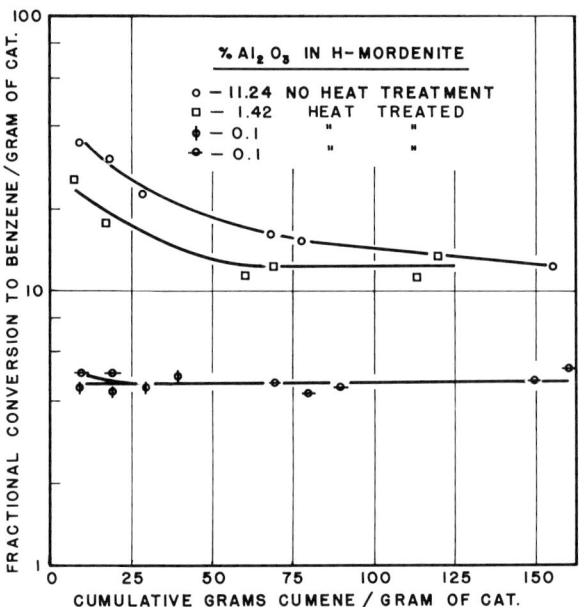

Figure 2. Cumene cracking activity

H_2 of the catalysts containing more alumina is within the limits of error of catalyst mass as described previously.

Arrhenius plots of nondeclining activity data calculated as first-order rate constants were fitted for the 0.1 and 1.42% Al_2O_3 catalysts. Benzene production on both catalysts can be fitted to the same activation energy of 15.8 Kcal/mole for operation in both H_2 and He in the range of 347°–450°C.

Beecher et al. (2) investigated 1/2% Pd on H-mordenite and 1/2% Pd on Al-deficient H-mordenite ("acid-leached . . . to give a Si/Al ratio several-fold greater than 5"). They showed that the acid treatment greatly improved catalytic activity for hydrocracking n-decane and decalin. Topchieva et al. (7) studied pulse microreactor cumene cracking of H-mordenites having Si/Al ratios of 5, 6.5, and 9. The data obtained (but not presented) implied that catalytic activity increased with increasing Si/Al ratio. Similar work by Weller and Brauer (8) in the range of 7.5 to 9 Si/Al corroborated the increased cracking activity of alumina-deficient mordenites, in this case for hexane cracking.

In the present work, Si/Al ratios varied from 6 to 600. Neither initial activity levels nor sustained operation showed an activity advantage for Al_2O_3 removal. Rather, activity appeared to decrease with increasing Si/Al at the 1% cumene conversion level, both with H_2 and with helium.

The materials selected for this preliminary investigation were chosen so that Si/Al ratios would be separated by about an order of magnitude— i.e., 6, 50, and 600. The range studied by the other investigators cited was between the first 2 of these points. This work does not preclude the possibility that the increase in activity with decreasing aluminum content observed by the others might exist also in that same range for other materials. Weller and Brauer (8) in fact did report a maximum in cracking activity at a Si/Al ratio of about 8.7. Work is continuing in this range.

Butene Isomerization. INITIAL DEACTIVATION. Conversion of 1-butene to the 2-butenes was much more rapid initially than in the steady state for all catalysts at all temperatures studied. Initial activity of the fully-leached material was much greater than that of the original H-mor-

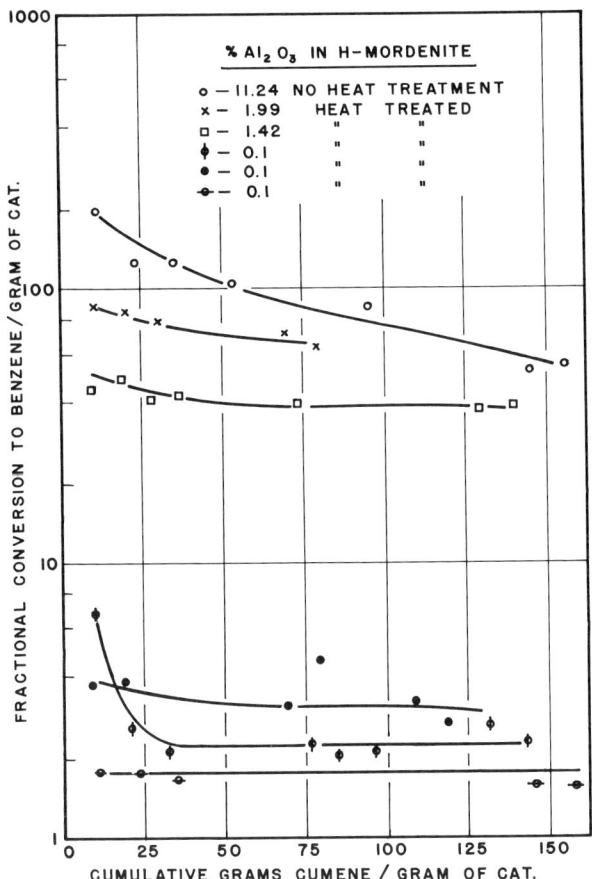

Figure 3. Cumene hydrocracking activity

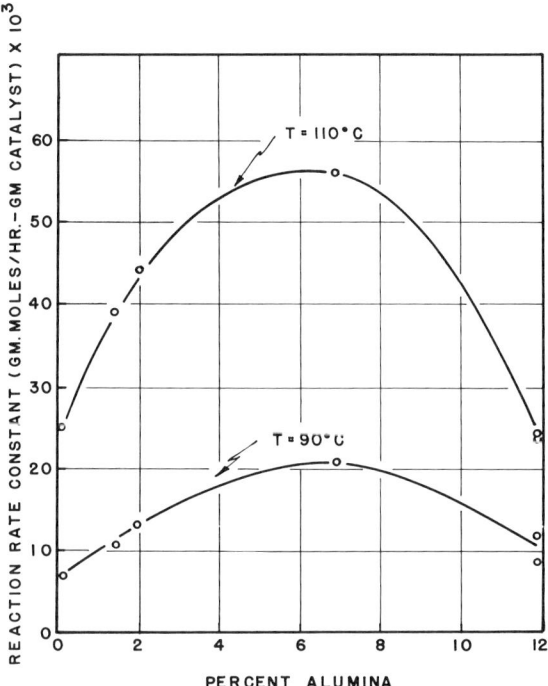

Figure 4. 1-Butene isomerization activity

denite. Approximately 40% of the 1-butene was converted at a temperature of 20°C by 0.75 gram of leached material after 4 minutes on stream. In contrast, only about 10% was converted under the same conditions over the untreated H-mordenite. Catalysts did not reach a steady state activity in the range 20°–70°C, but completely deactivated with time. Above 90°C, a steady-state activity was reached which persisted for many hours with only slight decay.

Approximately 75% of the lost activity could be recovered by purging for several hours with dry helium. It is believed that the lost activity results primarily from polymerization of the butenes to octenes, docecenes, and higher which cannot readily escape from the pores and thus prevent access to the inner active sites. On prolonged purging in a helium stream, these somewhat volatile polymers slowly diffuse out of the catalyst.

Above about 90°C, the volatility of the low molecular weight polymer byproducts is sufficient to permit them to diffuse slowly but steadily out of the pores and allow continuing isomerization to occur. The presence of volatile polymers in catalyst which had reached steady-state activity was confirmed by scanning deactivated catalyst at various temperatures in a mass spectrometer. At 25°C, C_8 hydrocarbons were observed; at 50°C, C_{12} compounds appeared; and at 75°C there were significant peaks as high as the C_{40} range.

STEADY STATE. The variation of reaction rate constant with percentage alumina at the steady state is shown in Figure 4. As the mordenite is decationized, the activity increases to a maximum and then falls with further leaching.

This may be explained as an initial activation of internal sites by the acid and heat treatment, followed by some loss in activity as the number of acid sites is reduced.

A marked effect on steady-state selectivity, measured as trans/cis, results from reduction in aluminum. The selectivity of the fully-leached catalyst is about 2.4, while that of the H-mordenite is about 1.3.

Apparent selectivity of the fully-leached freshly activated material was even more marked than that in the steady state. In the first samples from the fresh catalyst, no *cis*-2-butene was observed even though half of the 1-butene was converted.

At present, it is not known whether this is observed because the cis isomer is not formed at a rate measurable relative to trans or whether the cis is formed, then strongly adsorbed or quickly polymerized.

Conclusions

Aluminum can be leached from H-mordenite almost completely by a combination of thermal and acid treatment. No significant variation of the basic mordenite lattice occurs as a result of this treatment, even when the final Si/Al ratio is greater than 600.

Both catalytic and adsorptive properties of the H-mordenite are altered significantly as aluminum is removed, particularly as decationization nears completion.

As Si/Al is increased by leaching, adsorption capacity for cumene at ambient temperature and a given pressure is enhanced at very low pressures, but diminished at pressure above 0.15 torr.

Leaching of H-mordenite initially increases catalytic activity for 1-butene isomerization, but above a Si/Al ratio of about 15, the activity diminishes.

At high Si/Al ratios, removal of aluminum also results in a loss of cracking and hydrocracking activity for cumene. The disappearance of acid sites with decationization could account for loss of activity in both cracking and isomerization.

Studies are continuing with the objectives of further defining the results on adsorption and catalysis and explaining the phenomena observed in terms of structural changes.

Acknowledgment

Techniques for the preparation of the aluminum-deficient mordenites were developed under a contract with the Norton Co. The following students at Worcester Polytechnic Institute performed the experimental work: H. Chang, J. Tamboli, J. R. Pratt, D. Kremer, H. Bierenbaum, and S. Chiramongkol. J. R. Pratt and S. Chiramongkol were supported by a National Science Foundation grant for Undergraduate Research Participation. H. Bierenbaum was supported by an NDEA Title IV Fellowship. The cooperation of the following Norton Co. scientists in the analysis of the modified mordenites is appreciated: R. H. Benson, A. J. Regis, W. H. Gerdes. Contributions to the design and construction of apparatus by C. A. Keisling, and the collection of adsorption data by R. Knapik are acknowledged.

Literature Cited

(1) Barrer, R. M., Makki, M. B., *Can. J. Chem.* **1964**, 42, 1481.
(2) Beecher, R., Voorhies, A., Eberly, P., *Symp. New Chem. Hydrogen Process. Petrol.*, ACS, Chicago, Ill., September 11–15, **1967**.
(3) Belen'kaya, I. M., Dubinin, M. M., Krishtofori, I. I., *Izv. Akad. Nauk SSSR* **1967**, 10, 2164–71.
(4) Chang, H. D., Sand, L. B., in preparation.
(5) Frilette, V. J., Rubin, M. K., *J. Catalysis* **1965**, 4, 310.
(6) Sand, L. B., "Molecular Sieves," Society of the Chemical Industry Spec. Publ., 71–77, 1968.
(7) Topchieva, K. V., Romanovsky, B. V., Piquzova, L. I., Thoang, HoSi, Bizreh, Y. W., *Intern. Congr. Catalysis, 4th, Moscow, USSR,* **1968**.
(8) Weller, S. W., Brauer, J. M., *AIChE Annual Meeting, 62nd, Washington, D. C.,* November 16–20, **1969**.

RECEIVED February 25, 1970.

Discussion

Alfred E. Hirschler (Sun Oil Co., Marcus Hook, Pa.): I wonder if you could be a little more specific on the manner in which the thermal and acid leaching procedures were combined in obtaining the samples described?

L. B. Sand: A typical procedure was 700°C for two hours in a dry nitrogen purge followed by boiling in 6N HCl for several hours. Delineation of parameters can be found in H. D. Chang's PhD thesis on file at the W.P.I. library.

B. C. Gates: (University of Delaware, Newark, Del. 19711): You reported data for crystallite size distribution. How did you measure crystallite sizes? Did you attempt size-fractionation of the crystallites?

L. B. Sand: Crystallite sizes were measured with an optical microscope. No fractionation was attempted.

J. R. Katzer (University of Delaware, Newark, Del. 19711): You refer to the fact that diffusivities were measured and comment that you are attempting to explain the variations found on the basis of strongly adsorbed species blocking the diffusion process. Did you measure both adsorption and desorption rates, and if you did, what were the relative values of the two rates?

W. L. Kranich: Because of limited availability of sorption apparatus when this study was made, we did not take desorption data. In view of the results, we recognize the special value of desorption rates and hope to take such data in the future.

J. R. Katzer: We are using a similar explanation for our results on the desorption of benzene and cumene from H-mordenite and have reported this work at the Puerto Rico A.I.Ch.E. meeting in May, 1970. (*Ind. Eng. Chem. Fundamentals*, in press.) I think that this is an important factor. If you could do some desorption measurements, the results would be very helpful and would clarify your speculation. We showed quite clearly that benzene and cumene molecules cannot counterdiffuse in the H-mordenite pores at reasonable temperatures.

41

Investigation of Oxygen Mobility in Synthetic Zeolites by Isotopic Exchange Method

G. V. ANTOSHIN, KH. M. MINACHEV, E. N. SEVASTJANOV,
D. A. KONDRATJEV, and CHAN ZUI NEWY

N. D. Zelinsky Institute of Organic Chemistry, Academy of Sciences, Moscow, USSR

> *A study has been carried out on the O^{18} exchange between O_2 and following zeolites: NaX, NaY ($SiO_2/Al_2O_3 = 3.4$), NaY ($SiO_2/Al_2O_3 = 4.7$), Na-mordenite, and Cu and Ag forms of Y zeolite. The sodium forms have similar kinetic characteristics of oxygen mobility. Activation energies of exchange for these forms are about 45–50 kcal/mole. The replacement of Na by Cu in Y zeolite causes a large increase of oxygen mobility. The same effect, but less pronounced, has been found for silver.*

The present paper reports some data on the oxygen mobility of Na-mordenite, and NaX, and NaY zeolites, and that of silver and copper forms of zeolite Y with various degrees of replacement of sodium by the cations. These have been obtained by studying the zeolite–molecular oxygen isotopic exchange kinetics. Such data may prove valuable in gaining insight into the correlation dependence between zeolite chemical composition and their properties, such as catalytic activity and thermal stability (*1, 4, 6*). No data on isotopic exchange between O_2 and zeolites are available in the literature, except one work (*3*) in which the exchange between O_2^{18} and zeolite Linde 5A was not found at 250°C.

Experimental

Materials. The samples were commercially supplied zeolites NaX ($SiO_2/Al_2O_3 = 2.45$), NaY ($SiO_2/Al_2O_3 = 3.4$ and 4.7), and Na-mordenite ($SiO_2/Al_2O_3 = 10$) without binder. The crystal sizes of zeolites used were 1–2μ. Silver and copper forms of zeolite Y were prepared by cation exchange between the starting sodium containing zeolite Y with $SiO_2/Al_2O_3 = 3.4$ ratio and silver nitrate or copper acetate solution. Before experiments, the zeolites were washed by distilled water. Pelleted forms of zeolites 4 mm in diameter and 5 mm high were used. Surface areas (BET) of X and Y zeolites were *ca.* 700 m²/gram. The corresponding value for Na-mordenite was 300 m²/gram. Oxygen enriched in O^{18} was prepared by electrolysis of heavy-oxygen water. Oxygen with O^{18} concentration about 7% atm (equilibrium mixture of different isotopic species) was used in most of experiments.

Apparatus. A vacuum circular static unit was used to study the isotopic exchange. The temperature in the zeolite bed was maintained with accuracy of $\pm 1\,°C$. Omegatrone was employed to analyze the gas isotopic composition.

Procedure. Zeolite samples were treated in high vacuum (10^{-5} mm) at 700°C for 5 hours and then kept in oxygen at the same temperature for not less than 5 hours at a pressure of 5–10 mm. Isotopic exchange was studied at 600°–700°C under 5–60 mm pressure. The reaction vessel was fitted with traps and cooled with liquid nitrogen to prevent grease from getting into the zeolites. The capacity of the zeolite samples used was determined against nitrogen at the temperature of liquid nitrogen before and after the exchange experiments.

Results and Discussion

Preliminary experiments with zeolite NaY ($SiO_2/Al_2O_3 = 3.4$) have shown oxygen mobility (exchange between phases) to be measurable only at temperatures exceeding 600°C after vacuum-treating the zeolite showed varying increased activity. This was decreased to a constant value by maintaining the zeolite in oxygen. The data obtained allowed a choice of conditions for the pre-treatment of the samples. The O^{18} concentration increase in zeolite resulting from one exchange with the enriched gas was negligible. (The ratio between the number of oxygen atoms in the zeolite and that in the gas was about 200 in most of the experiments.) This made possible a series of experiments with the same zeolite sample to study the dependence of the exchange rate on oxygen temperature and pressure. After the series of experiments, the initial one was repeated under the starting conditions. The reproducibility of results was 10%.

Evaluation of the experimental data has shown the molecular oxygen–zeolite isotopic exchange kinetics to obey first order equations. For example, Figure 1 presents some kinetic data on the exchange between O_2 and sodium forms of zeolites studied at 620° and 700°C. These are treated according to the equation

$$R = - \frac{2.3 \cdot N_1 \cdot \log(1 - F)}{g \cdot (1 + \frac{N_1}{N_2}) \cdot \tau} \quad (1)$$

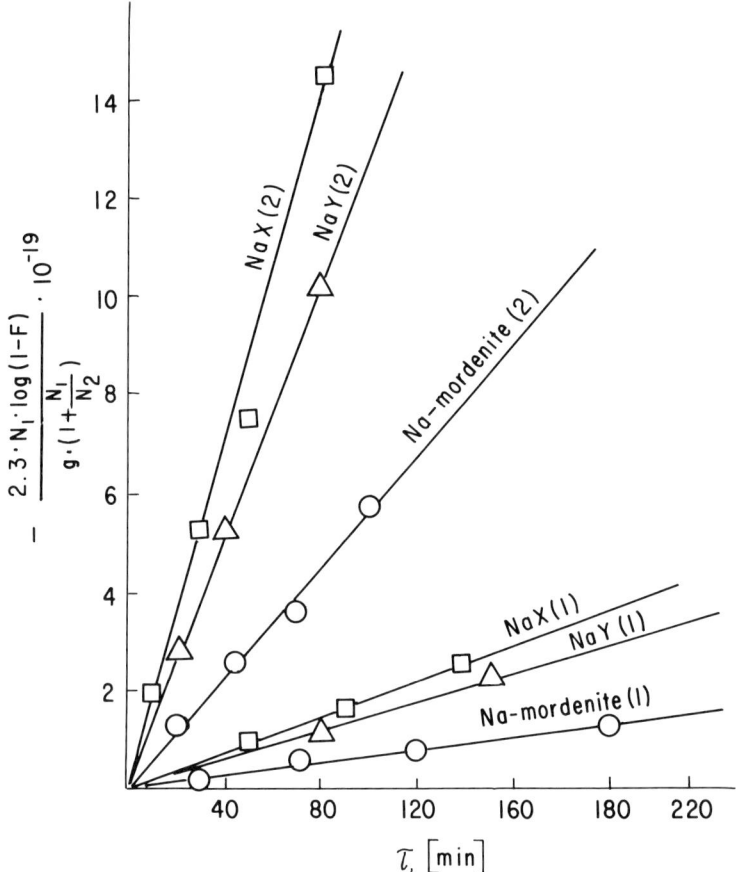

Figure 1. Kinetics of exchange between O_2 and sodium forms of zeolites studied at temperatures 620°C (1) and 700°C (2). Oxygen pressure was 5 mm Hg.

where R is the isotopic exchange rate constant, atoms/gram min; N_1 is the number of oxygen atoms in the gas; N_2 is the number of oxygen atoms in the zeolite; F is the degree of exchange; g is the zeolite specimen (grams); τ is the exchange time (min). The linear dependence shown in Figure 1 provides evidence that all oxygen of the zeolites studied is equivalent as to its exchange with the gas. These results will be discussed further.

Table I summarizes the main kinetic characteristics of the isotopic exchange between molecular oxygen and the zeolites studied: R is the rate constant of exchange at 670°C and 5-mm pressure, atoms/gram min; E is the apparent activation energy, kcal/mole; n is the reaction order

Table I. Kinetic Characteristics of O^{18} Exchange between Zeolites and Molecular Oxygen

Zeolite	$\frac{SiO_2}{Al_2O_3}$ Ratio	Z^a	$R \times 10^{-17}$ Atoms / Gram Min	E, Kcal / Mole	n
NaX	2.45	—	8.40	47	1.0
NaY	3.4	—	5.70	45	1.0
NaY	4.7	—	2.35	49	1.0
Na-mordenite	10	—	2.88	45	1.0
CuNaY	3.4	0.14	37.5	38	0.9
CuNaY	3.4	0.26	121	33	1.0
CuNaY	3.4	0.42	433	23	0.9
AgNaY	3.4	0.14	12.7	38	0.8
AgNaY	3.4	0.26	21.0	39	—

^{a} Z = degree of sodium replacement by the corresponding cation.

against oxygen pressure. As seen from the data presented in Table I, the oxygen mobility kinetic characteristics of the sodium forms of the zeolites studied are similar.

The replacement of sodium by copper causes a strong promoting effect. For example, the exchange rate constant is 6 times as high for a NaY sample with $Z = 0.14$ as for the starting zeolite, the exchange activation energy decreasing from 45 to 38 kcal/mole. When the degree of cationic exchange increases up to 0.42, the rate constant increases further and the activation energy decreases to 23 kcal/mole. The promoting effect also is caused when sodium is replaced by silver, but in this case the effect is less pronounced.

One may observe compensating dependence between the preexponential factor in the Arrhenius equation and the activation energy. This dependence is shown in Figure 2. As the activation energy in the sample with the exchange degree 0.42 decreases to 23 kcal/mole, the preexponential factor becomes 3 orders lower.

Some experiments with various cationic forms of zeolite Y at the highest temperatures have been carried out using the gas with O^{18} concentration about 40% atm, the content of various isotopic molecules of oxygen ($O^{16}O^{16}$, $O^{16}O^{18}$, and $O^{18}O^{18}$) in it being close to equilibrium. With sodium and silver forms, the equilibrium in the gas phase was retained, whereas with the copper forms a sharp deviation from equilibrium was observed at the initial stage of exchange. This indicated (2) that in the cases of silver and sodium forms the exchange with one ion of zeolite oxygen would be predominant, while in the case of copper forms it would be with 2 ions.

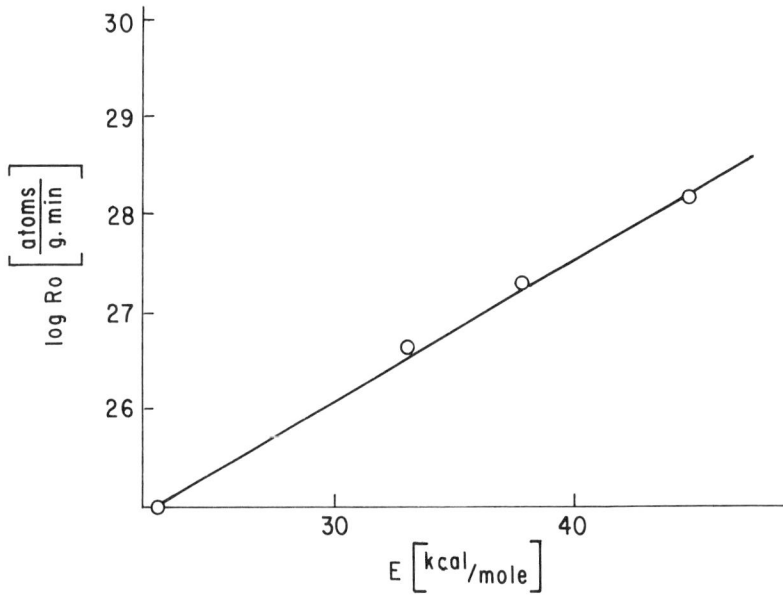

Figure 2. The compensatory dependence between parameters in the Arrhenius equation for the exchange between O_2 and the copper forms of zeolite Y

Thus, as follows from the analysis of the exchange kinetic characteristics, the substitution of copper for sodium causes a considerable change in the exchange mechanism.

The temperatures of the preliminary treatment and exchange experiments were sufficiently high that one might suggest that the structure of the samples under study should change. However, the data on nitrogen adsorption provided evidence in favor of retaining the structure.

Now, we are to consider the problem of zeolite oxygen equivalence with respect to exchange. Since most of the experiments have been carried out with the ratio $N_2/N_1 = 200$, that is, in a large excess of oxygen in the solid phase, the kinetics observed could be characteristic of the exchange of a small part of zeolite oxygen. Then the conclusion about the oxygen equivalence could be true only for a small part of the oxygen involved in the exchange rather than for the zeolite oxygen as a whole. However, in the experiment with NaY ($SiO_2/Al_2O_3 = 3.4$) at 670°C and at oxygen pressure of 5 mm, with N_2/N_1 ratio being 20.8, the rate constant value practically coincides with the corresponding value presented in Table I. Furthermore, a series of as many as 10–15 experiments with the same sample showed that the experimental values of equilibrium concentrations obtained during the exchange between new portions of gas and the same sample did not exceed 0.5–0.7% atm. These data allow one to assume that all the oxygen of zeolites would be involved and, hence, the conclusion about its equivalence with respect to exchange would be true for the zeolite oxygen as a whole.

The kinetic dependences observed might be caused by the exchange diffusion hindrance. It has been shown experimentally that the exchange is not limited by outer diffusion; the results obtained from the experiments with various rates of gas circulation coincide. Exchange probably is not limited by diffusion in zeolite pores. In fact, with as little as 14% of sodium replaced by copper, which does not practically affect the porous structure, the exchange rate constant of zeolite Y becomes more than 6 times as high.

The high activation energy values for sodium forms also provide evidence against diffusion hindrance. Thus, one cannot rely on the assumption that exchange is limited by the diffusion process to account for oxygen equivalence with respect to exchange. It is possible to explain it by making the following assumptions.

a) The limiting stage of exchange may be one of the adsorption or desorption stages, with all the adsorption–desorption sites being equivalent in their properties.

b) All oxygen ions of zeolite may be identical, no matter what their close surrounding is (for example, owing to the uniform distribution of the excess negative charge throughout the framework of zeolite).

The data available do not allow one to decide between these 2 assumptions. Study in this direction is under way.

Finally, a few words should be said about results that fell beyond the scope of this report. The reduced copper forms of zeolite Y possess a high catalytic activity in relation to the reaction $O_2^{16} + O_2^{18} = 2\,O^{16}O^{18}$ at temperatures as low as 100°–300°C. Thus, zeolites could be used as oxidizing catalysts.

To conclude, the zeolites studied rank as an intermediate between alumina and silica in their oxygen mobility (5); zeolites are less active in the exchange with O_2 than alumina and more active than silica.

Literature Cited

(1) Minachev, Kh. M., Garanin, V. I., Isakov, Ya. I., *Usp. Khim.* **1966**, 35, 2151.
(2) Muzykantov, V. S., Popovskii, V. V., Boreskow, G. K., *Kinetics Catalysis* **1964**, 5, 624.
(3) Saxena, S. C., Taylor, T. I., *J. Inorg. Nucl. Chem.* **1962**, 25, 261.
(4) Venuto, P. B., Landis, P. S., *Advan. Catalysis* **1968**, 18, 259.
(5) Winter, E. R. S., *J. Chem. Soc.* **1968**, (A) 12, 2889.
(6) Zhdanov, F. P., Egorova, E. M., "Chemistry of Zeolites," Nauka, Leningrad, 1968.

RECEIVED February 13, 1970.

INDEX

A

Acetaldehyde 399
Acetonitrile 399
Acidity, surface 391
Activation energies 372, 462
Activity coefficient, mean molal .. 423
Adsorption 4
   capacity 88, 99, 445
   stereospecific 480
Advances in zeolite science 1
Affinity order 430
Albite 150
Alkali
   aluminosilica gels 21
   basalt glass 334
   metal ion exchange 351
Alkalic lake waters 317
Alkaline earth ion exchange 358
Alkalinity 35
Alkylammonium ions 364, 426
Aluminosilicate
   crystallization of zeolitic 122
   framework 171, 221, 230, 353
   gels 26, 44
Aluminosilicophosphate zeolites .. 78
Aluminum-deficient large-port
   mordenites 502
Aluminum removal 273
Amido-zeolite Y 184
Ammonia 480, 484
Ammonium
   ion exchange 303
   zeolites 384
Amorphous
   precipitated silica 103
   structure 46
Analcime 77, 138, 156, 204, 280,
                  285, 304, 343
   composition 152
   structure 135
Analcite 54, 100, 140, 142, 329, 331
Anorthite 176
Apatite 77
Apparent phase transition effect .. 475
Applications
   in engineering sciences 1
   future 16
Arsenic 74
Atomic
   coordinates 268
   parameters 252, 260

Authigenic
   feldspar 304
   silicate minerals 279
Autocatalysis 30, 38

B

Barium form of zeolite A 436
Basaltic rock 284, 293
Beds of zeolite 279
Benzene 400
Beryllosilicate 135, 138
Beryllosodalite 136
Bikitaite 165
Binary ion exchange 405
Bond
   angles 255, 261
   distances 270
Bonding, cation–oxygen 189
Breccias 318
Brewsterite 164
Building units 210, 223
Burial metamorphic sequences ... 317
Butene-1 400
Butene isomerization 506

C

Calcic zeolites 320
Calcination 277, 394
Cancrinite 77, 159, 230, 365
   hydrate 124
Carbon monoxide adsorption 397
Carnegieite 117
Ca^{2+}/Sr^{2+}–chabazite system 406
Catalysts, zeolite 7
Catalytic activity 401
Cation
   distribution 432
   exchange 140, 143, 186, 350
   movement 219
   occupancy and position 185
   –oxygen bonding 189
   ratio 70
   zeolites 389
Cell-dimension measurement,
   techniques for 180
Cementation of clastics 312
Cenozoic 279
Centrifugation 50
Cesium-137, recovery and purification of 450
Chabazite 22, 77, 119, 150, 160,
             214, 286, 329, 366, 408
   -like 124

Channel geometry 159
Chemical structure of alkali
  aluminosilica gels 21
Chkalovite 138
Citrate 66
Classification scheme 11
Clinoptilolite 292, 294, 300, 305,
                        311, 334, 343, 364
Co(II)A synthetic zeolite 480
Coexisting phases 132
Cole–Cole plot 465
Complexing agent 63
Component concentrations 26
Concentration change 102
Conductivity, specific 460
Correlation time 473
Coulombic potentional energy 198
Cristobalite 49
Crystal
  chemical relationships 140
  size distribution 31
  structure 171, 238, 250, 259, 266
Crystallite laminae 45
Crystallization 20, 29, 37, 63,
                          104, 110, 122, 124
Crystallographic free diameters ... 161
Cumene 505, 507
Cyclopropane 480, 484

D

Dachiardite 164
Dacite glass 343
Deammination 386
Decomposition
  patterns, thermal 490
  of TTMA cations 494, 497
Decontamination factors 452
Deep bed geometry 272
Dehydrated specimens 186
Dehydration 183, 219
  incomplete 190
Diagenesis 303, 317
  zeolitic 335, 342
Dielectric study 456, 472
Diethanolglycinate 67
Differential thermal analysis 494
Diffusion 367
  equation 405
Diphosphate 66
Dipole relaxation 457
Disposition of radioactive waste .. 450
Distribution function 475

E

Edingtonite 158
Electric measurements 459
Electrical conductivity 456
Electron
  diffraction
    by lamellae 46
    patterns 232
  microprobe analyses 101
  microscopy 45

Electroselective effect 357
Engineering sciences, applications in 1
Epigenesis 317
Epistilbite 165
Erionite 22, 119, 160, 230, 288, 293
Ethylenediamine
  di(o-hydroxyphenylacetate) .. 67
Ethylenediaminetetraacetate 67
Ethylene oxide 484
Exchange
  cation 350
  process 475
Exchangeable cations 186

F

Facies, zeolite 317
Faujasite 66, 106, 118, 124, 166,
                   171, 194, 214, 472, 490
  low-silica 70
  synthetic 351, 426
  -type structures 171
  ultrastable 266
Fault planes 163
Feldspar 49, 178
  authigenic 304
Felspathoid 153, 214
Felsic glass 334
Ferrierite 165
Fickian diffusion process 369
Fission products 450
Fluidity of water adsorbed 473
Formation
  kinetic 56
  zeolite 51, 102
Framework
  aluminosilicate 353
  density 224
  distortion 219
  hydroxyls 181
  models 172
  ordering 215
  silicates 153
  zeolite 155
Frequency shift 217

G

Gallosilicate 69
  faujasite 199
Gallogermanate 135
Garronite 150
Gehlenite 177
Gel crystallization 78
Genesis of zeolites 303
Geosynclinal deposits, neogene ... 342
Gismondine 162, 239
Gismondite 250
Glass
  dacite 343
  semisynthetic 53
  synthetic and natural 51
  volcanic 334
  zeolitization
    of natural 59
    of synthetic 52

INDEX

Glassy tuffs 317
Gluconate 67
Gmelinite ... 77, 119, 159, 214, 331, 365
Graphical analysis 328
Greenschist 322, 327
Griphite 77
Growth, stages of zeolite 44

H

Harmotome 85, 238
Helvine 136
Herschelite 286
Heteroionic forms, stability of 436
Heulandite 163, 292, 311, 319, 329
High-silica zeolites 64
H_2O occupancies 257
HT mechanism 461
Hydrated species 191
Hydrocarbon oxidation 488
Hydrocracking 505
Hydrogen 484
  zeolites 384, 394
Hydronium ion exchange, degree of 492
Hydrothermal
  behavior 316
  metamorphism 318
  synthesis 137
Hydroxy
  cancrinite 152, 205
  sodalite 202
Hydroxyl
  group, structural 381
  stretching 382

I

Igneous rocks 287, 303
Induction period 29
Infrared
  assignments 212
  pattern 207
  spectra 202
  spectroscopic studies 380
  structural studies 201
Interaction with
  inorganic molecules 397
  organic molecules 399
  sulfur-containing molecules 398
  water 389
Interatomic distances 254, 260
Interstitial spaces, precepitation in 312
Intertetrahedral angles 177
Intracrystalline fluid 473
Intrastratal waters, salinity of 322
Ion
  migration 457
  sieve effects 426
Ion exchange
  alkali metal 351
  alkaline earth 358
  alkylammonium 364
  binary 405
  equilibria 366

Ion exchange (Continued)
  incomplete 189
  isotopic 367
  kinetics 367, 405
    particle-controlled 374
  model 415
  rare earth 356
  selectivity 422
Isomorphous
  replacement 135
  substitution 76
Isotope experiments 437
Isotopic
  exchange method 514
  ion exchange 367

K

Kehoeite 77, 135
K-F 124
Kinetics
  ion exchange 367, 405
  particle-controlled ion exchange 374
  of zeolite crystallization 29

L

Lacustrine rocks 302
Lamellae, electron diffraction by .. 46
Lattice collapse 436
Laumontite 156, 259, 319, 329, 343
Lawsonite 324
Leaching procedures 512
Least-squares distance refinement . 259
Leucite 144
Linde A 124, 166, 414
Linde B 237, 241
Linde L 161
Li-A 124
$Li_2O–Na_2O–Al_2O_3–SiO_2–H_2O$
  system 122
Lithium
  aluminate 124
  analcime 127
  metasilicate 124
  mordenite, synthesis of 127
  phillipsite 127
Low-silica faujasite 70
LT mechanism 463

M

Masugata 342
Mean molal activity coefficient ... 423
Mesolite 329
Metamorphism, hydrothermal 318
Metastable crystallization 149
Methylammonium cation exchange 490
Mid-infrared spectroscopy 202
Migration in large cavities 468
Mineral
  assemblages, zeolite 328
  phases 237
  zeolite 3
Mobility of exchangeable cations .. 456

Model
- ion exchange 415
- osmotic 418
- self-diffusion 369

Molecular jumps 473
Molecule distribution 190
Montmorillonite 305
Mordenite 77, 164, 204, 288, 301, 319, 343, 514
- aluminum-deficient large-port .. 502

Mullite transformation 500

N

Nakanosawa tuff 334
Na-mordenite 49
$Na_2O–Al_2O_3–SiO_2–H_2O$ system 102, 149
Na-P1 239
- and Na-P2 124

Natrolite 150, 157, 329
Natural glasses, zeolitization of ... 59
Neogene geosynclinal deposits 342
Nepheline 150, 177
Nephelinite 334
Niigata oil field 342
Nitrous oxide 480, 484
NMR relaxation times 473
Nomenclature 9, 171, 174, 194, 197
Nuclear fuels, reprocessing of 450
Nucleation rate 149

O

Obuchi 342
Occupancy parameters 270
Ocean basins, deep 317
Offretite 161, 230
- –erionite 214

OH stretching vibrations 389
Ordering distribution 153
Organic molecules, interaction with 399
Osmotic model 418
O–T–O angles 175
Overtone vibrations 229
Oxalate 66
Oxygen 484
- equivalence 518
- mobility in synthetic zeolites ... 514
- packing models 224
- removal 183

P

Paramagnetic
- impurities 473
- ions 478

Parameters
- atomic 252, 260
- occupancy 270
- thermal 268

Partial exchange, thermodynamic functions for 429
Particle-controlled ion exchange kinetics 374
Paulingite 167

Pauling's rules 188
pH, effect of 69
Phase relationships 130
Phillipsite 71, 85, 106, 108, 162, 214, 239, 249, 279, 288, 302
- –harmotone 77

Phosphate 74
- -containing reaction mixtures ... 64

Phosphorous substitution ... 76, 99, 100
- mechanisms of 95

Photosensitized redox reaction . 143, 146
Phytate 68
Pistacitic epidote 320
Plagioclase 333
- replacement of 313

Polycondensation mechanism 24
Polymerization 72
- state 105

Polytypism 166
Precipitation in interstitial spaces . 312
Predigestion, effect of 107
Prehnite 320, 324, 327
Prepolymerization of silicate 71
Pressure filtration method 367
Protonic acidity 495
Pulsed NMR apparatus 474
Pumpellyite 320, 324, 327
Purex waste 450
Pyroclastics 311, 327
Pyrogenic Crystals 305

Q

Quartz 329

R

Radial thermal displacement 256
Radioactive waste, disposition of .. 450
Raman spectra 222, 228
Rare earth ion exchange 356
Rate of crystal growth 34
Reactivity 380
Recovery and purification of cesium-137 450
Refinement 250, 259
Relaxation of water, NMR 473
Replacement of plagioclase 313
Reprocessing of nuclear fuels 450
Rhyolite 343
Rhyolitic glass 335
Rioltic pumices 53
Rock
- basaltic 284, 293
- igneous 287, 303
- lacustrine 302
- sedimentary 280, 285, 317
- tuffaceous 302
- volcanic 318, 320

S

Saline alkaline lake 304
Salinity of intrastratal waters 322
Scolecite 329

INDEX

Searlesite 304
Sedimentary
   deposits 279
   rocks 285, 317
Seeding 36, 150
Segregation veins 313
Selectivity
   coefficient 429
   data 416
   ion-exchange 422
   lack of 488
   thermodynamic 353, 365
Self diffusion 367
Semisynthetic glasses 53
Shape of aluminosilicate framework 175
Shimoigarashi 342
Sieve effects, ion 426
Silica source 72
Silica transport 275, 278
Silicate minerals, authigenic 279
Silicic
   glass, solution of 303
   vitric tuff 304
Silicon–aluminum
   distributions 252
   order–disorder 177
   ratio 198
   of clinoptilolites 338
Site occupancies 252
Smectite 47
Sodalite 77, 105, 158
   hydrate 124
   unit 173
Sodium
   aluminosilicate gels 102
   –aluminum Y 272
   chloride 138
Solubility 28
Sorption
   capacity 449
   complexes 193
Space group 259
Specific conductivity 460
Spectroscopic studies, infrared 380
Stability of heteroionic forms 436
Stacking
   fault 119
   sequences 117, 163
Stereoscopic drawings 155
Stereospecific adsorption 480
Stilbite 163, 329
Structural
   characteristics of zeolites 210
   chemistry 2
   hydroxyl groups 381
   studies 230, 351
      infrared 201
Structure
   of alkali aluminosilica gels,
      chemical 21
   amorphous 46
   crystal 250, 259, 266
   evolution of 443

Structure (Continued)
   group 214
   –infrared relationships 221
   types 156
Sulfate 66
Sulfur-containing molecules,
   interaction with 398
Surface
   acidity 391, 401
   properties 380
Symmetry 259
   changes .... , 144
Synthesis 140
   of a beryllosilicate 135
   and structural features 109
   of thermodynamically stable
      zeolites 149
   zeolite 3
Synthetic
   analcimes 131
   erionite 233
   faujasite 351, 426
   glasses, zeolitization of 52
   mordenites 128
   and natural glasses 51
   offretite 233
   phases 237
   zeolites 6, 102, 206
   A 436
   Co(II) A 480
   oxygen mobility in 514

T

Technical water glass solution 103
Temperature
   dependence 34
   jump method 367
Tertiary acidic tuff 311
Texture, evolution of 443
Thermal
   analyses 437
   decomposition 272, 490
   parameters 268
   stability 183, 440
   treatment 443
Thermodynamic
   functions for partial exchange .. 429
   selectivity 353, 365
Thermogravimetric analysis 493
Thiophene 398
Thomsonite 157, 329
T–O bond length 253, 263
T–O distances 175
T–O–T angles 254, 263
Transition metal ions 480
Transport of silica 275
TTMA cations, decomposition of 494, 497
Tuff
   glassy 317
   Nakanosawa 334
   silicic vitric 304
   tertiary acidic 311
   vitric 318, 343

Tuffaceous rocks	302
Tugtupite	136
Twinning	114

U

Ultrastable faujasites	266
Ultrastable zeolite	387

V

Vibration	
amplitude	256
OH stretching	389
water bending	382
Vibrational	
frequencies	208
spectra	222
Viseite	76, 135
Vitric	
materials	312
tuffs	318, 334, 343
Vitroclastic texture	305
Volcanic	
glass	59, 334
rocks	318, 320

W

Wairakite	319
Waste, disposition of radioactive	450
Water	480, 484
bending vibration	382
interaction with	389
NMR relaxation of	473

X

X-ray data	182
X-ray diffraction	
methods	172
patterns	48, 240
X-ray powder diffraction	439

Y

Yugawaralite	162

Z

Zeolite	
A	22, 31, 54, 68, 77, 90, 105, 202, 206, 355, 417, 436, 480
4A	443
B	71, 77, 92, 237, 241
B(P1)	203
C	205
D	204
facies	317
formation, process of	102
frameworks	76, 155
G	204
genesis of	303
L	71, 91, 205
mineral assemblages	328
N-A	206
P	54
P-A	80
P-B	81
P-C	78
P-G	80
P-L	80
P-R	80
properties	83
P-W	80
R	204
S	77, 203
T	205, 365
W	85, 205
X	22, 54, 68, 77, 171, 202, 218, 351, 426, 456, 514
13-X	473
Y	22, 65, 77, 171, 202, 218, 351, 426, 456, 490, 514
ZK-5	167, 203
ZSM-3	109
Ω	203
Zeolitic	
diagenesis	335, 342
zonation	348
Zeolitization	
of natural glasses	59
of synthetic glasses	52
Zeolon	205
Zonation of authigenic silicate	
minerals	305